Dynamical Mean-Field Theory for Strongly Correlated Materials

Volodymyr Turkowski

Dynamical Mean-Field Theory for Strongly Correlated Materials

Springer

Volodymyr Turkowski
Department of Physics
University of Central Florida
Orlando, FL, USA

ISBN 978-3-030-64903-6   ISBN 978-3-030-64904-3 (eBook)
https://doi.org/10.1007/978-3-030-64904-3

This Springer imprint is published by the registered company Springer Nature Switzerland AG
The registered company address is: Gewerbestrasse 11, 6330 Cham, Switzerland

To Lili

Contents

Chapter 1
Dynamical Mean-Field Theory for Strongly Correlated Materials

1.1 Introduction

In 1927, right after the principles of quantum mechanics were formulated [1–4], the first quantum theories of solids were proposed—Wolfgang Pauli applied the Fermi–Dirac statistics to describe paramagnetism in metals [5], and Arnold Sommerfeld generalized the classical Drude model of conductivity of free electrons in metals on the quantum case [6] (on the early history of quantum theory of solids, see review paper [7]). Next year, Felix Bloch constructed a general theory of electrons in lattices [8] that led to a basic understanding of the electronic properties of metals. In particular, Bloch's model explained the reason for finite resistivity of metals—it is caused by scattering of electrons on impurities or other "imperfections." After metals, the basics of the electronic properties of semiconductors were explained when Rudolf Peierls introduced [9] and Leon Brillouin applied [10, 11] the idea of gap states in solids. These works were followed by Alan Wilson's papers [12, 13] in which he has formulated the band theory of solids using the concept of empty and filled energy bands. This concept allowed to explain the principal difference between metals and semiconductors—in semiconductors, contrary to metals, all momentum states in the top occupied (valence) band are filled, thus the electrons cannot change their momentum when an electric field is applied and hence the semiconductors do not conduct electricity (which is essentially true in the idealistic case of crystals with no impurities and/or at zero temperature, and when the applied field is not very strong).

The serene sky of solid-state theory was seriously perturbed in approximately 4 years, at an international conference in Bristol in 1937, when Jan Hendrik de Boer and Evert Verwey presented data on conductivity of NiO, CoO, and several other transition-metal oxides [14]. As it followed from their data, most of the systems (except Fe_3O_4) were insulators, despite the fact that the d-bands of the materials were

To Lili

V. Turkowski, *Dynamical Mean-Field Theory for Strongly Correlated Materials*,
https://doi.org/10.1007/978-3-030-64904-3_1

only partially filled, i.e., from the band-theory point of view they supposed to be metals, similar to Ni and their other transition-metal "cousins" with non-fully occupied d-orbitals. It was immediately realized that the reason for insulating state of transition-metal oxides must be different from the one based on the fully occupied bands. In particular, in the following discussion, Wilson suggested that it might be that the band theory is not applicable to transition-metal oxides if one assumes that the potential barrier for the interatomic tunneling of the electrons in these nonconducting systems is very high. Indeed, according to his arguments, it is difficult to talk about bands when the kinetic (hopping) energy of the electrons is very small, so not bands, but rather sharp energy levels should appropriately describe the electronic structure of transition-metal oxides. On the other hand, it was realized that the distances between the Ni atoms in Ni and NiO are similar, thus the high interatomic potential barrier might be not a sufficient argument to explain the insulating state in NiO. Peierls suggested that the reason for the insulating state in the transition-metal oxides is a strong repulsion between electrons (or in modern language, strong electron correlations), when they meet on the d-orbitals of Ni and other transition-metal atoms, which prevents them from easy hopping on the transition-metal sublattice. As a Chair of the Conference, Mott documented these "prophesying" thoughts in the Conference proceedings paper [15]. Thus, one can put "Bristol, UK; 1937" as the place and the year of birth of the field of physics of strongly correlated materials (for a more detailed historic over-review of the early developments of the field, see, e.g., paper [16]).

In late 1940s, after a long break during which most of the physicists were engaged in applied defense research forced by the Second World War, Neville Mott has returned to the problem of conductivity in transition-metal materials [17–21]. Mott's rather phenomenological but physically insightful analysis was a very important step in laying the foundations of the principles of physics of strongly correlated systems. Thus, let us summarize some of his important ideas and findings.

- In paper [17], analyzing the properties of NiO, he argued that the atomic (London–Heitler), not band (Bloch), picture is more relevant for this material since a quite large energy is necessary to move electron from one Ni ion to another. It was also shown that when the lattice constant is large enough (in particular, when no pressure is applied), the state with no electron–hole pairs (i.e., an insulating state) is the ground state of the system.
- In paper [19], he analyzed metallization of insulators when the lattice constant decreases. He noticed that under the insulator-to-metal transition, the electron wave functions undergo an important transformation—it acquires a finite imaginary part, i.e., a nonzero phase (since metallicity assumes a nonzero current that is proportional to the gradient of the phase of the wavefunction). Quite remarkably, in this work he operated with the bandwidth and the atomic ionization as basic physical quantities of the considered model and did not take into account the local Coulomb repulsion.
- He has further elaborated the connection between the insulator-to-metal transition and the transformation of the electronic wave function in work [20]. As he noted,

at large interatomic distances one can use atomic functions as the basis and construct out of them the total, Slater-determinant, wave function that has to give zero conductivity. He continued with the question—what will happen when the interatomic distance will decrease? Naturally, the electrons will start to hop from site to site and the Slater-determinant wavefunction will not be an appropriate tool anymore. Indeed, the Slater determinant does not reflect whether an electron on given site meets another electron or not (i.e., whether this site is also occupied by another electron with the opposite spin). Obviously, the Slater-determinant ("exchange") approximation is not correct in this case. By that time, Wigner, Bohm, Pines, and others already attempted to include this phenomenon of electron correlations, i.e., the tendency of electrons with opposite spins to keep away from each other, into the picture. In particular, it was concluded that at large distances between the atoms one needs to use a correlated wave function Ψ_c (that includes correlation effects in zero-order approximation) instead of the "non-correlated" Slater wave function Ψ_{nc}. Mott argued that at the insulator-to-metal transition, when the interatomic distance decreases, the transformation of the wave function from Ψ_c to Ψ_{nc} is sharp—either every or no one electron can move!

- Another important Mott's concept relevant to the insulator-to-metal transition is polar states [20]: to provide a finite conductivity the total wave function must include free carriers, i.e., the metallic system should have polar states. Polar states have to come from overlapping wave functions; otherwise, there will be no charge transfer to create polarization. On the other hand, presence of polar states does not guarantee finite conductivity (i.e., polar states are a necessary but not a sufficient condition for finite conductivity).
- He analyzed [20] the steps of the Slater's explanation [22] of the conductivity quench at the metal–insulator transition (MIT) based on the assumption that the insulating lattice is antiferromagnetic (AFM). According to Slater, in the Hartree–Fock (HF) approximation any two Bloch electrons repel each other less when their spins are parallel. This corresponds to the presence of an effective superlattice field in the system that induces a split of the first Brillouine zone into two zones—a full and an empty one. Mott argued that this argument is wrong due to the following reason—according to the Slater scenario, the activation energy to generate the current and the energy needed to destroy the AFM order have the same order of magnitude. In reality, the last ("AFM") energy is the exchange energy defined by the overlap integral of the electron wave functions on the neighboring atoms, and the first (activation) energy is equal to the ionization energy potential I minus the electron affinity E, and it is not obvious that these two energies are similar.
- To estimate the critical interatomic distance when the insulator-to-metal transition occurs, he used a model of hydrogen atoms on a cubic lattice with the lattice parameter b [20]. He obtained analytical result for the order of magnitude of the interatomic hopping parameter (hopping energy of electrons):

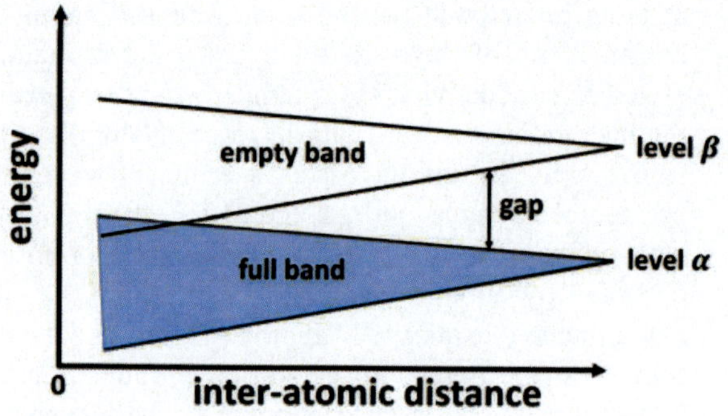

Fig. 1.1 Cartoon presentation of a possible mechanism of the insulator-to-metal transition with the interatomic distance decrease: when the distance reaches the critical value, the full and the empty bands begin to overlap and the system becomes metallic

$$\gamma = \int \psi^*(r)\frac{e^2}{r}\psi(r)d^3r = \frac{me^4}{2\hbar^2}\left(1+\frac{b}{a_H}\right)e^{-\frac{b}{a_H}}, \tag{1.1}$$

where a_H is the Bohr radius. This result also gives the formula for the bandwidth W that is equal 6γ (W gives the range of the kinetic energy of electrons: it can be between 0 and 6 γ). He also obtained an estimated formula for the average repulsion energy of electrons U (as the average potential within the hydrogen atom):

$$U = \frac{4\pi e^2 a_H^2}{b^3}. \tag{1.2}$$

Next, he assumed that the transition occurs when the bottom of the conduction band (energy of an extra electron in the system) touches the top of the valence band, i.e., when 6 γ + $U = I - E$. Calculating the ionization potential I and estimating the affinity to be E~0.1 · I, he obtained the value of the critical interatomic distance $b = 4.5a_H$. Quite remarkably, in this simple approximation one gets for the critical repulsion energy $U = 3.75$ eV and the critical bandwidth $W = 4.95$ eV. Thus, the ratio of the repulsion energy to the bandwidth at the transition point is $\frac{U}{W} = 0.76$, or $\frac{U}{\gamma} \approx 4.55$, which has the right order of magnitude confirmed by later, more sophisticated, calculations (see, e.g., Ref. [23]). Thus, Mott had rightly predicted that the MIT transition happens when the local Coulomb repulsion energy is of the order of the bandwidth (kinetic energy).

- He suggested that with the interatomic distance decreasing the insulating-state atomic levels transform into filled and empty bands ("zones"), and that at some distance when the bands start to overlap a transition to a metallic state takes place [21]. This is an important concept describing one of the possible scenarios of the insulator-to-metal transition in systems with localized orbitals. It is illustrated in Fig. 1.1.

As an important step to improve the theory, Mott suggested [20] that it is necessary to include "the effect on energy of correlation in the case of narrow bands." Besides, most of his analysis was done in terms of single-electron (orbital, or Bloch) wavefunctions, which is not the most suitable approach to study the correlation and other many-body effects. Only with advances of the many-body theory, the path to rigorously include the electron correlation effects in description of solids was opened.

Before moving to the discussion of the many-body studies of the spectral properties of strongly correlated materials, it is important to mention another important and challenging problem the solid-state theorists faced from the early times—ferromagnetism in Ni and other transition metals. The first attempt to explain the nature of ferromagnetism on the basis of quantum theory was performed by Heisenberg who used his concepts of exchange interaction and exchange energy. Based on these concepts, he formulated the famous Heisenberg model for site-fixed spins [24]. Unfortunately, the exchange parameter for this model is negative in overwhelming cases, i.e., it has the sign opposite to the one needed to produce ferromagnetism. Quite reasonably, Bloch argued [25] that static localized spins are not enough to describe ferromagnetism in metals (e.g., in Fe, Ni, and Co), where itinerant electrons play an important role and details of the material band structure can also matter. In the Bloch model, the main element is the plane waves traveling through the lattice. Despite the physical soundness of the model, in many cases it gives too large lattice constants to produce ferromagnetism as compared to the experimental values. An important step in developing theory of ferromagnetism was made by Stoner who showed that magnetism in itinerant electron systems can exist when it is energetically favorable for electrons to occupy the band with one spin orientation [26]. Almost at the same time, Wigner [27] analyzed the effects of the correlation energy in ferromagnetism and showed that inclusion of these effects significantly modifies the results of the Bloch theory, in particular it excludes ferromagnetism in many cases.

Thus, it was soon realized that, similar to the spectral properties, ferromagnetism in transition-metal systems is a complex phenomenon defined by a delicate balance between the correlation and many-body band structure effects. The most important and fundamental question was—are the d-electrons localized, i.e., do they provide static (local) fixed magnetic moments for ferromagnetism? As a possible answer, Slater suggested [28] that the (Hund's rule) intra-atomic exchange might be responsible for ferromagnetism in metals, i.e., the hopping of electrons or holes might generate ordering of preformed atomic moments and hence correlation effects. Similar to the spectral and transport studies, the events of WWII had postponed the research on transition-metal magnetism for approximately a decade. After the War, the ideas of Slater and others were elaborated and further extended by himself, van Vleck, Wolfarth, and several other researchers. The status of the field in early 1950s was summarized in several review papers [29–31].

Hence, as it was mentioned above, in the late 1950s the relative role of the kinetic (band) and correlation (local Coulomb repulsion) energy effects in the spectral, transport, and magnetic properties of transition metal systems remained unclear. With advances of many-body theory (summarized in early books [32–34]), the community has acquired an opportunity to solve these problems by using rigorous and accurate tools.

To describe ferromagnetism in d-electron systems, where both kinetic and Coulomb repulsion energy effects are important, Hubbard [35] and independently Gutzwiller [36] and Kanamori [37] introduced a first simple many-body microscopic model, later named the Hubbard model. The second quantized Hamiltonian of the model includes the kinetic energy (described by the nearest-site hopping parameter t) and local Coulomb repulsion (with the on-site repulsion energy U) terms. The main expectation from the model was that it will help to prove that correlation effects make electrons, and hence their magnetic moments, localized ("frozen") on the atomic sites. Another goal was to understand why these magnetic moments point preferably in one direction, i.e., why the systems like Ni are ferromagnets? Despite rather modest early successes in describing ferromagnetism (in fact, for a long time, the Hubbard model was regarded as a generic model of antiferromagnetism, not ferromagnetism), the first solutions of the model, especially by Hubbard [35, 38–42] and Gutzwiller [36, 43, 44] (who introduced a trial wave function approach with suppressed double occupancy), gave useful insights on the properties of strongly correlated materials, most notably on possible features of the MIT. It is interesting that Hubbard almost immediately came to the conclusion that the electron self-energy obtained from the solution of the Hubbard model has to be frequency-dependent, i.e., dynamical effects should play an important role in properties of strongly correlated systems, similarly to the case of impurity models where the corresponding (frequency-dependent effective potential) solution is called coherent-potential approximation (CPA) [45–47]. Thus, both Hubbard and CPA solutions might be regarded as the first forerunners of the dynamical mean-field theory (DMFT), the theory where the dynamical effects play a crucial role.

Now, let us mention some early (pre-DMFT) original approaches to the Hubbard model. Spectral and other properties of the model were analyzed with the so-called Hubbard operator formalism, by using the Hubbard operators that correspond to creation and annihilation of different types of states—empty site, singly- and doubly-occupied sites, a pair of sites with different occupancies, etc. In this approach, one needs to apply a truncation scheme to get a finite set of the equations of motion for the operators, derived from different sum rules and other constraints on the electron Green's functions (GFs) (see, e.g., Ref. [48] and for the references–paper [49]). Another important methodology is the path-integral slave-boson saddle-point solutions [50–57], where one uses the double occupancy and other boson operators and exploits constraint on possible site occupancies and other constraints. This approach was actively used to study the properties of heavy fermion materials. In the one-dimensional case, Bethe ansatz approach was used to solve the Hubbard model exactly by Lieb and Wu [58]. An important result they obtained is the absence of MIT in the 1D case, i.e., at arbitrary small local Coulomb repulsion between the electrons the system described by the Hubbard model is in insulating state.

In the magnetic case, a promising result was obtained by Nagaoka [59], who proved that when the number of electrons in the (bulk) system is smaller than the number of sites by one (the case of one hole) and $U = \infty$, the ferromagnetic (FM) state is the ground state of the system. Much later, another type of magnetic order in the Hubbard model—ferrimagnetic order—was found by Lieb [60] who proved that for the bipartite lattice with different number of nonequivalent sites at half-filling the total spin of the system is different from zero.

All sophisticated and rather powerful approximate methods described above have at least one common defect—they cannot be used in a controllable manner (except the 1D Bethe ansatz approach). Most importantly, the accuracy of the solutions obtained with these approaches cannot be controlled in the most difficult case of intermediate values of U, i.e., when $U \sim W$, and when neither U/W nor W/U perturbation theory can be used. As it will be shown in next chapters, the case $U \sim W$ is probably the most interesting, since it corresponds to many materials with very unusual properties. Fortunately, it was possible to formulate a universal "all-U" approach after the 1989 fundamental work by Vollhardt and Metzner [61] who showed that in the high-dimensional, $d \to \infty$ (or equivalently large atomic coordination number) limit, one can neglect the momentum dependence of the electron self-energy, i.e., the self-energy is local in space. This finding happened after a renewal of interest to strongly correlated materials caused by experimental discovery of high-temperature superconductivity [62] and Anderson's suggestion that strong electron–electron correlations are responsible for this phenomenon [63]. Vollhardt and Metzner's result was used by George and Kotliar [64] and by Jarrell [65] to build a DMFT formalism that allows one to solve the problem exactly taking advantage of the momentum-independent electron self-energy. The key step in formulating DMFT was mapping the problem on one-site problem, or the problem of an impurity in an effective self-consistently calculated (bath) field of other electrons (in DMFT, it is sufficient to consider one site, since all sites are equivalent in the infinite coordination number approximation). Since the local self-energy is frequency-dependent, it means that the DMFT solution takes into account local-in-space time-resolved interactions between electrons.

In early 1990s, first DMFT solutions for the one-band Hubbard model were obtained. They provided new unexpected insights on the MIT in correlated materials [64, 65] and established the conditions for the FM ground state in the Hubbard model [66–68] (see also Refs. [69–72] for over-review).

As a next step, the DMFT approach was generalized on the case with electron–phonon interaction [69, 73, 74], systems with impurities [75–77], the multi-orbital case [78–81], and the nonequilibrium case [82–84]. Availability of the "minimal" (or "zero-order") theory for strongly correlated system with local self-energy also opened the door to generalized solutions that include missing in DMFT momentum-dependence of the self-energy by using, e.g., $\frac{1}{d}$-perturbation theory, popular now dynamical vertex and other approximations (see, e.g., Refs. [85, 86], review paper [87], and Chap. 9 below).

The role of DMFT in condensed matter and material science communities further increased after the theory merged with the *ab initio* density functional theory (DFT) [88, 89], resulting in a powerful DFT + DMFT approach to describe complex real systems [90, 91]. DFT + DMFT combines the *ab initio* accuracy of DFT and the strength of the many-body DMFT to describe the effects of strong electron–electron correlations. In DFT + DMFT approach, one first calculates the electronic properties of the system within LDA or another available DFT exchange–correlation (XC) potential, and then takes into account the effects of strong correlations by solving the corresponding Hubbard model for the "non-correlated" DFT Kohn–Sham electrons (at current stage, both problems are solved self-consistently together). The DFT + DMFT has an important advantage comparing to the static approximations (with frequency-independent self-energy), like DFT + U [92]—it takes into account the above-mentioned time-resolved interaction effects, which appear to be crucial to describe properties of strongly correlated materials, e.g., the density of states (DOS) of such a complex material as Pu [93]. During the last two decades, DFT + DMFT was successfully applied to many different materials, from bulk systems [70, 71] to nanostructures [94–98].

At the same time, ability of DFT + DMFT to describe complex materials is still rather limited due to a large number of nonequivalent atoms in these systems. Therefore, it is desirable to have an *ab initio* DFT approach with an XC potential that properly (contrary to LDA and GGA and other standard potentials used in DFT) takes into account the effects of strong electron–electron correlations. Indeed, DFT being a theory of one variable—charge density, is technically much simpler than DMFT and other many-body theories that require to find multivariable GFs. There are different ways to proceed in constructing such an XC potential from DMFT or another many-body theory. One possibility is to calculate the XC energy of the Hubbard model as function of the particle density n and then differentiate the result with respect to the particle density (since the XC potential is a derivative of the XC energy with respect to n). Another way is to derive the XC potential by using the Sham–Schlueter equation methodology [99–101] that connects the XC potential with the many-body electron self-energy. A significant progress in the first directions was made by several groups [102–107] who used either the Bethe ansatz approach or DMFT solutions to calculate the XC energy of the Hubbard model. Efforts in both directions were recently undertaken also in the time-dependent case, where several groups derived the XC potentials (or XC kernels in the linear-response regime; see, e.g., Refs. [107–109]) for the time-dependent DFT (TDDFT) [110] to study nonequilibrium properties of materials. These *ab initio* tools, in particular, the ones that employ DMFT, show a promise to help in deeper understanding of the physical properties of very complex strongly correlated systems, in- and out-of-equilibrium.

The main goals of this book are: (1) to formulate DMFT and to summarize current understanding of different physical, including nonequilibrium, properties of the Hubbard model solved with DMFT; (2) to give details on how one can go beyond DMFT to take into account momentum dependence of the self-energy; (3) to formulate DFT + DMFT approach and to show its applications to describe properties of strongly correlated materials; (4) to give details on how to construct XC potentials and kernels for *ab initio* DFT and TDDFT approaches by using the DMFT solution of the Hubbard model; and (5) to show how these functionals perform in the static and time-dependent cases. The first two items of this program form the first part of the book—A Many-Body Theory, and the last three—the second part, A Computational Tool.

Finally, I use the opportunity to thank my Teachers, Colleagues, Family, and Friends for their inspiration and support. I acknowledge partial support from DOE Grant No. DE-FG02-07ER46354. I also would like to thank the Springer Nature team for the help, especially Arun Pandian KJ, Michael Luby, and Brian Halm who kindly extended the deadline for submission of the book for 7 weeks, which was very helpful in these uneasy pandemic times.

References

1. Heisenberg, W.: Z. Phys. **33**, 879 (1925)
2. Born, M., Heisenberg, W., Jordan, P.: Z. Phys. **35**, 557 (1926)
3. Schrödinger, E.: Ann. Phys. **384**, 361 (1926)
4. Schrödinger, E.: Ann. Phys. **384**, 489 (1926)
5. Pauli, W.: Z. Phys. **41**, 81 (1927)
6. Sommerfeld, A.: Naturwiss. **15**, 825 (1927)
7. Hoddeson, L., Baym, G., Eckert, M.: Rev. Mod. Phys. **59**, 287 (1987)
8. Bloch, F.: Z. Phys. **52**, 555 (1928)
9. Peierls, R.E.: Ann. Phys. (Leipzig). **4**, 121 (1930)
10. Brillouin, L.: C. R. Acad. Sci. **191**, 198 (1930)
11. Brillouin, L.: C. R. Acad. Sci. **191**, 292 (1930)
12. Wilson, A.H.: Proc. R. Soc. London Ser. A. **133**, 458 (1931)
13. Wilson, A.H.: Proc. R. Soc. London Ser. A. **134**, 277 (1931)
14. de Boer, J.H., Verwey, E.J.W.: Proc. Phys. Soc. **49**, 59 (1937)
15. Mott, N.F., Peierls, R.: Proc. Phys. Soc. **49**, 72 (1937)
16. Brandow, B.H.: Adv. in Phys. **26**, 651 (1977)
17. Mott, N.F.: Proc. Phys. Soc. A. **62**, 416 (1949)
18. Mott, N.F.: Prog. Metal Phys. **3**, 76 (1952)
19. Mott, N.F.: Can. J. Phys. **34**, 1356 (1956)
20. Mott, N.F.: Supl. to Nuovo Cim. **7**, 312 (1958)
21. Mott, N.F.: Phil. Mag. **6**, 281 (1961)
22. Slater, J.C.: Phys. Rev. **82**, 568 (1951)
23. Staudt, R., Dzierzawa, M., Muramatsu, A.: Eur. Phys. J. B. **17**, 411 (2000)
24. Heisenberg, W.: Z. Phys. **49**, 619 (1928)
25. Bloch, F.: Z. Phys. **57**, 545 (1929)
26. Stoner, E.C.: Magnetism and Matter. Methuen, London (1934)
27. Wigner, E.: Phys. Rev. **46**, 1002 (1934)
28. Slater, J.C.: Phys. Rev. **49**, 537 (1936)
29. Slater, J.C.: Rev. Mod. Phys. **25**, 199 (1953)
30. Van Vleck, J.H.: Rev. Mod. Phys. **25**, 220 (1953)
31. Wohlfarth, E.P.: Rev. Mod. Phys. **25**, 211 (1953)
32. Pines, D.: The Many-Body Problem. W.A. Benjamin, Inc., New York (1961)
33. Abrikosov, A.A., Dzyaloshinski, I.E., Gorkov, L.P.: Methods of Quantum Field Theory in Statistical Physics. Prentice Hall Press, Upper Saddle River (1963)
34. Nozières, P.: Le problème à N corps: propriétés générales des gaz de fermions. Dunod, Paris (1963)
35. Hubbard, J.: Proc. Roy. Soc. Lond. Ser. A. **276**, 238 (1963)
36. Gutzwiller, M.C.: Phys. Rev. Lett. **10**, 159 (1963)
37. Kanamori, J.: Prog. Theor. Phys. **30**, 275 (1963)
38. Hubbard, J.: Proc. Roy. Soc. Lond. Ser. A. **277**, 237 (1964)

39. Hubbard, J.: Proc. Roy. Soc. Lond. Ser. A. **281**, 401 (1964)
40. Hubbard, J.: Proc. Roy. Soc. Lond. Ser. A. **285**, 542 (1965)
41. Hubbard, J.: Proc. Roy. Soc. Lond. Ser. A. **286**, 82 (1966)
42. Hubbard, J.: Proc. Roy. Soc. Lond. Ser. A. **286**, 100 (1966)
43. Gutzwiller, M.C.: Phys. Rev. **134**, A923 (1964)
44. Gutzwiller, M.C.: Phys. Rev. **137**, A1726 (1965)
45. Soven, P.: Phys. Rev. **156**, 809 (1967)
46. Soven, P.: Phys. Rev. **178**, 1136 (1969)
47. Hirooka, S., Shimizu, M.: J. Phys. Soc. Jpn. **43**, 70 (1977)
48. Roth, L.M.: Phys. Rev. **184**, 451 (1969)
49. Mancini, F., Avella, A.: Adv. Phys. **53**, 537 (2004)
50. Barnes, S.E.: J. Phys. F. **6**, 1375 (1976)
51. Barnes, S.E.: J. Phys. F. **7**, 2637 (1977)
52. Coleman, P.: Phys. Rev. B. **28**, 5255 (1983)
53. Read, N., Newns, D.M.: J. Phys. C. **16**, 3273 (1983)
54. Coleman, P.: Phys. Rev. B. **29**, 3035 (1984)
55. Kotliar, G., Ruckenstein, A.E.: Phys. Rev. Lett. **57**, 1362 (1986)
56. Coleman, P.: Phys. Rev. B. **35**, 5072 (1987)
57. Read, N., Newns, D.M.: Adv. Phys. **36**, 799 (1987)
58. Lieb, E.H., Wu, F.Y.: Phys. Rev. Lett. **20**, 1445 (1968).; Erratum Phys. Rev. Lett. **21**, 192 (1968)
59. Nagaoka, Y.: Phys. Rev. **147**, 392 (1966)
60. Lieb, E.H.: Phys. Rev. Lett. **62**, 1201 (1989)
61. Metzner, W., Vollhardt, D.: Phys. Rev. Lett. **62**, 324 (1989)
62. Bednorz, J.G., Müller, K.A.: Z. Phys. B. **64**, 189 (1986)
63. Anderson, P.W.: Science. **235**, 1196 (1987)
64. Georges, A., Kotliar, G.: Phys. Rev. B. **45**, 6479 (1992)
65. Jarrell, M.: Phys. Rev. Lett. **69**, 168 (1992)
66. Obermeier, T., Pruschke, T., Keller, J.: Phys. Rev. B. **56**, 8479 (1997)
67. Ulmke, M.: Eur. Phys. J. B. **1**, 301 (1998)
68. Vollhardt, D., Blumer, N., Held, K., Kollar, M.: Adv. Solid State Phys. **38**, 383 (1999)
69. Georges, A., Kotliar, G., Krauth, W., Rozenberg, M.J.: Rev. Mod. Phys. **68**, 13 (1996)
70. Kotliar, G., Savrasov, S.Y., Haule, K., Oudovenko, V.S., Parcollet, O., Marianetti, C.: Rev. Mod. Phys. **78**, 865 (2006)
71. Held, K., Nekrasov, I., Keller, G., Eyert, V., Blümer, N., McMahan, A., Scalettar, R., Pruschke, T., Anisimov, V., Vollhardt, D.: Phys. Status Solidi (b). **243**, 2599 (2006)
72. Katsnelson, M.I., Irkhin, V.Y., Chioncel, L., Lichtenstein, A.I., de Groot, R.A.: Rev. Mod. Phys. **80**, 315 (2008)
73. Freericks, J.K., Jarrell, M., Scalapino, D.J.: Phys. Rev. B. **48**, 6302 (1993)
74. Freericks, J.K., Jarrell, J.K.M., Scalapino, D.J.: Europhys. Lett. **25**, 37 (1994)
75. Dobrosavljević, V., Kotliar, G.: Phys. Rev. Lett. **71**, 3218 (1993)
76. Dobrosavljević, V., Kotliar, G.: Phys. Rev. B. **50**, 1430 (1994)
77. Dobrosavljevic, V., Pastor, A.A., Nikolić, B.K.: Europhys. Lett. **62**, 76 (2003)
78. Georges, A., Kotliar, G., Krauth, W.: Z. Phys. B: Condens. Matter. **92**, 313 (1993)
79. Caffarel, M., Krauth, W.: Phys. Rev. Lett. **72**, 1545 (1994)
80. Maier, Th., Zolfl, M.B., Pruschke, Th., Keller, J.: Eur. Phys. J. B. **7**, 377 (1999)
81. Florens, S., Georges, A., Kotliar, G., Parcollet, O.: Phys. Rev. B. **66**, 205102 (2002)
82. Schmidt, P., and Monien, H.: preprint arXiv:cond-mat/0202046v1 (2002)
83. Freericks, J.K., Turkowski, V., Zlatić, V.: Phys. Rev. Lett. **97**, 266408 (2006)
84. Aoki, H., Tsuji, N., Eckstein, M., Kollar, M., Oka, T., Werner, P.: Rev. Mod. Phys. **86**, 779 (2014)
85. Schiller, A., Ingersent, K.: Phys. Rev. Lett. **75**, 113 (1995)
86. Toschi, A., Katanin, A.A., Held, K.: Phys. Rev. B. **75**, 045118 (2007)

87. Rohringer, G., Hafermann, H., Toschi, A., Katanin, A.A., Antipov, A.E., Katsnelson, M.I., Lichtenstein, A.I., Rubtsov, A.N., Held, K.: Rev. Mod. Phys. **90**, 025003 (2018)
88. Hohenberg, P., Kohn, W.: Phys. Rev. **136**, B864 (1964)
89. Kohn, W., Sham, L.J.: Phys. Rev. **140**, A1133 (1965)
90. Anisimov, V., Poteryaev, A., Korotin, M., Anokhin, A., Kotliar, G.: J. Phys. Condens. Matter. **9**, 7359 (1997)
91. Lichtenstein, A., Katsnelson, M.: Phys. Rev. B. **57**, 6884 (1998)
92. Anisimov, V.I., Aryasetiawan, F., Lichtenstein, A.I.: J. Phys. Condens. Matter. **9**, 767 (1997)
93. Savrasov, S.Y., Kotliar, G., Abrahams, E.: Nature. **410**, 793 (2001)
94. Boukhvalov, D.W., Vergara, L.I., Dobrovitskii, V.V., Katsnelson, M.I., Lichtenstein, A.I., Kögerler, P., Musfeldt, J.L., Harmon, B.N.: Phys. Rev. B. **77**, 180402 (2008)
95. Valli, A., Sangiovanni, G., Gunnarsson, O., Toschi, A., Held, K.: Phys. Rev. Lett. **104**, 246402 (2010)
96. Jacob, D., Haule, K., Kotliar, G.: Phys. Rev. B. **82**, 195115 (2010)
97. Turkowski, V., Kabir, A., Nayyar, N., Rahman, T.S.: J. Chem. Phys. **136**, 114108 (2012)
98. Chen, J., Millis, A.J., Marianetti, C.A.: Phys. Rev. B. **91**, 241111 (2015)
99. Sham, L., Schlüter, M.: Phys. Rev. Lett. **51**, 1888 (1983)
100. Sham, L.: Phys. Rev. B. **32**, 3876 (1985)
101. van Leeuwen, R.: Phys. Rev. Lett. **76**, 3610 (1996)
102. Lima, N.A., Silva, M.F., Oliveira, L.N., Capelle, K.: Phys. Rev. Lett. **90**, 146402 (2003)
103. Xianlong, G., Polini, M., Tossi, M.P., Campo Jr., V.L., Capelle, K., Rigol, M.: Phys. Rev. B. **73**, 165120 (2006)
104. Verdozzi, C.: Phys. Rev. Lett. **101**, 166401 (2008)
105. Li, W., Xianlong, G., Kollath, C., Polini, M.: Phys. Rev. B. **78**, 195109 (2008)
106. Bergeld, J.P., Liu, Z.-F., Burke, K., Staford, C.A.: Phys. Rev. Lett. **108**, 066801 (2012)
107. Karlsson, D., Privitera, A., Verdozzi, C.: Phys. Rev. Lett. **106**, 116401 (2011)
108. Aryasetiawan, F., Gunnarsson, O.: Phys. Rev. B. **66**, 165119 (2002)
109. Turkowski, V., Rahman, T.S.: J. Phys. Condens. Matter. **29**, 455601 (2017)
110. Runge, E., Gross, E.K.: Phys. Rev. Lett. **52**, 997 (1984)

Part I
A Many-Body Theory

Chapter 2
One-Band Hubbard Model: Definition and Early Solutions

2.1 Many-Body Problem and the Hubbard Hamiltonian

In this section, we derive the Hubbard Hamiltonian in the second-quantization form that will play a central role in this book (for details, see, e.g., Ref. [1]; a part of the presentation in this chapter follows the presentation in this manuscript). For this, we begin from the most general (first-quantized) form of a solid-system Hamiltonian with N electrons (e) and L ions (I):

$$\mathcal{H} = \sum_{i=1}^{L} \frac{\vec{P}_i^2}{2M_i} + \sum_{i<j=1}^{L} V_{I-I}\left(\vec{R}_i - \vec{R}_j\right) + \sum_{i=1}^{N} \frac{\vec{p}_i^2}{2m} + \sum_{i<j=1}^{N} V_{e-e}\left(\vec{r}_i - \vec{r}_j\right)$$
$$+ \sum_{i=1}^{N} \sum_{j=1}^{L} V_{e-I}\left(\vec{r}_i - \vec{R}_j\right), \tag{2.1}$$

where $\vec{R}_i$ and $\vec{P}_\mathrm{i}$ are the coordinate and momentum operators and M_i is the mass of the ions; similarly, $\vec{r}_i$, $\vec{p}_i$, and m are the corresponding quantities for the electrons. In Hamiltonian (2.1), the first and the third terms are the kinetic energy operators of the ions and electrons,

$$V_{I-I}\left(\vec{R}_i - \vec{R}_j\right) = \frac{e^2 Z_i Z_j}{\left|\vec{R}_i - \vec{R}_j\right|} \tag{2.2}$$

is the ion–ion interaction potential (e is the charge of electron; Z_i and Z_j are the atomic numbers of the ions on cites $\vec{R}_i$ and $\vec{R}_j$), and similarly,

V. Turkowski, *Dynamical Mean-Field Theory for Strongly Correlated Materials*,
https://doi.org/10.1007/978-3-030-64904-3_2

$$V_{e-e}\left(\vec{r}_i - \vec{r}_j\right) = \frac{e^2}{\left|\vec{r}_i - \vec{r}_j\right|} \tag{2.3}$$

and

$$V_{e-I}\left(\vec{r}_i - \vec{R}_j\right) = -\frac{e^2 Z_j}{\left|\vec{r}_i - \vec{R}_j\right|} \tag{2.4}$$

are the electron–electron and the electron–ion interaction potentials.

Since ions are much heavier than electrons, the electronic subsystem (our main interest) can be analyzed separately, in accordance with the Born–Oppenheimer approximation. Namely, one can use the following electronic Hamiltonian in the field of “frozen” heavy ions:

$$H = \sum_{i=1}^{N}\left(\frac{\vec{p}_i^{\,2}}{2m} + V_{\text{ion}}\left(\vec{r}_i\right)\right) + \sum_{i<j=1}^{N} V_{e-e}\left(\vec{r}_i - \vec{r}_j\right), \tag{2.5}$$

where

$$V_{\text{ion}}\left(\vec{r}_i\right) = \sum_{j=1}^{L} V_{e-I}\left(\vec{r}_i - \vec{R}_j\right) \tag{2.6}$$

is the periodic potential of ions.

The problem described by the Hamiltonian (2.5) can be simplified by splitting the Hamiltonian into one- and two-particle parts in such a way that the matrix elements of the last part are small, i.e.

$$H = \sum_{i=1}^{N} h_1\left(\vec{r}_i\right) + \sum_{i<j=1}^{N} U\left(\vec{r}_i - \vec{r}_j\right), \tag{2.7}$$

where

$$h_1\left(\vec{r}_i\right) = \frac{\vec{p}_i^{\,2}}{2m} + V\left(\vec{r}_i\right) \tag{2.8}$$

is one-particle Hamiltonian with an effective single-particle potential $V\left(\vec{r}_i\right)$. This potential is sum of the ionic potential $V_{\text{ion}}\left(\vec{r}_i\right)$ and of an effective auxiliary potential $V_{\text{aux}}\left(\vec{r}_i\right)$ that describes the interaction of the electron with all other electrons in the systems:

$$V\left(\vec{r}_i\right) = V_{\text{ion}}\left(\vec{r}_i\right) + V_{\text{aux}}\left(\vec{r}_i\right). \tag{2.9}$$

To adjust the introduced effective single-particle Hamiltonian (2.8), the electron interaction in the original Hamiltonian (2.5) must be transformed to a new function $U\left(\vec{r}_i - \vec{r}_j\right)$ in the following way:

$$\begin{aligned} V_{e-e}\left(\vec{r}_i - \vec{r}_j\right) &\to U\left(\vec{r}_i - \vec{r}_j\right) \\ &= V_{e-e}\left(\vec{r}_i - \vec{r}_j\right) - \frac{1}{N-1}\left(V_{\text{aux}}\left(\vec{r}_i\right) + V_{\text{aux}}\left(\vec{r}_j\right)\right). \end{aligned} \tag{2.10}$$

Then, the next goal is to get such $V_{\text{aux}}\left(\vec{r}_i\right)$ that the matrix elements of the effective potential $U\left(\vec{r}_i - \vec{r}_j\right)$ with respect to the eigenstates of $h_1\left(\vec{r}_i\right)$ are as small as possible. In other words, it is necessary to find an effective single-particle Hamiltonian (2.8) that is a good approximation to the exact Hamiltonian (2.5). Provided a good approximation for $h_1\left(\vec{r}_i\right)$ is found, one can solve the eigenproblem:

$$h_1\left(\vec{r}_i\right)\varphi^{\alpha}_{\vec{k}}\left(\vec{r}\right) = \varepsilon_{\alpha}\left(\vec{k}\right)\varphi^{\alpha}_{\vec{k}}\left(\vec{r}\right), \tag{2.11}$$

where $\varphi^{\alpha}_{\vec{k}}\left(\vec{r}\right)$ and $\varepsilon_{\alpha}\left(\vec{k}\right)$ are the quasiparticle Bloch eigenfunctions and eigenenergies/bands dispersion ($\vec{k}$ is the electron momentum and α is the band number). Once the problem (2.8) is solved, the missing many-body effects can be calculated by perturbation theory in $U\left(\vec{r}_i - \vec{r}_j\right)$.

In the case of on-site electron–electron interactions, it is more convenient to use the basis of the Wannier local atomic functions $\phi^{\alpha}\left(\vec{r} - \vec{R}_i\right)$ (centered around the site $\vec{R}_i$) instead of $\varphi^{\alpha}_{\vec{k}}\left(\vec{r}\right)$. The Bloch eigenfunctions can be expanded in the Wannier function basis as

$$\varphi^{\alpha}_{\vec{k}}\left(\vec{r}\right) = \frac{1}{\sqrt{L}}\sum_i e^{i\vec{k}\vec{R}_i}\phi^{\alpha}\left(\vec{r} - \vec{R}_i\right), \tag{2.12}$$

and inversely,

$$\phi^{\alpha}\left(\vec{r} - \vec{R}_i\right) = \frac{1}{\sqrt{L}}\sum_{\vec{k}} e^{-i\vec{k}\vec{R}_i}\varphi^{\alpha}_{\vec{k}}\left(\vec{r}\right). \tag{2.13}$$

At this stage, we are ready to build the second quantized Hamiltonian of the strongly correlated system. For this, we introduce standard creation $\hat{c}^{+}_{\alpha\vec{k},\sigma}$ and annihilation $\hat{c}_{\alpha\vec{k},\sigma}$ operators of the electron with momentum $\vec{k}$ and spin σ in the state described by $\varphi^{\alpha}_{\vec{k}}\left(\vec{r}\right)$. These operators satisfy the following anticommutation relations:

$$\left\{\hat{c}_{\alpha\vec{k},\sigma}, \hat{c}_{\beta\vec{q},b}\right\} \equiv \hat{c}_{\alpha\vec{k},\sigma}\hat{c}_{\beta\vec{q},\sigma'} + \hat{c}_{\beta\vec{q},\sigma'}\hat{c}_{\alpha\vec{k},\sigma} = 0, \tag{2.14}$$

$$\left\{\hat{c}^{+}_{\alpha\vec{k},\sigma}, \hat{c}^{+}_{\beta\vec{q},\sigma'}\right\} \equiv \hat{c}^{+}_{\alpha\vec{k},\ \sigma}\hat{c}^{+}_{\beta\vec{q},\sigma'} + \hat{c}^{+}_{\beta\vec{q},\sigma'}\hat{c}^{+}_{\alpha\vec{k},\sigma} = 0, \tag{2.15}$$

$$\left\{\hat{c}_{\alpha\vec{k},\sigma}, \hat{c}^{+}_{\beta\vec{q},\sigma'}\right\} \equiv \hat{c}_{\alpha\vec{k},\sigma}\hat{c}^{+}_{\beta\vec{q},\sigma'} + \hat{c}^{+}_{\beta\vec{q},\sigma'}\hat{c}_{\alpha\vec{k},\sigma} = \delta_{\alpha\beta}\delta_{\vec{k}\vec{q}}\delta_{\sigma\sigma'} \tag{2.16}$$

and together with the Bloch eigenfunction can be used to construct the second quantized electron wavefunction:

$$\hat{\Psi}_{\sigma}\left(\vec{r}\right) = \sum_{\vec{k},\alpha}\varphi^{\alpha}_{\vec{k}}\left(\vec{r}\right)\hat{c}_{\alpha\vec{k},\sigma} \tag{2.17}$$

Similarly, for the conjugated function one gets

$$\hat{\Psi}^{+}_{\sigma}\left(\vec{r}\right) = \sum_{\vec{k},\alpha}\varphi^{\alpha *}_{\vec{k}}\left(\vec{r}\right)\hat{c}^{+}_{\alpha\vec{k},\sigma}. \tag{2.18}$$

As the next step, one can introduce more convenient electron site-representation operators connected with the momentum-representation operators as:

$$\hat{c}_{\alpha i,\sigma} = \frac{1}{\sqrt{L}}\sum_{\vec{k}} e^{i\vec{k}\vec{R}_i}\hat{c}_{\alpha\vec{k},\sigma}, \tag{2.19}$$

$$\hat{c}^{+}_{\alpha i,\sigma} = \frac{1}{\sqrt{L}}\sum_{\vec{k}} e^{-i\vec{k}\vec{R}_i}\hat{c}_{\alpha\vec{k},\sigma}. \tag{2.20}$$

Using the last two relations, Eq. (2.13) and its complex-conjugated version, one can express the wavefunction operators (2.17) and (2.18) in terms of the "site" operators:

$$\hat{\Psi}_{\sigma}\left(\vec{r}\right) = \sum_{i,\alpha}\phi^{\alpha}\left(\vec{r} - \vec{R}_i\right)\hat{c}_{\alpha i,\sigma}, \tag{2.21}$$

$$\widehat{\Psi}_{\sigma}^{+}\left(\vec{r}\right) = \sum_{i,\alpha} \phi^{\alpha *}\left(\vec{r} - \vec{R}_i\right) \widehat{c}_{\alpha i,\sigma}^{+}. \tag{2.22}$$

The second quantized version of the Hamiltonian (2.7) is

$$\widehat{H} = \sum_{\sigma=\uparrow,\downarrow} \int d\vec{r}\,\widehat{\Psi}_{\sigma}^{+}\left(\vec{r}\right) h_1\left(\vec{r}\right) \widehat{\Psi}_{\sigma}\left(\vec{r}\right) + \frac{1}{2} \sum_{\sigma,\sigma'=\uparrow,\downarrow} \int d\vec{r} \int d\vec{r}'\, \widehat{\Psi}_{\sigma}^{+}\left(\vec{r}\right) \widehat{\Psi}_{\sigma'}^{+}\left(\vec{r}'\right) U\left(\vec{r} - \vec{r}'\right) \widehat{\Psi}_{\sigma'}\left(\vec{r}'\right) \widehat{\Psi}_{\sigma}\left(\vec{r}\right), \tag{2.23}$$

or using Eqs. (2.21) and (2.22),

$$\widehat{H} = -\sum_{i,j,\alpha,\sigma} t_{ij}^{\alpha} \widehat{c}_{\alpha i,\sigma}^{+} \widehat{c}_{\alpha j,\sigma} + \frac{1}{2} \sum_{i,j,k,l\alpha,} \sum_{\beta,\gamma,\delta\sigma,\sigma'} \sum U_{ijkl}^{\alpha\beta\gamma\delta} \widehat{c}_{\alpha i,\sigma}^{+} \widehat{c}_{\beta j,\sigma'}^{+} \widehat{c}_{\gamma k,\sigma'} \widehat{c}_{\delta l,\sigma}, \tag{2.24}$$

where t_{ij}^{α} and defined $U_{ijkl}^{\alpha\beta\gamma\delta}$ below (Eqs. (2.25) and (2.26)).

It must be emphasized that we have not made any approximation so far, and that Hamiltonian (2.24) is the exact many-electron Hamiltonian in the site representation. The problem described by this Hamiltonian cannot be solved exactly due to an extremely large (to be precise, infinite) number of the kinetic energy terms with the hopping

$$t_{ij}^{\alpha} = -\int d\vec{r}\; \phi^{\alpha *}\left(\vec{r} - \vec{R}_i\right) h_1\left(\vec{r}\right) \phi^{\alpha}\left(\vec{r} - \vec{R}_j\right), \tag{2.25}$$

and the electron–electron interaction terms with the repulsion energy

$$U_{ijkl}^{\alpha\beta\gamma\delta} = \int d\vec{r} \int d\vec{r}'\, \phi^{\alpha *}\left(\vec{r} - \vec{R}_i\right) \phi^{\beta *}\left(\vec{r} - \vec{R}_j\right) U\left(\vec{r} - \vec{r}'\right) \phi^{\gamma}\left(\vec{r}' - \vec{R}_k\right) \phi^{\delta}\left(\vec{r}' - \vec{R}_l\right) \tag{2.26}$$

parameters (e.g., even in the rather simple case of a finite ten-site one-orbital system, Eq. (2.24) includes ~$L^4 = 10{,}000$ interaction terms!).

Thus, Hamiltonian (2.24) has to be simplified by using some approximation(s). A quite natural approximation is to neglect the nonlocal-in-space interaction energies (2.26), i.e., to put

$$U_{ijkl}^{\alpha\beta\gamma\delta} = \delta_{ij}\delta_{ik}\delta_{il} U_{iiii}^{\alpha\beta\gamma\delta} \equiv \delta_{ij}\delta_{ik}\delta_{il} U_{i}^{\alpha\beta\gamma\delta}. \tag{2.27}$$

Indeed, since the Wannier functions $\phi^{\alpha*}\left(\vec{r}-\vec{R}_i\right)$ are strongly localized around the corresponding atoms, the functions (2.26) with $i=j=k=l$ are much larger as compared to all others with two or more different site indices. For instance, the next after (2.27) largest matrix elements—the nearest-neighbor and the next-nearest neighbor elements—are typically ~3 and ~10 times smaller than $U_i^{\alpha\beta\gamma\delta}$, correspondingly. Using this approximation, the multi-orbital Hamiltonian (2.24) can be transformed to

$$\widehat{H}=-\sum_{i,j,\alpha,\sigma}t_{ij}^{\alpha}\widehat{c}_{\alpha i,\sigma}^{+}\widehat{c}_{\beta j,\sigma}+\frac{1}{2}\sum_{i}\sum_{\alpha,\beta,\gamma,\delta}\sum_{\sigma,\sigma'}U_i^{\alpha\beta\gamma\delta}\widehat{c}_{\alpha i,\sigma}^{+}\widehat{c}_{\beta i,\sigma'}^{+}\widehat{c}_{\gamma i,\sigma'}\widehat{c}_{\delta i,\sigma}. \tag{2.28}$$

To simplify notations, we introduce the electron spin-orbital density-matrix operators

$$\widehat{n}_{\alpha\beta i,\sigma}=\widehat{c}_{\alpha i,\sigma}^{+}\widehat{c}_{\beta i,\sigma} \tag{2.29}$$

(the diagonal in the band index operators $\widehat{n}_{\alpha\alpha i,a}\equiv\widehat{n}_{\alpha i,a}$ are the density operators for the electron on the site i in the state (α,σ)).

Using definition (2.29), the Hamiltonian (2.28) can be rewritten as

$$\begin{aligned}\widehat{H}=&-\sum_{i,j,\alpha,\sigma}t_{ij}^{\alpha}\widehat{c}_{\alpha i,\sigma}^{+}\widehat{c}_{\beta j,\sigma}-\mu\sum_{i,\alpha,\sigma}\widehat{n}_{\alpha i,\sigma}+\frac{1}{2}\sum_{i}\sum_{\alpha,\beta,\gamma,\delta}\sum_{\sigma,\sigma'}U_i^{\alpha\beta\gamma\delta}\widehat{n}_{\alpha\delta i,\sigma}\,\widehat{n}_{\beta\gamma i,\sigma'}\\&-\frac{1}{2}\sum_{i}\sum_{\alpha,\beta,\gamma}\sum_{\sigma}U_i^{\alpha\beta\gamma\beta}\widehat{n}_{\alpha\gamma i,\sigma},\end{aligned} \tag{2.30}$$

where in order to be able to analyze the cases with different number of electrons, we have added the standard chemical potential term $-\mu\sum_{i,\alpha,\sigma}\widehat{n}_{\alpha i,\sigma}$.

The Hamiltonian (2.30) is more convenient for doing further approximations as compared to the Hamiltonian (2.28), since the density-matrix operator representation helps to better understand the physical meaning of different terms and to identify the largest terms in the Hamiltonian.

In the Hubbard model approximation, the following terms of the interaction part of the Hamiltonian (2.30) (the first term in the second line) are taken into account:

- Repulsion between two electrons on the same orbital with opposite spin ($\sim\delta_{\alpha\delta}\delta_{\beta\gamma}\delta_{\alpha\beta}$ at $\sigma'=\overline{\sigma}$), $\frac{1}{2}U_i^{\alpha\alpha\alpha\alpha}\widehat{n}_{\alpha i,\sigma}\widehat{n}_{\alpha i,\overline{\sigma}}$;
- Repulsion between two electrons on different orbitals with opposite spins ($\sim\delta_{\alpha\delta}\delta_{\beta\gamma}(1-\delta_{\alpha\beta})$ at $\sigma'=\overline{\sigma}$), $\frac{1}{2}U_i^{\alpha\beta\beta\alpha}\widehat{n}_{\alpha i,\sigma}\widehat{n}_{\beta i,\overline{\sigma}}$;
- Repulsion between two electrons on different orbitals with the same spin ($\sim\delta_{\alpha\delta}\delta_{\beta\gamma}(1-\delta_{\alpha\beta})$ at $\sigma=\sigma$), $\frac{1}{2}U_i^{\alpha\beta\beta\alpha}\widehat{n}_{\alpha i,\sigma}\widehat{n}_{\beta i,\sigma}$.

In addition to these, sometimes two other (spin-flipping terms) are included:

– Simultaneous flip of spins of two opposite-spin electrons on different orbitals ($\sim\delta_{\alpha\gamma}\delta_{\beta\delta}(1-\delta_{\alpha\beta})$ at $\sigma'=\overline{\sigma}$), $-\frac{1}{2}U_i^{\alpha\beta\alpha\beta}\widehat{c}^{+}_{\alpha i,\sigma}\widehat{c}_{\alpha i,\overline{\sigma}}\widehat{c}^{+}_{\beta i,\overline{\sigma}}\widehat{c}_{\beta i,\sigma}$;
– Pair transition of two electrons with opposite spins from one orbital to another ($\sim\delta_{\alpha\beta}\delta_{\gamma\delta}(1-\delta_{\alpha\gamma})$ at $\sigma'=\overline{\sigma}$), $-\frac{1}{2}U_i^{\alpha\alpha\gamma\gamma}\widehat{c}^{+}_{\alpha i,\sigma}\widehat{c}^{+}_{\alpha i,\overline{\sigma}}\widehat{c}_{\gamma i,\sigma}\widehat{c}_{\gamma i,\overline{\sigma}}$.

In other words, only the terms with no more than two different orbital indices in $U_i^{\alpha\beta\gamma\delta}$ (i.e., the terms with the largest overlap of the wave functions in the integral that defines interactions (2.26)) are taken into account. Accordingly, with this level of accuracy the remaining last term in Eq. (2.30) can be approximated as

$$-\frac{1}{2}\sum_i\sum_{\alpha,\beta,\gamma}\sum_\sigma U_i^{\alpha\beta\gamma\beta}\widehat{n}_{\alpha\gamma i,\sigma}\approx-\frac{1}{2}\sum_i\sum_{\alpha,\beta}\sum_\sigma U_i^{\alpha\beta\alpha\beta}\widehat{n}_{\alpha i,\sigma}.$$

Thus, we obtain the most general form of the Hubbard Hamiltonian [2–4]:

$$\begin{aligned}\widehat{H}=&-\sum_{i,j,\alpha,\sigma}t_{ij}^{\alpha}\widehat{c}^{+}_{\alpha i,\sigma}\widehat{c}_{\beta j,\sigma}-\mu\sum_{i,\alpha,\sigma}\widehat{n}_{\alpha i,\sigma}+\frac{1}{2}\sum_i\sum_\sigma U_i^{\alpha\alpha\alpha\alpha}\widehat{n}_{\alpha i,\sigma}\widehat{n}_{\alpha i,\overline{\sigma}}\\&+\frac{1}{2}\sum_i\sum_{\alpha\neq\beta,\sigma}U_i^{\alpha\beta\beta\alpha}\widehat{n}_{\alpha i,\sigma}\widehat{n}_{\beta i,\overline{\sigma}}+\frac{1}{2}\sum_i\sum_{\alpha\neq\beta,\sigma}U_i^{\alpha\beta\beta\alpha}\widehat{n}_{\alpha i,\sigma}\widehat{n}_{\beta i,\sigma}\\&-\frac{1}{2}\sum_i\sum_{\alpha\neq\beta,\sigma}U_i^{\alpha\beta\alpha\beta}\widehat{c}^{+}_{\alpha i,\sigma}\widehat{c}_{\alpha i,\overline{\sigma}}\widehat{c}^{+}_{\beta i,\overline{\sigma}}\widehat{c}_{\beta i,\sigma}\\&-\frac{1}{2}\sum_i\sum_{\alpha\neq\beta,\sigma}U_i^{\alpha\alpha\beta\beta}\widehat{c}^{+}_{\alpha i,\sigma}\widehat{c}^{+}_{\alpha i,\overline{\sigma}}\widehat{c}_{\beta i,\sigma}\widehat{c}_{\beta i,\overline{\sigma}}-\frac{1}{2}\sum_i\sum_{\alpha,\beta}\sum_\sigma U_i^{\alpha\beta\alpha\beta}\widehat{n}_{\alpha i,\sigma}.\end{aligned}\tag{2.31}$$

Usually, the site- and orbital-dependent interaction matrix elements are approximated by their averages: $U_i^{\alpha\alpha\alpha\alpha}\approx U$ (intra-orbital Coulomb repulsion), $U_i^{\alpha\beta\beta\alpha}\approx U'$ (inter-orbital Coulomb repulsion), $U_i^{\alpha\beta\alpha\beta}\approx J$ (Hund exchange energy) and $U_i^{\alpha\alpha\beta\beta}\approx\widetilde{J}$ (pair hopping energy). This gives

$$\begin{aligned}\widehat{H}=&-\sum_{i,j,\alpha,\sigma}t_{ij}^{\alpha}\widehat{c}^{+}_{\alpha i,\sigma}\widehat{c}_{\beta j,\sigma}-\mu\sum_{i,\alpha,\sigma}\widehat{n}_{\alpha i,\sigma}+\frac{1}{2}U\sum_i\sum_\sigma\widehat{n}_{\alpha i,\sigma}\widehat{n}_{\alpha i,\overline{\sigma}}\\&+\frac{1}{2}U'\sum_i\sum_{\alpha\neq\beta,\sigma}\widehat{n}_{\alpha i,\sigma}\widehat{n}_{\beta i,\overline{\sigma}}+\frac{1}{2}U'\sum_i\sum_{\alpha\neq\beta,\sigma}\widehat{n}_{\alpha i,\sigma}\widehat{n}_{\beta i,\sigma}\\&-\frac{1}{2}J\sum_i\sum_{\alpha\neq\beta,\sigma}\widehat{c}^{+}_{\alpha i,\sigma}\widehat{c}_{\alpha i,\overline{\sigma}}\widehat{c}^{+}_{\beta i,\overline{\sigma}}\widehat{c}_{\beta i,\sigma}-\frac{1}{2}\widetilde{J}\sum_i\sum_{\alpha\neq\beta,\sigma}\widehat{c}^{+}_{\alpha i,\sigma}\widehat{c}^{+}_{\alpha i,\overline{\sigma}}\widehat{c}_{\beta i,\sigma}\widehat{c}_{\beta i,\overline{\sigma}}\\&-\frac{1}{2}J\sum_i\sum_\alpha\sum_\sigma\widehat{n}_{\alpha i,\sigma}.\end{aligned}\tag{2.32}$$

The terms in the third line significantly complicate the calculations, since they cannot be expressed in terms of the density operators. Fortunately, in many cases they are much smaller as compared to the other interaction terms and can be neglected. Next, the last term in (2.32) can be incorporated into the chemical potential term (leading to a renormalization of the chemical potential). Therefore, usually the last three terms in the Hubbard Hamiltonian (2.32) are neglected. The remaining terms can be rewritten by using modified interaction parameters in the Kanamori form [3]:

$$\begin{aligned}\widehat{H} = & -\sum_{i,j,\alpha,\sigma} t_{ij}^{\alpha}\widehat{c}_{\alpha i,\sigma}^{+}\widehat{c}_{\alpha j,\sigma} - \mu\sum_{i}\sum_{\alpha,\sigma}\widehat{n}_{\alpha i,\sigma} + \frac{U}{2}\sum_{i}\sum_{\alpha,\sigma}\widehat{n}_{\alpha i,\sigma}\widehat{n}_{\alpha i,\bar{\sigma}} \\ & + (U-2J)\frac{1}{2}\sum_{i}\sum_{\alpha\neq\beta,\sigma}\widehat{n}_{\alpha i,\sigma}\widehat{n}_{\beta i,\bar{\sigma}} + (U-3J)\frac{1}{2}\sum_{i}\sum_{\alpha\neq\beta,\sigma}\widehat{n}_{\alpha i,\sigma}\widehat{n}_{\beta i,\sigma}.\end{aligned} \tag{2.33}$$

Indeed, since the inter-orbital same-spin interaction energy of two electrons is smaller than the corresponding energy in the case of opposite spins (due to Pauli principle), it is meaningful to use different interaction parameters for these quantities, $U - 2J$ and $U - 3J$.

Finally, since the inter-orbital interactions are usually significantly smaller than the intra-orbital interaction U, in the simplest case only the interaction term $U\sum_{i}\widehat{n}_{\alpha i,\uparrow}\widehat{n}_{\alpha i,\downarrow}$ is included in the Hubbard model. For the kinetic part of the Hamiltonian (2.33), most often only the terms corresponding to the nearest-neighbor (hopping parameter t), next nearest-neighbor (t') and next-next nearest neighbor (t'') hoppings are included. This is a well-justified approximation, since the hopping parameters rapidly decrease with increasing of the inter-atomic distance l (e.g., for d-electrons $t \sim 1/l^{5/2}$). Due to this reason, in many cases, it is also put $t' = t'' = 0$.

In the next section, we will analyze solutions of the approximated one-band Hubbard model that includes only terms with t and U, defined above. The corresponding Hamiltonian has a rather simple form:

$$\widehat{H} = -t\sum_{i,j,\sigma}\widehat{c}_{i,\sigma}^{+}\widehat{c}_{j,\sigma} - \mu\sum_{i,\sigma}\widehat{c}_{i,\sigma}^{+}\widehat{c}_{i,\sigma} + U\sum_{i}\widehat{n}_{i,\uparrow}\widehat{n}_{i,\downarrow}. \tag{2.34}$$

In many cases, it is convenient to write down the noninteracting part of this Hamiltonian in the momentum representation:

$$-t\sum_{i,j,\sigma}\widehat{c}_{i,\sigma}^{+}\widehat{c}_{j,\sigma} - \mu\sum_{i,\sigma}\widehat{c}_{i,\sigma}^{+}\widehat{c}_{i,\sigma} \rightarrow \sum_{\vec{k},\sigma}\left(\varepsilon\left(\vec{k}\right) - \mu\right)\widehat{c}_{\vec{k},\sigma}^{+}\widehat{c}_{\vec{k},\sigma},$$

where $\varepsilon\left(\vec{k}\right)$ is the free-electron spectrum. In particular, in the case of d-dimensional hypercubic lattice with the lattice parameter a,

$$\varepsilon\left(\vec{k}\right) = -2t\sum_{i=1}^{d}\cos\left(ak_i\right). \tag{2.35}$$

2.2 Early Solutions of the Hubbard Model

Despite formal simplicity of the Hamiltonian (2.34), the problem cannot be solved exactly, except the one-dimensional case [5]. Before introducing the DMFT approach to solve the Hubbard model, in this part of Chap. 2 we give a summary on early, "pre-DMFT" (1963–1991), approximate approaches to the Hubbard model and discuss their strengths and weaknesses. We begin with the simplest second-order perturbation theory small- and large-U solutions and proceed with the solutions applicable to an arbitrary value of U (for a more detailed summary of early approaches, see book [1]).

2.2.1 *Small* U *Solution*

We begin with a simple second-order perturbation-theory solution of the Hubbard model in the case of small values of U. To analyze the spectral properties of the system, it is convenient to calculate the electron retarded GF

$$
\begin{aligned}
G^R_{ij,\sigma}(t-t') &= -i\theta(t-t')\left\langle\left\{\hat{c}_{i,\sigma}(t), \hat{c}^+_{j,\sigma}(t')\right\}\right\rangle \\
&\equiv -i\theta(t-t')\left(\left\langle\hat{c}_{i,\sigma}(t)\hat{c}^+_{j,\sigma}(t')\right\rangle + \left\langle\hat{c}^+_{j,\sigma}(t')\hat{c}_{i,\sigma}(t)\right\rangle\right),
\end{aligned} \tag{2.36}
$$

where $\langle\ldots\rangle$ means the quantum statistical average and $\theta(t - t')$ is the standard Heavyside's step function. The retarded GF satisfies the following equation of motion:

$$
\begin{aligned}
\left(i\frac{d}{dt}+\mu\right)G^R_{ij,\sigma}(t-t') &= \delta_{ij}\delta(t-t') - \sum_l t_{i,l}G^R_{lj,\sigma}(t-t') + U(-i)\theta(t-t') \\
&\times\left\langle\left\{\hat{c}_{i,\sigma}(t)\hat{n}_{i,\bar{\sigma}}(t), \hat{c}^+_{j,\sigma}(t')\right\}\right\rangle.
\end{aligned} \tag{2.37}
$$

The last equation can be obtained by differentiating Eq. (2.36) with respect to time and using the Heisenberg equation of motion for the annihilation operators

$$
i\frac{d\hat{c}_{i,\sigma}}{dt} = \left[\hat{c}_{i,\sigma}, \hat{H}\right], \tag{2.38}
$$

where $\hat{H}$ is the Hamiltonian (2.34).

In the noninteracting case ($U = 0$), Eq. (2.37) can be easily solved. The solution in the momentum–frequency representation has the following form:

$$G_{\sigma}^{R(0)}\left(\vec{k},\omega\right)=\frac{1}{\omega-\varepsilon\left(\vec{k}\right)+\mu+i\delta}, \tag{2.39}$$

where $\delta=\ \mathrm{sgn}\left(\varepsilon\left(\vec{k}\right)-\mu\right)$. To calculate the second order in U correction to the GF (2.39), we use the following strategy. First, we write down the Dyson equation for the electron's GF in the site–frequency representation:

$$G_{\sigma ij}^{R}(\omega)=G_{\sigma ij}^{R(0)}(\omega)+G_{\sigma il}^{R(0)}(\omega)\Sigma_{\sigma lm}^{R}(\omega)G_{\sigma mj}^{R}(\omega), \tag{2.40}$$

where $\Sigma_{\sigma lm}^{R}(\omega)$ is the free-electron self-energy (here and below the sum over same indices in the products of functions is assumed). Next, we expand the interacting GF and the self-energy up to the U^2 (second order in U) terms:

$$G_{ij}^{(0)}+G_{ij}^{(1)}+G_{ij}^{(2)}=G_{ij}^{(0)}+G_{il}^{(0)}\left[\Sigma_{lm}^{(1)}+\Sigma_{lm}^{(2)}\right]\left[G_{mj}^{(0)}+G_{mj}^{(1)}+G_{mj}^{(2)}\right] \tag{2.41}$$

(we omit the frequency variable and the index R to shorten notations; below we drop the index R in front of the GF and other functions, it will be assumed they are retarded unless specified), and write down the linear and the second order in U equations:

$$G_{ij}^{(1)}=G_{il}^{(0)}\Sigma_{lm}^{(1)}G_{mj}^{(0)}, \tag{2.42}$$

$$\begin{aligned}G_{ij}^{(2)}&=G_{il}^{(0)}\Sigma_{lm}^{(2)}G_{mj}^{(0)}+G_{il}^{(0)}\Sigma_{lm}^{(1)}G_{mj}^{(1)}\\&\equiv G_{il}^{(0)}\Sigma_{lm}^{(2)}G_{mj}^{(0)}+G_{il}^{(0)}\Sigma_{lm}^{(1)}G_{mp}^{(0)}\Sigma_{ps}^{(1)}G_{sj}^{(0)}.\end{aligned} \tag{2.43}$$

Then, one can calculate the first-order and the second-order *GFs* $G_{ij}^{(1)}$ and $G_{ij}^{(2)}$ by using the many-body perturbation theory:

$$\begin{aligned}iG_{\sigma ij}(t_1,t_2)&=\frac{1}{Z}\left\langle\widehat{c}_{\sigma i}(t_1)\widehat{c}_{\sigma j}^{+}(t_2)e^{-i\int H_U(t)dt}\right\rangle\\&=\left\langle\widehat{c}_{\sigma i}(t_1)\widehat{c}_{\sigma j}^{+}(t_2)\right\rangle+\left\langle\widehat{c}_{\sigma i}(t_1)\widehat{c}_{\sigma j}^{+}(t_2)(-i)\int H_U(t)dt\right\rangle\\&+\ \left\langle\widehat{c}_{\sigma i}(t_1)\widehat{c}_{\sigma j}^{+}(t_2)(-i)^2\frac{1}{2}\int H_U(t)dt\int H_U(t')dt'\right\rangle+\ldots,\end{aligned} \tag{2.44}$$

where

$$Z = \left\langle e^{-i\int H_U(t)dt} \right\rangle$$
$$= 1 - i\left\langle \int H_U(t)dt \right\rangle + (-i)^2 \frac{1}{2} \left\langle \int H_U(t)dt \int H_U(t')dt' \right\rangle + \ldots \quad (2.45)$$

is the partition function and the averages in the last two lines in Eq. (2.44) include only connected diagrams.

After $G_{ij}^{(1)}$ and $G_{ij}^{(1)}$ are calculated, one can obtain $\Sigma_{lm}^{(1)}$ and $\Sigma_{lm}^{(2)}$ from Eqs. (2.42) and (2.43). This gives for the second-order self-energy

$$\Sigma_{ij\sigma}(t - t') = \delta_{ij}\delta(t - t')Un_{\bar{\sigma}} + \delta_{ij}U^2 G_{ii\sigma}^{(0)}(t - t')G_{ii\bar{\sigma}}^{(0)}(t' - t)G_{ii\bar{\sigma}}^{(0)}(t - t'), \quad (2.46)$$

or in the momentum–frequency representation,

$$\Sigma_\sigma\left(\omega, \vec{k}\right) = Un_{\bar{\sigma}}$$
$$+ U^2 \int \frac{d\omega'}{(2\pi)} \int \frac{d\omega''}{(2\pi)} G_\sigma^{(0)\mathrm{loc}}(\omega')G_{ii\bar{\sigma}}^{(0)\mathrm{loc}}(\omega'')G_{ii\bar{\sigma}}^{(0)\mathrm{loc}}(\omega' + \omega'' - \omega), \quad (2.47)$$

where the functions under the integral are local GFs

$$G_\sigma^{(0)\mathrm{loc}}(\omega) \equiv G_{ii\sigma}^{(0)\mathrm{loc}}(\omega) = \sum_{\vec{k}} G_\sigma^{(0)}\left(\omega, \vec{k}\right). \quad (2.48)$$

Though it is possible to show that the second order in U solution gives both MIT and ferromagnetism, the first phenomenon occurs at large values of $U(\gg t)$ (depending on the type of lattice), far beyond the region of applicability of the second-order perturbation theory.

2.2.2 *Large* U *Solution*

In a similar to the previous subsection way, one can obtain the second order in hopping parameter solution for the GF, with the atomic-limit ($t = 0$) GF as zero-order approximation. The last function can be easily obtained:

$$G_{ij,\sigma}^{(at)}(t - t') = -i\theta(t - t')\delta_{ij}\left[e^{-i(U-\mu)t}\langle \hat{n}_{i,\bar{\sigma}}\rangle + e^{i\mu t}(1 - \langle \hat{n}_{i,\bar{\sigma}}\rangle)\right], \quad (2.49)$$

or in the frequency domain:

$$G_{ij,\sigma}^{(at)}(\omega) = \delta_{ij}\frac{1-\langle \widehat{n}_{i,\bar{\sigma}}\rangle}{\omega+\mu+i\delta} + \delta_{ij}\frac{\langle \widehat{n}_{i,\bar{\sigma}}\rangle}{\omega+\mu-U+i\delta}. \tag{2.50}$$

Since in the atomic limit in the momentum representation $G_\sigma^{(at)}(\omega) = \frac{1}{\omega+\mu-\Sigma_\sigma^{(at)}(\omega)}$, one can obtain the atomic self-energy from (2.50):

$$\Sigma_\sigma^{(at)}(\omega) = Un_{\bar{\sigma}} + \frac{U^2 n_{\bar{\sigma}}(1-n_{\bar{\sigma}})}{\omega - U(1-n_{\bar{\sigma}})} \tag{2.51}$$

(we consider spatially homogeneous case and use the notation $n_{\bar{\sigma}} = \langle \widehat{n}_{i,\bar{\sigma}}\rangle$).

The t^2-correction to Eq. (2.51) can be obtained by repeating the steps between Eqs. (2.39) and (2.46). The only difference from Sect. 2.2.2 is that instead of the interaction part of the Hamiltonian $H_U(t)$, in the present case one uses the kinetic part $H_t(t)$ as the perturbation. The final result for the self-energy is

$$\Sigma_{\sigma ml}(t-t') = \delta_{ml}\Sigma_\sigma^{(at)}(t-t') - \delta(t-t')t_{ml} + \sum_{p,d,b} t_{mp}G_\sigma^{(at)}(t-t')t_{pm}, \tag{2.52}$$

or in the momentum–frequency representation:

$$\Sigma_\sigma\left(\omega,\vec{k}\right) = Un_{\bar{\sigma}} + \frac{U^2 n_{\bar{\sigma}}(1-n_{\bar{\sigma}})}{\omega - U(1-n_{\bar{\sigma}})} - t\left(\vec{k}\right) + t^2\left(\vec{k}\right)G_\sigma^{(at)}(\omega), \tag{2.53}$$

where $t\left(\vec{k}\right)$ is the momentum Fourier transform of the hopping matrix t_{ij}, connected with the free-electron spectrum as $\varepsilon\left(\vec{k}\right) = t\left(\vec{k}\right) - \mu$ (see Eq. (2.35)).

Similar to the small U case, the solution (2.53) gives both an MIT and an FM state. However, the transition happens at rather large values of U, thus higher order in hopping terms in the expansion are needed to get the transition at smaller Us.

2.2.3 *Hartree and Hartree–Fock Solutions*

The solutions obtained in the previous two subsections are valid in the limits of small and large Us. From now on until the end of Sect. 2.2, we will discuss early solutions of the Hubbard model valid for all values of U (non-perturbative solutions). We begin with the simplest approaches that give such "universal" solution—the Hartree (H) and the Hartree–Fock (HF) mean-field approximations (see, e.g., Ref. [1]). The key step in these approaches is to approximate the double-occupancy (interaction) operator in the Hubbard Hamiltonian $\widehat{D}_i = \widehat{c}_{i\uparrow}^+\widehat{c}_{i\uparrow}\widehat{c}_{i\downarrow}^+\widehat{c}_{i\downarrow}$ by a sum of two bilinear operators:

$$\widehat{D}_i \approx \widehat{D}_i^H + \widehat{D}_i^F, \tag{2.54}$$

where

$$\widehat{D}_i^H = \left\langle \widehat{c}_{i\uparrow}^+ \widehat{c}_{i\uparrow} \right\rangle \widehat{c}_{i\downarrow}^+ \widehat{c}_{i\downarrow} + \widehat{c}_{i\uparrow}^+ \widehat{c}_{i\uparrow} \left\langle \widehat{c}_{i\downarrow}^+ \widehat{c}_{i\downarrow} \right\rangle - \left\langle \widehat{c}_{i\uparrow}^+ \widehat{c}_{i\uparrow} \right\rangle \left\langle \widehat{c}_{i\downarrow}^+ \widehat{c}_{i\downarrow} \right\rangle \tag{2.55}$$

is the Hartree and

$$\widehat{D}_i^F = -\left\langle \widehat{c}_{i\uparrow}^+ \widehat{c}_{i\downarrow} \right\rangle \widehat{c}_{i\downarrow}^+ \widehat{c}_{i\uparrow} - \widehat{c}_{i\uparrow}^+ \widehat{c}_{i\downarrow} \left\langle \widehat{c}_{i\downarrow}^+ \widehat{c}_{i\uparrow} \right\rangle + \left\langle \widehat{c}_{i\uparrow}^+ \widehat{c}_{i\downarrow} \right\rangle \left\langle \widehat{c}_{i\downarrow}^+ \widehat{c}_{i\uparrow} \right\rangle \tag{2.56}$$

is the Fock term (the last constant terms in (2.55) and (2.56) are added to assure that the averages of the approximate $\widehat{D}^H + \widehat{D}^F$ and of the exact $\widehat{D}$ operators are equal).

One can write down the approximate operator (2.56) in a more symmetric form by using the generators of the spin-rotation algebra – spin operators $\widehat{S}^+ = \widehat{S}^x + i\widehat{S}^y = \widehat{c}_\uparrow^+ \widehat{c}_\downarrow$, $\widehat{S}^+ = \widehat{S}^x - i\widehat{S}^y = \widehat{c}_\downarrow^+ \widehat{c}_\uparrow$ and $\widehat{S}^z = \frac{1}{2}\left(\widehat{c}_\uparrow^+ \widehat{c}_\uparrow - \widehat{c}_\downarrow^+ \widehat{c}_\downarrow\right)$, where $\widehat{S}^i = \left(\widehat{c}_\uparrow^+, \widehat{c}_\downarrow^+\right) \frac{1}{2} \widehat{\sigma}^i \begin{pmatrix} \widehat{c}_\uparrow \\ \widehat{c}_\downarrow \end{pmatrix}$ $(i = x, y, z)$ and

$$\widehat{\sigma}^x = \begin{pmatrix} 0 & 1 \\ 1 & 0 \end{pmatrix}, \widehat{\sigma}^y = \begin{pmatrix} 0 & -i \\ i & 0 \end{pmatrix}, \widehat{\sigma}^z = \begin{pmatrix} 1 & 0 \\ 0 & -1 \end{pmatrix}$$

are the Pauli matrices (we skip the site indices in the spin operators for sake of simplicity of the notations).

Using the spin operators one can show that

$$\widehat{D}_i^H = \langle \widehat{n}_{i\uparrow} \rangle \widehat{n}_{i\downarrow} + \widehat{n}_{i\uparrow} \langle n_{i\downarrow} \rangle - \langle \widehat{n}_{i\uparrow} \rangle \langle n_{i\downarrow} \rangle, \tag{2.57}$$

$$\widehat{D}_i^F = -\left\langle \widehat{S}_i^+ \right\rangle \widehat{S}_i^- - \widehat{S}_i^+ \left\langle \widehat{S}_i^- \right\rangle + \left\langle \widehat{S}_i^+ \right\rangle \left\langle \widehat{S}_i^- \right\rangle. \tag{2.58}$$

Obviously, general solution of the HF problem is still a nontrivial task due to the site-dependence of the Hartree and HF operators. In particular, this dependence may result in different spatially nonhomogeneous (spin- and charge-) orders in the system. One obvious example of such an order is an AFM state. Here we present results for the spin GFs in the simplest, spatially homogeneous (when the averages in Eqs. (2.57) and (2.58) are site-independent), case:

$$G_\sigma\left(\omega, \vec{k}\right) = \frac{1}{\omega - \left(\varepsilon\left(\vec{k}\right) - \mu\right) - Un_{\bar{\sigma}} - \frac{U|S|^2}{\omega - \left(\varepsilon_\sigma\left(\vec{k}\right) - \mu\right) - Un_\sigma} + i\delta}, \tag{2.59}$$

where we introduced $S = \left\langle \widehat{S}_i^- \right\rangle = \left\langle \widehat{S}_i^+ \right\rangle^*$.

The last equation is the solutions for the GFs in the HF approximation. It includes also a partial case of the Hartree approximation that corresponds to $S = 0$. Despite its simplicity, the Hartree solution, with the spectrum

$$E_\sigma\left(\vec{k}\right) = \varepsilon\left(\vec{k}\right) - \mu + U\langle\widehat{n_{\bar{\sigma}}}\rangle, \tag{2.60}$$

can describe FM state when the spin-up and spin-down densities are sufficiently different, $n_\downarrow \neq n_\uparrow$, i.e. when one-spin band is lower in energy than the other and the electrons prefer to occupy the lower energy spin states. On the other hand, Hartree approximation does not give MIT. This can be seen directly, e.g., by analyzing the case of half-filling, where $n_\downarrow = n_\uparrow$ and both spin spectra are the same and equally shifted by U, i.e., one cannot get a gap in the spectrum with increasing U.

The general HF solution (2.59) is much richer comparing to the Hatree one. Besides possible ferromagnetism, the HF spectrum consists of two different bands for each spin (there are two poles in the GF), which split at large values of U that corresponds to an MIT. Namely, the spectra of the two bands are

$$E_\pm\left(\vec{k}\right) = \varepsilon\left(\vec{k}\right) - \mu + \frac{U}{2}(n_\uparrow + n_\downarrow) \\ \pm\sqrt{\left(\varepsilon\left(\vec{k}\right) - \mu + \frac{U}{2}(n_\uparrow + n_\downarrow)\right)^2 - U^2 n_\uparrow n_\downarrow + U^2|S|^2}. \tag{2.61}$$

Using these dispersions, the GFs can be written as

$$G\left(\omega, \vec{k}\right) = \frac{Z_+\left(\vec{k}\right)}{\omega - E_+\left(\vec{k}\right)} + \frac{Z_-\left(\vec{k}\right)}{\omega - E_-\left(\vec{k}\right)}, \tag{2.62}$$

where

$$Z_\pm\left(\vec{k}\right) = \frac{1}{2} \pm \frac{U(-n_\uparrow + n_\downarrow)}{4\sqrt{\left(\varepsilon\left(\vec{k}\right) - \mu + \frac{U}{2}(n_\uparrow + n_\downarrow)\right)^2 - U^2 n_\uparrow n_\downarrow + U^2|S|^2}}, \tag{2.63}$$

are the corresponding quasiparticle weights.

The HF phase diagram includes both FM and AFM phases (in particular, the HF solution has the FM ground state when the number of electrons is smaller than 1 (below half-filling) and U is infinitely large). Also, the HF solution gives an MIT. However, similar to the Hartree approximation, this solution is not very helpful for understanding physics of strongly correlated systems since it basically misses electron–electron correlation effects.

2.2.4 Hubbard-I Solution

In the first solution of his model Hubbard used a different, more refined as compared the HF approximation, approach [2]. He considered the exact retarded GF in Eq. (2.37),

$$\left(i\frac{d}{dt}+\mu\right)G_{ij,\sigma}(t-t') = \delta_{ij}\delta(t-t') - \sum_l t_{il}G_{lj,\sigma}(t-t') + U\Gamma_{ij,\sigma}(t-t'), \quad (2.64)$$

where

$$\Gamma_{ij,\sigma}(t-t') = (-i)\theta(t-t')\left\langle\left\{\hat{c}_{i,\sigma}(t)\hat{n}_{i,\bar{\sigma}}(t), \hat{c}^+_{j,\sigma}(t')\right\}\right\rangle \quad (2.65)$$

is a new retarded GF for the composite operator

$$\hat{c}_{i,\sigma}(t)\hat{n}_{i,\bar{\sigma}}(t) \quad (2.66)$$

(for a doubly occupied (excited) atomic state). This GF satisfies the following equation of motion:

$$\begin{aligned} i\frac{d}{dt}\Gamma_{ij,\sigma}(t-t') &= \delta_{ij}\delta(t-t')\langle\hat{n}_{i,\bar{\sigma}}(t)\rangle + (U-\mu)\Gamma_{ij,\sigma}(t-t') \\ &+ i\theta(t-t')\sum_m t_{im}\left\langle\left\{\hat{c}_{m,\sigma}\hat{n}_{i,\bar{\sigma}}, \hat{c}^+_{j,\sigma}(t')\right\}\right\rangle \\ &- i\theta(t-t')\sum_m t_{li}\left\langle\left\{\hat{c}_{i,\sigma}(t)\hat{c}^+_{l,\bar{\sigma}}\hat{c}_{i\bar{\sigma}}, \hat{c}^+_{j,\sigma}(t')\right\}\right\rangle \\ &+ i\theta(t-t')\sum_m t_{im}\left\langle\left\{\hat{c}_{i,\sigma}(t)\hat{c}^+_{i,\bar{\sigma}}\,\hat{c}_{m,\bar{\sigma}}, \hat{c}^+_{j,\sigma}(t')\right\}\right\rangle. \end{aligned} \quad (2.67)$$

Since the system of Eqs. (2.64) and (2.67) cannot be solved exactly, Hubbard decoupled the opposite spin states in the last equation by approximating in the last three terms: $\hat{c}_{m,\sigma}\hat{n}_{i,\bar{\sigma}} \approx \hat{c}_{m,\sigma}\langle\hat{n}_{i,\bar{\sigma}}\rangle$, $\hat{c}_{i,\sigma}(t)\hat{c}^+_{l,\bar{\sigma}}\hat{c}_{i\bar{\sigma}} \approx \hat{c}_{i,\sigma}(t)\langle\hat{c}^+_{l,\bar{\sigma}}\hat{c}_{i\bar{\sigma}}\rangle$, $\hat{c}_{i,\sigma}(t)\hat{c}^+_{i,\bar{\sigma}}\,\hat{c}_{m,\bar{\sigma}} \approx \hat{c}_{i,\sigma}(t)\langle\hat{c}^+_{i,\bar{\sigma}}\,\hat{c}_{m,\bar{\sigma}}\rangle$. This, Hubbard-I, approximation allows one to solve the resulting system of equations exactly. Namely, it is easy to show that the approximated last two terms in Eq. (2.67) cancel, and the equation becomes

$$\begin{aligned} i\frac{d}{dt}\Gamma_{ij,\sigma}(t-t') &= \delta_{ij}\delta(t-t')\langle\hat{n}_{i,\bar{\sigma}}(t)\rangle + (U-\mu)\Gamma_{ij,\sigma}(t-t') \\ &- \langle\hat{n}_{i,\bar{\sigma}}\rangle\sum_m t_{im}G_{mj,\sigma}(t-t'). \end{aligned} \quad (2.68)$$

The system of Eqs. (2.64) and (2.68) can be easily solved, resulting in the following two-pole electron GF in the momentum–frequency representation:

$$G_\sigma\left(\omega, \vec{k}\right) = \frac{Z_{\sigma+}\left(\vec{k}\right)}{\omega - E_{\sigma+}\left(\vec{k}\right)} + \frac{Z_{\sigma-}\left(\vec{k}\right)}{\omega - E_{\sigma-}\left(\vec{k}\right)}, \tag{2.69}$$

where

$$E_{\sigma\pm}\left(\vec{k}\right) = \frac{1}{2}\left[\varepsilon\left(\vec{k}\right) + U \pm \sqrt{\left(\varepsilon\left(\vec{k}\right) - U\right)^2 + 4\varepsilon\left(\vec{k}\right)Un_{\bar{\sigma}}}\right] \tag{2.70}$$

and

$$Z_{\sigma\pm}\left(\vec{k}\right) = \pm\left[\frac{E_{\sigma\pm}\left(\vec{k}\right) - U(1 - n_{\bar{\sigma}})}{E_{\sigma+}\left(\vec{k}\right) - E_{\sigma-}\left(\vec{k}\right)}\right] \tag{2.71}$$

are the band dispersions and the corresponding quasiparticle weights, correspondingly (we did not include the chemical potential in the solution, assuming that it shifts the free-electron spectrum).
From

$$G_\sigma\left(\omega, \vec{k}\right) = \frac{1}{\omega - \varepsilon\left(\vec{k}\right) - \Sigma_\sigma\left(\omega, \vec{k}\right)} \tag{2.72}$$

one can obtain the expression for the electron self-energy in the Hubbard-I approximation:

$$\Sigma_\sigma\left(\omega, \vec{k}\right) = Un_{\bar{\sigma}} + \frac{U^2 n_{\bar{\sigma}}(1 - n_{\bar{\sigma}})}{\omega - U(1 - n_{\bar{\sigma}})}, \tag{2.73}$$

which is nothing else but the atomic-limit self-energy (2.51)!

Therefore, the Hubbard-I approximation corresponds to the approximation of the electron self-energy by its atomic limit expression. Importantly, the solution (2.69) has two – lower and upper – (Hubbard) bands, typical for strongly correlated systems. Unfortunately, this solution gives and insulating (gapped) state for any U, i.e., Hubbard-I approximation does give an insulator-to-metal transition with changing U/t. Also, it gives FM phase only in special cases of the free-electron density of states (DOS). Another important shortcoming of the Hubbard-I solution is violated electron–hole symmetry, which results in many flaws of the solution.

2.2.5 *Hubbard-II (Falicov–Kimball) Solution*

One of the reasons for failures of the Hubbard-I solution is that it is based on an assumption of "frozen" (fixed) on-site spin density. To improve the solution, in particular to describe MIT, Hubbard proposed to take into account the time-resolved interactions of electrons [6]. To do this, it is convenient to assume the electrons with one of the spins ($\overline{\sigma}$) do not move (the analysis is significantly simplified in this case). This approximation, used also by Falicov and Kimball in their famous model [7], can be called Hubbard-II approximation. It corresponds to the following situation: any site the spin-σ electron hops on is either empty or occupied by another electron with spin $\overline{\sigma}$. Thus, the energy of the σ electron will fluctuate in time (between, roughly speaking, ε_0 (atomic level energy) and $\varepsilon_0 + U$). This effect can be described by a frequency- (or equivalently, time-) dependent correction $\Delta_\sigma(\omega)$ to the Hubbard-I self-energy:

$$\Sigma_\sigma(\omega) \to \Sigma_\sigma^{\text{atom}}(\omega - \Delta_\sigma(\omega)) = Un_{\overline{\sigma}} + \frac{U^2 n_{\overline{\sigma}}(1 - n_{\overline{\sigma}})}{\omega - \Delta_\sigma(\omega) - U(1 - n_{\overline{\sigma}})}. \tag{2.74}$$

Since this approximation corresponds to the motion of electron in a random field of the "frozen" opposite-spin electrons, it can be called the "alloy-analogy" approximation (AAA), (see, e.g., book [1]; we discuss the relation of the dynamic solutions of the alloy problems and of the Hubbard model in the next chapter).

As the next step, Hubbard used a constraint that the local GF

$$G_\sigma(\omega) = \sum_{\vec{k}} \frac{1}{\omega - \left(\varepsilon\left(\vec{k}\right) - \mu\right) - \Sigma_\sigma\left(\omega, \vec{k}\right)} \tag{2.75}$$

depends on the field $\Delta_\sigma(\omega)$ as:

$$\begin{aligned} G_\sigma(\omega) &= \frac{1}{\omega + \mu - \Delta_\sigma(\omega) - \Sigma_\sigma\left(\omega, \vec{k}\right)} \\ &\equiv \frac{1}{\omega - \Delta_\sigma(\omega) + \mu} + \frac{1}{\omega - \Delta_\sigma(\omega) + \mu - U}. \end{aligned} \tag{2.76}$$

The system of Eqs. (2.74)–(2.76) can be solved exactly. The most important improvement of the Hubbard-I solsution brought by the Hubbard-II approximation is the ability of the last solution to describe the MIT at finite U. In particular, in the case of Bethe lattice, the transition happens when the value of U is equal to the value of the bandwidth W.

2.2.6 *Hubbard-III Solution*

The next level of accuracy of the solution for the Hubbard model can be achieved with the Hubbard-III approximation by assuming that electrons with both spins can move [8]. In this case, the expression for the dynamical field $\Delta_\sigma(\omega)$ has to be generalized by adding two terms that take into account motion of the opposite spin electrons: (i) the term $\Delta_{\overline{\sigma}}(\omega)$ that describes the correction to the electron self-energy due to the process when the σ electron comes to and the $\overline{\sigma}$ electron leaves the site and (ii) the term $-\Delta_{\overline{\sigma}}(\omega - U)$ corresponding to the process when the σ electron comes to and the hole $\overline{\sigma}$ leaves the site (the frequency dependence $-\Delta_{\overline{\sigma}}(\omega - U)$ comes from the electron-hole symmetry). Therefore, the resulting total dynamical correction to the atomic self-energy is

$$\Delta_\sigma^{\text{tot}}(\omega) = \Delta_\sigma(\omega) + \Delta_{\overline{\sigma}}(\omega) - \Delta_{\overline{\sigma}}(\omega - U), \tag{2.77}$$

which is used instead of $\Delta_\sigma(\omega)$ in the corrected self-energy (2.74). Thus, in the Hubbard-III approximation one needs to solve the system of Eqs. (2.74)–(2.76) with $\Delta_\sigma(\omega) \rightarrow \Delta_\sigma^{\text{tot}}(\omega)$ for both spins.

There are some notable differences between the Hubbard-II and Hubbard-III solutions. For example, for the system with semielliptic DOS in PM case (when $\Delta_\sigma^{\text{tot}}(\omega) = 3\Delta_\sigma(\omega)$) the last approximation gives a lower critical U for the MIT. The most important improvement brought by the Hubbard-III approximation is independent noninteracting bands at large Us, in agreement with the Mott's statement that charge excitations move freely when the spin background is neglected.

2.2.7 *Projection Technique Solutions: Gutzwiller Approximation and the t–J Model*

The solution of the Hubbard model can be in some cases forced to a physically meaningful "direction" by using different constraints, i.e., by taking into account only certain subspaces of the solution space. One of the examples is the case of large values of U when the double occupancy of the sites is suppressed. To take this effect into account, Gutzwiller proposed [9, 10] the following wave function relevant to the Hubbard model and other systems with a local interaction:

$$|\Psi\rangle_g = g^{\widehat{D}}|\Psi_0\rangle, \tag{2.78}$$

where $|\Psi_0\rangle$ is the ground-state (one-electron product) wave function, e.g., $|\Psi_0\rangle = \prod_{|\vec{k}|<k_F,\sigma} \widehat{c}^+_{\vec{k}\sigma}|0\rangle$ in the PM case, g is a variational parameter and $\widehat{D} = \sum_i \widehat{c}^+_{i\uparrow}\widehat{c}_{i\uparrow}\widehat{c}^+_{i\downarrow}\widehat{c}_{i\downarrow} \equiv \sum_i \widehat{D}_i$ is the double occupancy operator.

The Gutzwiller operator $g^{\widehat{D}}$ can be rewritten as $g^{\widehat{D}} \equiv e^{ln(g)\sum_i \widehat{D}_i} = \prod_i \left(1-(1-g)\widehat{D}_i\right)$. As it follows from the last equation, when $g \to 0$ the operator $g^{\widehat{D}}$ reduces to the Gutzwiller projector on states with no double occupancy:

$$\widehat{P}_{D=0} = lim_{g\to 0} g^{\widehat{D}} = \prod_i \left(1-\widehat{D}_i\right). \tag{2.79}$$

At finite gs and $g < 1$, the Gutzwiller operator $g^{\widehat{D}} = \prod_i \left(1-(1-g)\widehat{D}_i\right)$ reduces the number of doubly occupied states.

To find the Gutzwiller wave function $|\Psi\rangle_g$, one needs to obtain the value of g (and in general case, other possible parameters) in $|\Psi_0\rangle$ by minimizing the ground-state energy

$$E = \frac{{}_g\left\langle \Psi|\widehat{H}|\Psi\right\rangle_g}{{}_g\langle\Psi|\Psi\rangle_g}. \tag{2.80}$$

Once $|\Psi\rangle_g$ is obtained, all expectation values of the physical quantities of interest can be calculated:

$$A = \frac{{}_g\left\langle \Psi|\widehat{A}|\Psi\right\rangle_g}{{}_g\langle\Psi|\Psi\rangle_g}. \tag{2.81}$$

Quite remarkably, using this approach Brinkman and Rice were capable to obtain an MIT in the infinite-dimensional case at half-filling [11]. In fact, the Gutzwiller approximation for the wave function is exact in the limit of infinite dimensions. The authors of the last paper have shown that the critical interaction at which the transition occurs is $U = 8\overline{T}$, where $\overline{T}$ is the average kinetic energy of the Fermi sea per site. Unfortunately, as it follows from the results of the calculations within the Gutzwiller approach, such a transition does not exist in finite dimensions.

Projection operator on the subspace of states with no doubly occupancy (2.78), $\widehat{P}_{D=0} \equiv \widehat{P}$, can be also used to derive the t-J model [12–14], an approximation to the Hubbard model at large Us. The advantage of the t-J model is that it has a small parameter $\left(\frac{t}{U}\right)$ allowing perturbative analysis (the price paid for this is complex operators in the zero-order "noninteracting" Hamiltonian (operators of singly occupied sites, instead of simple electron creation and annihilation operators)).

To derive the t-J model, one can start with derivation of the equation for the projected wave function $\widehat{P}|\Psi\rangle$. Once the corresponding equation, that can be written as $\widehat{\mathcal{H}}\widehat{P}|\Psi\rangle = E\widehat{P}|\Psi\rangle$, is found it will immediately give the effective Hamiltonian $\widehat{\mathcal{H}}$ for the projected subspace. To find this Hamiltonian, it is convenient to "move" the Hubbard Hamiltonian to the subspace of singly occupied sites by using $\widehat{P}$ and another operator $\widehat{Q} = 1 - \widehat{P}$. Both $\widehat{P}$ and $\widehat{Q}$ operators are projection operators that satisfy $\widehat{P}^2 = \widehat{P}$, $\widehat{Q}^2 = \widehat{Q}$, $\widehat{P}\widehat{Q} = \widehat{Q}\widehat{P} = 0$. Then, to obtain the equation for the projected wave function $\widehat{P}|\Psi\rangle$ (see, e.g., book [15]), we rewrite the original equation for the Hubbard model wave function

$$\widehat{H}|\Psi\rangle = E|\Psi\rangle \tag{2.82}$$

in the following form:

$$\widehat{H}\left(\widehat{P} + \widehat{Q}\right)|\Psi\rangle = E\left(\widehat{P} + \widehat{Q}\right)|\Psi\rangle. \tag{2.83}$$

It is important to note that the state $\widehat{P}|\Psi\rangle$ is an element of the low-energy subspace Hamiltonian $\widehat{\mathcal{H}}$, while the states $\widehat{Q}|\Psi\rangle$ have much higher, ~U, energies. Then, multiplying Eq. (2.83) from left by $\widehat{Q}$ and performing some trivial transformations, one can obtain $\widehat{Q}|\Psi\rangle = -\left(\widehat{Q}\widehat{H}\widehat{Q} - E\right)^{-1}\widehat{Q}\widehat{H}\widehat{P}|\Psi\rangle$. Substitution of this expression back into Eq. (2.83) gives $\left(\widehat{H} - \widehat{H}\left(\widehat{Q}\widehat{H}\widehat{Q} - E\right)^{-1}\widehat{Q}\widehat{H}\right)\widehat{P}|\Psi\rangle = E\left(\widehat{P} + \widehat{Q}\right)|\Psi\rangle$. Next, multiplying the last equation from left by $\widehat{P}$ and transforming the result by using $\widehat{P}\widehat{H}\widehat{P} = \widehat{P}\widehat{H}_U\widehat{P} = 0$ and $\widehat{Q}\widehat{P} = 0$ gives (still exact!) equation for the projected wave function $\widehat{P}|\Psi\rangle$:

$$-\widehat{P}\widehat{H}\widehat{Q}\left[\left(\widehat{Q}\widehat{H}\widehat{Q} - E\right)^{-1}\right]\left(\widehat{Q}\widehat{H}\widehat{P}\right)\widehat{P}|\Psi\rangle \equiv \widehat{\widehat{\mathcal{H}}}\widehat{P}|\Psi\rangle = E\widehat{P}|\Psi\rangle. \tag{2.84}$$

To obtain a low-energy (second order in hopping) approximation for the Hamiltonian $\widehat{\widehat{\mathcal{H}}}$ we take into account the fact that operators $\widehat{P}\widehat{H}\widehat{Q}$ and $\widehat{Q}\widehat{H}\widehat{P}$ are of the second order in hopping and $\left(\widehat{Q}\widehat{H}\widehat{Q} - E\right)^{-1}$ is relevant only to high-energy doubly-occupied states when $\left(\widehat{Q}\widehat{H}\widehat{Q} - E\right)^{-1} \sim \frac{1}{U-E} \approx \frac{1}{U}$ (since the projected state energies E are much lower than U), and it can be substituted with $\frac{1}{U}$. Therefore, Eq. (2.84) can be approximated by

$$-\frac{\left(\widehat{P}\widehat{H}\widehat{Q}\right)\left(\widehat{Q}\widehat{H}\widehat{P}\right)}{U}\widehat{P}|\Psi\rangle = E\widehat{P}|\Psi\rangle, \tag{2.85}$$

i.e., the searched low-energy Hamiltonian of the Hubbard model at large values of U is

$$\widehat{\mathcal{H}} = -\frac{\left(\widehat{P}\widehat{H}\widehat{Q}\right)\left(\widehat{Q}\widehat{H}\widehat{P}\right)}{U}, \tag{2.86}$$

where $\widehat{P}$ (and hence $\widehat{Q} = 1 - \widehat{P}$) is defined in Eq. (2.79). The explicit form of Eq. (2.86) in terms of the electron operators is

$$\widehat{\mathcal{H}} = -\sum_{i,j,\sigma} t_{ij}(1 - \widehat{n}_{i,\overline{\sigma}})\widehat{c}^{+}_{i,\sigma}\widehat{c}_{j,\sigma}(1 - \widehat{n}_{j,\overline{\sigma}}) - \mu\sum_{i,\sigma}\widehat{c}^{+}_{i,\sigma}\widehat{c}_{i,\sigma} + \sum_{i,j}\frac{2|t_{ij}|^2}{U}$$
$$\times\left(\widehat{\vec{S}}_i\cdot\widehat{\vec{S}}_j - \frac{1}{4}\widehat{n}_i\widehat{n}_j\right) - \sum_{i\neq k\neq j}\frac{t_{ik}t_{kj}}{U}\sum_{\sigma,\sigma'}(2\delta_{\sigma\sigma'} - 1)\widehat{c}^{+}_{i,\sigma}\widehat{c}^{+}_{k,\overline{\sigma}}\widehat{c}_{k,\overline{\sigma}'}\widehat{c}_{j,\sigma'} \tag{2.87}$$

Neglecting the second-line term (small comparing to the previous term) in this equations results in the well-known t-J model Hamiltonian [12–14]. Again, since the t-J model is valid at large values of U, besides difficulties in its accurate solution, it is also difficult to get a quantitative understanding of its applicability to the most nontrivial and interesting case of intermediate Us.

2.2.8 *Composite Operator Approaches Solution: Sum Rules and Other Constraints*

In previous subsections (see, e.g., Eqs. (2.66) and (2.87)), we have used composite (Hubbard) operators instead of the original electron operators to describe different on-site states. For example, $\widehat{X}_{i,\overline{\sigma}2} = \widehat{c}_{i,\sigma}\widehat{n}_{i,\overline{\sigma}}$ corresponds to the annihilation of an electron on a doubly occupied site ($\overline{\sigma}2$ indicates transition from the doubly occupied (two-electron) to the singly occupied $\overline{\sigma}$-electron state) and $\widehat{X}_{i,\sigma 0} = (1 - \widehat{n}_{i,\overline{\sigma}})\widehat{c}^{+}_{i,\sigma}$ – to creation of an electron on an empty site. There are many other such local and nonlocal operators that describe different electronic states/excitations, e.g., $\widehat{X}_{i,2\sigma} = \widehat{n}_{i,\overline{\sigma}}\widehat{c}^{+}_{i,\sigma}$ or $\left(1 - \widehat{n}_{j,\overline{\sigma}}\right)\widehat{c}^{+}_{i,\sigma}$ (at $i \neq j$), etc. Depending on the considered physical problem, different set of operators can be used (single site-occupancy constraint, two sublattices with different spin orientations, etc.). The system of equations for the GFs for these operators can be solved exactly in some (very limited) cases (like in Sect. 2.2.4 in the case with no hopping). In general, the equations will include higher-order operators with unknown averages. In some cases, one can truncate the system of equations by applying physically justified approximation for the unknown operators expressing them in terms of simpler operators (like in the HF approximation). However, one can try to use a different way to solve the problem by taking into account constraints for the GFs. Namely, one can force the GF to satisfy exact sum rules that will give new equation(s) for the unknown averages (see, e.g., Ref. [16] and for an over-review—paper [17]). For example, the number of electrons per site can be calculated as

$$n = \sum_{\sigma} \langle \widehat{c}_{i\sigma}^{+}(t)\widehat{c}_{i\sigma}(t) \rangle. \tag{2.88}$$

This equation can be written in terms of the time-ordered electron GF $G^T_{ij,\sigma}(t-t') = -i\left\langle \widehat{T}\widehat{c}_{i,\sigma}(t)\widehat{c}^{+}_{j,\sigma}(t')\right\rangle \equiv -i\theta(t-t')\left\langle \widehat{c}_{i,\sigma}(t)\widehat{c}^{+}_{j,\sigma}(t')\right\rangle + i\theta(t'-t)$ $\left\langle \widehat{c}^{+}_{j,\sigma}(t')\widehat{c}_{i,\sigma}(t)\right\rangle$:

$$n_{i\sigma} = \langle \widehat{c}_{i,\sigma}^{+}(0)\widehat{c}_{i,\sigma}(0) \rangle = i\sum_{\sigma} G^T_{ii,\sigma}(\delta \to 0^-). \tag{2.89}$$

On the other hand, since the electron operators can be expressed in terms of the Hubbard operators: $\widehat{c}_{i,\sigma} = \widehat{c}_{i,\sigma}\widehat{n}_{i,\bar{\sigma}} + \widehat{c}_{i,\sigma}(1-\widehat{n}_{i,\bar{\sigma}}) = \widehat{X}_{i,\bar{\sigma}2} + \widehat{X}_{i,0\sigma}$ and $\widehat{c}^{+}_{i\sigma} = \left(\widehat{X}_{i,\bar{\sigma}2} + \widehat{X}_{i,0\sigma}\right)^{+} = \widehat{X}_{i,2\bar{\sigma}} + \widehat{X}_{i,\sigma0}$, one can express the time-ordered electron GF in terms of the GFs for the X-operators: $G^T_{AB}(t-t') = -i\left\langle \widehat{T}\widehat{X}_A(t)\widehat{X}^{+}_B(t')\right\rangle$. Then, Eq. (2.89) becomes an equation for the unknown averages in $G^T_{AB}(t-t')$.

Unfortunately, it is often difficult to judge about the accuracy of such approximations since usually the number of the constraint equations is smaller than the number of unknown averages and one needs to do some uncontrolled approximation for one or several operators. Also, it is not easy to apply this approach in the nonhomogeneous cases (even for small-size periodic clusters, where one has rather complex nonlocal operators).

2.2.9 *Slave-Boson Solutions*

An alternative approach to solve the Hubbard model is to use the saddle-point expansion of the action in terms of fluctuations by applying the path-integral formalism. In detail, one can write down the path integral formula for the partition function

$$Z = \int D[\psi_\uparrow] D[\psi_\downarrow] e^{-S_0 - S_U}, \tag{2.90}$$

where $\psi_\uparrow$ and $\psi_\downarrow$ are Grassman variables and

$$S_0 = \int_0^\beta d\tau \sum_{\vec{R}_i, \sigma} \psi_\sigma^*\left(\vec{R}_i, \tau\right) [\partial_\tau - \mu] \psi_\sigma\left(\vec{R}_i, \tau\right)$$
$$- \int_0^\beta d\tau \sum_{i,j,\sigma} t_{ij} \psi_\sigma^*\left(\vec{R}_i, \tau\right) \psi_\sigma\left(\vec{R}_j, \tau\right) \tag{2.91}$$

is the free-electron part and

$$S_U = U \int_0^\beta d\tau \sum_i \psi_\uparrow^*\left(\vec{R}_i, \tau\right) \psi_\downarrow\left(\vec{R}_i, \tau\right) \psi_\downarrow^*\left(\vec{R}_i, \tau\right) \psi_\uparrow\left(\vec{R}_i, \tau\right) \tag{2.92}$$

is the interacting part of the action (τ is imaginary time and β is the inverse temperature).

The action with fourth-order interacting part can be transformed into an integrable quadratic action by using the Hubbard–Stratonovich transformation (see, e.g., Ref. [1]) by employing new real auxiliary charge $\phi_c\left(\vec{R}_i, \tau\right)$ and spin $\phi_s\left(\vec{R}_i, \tau\right)$ fields,

$$Z = \int D[\psi_\uparrow] D[\psi_\downarrow] D[\phi_c] D[\phi_s]$$
$$\times \exp\left[-S_0 - \int_0^\beta d\tau \sum_i \left[\frac{1}{U}(\phi_c^2 + \phi_s^2) - i\phi_c\left(\psi_\uparrow^* \psi_\uparrow + \psi_\downarrow^* \psi_\downarrow\right) - \phi_s\left(\psi_\uparrow^* \psi_\uparrow - \psi_\downarrow^* \psi_\downarrow\right)\right]\right]. \tag{2.93}$$

Integration over the Grassman fields in the last equation can be easily performed giving $Z = \int D[\phi_c] D[\phi_s] e^{-\mathcal{A}[\phi_c, \phi_s]}$, where $\mathcal{A}[\phi_c, \phi_s]$ is the effective action, and the saddle-point configuration of the charge and spin (bosonic) fields can be found by minimizing the effective action. Then, one can easily find the field-dependent electron GF. To improve the accuracy of the solution, one can take into account corrections due to the fluctuations of the auxiliary fields.

Instead of the procedure above, one can use more physically transparent approaches with constrained bosonic fields (slave bosons) that, for example, suppress the double occupancy. Probably, the most popular slave-boson approach to the Hubbard model is the approach by Kotliar and Ruckenstein (KR) [18] with the following set of bosonic field operators: empty site (hole) field $\widehat{e}_i$, single-electron on-site fields $\widehat{s}_{i\uparrow}$ and $\widehat{s}_{i\downarrow}$ and doubly occupied-site field $\widehat{d}_i$ operators. There are two constraints on the fields in the KR approximation. The first constraint is

$$\widehat{n}_i^e + \widehat{n}_{i\uparrow}^s + \widehat{n}_{i\downarrow}^s + \widehat{n}_i^d = 1, \tag{2..94}$$

a restriction on possible occupancies of the site. The second one implies that the site is either singly or doubly occupied by electron(s) with a given spin:

$$\widehat{n}_{i\sigma}^s + \widehat{n}_i^d = \widehat{c}_{i\sigma}^+ \widehat{c}_{i\sigma}. \tag{2.95}$$

The KR Hamiltonian has the following form

$$\widehat{H}^{\mathrm{KR}} = \sum_{i,j,\sigma} t_{ij} \widehat{z}_{i\sigma}^+ \widehat{z}_{j\sigma} \widehat{c}_{i\sigma}^+ \widehat{c}_{j\sigma} + U \sum_i \widehat{d}_i^+ \widehat{d}_i, \tag{2.96}$$

where

$$\widehat{z}_{i\sigma} = \widehat{e}_i^+ \widehat{s}_{i\sigma} + \widehat{s}_{i\bar{\sigma}}^+ \widehat{d}_i \tag{2.97}$$

are boson hopping operators.

Instead of the bosonic operators (2.97) it is more convenient to use transformed (equivalent) operators

$$\widehat{\widehat{z}}_{i\sigma} = \frac{1}{\sqrt{1 - \widehat{n}_{i\sigma}^s - \widehat{n}_i^d}} \widehat{z}_{i\sigma} \frac{1}{\sqrt{1 - \widehat{n}_i^e - \widehat{n}_{i\bar{\sigma}}^s}}, \tag{2.98}$$

since in this case, one can reproduce the exact $U = 0$ and the FM solutions in the saddle-point approximation.

In the KR mean-field approximation, one substitutes the annihilation operators by

$$\widehat{d}_i \to \left\langle \widehat{d}_i \right\rangle_0 = \sqrt{\bar{d}_i}, \quad \widehat{s}_{i\sigma} \to \langle \widehat{s}_{i\sigma} \rangle_0 = \sqrt{n_{i\sigma 0} - \bar{d}_i}, \; \widehat{e}_i \to \langle \widehat{e}_i \rangle_0 = \sqrt{1 - n_{i0} + \bar{d}_i}, \tag{2.99}$$

and the same for the corresponding creation operators (since the averages are real). This gives

$$\begin{aligned} \left\langle \widehat{\widehat{z}}_{R\sigma} \right\rangle_0 &= \sqrt{\frac{\sqrt{(1 - n_{R0} + \bar{d}_R)(n_{R\sigma 0} - \bar{d}_R)} + \sqrt{\bar{d}_R (n_{R-\sigma 0} - \bar{d}_R)}}{\sqrt{n_{R\sigma 0}(1 - n_{R\sigma 0})}}} \\ &\equiv \sqrt{q_{R\sigma}}, \end{aligned} \tag{2.100}$$

and the Hamiltonian reduces to a simple single-particle one:

$$\widehat{H}^{\mathrm{KR}} = \sum_{i,j,\sigma} t_{ij}\sqrt{q_{i\sigma}}\sqrt{q_{j\sigma}}\widehat{c}_{i\sigma}^{+}\widehat{c}_{j\sigma} + U\sum_{i}\overline{d}_i. \tag{2.101}$$

Due to constraints, there are only three independent parameters in the theory—$n_{i\uparrow 0}$, $n_{i\downarrow 0}$, and $\overline{d}_i$, that can be found by minimizing the free energy of the Hamiltonian (2.101). Quite remarkably, the ground-state energy for this Hamiltonian is the same as the one obtained with the Gutzwiller wave functions in infinite dimensions (Sect. 2.2.7).

Another popular slave-boson approach was formulated in works of Barnes, Coleman, Read, and Newns (BCRN) [19–25]. It is valid in the limit $U = \infty$, i.e., when the double occupancy is suppressed, which is fixed by the constraint

$$n_{i\uparrow} + n_{i\downarrow} + \widehat{b}_i^{+}\widehat{b}_i = 1 \tag{2.102}$$

($\widehat{b}_i^{+}$ and $\widehat{b}_i$ are the boson-hole creation and annihilation operators). The corresponding effective BCNR Hamiltonian is

$$\widehat{H}^{\mathrm{BSRN}} = \sum_{i,j,\sigma} t_{ij}\,\widehat{b}_{i\sigma}\,\widehat{c}_{i\sigma}^{+}\widehat{c}_{j\sigma}\widehat{b}_j^{+}. \tag{2.103}$$

To reduce the problem (2.103) to a problem of noninteracting fermions, similar to the KR approximation, in BCNR one needs to substitute the boson operators by their averages that can be found by minimizing the energy of the system.

There are several shortcomings in the slave-boson approximations, most notably its limited applicability to the cases with intermediate Us and impossibility to describe the non-Fermi Liquid phases (for more details, see book [1]).

References

1. Gebhard, F.: The Mott Metal-Insulator Transition. Models and Methods. Springer, Berlin (1997)
2. Hubbard, J.: Proc. Roy. Soc. London Ser. A. **276**, 238 (1963)
3. Kanamori, J.: Prog. Theor. Phys. **30**, 275 (1963)
4. Gutzwiller, M.C.: Phys. Rev. Lett. **10**, 159 (1963)
5. Lieb, E.H., Wu, F.Y.: Phys. Rev. Lett. **20**, 1445 (1968).; Erratum Phys. Rev. Lett. **21**, 192 (1968)
6. Hubbard, J.: Proc. Roy. Soc. London Ser. A. **277**, 237 (1964)
7. Falicov, L.M., Kimball, J.C.: Phys. Rev. Lett. **22**, 997 (1969)
8. Hubbard, J.: Proc. Roy. Soc. London Ser. A. **281**, 401 (1964)
9. Gutzwiller, M.C.: Phys. Rev. **134**, A923 (1964)
10. Gutzwiller, M.C.: Phys. Rev. **137**, A1726 (1965)
11. Brinkman, W.F., Rice, T.M.: Phys. Rev. B. **2**, 4302 (1970)
12. Harris, A.B., Lange, R.V.: Phys. Rev. **157**, 295 (1967)
13. Chao, K.A., Spalek, J., Oles, A.M.: J. Phys. C. **10**, L271 (1977)

14. Gros, C., Joynt, R., Rice, T.M.: Phys. Rev. B. **36**, 381 (1987)
15. Nagaosa, N.: Quantum field theory in strongly correlated electron systems. Springer, Berlin (1999)
16. Roth, L.M.: Phys. Rev. **184**, 451 (1969)
17. Mancini, F., Avella, A.: Adv. Phys. **53**, 537 (2004)
18. Kotliar, G., Ruckenstein, A.E.: Phys. Rev. Lett. **57**, 1362 (1986)
19. Barnes, S.E.: J. Phys. F. **6**, 1375 (1976)
20. Barnes, S.E.: J. Phys. F. **7**, 2637 (1977)
21. Coleman, P.: Phys. Rev. B. **28**, 5255 (1983)
22. Read, N., Newns, D.M.: J. Phys. C. **16**, 3273 (1983)
23. Coleman, P.: Phys. Rev. B. **29**, 3035 (1984)
24. Coleman, P.: Phys. Rev. B. **35**, 5072 (1987)
25. Read, N., Newns, D.M.: Adv. Phys. **36**, 799 (1987)

Chapter 3
Dynamical Mean-Field Theory

3.1 Local Self-Energy Approximation

3.1.1 Limit of Infinite Dimensions in Statistical Physics

In statistical physics, it is well-known for a long time that one can obtain a simple solution of the problem by artificially increasing dimensionality d of the system. In particular, the critical exponents for phase transitions get their mean-field value when the dimensionality is higher than the critical value (see, e.g., Ref. [1] and references therein). One of the simplest examples is the Ising model in infinite dimensions, where the spatial fluctuations are suppressed and the mean-field solution is exact. It is also known that to get a meaningful result in the case of infinite-dimensional Ising model, the coupling constant must be scaled as

$$J = \frac{J^*}{z}, \tag{3.1}$$

where z is the coordination number ($z = 2d$). In detail (see review [2]), let us consider the Ising model with the Hamiltonian

$$H = -\sum_{i,j} J_{ij} S_i S_j - h \sum_i S_i, \tag{3.2}$$

where S_i is spin on site i, J_{ij} is the FM coupling constant such that $J_{ij} = J > 0$ when i and j are nearest-neighbor sites and $J_{ij} = 0$ otherwise, and h is an external magnetic field. In the mean-field (Weiss) approximation, one can linearize the Hamiltonian (3.2) by approximating

V. Turkowski, *Dynamical Mean-Field Theory for Strongly Correlated Materials*,
https://doi.org/10.1007/978-3-030-64904-3_3

$$J_{ij}S_iS_j \approx J_{ij}S_i\langle S_j\rangle, \tag{3.3}$$

where $\langle S_j\rangle = m_j$ is the average magnetic moment on site j. In this approximation, all sites are equivalent with the same value of magnetization m. Thus, the effective Hamiltonian for the problem (3.2) has the following form:

$$H = -(h + zJm)\sum_i S_i. \tag{3.4}$$

The effective field

$$h + zJm \equiv h_{\text{eff}} \tag{3.5}$$

in the last equation is homogeneous in space, thus the on-site magnetization can be calculated by using the one-site model Hamiltonian

$$H_{\text{eff}}(S_0) = -(h + zJm)S_0 \tag{3.6}$$

as

$$m = \frac{\sum_{S_0=\pm 1} S_0 e^{-H_{\text{eff}}(S_0)}}{\sum_{S_0=\pm 1} e^{-H_{\text{eff}}(S_0)}}.$$

This gives the equation for the magnetization m:

$$m = \tanh\left[\frac{h + zJm}{T}\right]. \tag{3.7}$$

The mean-field equation (3.7) for the site magnetization becomes exact when z $(d) \to \infty$, i.e., when the spatial fluctuations are negligible and the mean-field theory is exact. It is obvious from Eq. (3.7) that in order to get a meaningful result at $z \to \infty$, the coupling constant should scale according to Eq. (3.1). Indeed, such a scaling gives a finite entropy and energy per site, and as result—a finite critical temperature of the FM transition [2].

Let us further analyze the meaning of the infinite-dimensional effective one-site model (3.6). When the number of the nearest to the site 0 neighbor sites z becomes large one can regard these neighbor sites as a bath for the site 0. Thus, in infinite dimensions the Ising problem reduces to *one-site problem in presence of a bath.* Since all sites in the case $z \to \infty$ are identical (with increasing z, the spatial non-homogeneity effects disappear), the mean-field theory is the exact theory in infinite dimensions. The same is valid for other statistical models. In the case of Ising model, the statement was proven in works [3, 4].

One can show that the scaled spin-lattice problem in infinite dimensions reduces to one-site problem also in the case of vector spins. For example, in the case of Heisenberg model, Gebhard [5] proved it by using the following reasoning. Let us introduce a spin operator that is the average of the spin operators on the nearest-to-site i neighbor sites:

$$\widehat{h}_i^{\alpha} = \frac{1}{z}\sum_j \widehat{S}_j^{\alpha}$$

($\alpha = x, y, z$ is the spin component index). Then, it is possible to show that commutators of these operators satisfy:

$$\left[\widehat{h}_i^{\alpha}, \widehat{h}_j^{\beta}\right] = O\left(\frac{1}{z}\right),$$

i.e., in the infinite-dimensional case, these operators commute with each other. Moreover, they also commute with the Hamiltonian in the lowest order in $\frac{1}{z}$. Thus, in infinite dimensions they describe conserved quantities and can be substituted in the Heisenberg Hamiltonian by their averages, similar to the Ising model case, Eq. (3.2). In other words, at $z \to \infty$ the mean-field approximation with suppressed spatial fluctuations is exact also in the case of the Heisenberg model.

3.1.2 Noninteracting Hubbard Model in Infinite Dimensions

Now, let us move to the infinite-dimensional Hubbard model on the hypercubic lattice, where the coordination number is also proportional to the dimensionality of the system ($z = 2d$) and try to analyze its physics by rescaling the hopping parameter t, in analogy with the Ising model case. To get the form of possible rescaling of t, we consider the one-band Hubbard model with the lattice parameter a and the nearest-neighbor hopping parameter that depends on the dimensionality of the system, $t = t(d)$:

$$H = -t(d)\sum_{i,j,\sigma}\left(c_{i\sigma}^{+}c_{j\sigma} + h.c.\right) + U\sum_{i,j} n_{i\uparrow}n_{i\downarrow}. \tag{3.8}$$

It is natural to expect that the hopping parameter in the Hubbard model is inversely proportional to the dimensionality of the system, similar to the Ising and the Heisenberg models. Thus, we postulate that the hopping parameter renormalizes as:

$$t = \frac{t^*}{(2d)^n}, \tag{3.9}$$

where n is a positive, not necessarily integer, number (2 in the denominator is introduced in order to have similar to the spin models relation $t{\sim}1/z^n$).

Now, let us find a physically meaningful value for n at $d \to \infty$. For this, we begin with rewriting the Hamiltonian (3.8) in momentum representation. After substitution of the expression for the hopping (3.9) into Eq. (3.8), one can show that the noninteracting (kinetic) part of the Hubbard Hamiltonian in the momentum representation

$$H_0 = \sum_{\vec{k},\sigma} \varepsilon\left(\vec{k}\right) c^+_{k\sigma} c_{k\sigma} \tag{3.10}$$

describes the electron system with the following spectrum:

$$\varepsilon\left(\vec{k}\right) = -\frac{2t^*}{(2d)^n} \sum\nolimits_{i=1}^{d} \cos\left(ak_i\right), \tag{3.11}$$

where momenta are defined in the first Brillouin zone: $-\frac{\pi}{a} \le k_i \le \frac{\pi}{a}$ $(i = 1, \ldots, d)$. Next, let us calculate the DOS that corresponds to the spectrum (3.11):

$$\rho(\varepsilon) = \int_{-\frac{\pi}{a}}^{\frac{\pi}{a}} \ldots \int_{-\frac{\pi}{a}}^{\frac{\pi}{a}} \delta\left(\varepsilon - \varepsilon\left(\vec{k}\right)\right) \frac{d^d k}{(2\pi)^d}. \tag{3.12}$$

Using the exponent representation for the delta-function $\delta(x) = \int_{-\infty}^{\infty} \frac{d\alpha}{2\pi} e^{i\alpha x}$, one gets from Eq. (3.12):

$$\rho(\varepsilon) = \int_{-\infty}^{\infty} \frac{d\alpha}{2\pi} e^{i\alpha\varepsilon} \left[\int_{-\frac{\pi}{a}}^{\frac{\pi}{a}} \exp\left[i\alpha \frac{2t^*}{(2d)^n} \cos\left(ak\right) \right] \frac{dk}{2\pi} \right]^d.$$

To perform the integration in the large square bracket in the last expression, let us assume for a moment that the hopping parameter $t = \frac{t^*}{(2d)^n}$ is small (this is a natural assumption, since t^* is finite and d is very large). Then, up to the second order in t one has

$$\int_{-\frac{\pi}{a}}^{\frac{\pi}{a}} e^{i\alpha \frac{2t^*}{(2d)^n} \cos(ak)} \frac{dk}{2\pi} \approx \frac{1}{2\pi a} \int_{-\pi}^{\pi} \left(1 + i\alpha \frac{2t^*}{(2d)^n} \cos(x) - \frac{1}{2}\alpha^2 \frac{4t^{*2}}{(2d)^{2n}} \cos^2(x) + O\left(\frac{1}{d^{4n}}\right) \right) dx$$
$$= \frac{1}{a}\left(1 - \alpha^2 \frac{t^{*2}}{(2d)^{2n}} + O\left(\frac{1}{d^{4n}}\right) \right) \approx \frac{1}{a} \exp\left[-\alpha^2 \frac{t^{*2}}{(2d)^{2n}} + O\left(\frac{1}{d^{4n}}\right) \right],$$

and therefore

$$\rho(\varepsilon) \approx \frac{1}{a^d} \int_{-\infty}^{\infty} \frac{d\alpha}{2\pi} e^{i\alpha\varepsilon} e^{-\alpha^2 \frac{t^{*2}}{2(2d)^{2n-1}}}.$$

The integration in the last expression can be easily performed and one gets

$$\rho(\varepsilon) \approx \frac{1}{a^d} \frac{(2d)^{n-\frac{1}{2}}}{\sqrt{2\pi} t^*} e^{-\frac{\varepsilon^2}{2t^{*2}}(2d)^{2n-1}}. \quad (3.13)$$

As it follows from Eq. (3.13), the DOS in the infinite-dimensional system is finite only when $n = 1/2$, i.e., when

$$t = \frac{t^*}{\sqrt{2d}}, \quad (3.14)$$

which is similar to the Ising model renormalization, Eq. (3.1) (in the Hubbard model, the prefactor $\frac{1}{a^d}$ in Eq. (3.13) cancels with the volume prefactor a^d when the summation over momentum is performed). This gives

$$\rho(\varepsilon) = \frac{1}{\sqrt{2\pi} t^*} e^{-\frac{\varepsilon^2}{2t^{*2}}}. \quad (3.15)$$

Then, a natural question arises: since at $d \to \infty$ the original ("bare") hopping parameter t disappears after renormalization (3.14), does the kinetic energy also approaches zero in this case, i.e., do we end up at the atomic limit, with no kinetics? Fortunately, the answer is "no." Straightforward calculation of the average kinetic energy $\langle\varepsilon\rangle$ by using the DOS (3.15) gives a finite value for this quantity:

$$\langle\varepsilon\rangle = \int_{-\infty}^{\epsilon_F} \varepsilon\rho(\varepsilon) d\varepsilon = \frac{1}{t^*\sqrt{d}} \int_{-\infty}^{\epsilon_F} \varepsilon e^{-\frac{\varepsilon^2}{t^{*2}}} d\varepsilon = -\frac{2\sqrt{2}}{\sqrt{\pi}} t^* e^{-\frac{\epsilon_F^2}{2t^{*2}}},$$

where ϵ_F is the Fermi energy of the system at $T = 0$. The average kinetic energy in the last equation is negative, since we included the chemical potential into the Hamiltonian (3.8) (see, e.g., Ref. [1]).

One can also give another argument in favor of nonzero kinetic energy in the infinite-dimensional case [1]. Namely, kinetic energy

$$\varepsilon\left(\vec{k}\right) = -\frac{\sqrt{2}t^*}{\sqrt{d}} \sum_{i=1}^{d} \cos\left(ak_i\right) \quad (3.16)$$

is defined by a sum of random numbers that oscillate between -1 and 1. Thus, in the limit $d \to \infty$, the function $\left(\sum_{i=1}^{d} \cos\left(ak_i\right)\right)^2$ grows as d and hence $\sum_{i=1}^{d} \cos\left(ak_i\right) \sim \sqrt{d}$.

Then, substitution of the last estimation into Eq. (3.16) gives a finite kinetic energy, $\varepsilon \sim t^*$.

Thus, we have shown that for the Gaussian DOS, the kinetic energy is finite. At the end of this subsection, let us prove the statement opposite to this. Since the kinetic energy is finite, its square is also finite. Indeed, using (3.16), one gets

$$\langle \varepsilon^2 \rangle \equiv a^d \int_{-\frac{\pi}{a}}^{\frac{\pi}{a}} \dots \int_{-\frac{\pi}{a}}^{\frac{\pi}{a}} \varepsilon^2 \left(\vec{k}\right) \frac{d^d k}{(2\pi)^d} = t^{*2} < \infty. \tag{3.17}$$

In this case, as it follows from the central limiting theorem, the DOS of the system has the Gaussian form, in agreement with Eq. (3.15) (the obtained result corresponds to a finite bandwidth). Thus, we have proved the opposite statement: finiteness of the square of the average kinetic energy, and hence of the kinetic energy, corresponds to the Gaussian DOS.

3.1.3 *Interacting Hubbard Model in Infinite Dimensions: Locality of the Diagrams*

We have analyzed the case of free electrons on the infinite-dimensional hypercubic lattice and found that the hopping parameter has to be renormalized, which results in the Gaussian DOS. As the next step, let us turn on the Coulomb repulsion beginning from the case of weak interaction and analyze the expression for the correlation energy per site $\left(U \frac{1}{N} \sum_i \frac{1}{Z} \langle n_{i\uparrow} n_{i\downarrow} \rangle \right)$ in the second order in U approximation (see Sect. 2.2.1) following the steps of the original analysis by Metzner and Vollhardt [6]. The equation for the above energy is

$$E_2 = U^2 \int \frac{d^d k}{(2\pi)^d} \int \frac{d^d k'}{(2\pi)^d} \int \frac{d^d q}{(2\pi)^d}$$
$$\times \frac{f_{0\uparrow}\left(\vec{k}\right) f_{0\downarrow}\left(\vec{k'}\right) \left(1 - f_{0\uparrow}\left(\vec{k} + \vec{q}\right)\right) \left(1 - f_{0\downarrow}\left(\vec{k'} - \vec{q}\right)\right)}{\varepsilon\left(\vec{k}\right) + \varepsilon\left(\vec{k'}\right) - \varepsilon\left(\vec{k} + \vec{q}\right) - \varepsilon\left(\vec{k'} - \vec{q}\right)}, \tag{3.18}$$

where $f_{0\sigma}\left(\vec{k}\right)$ is the zero-temperature Fermi function for the electron with spin $\sigma = \uparrow, \downarrow$. To simplify the analysis of Eq. (3.18), it is convenient to express the denominator in the expression under the integral in terms of the following integral:

$$\frac{1}{\varepsilon\left(\vec{k}\right)+\varepsilon\left(\vec{k}'\right)-\varepsilon\left(\vec{k}+\vec{q}\right)-\varepsilon\left(\vec{k}'-\vec{q}\right)}$$
$$=\int_0^\infty d\alpha e^{-\left(-\varepsilon\left(\vec{k}\right)-\varepsilon\left(\vec{k}'\right)+\varepsilon\left(\vec{k}+\vec{q}\right)+\varepsilon\left(\vec{k}'-\vec{q}\right)\right)\alpha} \tag{3.19}$$

(the integral on the right-hand side of the last equation converges since $-\varepsilon\left(\vec{k}\right)-\varepsilon\left(\vec{k}'\right)+\varepsilon\left(\vec{k}+\vec{q}\right)+\varepsilon\left(\vec{k}'-\vec{q}\right)>0$, which is guaranteed by the Fermi functions in Eq. (3.18)). Next, let us substitute Eq. (3.19) into Eq. (3.18) and express the functions $f_{0\uparrow}(k)e^{\varepsilon\left(\vec{k}\right)\alpha}$, $f_{0\downarrow}\left(\vec{k}'\right)e^{\varepsilon\left(\vec{k}'\right)\alpha}$, $\left(1-f_{0\uparrow}\left(\vec{k}+\vec{q}\right)\right)e^{-\varepsilon\left(\vec{k}+\vec{q}\right)\alpha}$, and $\left(1-f_{0\downarrow}\left(\vec{k}'-\vec{q}\right)\right)e^{-\varepsilon\left(\vec{k}'-\vec{q}\right)\alpha}$ by their spatial Fourier transforms:

$$f_{0\sigma}\left(\vec{k}\right)e^{\varepsilon\left(\vec{k}\right)\alpha}=\sum_{\vec{R}}e^{i\vec{k}\vec{R}}F_\sigma^+\left(\alpha,\vec{R}\right),\ \left(1-f_{0\sigma}\left(\vec{k}\right)\right)e^{-\varepsilon\left(\vec{k}\right)\alpha}=\sum_{\vec{R}}e^{i\vec{k}\vec{R}}F_\sigma^-\left(\alpha,\vec{R}\right).$$

This gives

$$E_2=U^2\int_0^\infty d\alpha\int\frac{d^dk}{(2\pi)^d}\int\frac{d^dk'}{(2\pi)^d}\int\frac{d^dq}{(2\pi)^d}\left(\sum_{\vec{R}_1}e^{i\vec{k}\vec{R}_1}F_\uparrow^+\left(\alpha,\vec{k}_1\right)\right)$$
$$\times\left(\sum_{\vec{R}_2}e^{i\vec{k}'\vec{R}_2}F_\downarrow^+\left(\alpha,\vec{R}_2\right)\right)\times\left(\sum_{\vec{R}_3}e^{i\left(\vec{k}+\vec{q}\right)\vec{R}_3}F_\uparrow^-\left(\alpha,\vec{R}_3\right)\right)$$
$$\times\left(\sum_{\vec{R}_4}e^{i\left(\vec{k}'-\vec{q}\right)R_4}F_\downarrow^-\left(\alpha,\vec{R}_4\right)\right).$$

Finally, insertion of the momentum conservation condition factor in the form $\delta\left(\vec{k}+\vec{k}'-\left(\vec{k}+\vec{q}\right)-\left(\vec{k}'-\vec{q}\right)\right)=\sum_{\vec{R}}e^{i\left(\vec{k}+\vec{k}'-\left(\vec{k}+\vec{q}\right)-\left(\vec{k}'-\vec{q}\right)\right)\vec{R}}$ under the integrals in the last expression and integration over momenta gives:

$$E_2=-U^2\int_0^\infty d\alpha\sum_{\vec{R}}F_\uparrow^+\left(\alpha,\vec{R}\right)F_\downarrow^+\left(\alpha,\vec{R}\right)F_\uparrow^-\left(\alpha,\vec{R}\right)F_\downarrow^-\left(\alpha,\vec{R}\right). \tag{3.20}$$

As Metzner and Vollhardt argued, since $F^{\pm}_{\uparrow(\downarrow)}\left(\alpha, \vec{R}\right)$ count the number of nearest neighbors with respect to site R and they are finite, $F^{\pm}_{\uparrow(\downarrow)}\left(\alpha, \vec{R} \neq 0\right)$ should be of order or smaller than $\frac{1}{\sqrt{d}}$ when $d \to \infty$. Thus, in the case of infinite dimension, the terms with $\vec{R} \neq 0$ vanish and only one site $\left(\vec{R} = 0\right)$ term contributes to the right-hand side of Eq. (3.20).

To further simplify Eq. (3.20), one can use the following expressions:

$$F^{+}_{\uparrow}(\alpha, R = 0) = \sum_k f_{0\uparrow}(k) e^{\varepsilon(k)\alpha} = \int_{-\infty}^{\varepsilon_F} e^{-\varepsilon^2} e^{\varepsilon\alpha} d\varepsilon = e^{\frac{\alpha^2}{4}} \int_{-\infty}^{\varepsilon_F - \frac{\alpha}{2}} e^{-x^2} dx,$$

$$F^{-}_{\uparrow}(\alpha, R = 0) = \sum_k \left(1 - f_{0\uparrow}(k)\right) \mathrm{e}^{-\varepsilon(k)\alpha} = \int_{\varepsilon_F}^{\infty} e^{-\varepsilon^2} e^{-\varepsilon\alpha} d\varepsilon = e^{\frac{\alpha^2}{4}} \int_{\varepsilon_F + \frac{\alpha}{2}}^{\infty} e^{-x^2} dx$$

(and similar for the spin-down functions). This gives,

$$E_2 = -U^2 \int_0^{\infty} d\alpha e^{\alpha^2} \prod_{\sigma} P\left(\varepsilon_{F\sigma} - \frac{\alpha}{2}\right) \prod_{\sigma} \left(\sqrt{\pi} - P\left(\varepsilon_{F\sigma} + \frac{\alpha}{2}\right)\right), \quad (3.21)$$

where P is the Gaussian probability functions.

Thus, as it follows from Eq. (3.20), only diagrams with local-in-space vertices contribute to the correlation energy in the second order in U approximation. In more detail, the integration over momenta in the original Eq. (3.18) was transformed into the integration over mutually independent (for each site) energies, Eq. (3.20). This was possible due to the fact that in infinite dimensions, the energies in Eq. (3.18) are randomized by Umklapp scatterings (change of the lattice momenta). In other words, one may state that the vortex momenta in infinite dimensions are arbitrary and the "correlated sector" of physics does not depend on the momenta. In particular, going ahead we mention that one of the consequences of the last statement is that the electron self-energy is momentum-independent, or equivalently, is local in space. On the other hand, locality of the self-energy in space means that spatial fluctuations can be neglected. As we show below, locality of the vertices in infinite dimensions holds also in higher orders of the perturbation theory.

Metzner and Vollhardt showed that the dependence of the second-order correlation energy of the infinite-dimensional Hubbard model on the electron density is very close to the 3D result (Fig. 3.1), which suggests that neglecting conservation of the vertices momenta and momentum dependence of the electron self-energy must be a good approximation for bulk systems as well.

Though valid in the case of weak interactions, this was a breakthrough result that not only led to a significant simplification of the calculations (neglecting the

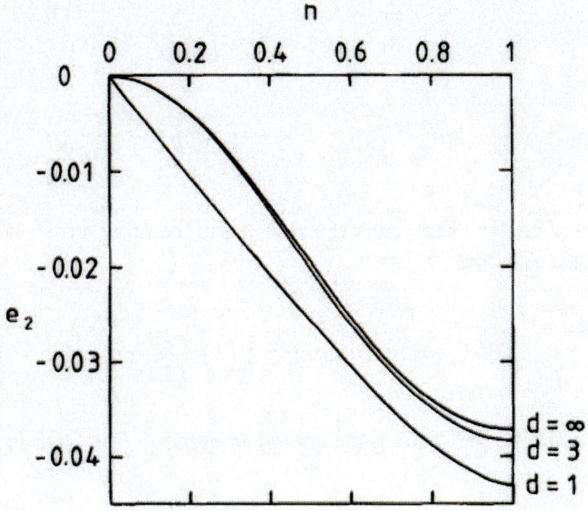

Fig. 3.1 The second-order correlation energy per site as function of the density of electrons in the case of different dimensions. The energy is divided by $U^2/\langle T\rangle$, where $\langle T\rangle$ is the average kinetic energy at $n\uparrow = n\downarrow = 1/2$. (Reprinted Fig. 3.1 with permission from [(W. Metzner and D. Vollhardt, Phys. Rev. Lett. 62, 324 (1989)]. Copyright 1989 by the American Physical Society. https://doi.org/10.1103/PhysRevLett.62.324)

momentum dependence of the electron self-energy), but revealed a fundamental property of the correlated systems: in the case with high atomic coordination number, the most important effects of electron–electron correlations come from the local-in-space time-resolved interactions between the electrons (described by the frequency- or time-dependent electron self-energy).

Mueller-Hartmann demonstrated [7] that conclusion about the momentum independence (locality in space) of the vertex functions is valid for any higher order in U expansion as well. Here, we go through the main steps of his proof. To show that in infinite dimensions in any order of the perturbation theory the vertex functions also have Gaussian dependence on the energies, similar to Eq. (3.21), let us consider an arbitrary four-energy vertex function (generalization of his proof on the case of higher-order vertices is straightforward):

$$\Gamma(\varepsilon_1, \varepsilon_2, \varepsilon_3, \varepsilon_4) = \int \prod_{l=1}^{4} \frac{d^d k_l}{(2\pi)^d} \delta\left(\varepsilon_l - \varepsilon\left(\vec{k}_l\right)\right) \sum_{\vec{R}} e^{i\vec{R}\left(\vec{k}_1 - \vec{k}_2 + \vec{k}_3 - \vec{k}_4\right)}. \qquad (3.22)$$

The last equation is valid in the case of momentum-independent interaction (the case of the Hubbard model), so the momenta enter in the vertex function only in the form of the energy function $\varepsilon\left(\vec{k}\right)$ and of the factor $\sum_{\vec{R}} e^{i\vec{R}\left(\vec{k}_1 - \vec{k}_2 + \vec{k}_3 - \vec{k}_4\right)}$ that guarantees the vortex momentum conservation (up to the reciprocal (G-) vectors). Next, let us perform the Fourier transformation of the function (3.22):

$$\Gamma(\alpha_1,\alpha_2,\alpha_3,\alpha_4) = \int \prod_{l=1}^{4} e^{i\alpha_l\varepsilon_l}\Gamma(\varepsilon_1,\varepsilon_2,\varepsilon_3,\varepsilon_4)d\varepsilon_l$$
$$= \sum_R \prod_{l=1}^{4} \int \frac{d^d k_l}{(2\pi)^d} e^{i\alpha_l\varepsilon\left(\vec{k}_l\right) - i(-1)^l \vec{k}_l \vec{R}} \tag{3.23}$$

and analyze separately the $\vec{R}=0$ and the other terms. For $\vec{R}=0$, one gets from the last equation

$$\Gamma_{R=0}(\alpha_1,\alpha_2,\alpha_3,\alpha_4) = \prod_{l=1}^{4} \int \frac{d^d k_l}{(2\pi)^d} e^{i\alpha_l\varepsilon\left(\vec{k}_l\right)} = \prod_{l=1}^{4} \int \rho(\varepsilon) e^{i\alpha_l\varepsilon} d\varepsilon,$$

and after performing the energy integration (the DOS is given in Eq. (3.15)),

$$\Gamma_{R=0}(\alpha_1,\alpha_2,\alpha_3,\alpha_4) = \prod_{l=1}^{4} e^{-\frac{\alpha_l^2 t^{*2}}{2}}. \tag{3.24}$$

Now, let us consider the finite-R terms in Eq. (3.23) and show that they are of order $\frac{1}{d^{1/2}}$ or smaller. To do this, we expand the exponential factors in Eq. (3.23) in powers of α_l:

$$e^{i\alpha_l\varepsilon\left(\vec{k}_l\right) - i(-1)^l \vec{k}_l \vec{R}} \approx \left(1 + i\alpha_l\varepsilon\left(\vec{k}_l\right) + \frac{1}{2}\left(i\alpha_l\varepsilon\left(\vec{k}_l\right)\right)^2 + \dots\right) e^{-i(-1)^l \vec{k}_l \vec{R}},$$

and first consider the terms with the nearest neighbor to 0 sites $\vec{R}$ (we use notation $\Gamma_{\left|\vec{R}\right|=a}$ for these terms). Then,

$$\Gamma_{\left|\vec{R}\right|=a}(\alpha_1,\alpha_2,\alpha_3,\alpha_4) \approx \sum_{\left|\vec{R}\right|=a} \prod_{l=1}^{4} \int \frac{d^d k_l}{(2\pi)^d}$$
$$\times \left(1 + i\alpha_l\varepsilon\left(\vec{k}_l\right) + \frac{1}{2}\left(i\alpha_l\varepsilon\left(\vec{k}_l\right)\right)^2 + \dots\right) e^{-i(-1)^l \vec{k}_l \vec{R}}$$
$$= \prod_{l=1}^{4} \int \frac{d^d k_l}{(2\pi)^d}$$
$$\times \left(1 + i\alpha_l\varepsilon\left(\vec{k}_l\right) + \frac{1}{2}\left(i\alpha_l\varepsilon\left(\vec{k}_l\right)\right)^2 + \dots\right)$$
$$\times \sum_{j=1}^{d} 2\cos\left(a k_l^j\right), \tag{3.25}$$

where k_l^j are components of the momentum $\vec{k}_l = \left(k_l^1, \ldots, k_l^d\right)$ (we omit the vector symbol, to simplify the notations). Then, substituting the expression for energy $\varepsilon\left(\vec{k}_l\right)$ (given by Eq. (3.11) at $n = 1/2$), into the last line in Eq. (3.25) and integrating over momenta, it is easy to show that in the lowest order in $\frac{1}{d^{1/2}}$ approximation

$$\Gamma_{|R|=a}(\alpha_1, \alpha_2, \alpha_3, \alpha_4) = \prod_{l=1}^{4} \left[i\alpha_l \frac{2t^*}{(2d)^{1/2}} \right]^d . \tag{3.26}$$

Similarly, one can also show that the higher order in α and larger-R terms in the formula for the vertex give smaller (higher in powers of $\frac{1}{d}$) contribution to (3.23), as compared to (3.26). Therefore, one gets

$$\begin{aligned} \Gamma(\alpha_1, \alpha_2, \alpha_3, \alpha_4) &= \prod_{l=1}^{4} e^{-\frac{\alpha_l^2 t^{*2}}{2}} \left(1 + \left[i\alpha_l \frac{2t^*}{(2d)^{\frac{1}{2}}} \right]^d + o\left(\frac{1}{d^2}\right) \right) \\ &\approx \prod_{l=1}^{4} e^{-\frac{\alpha_l^2 t^{*2}}{2}} e^{\prod_{j=1}^{4} \frac{\alpha_j t^{*4}}{2d} + o\left(\frac{1}{d^2}\right)} . \end{aligned} \tag{3.27}$$

Performing the inverse to Eq. (3.23) Fourier transform of Eq. (3.27) gives

$$\begin{aligned} \Gamma(\varepsilon_1, \varepsilon_2, \varepsilon_3, \varepsilon_4) &= \int \prod_{l=1}^{4} e^{-i\alpha_l \varepsilon_l} \Gamma(\alpha_1, \alpha_2, \alpha_3, \alpha_4) d\alpha_l \\ &= \prod_{l=1}^{4} \rho(\varepsilon_l) e^{\prod_{j=1}^{4} \frac{\varepsilon_j}{2dt^{*4}} + o\left(\frac{1}{d^2}\right)} \approx \prod_{l=1}^{4} \rho(\varepsilon_l). \end{aligned} \tag{3.28}$$

Thus, we have demonstrated that in infinite dimensions in all orders of the perturbation theory, the momentum conservation multipliers (Laue functions) $\sum_{\vec{R}} e^{i\vec{R}\left(\vec{k}_1 - \vec{k}_2 + \vec{k}_3 - \vec{k}_4\right)}$ in the vertices (3.22) can be substituted by 1, i.e., the momentum conservation in the vertices can be neglected (for more details, we refer the reader to work [7]).

Let us analyze the nature of the last result in terms of the Feynman diagrams. Since the leading effect of the momentum conservation comes from the nearest to $\vec{R} = 0$ neighbor sites, each nonlocal contribution to the "momentum–non-conserving vertex" gives correction of order d. On the other hand, as we showed above, expansion in power of α_l for these neighbor terms results in factor $\frac{1}{\sqrt{d}}$ for each momentum to be integrated, Eq. (3.26). In other words, each internal line gives a correction of order $\frac{1}{\sqrt{d}}$. One can also prove the statement in the last sentence by using another argument [8]. Namely, since the average kinetic energy

$$E_{kin,\sigma} = \sum_{i,j} t_{ij} \langle c_{i\sigma}^{+} c_{j\sigma} \rangle = \sum_{i,j} t_{ij} \int_{-\infty}^{\infty} \frac{d\omega}{2\pi i} G_{ij\sigma}(\omega) e^{i\omega \cdot 0^{+}} \tag{3.29}$$

is finite and the rescaled hopping parameters satisfy

$$t_{ij} \sim \frac{1}{d^{|R_i - R_j|/2}}, \tag{3.30}$$

one should have

$$G_{ij}(\omega) \sim \frac{1}{d^{|\vec{R}_i - \vec{R}_j|/2}}, \quad (i \neq j) \tag{3.31}$$

(naturally, on the same site $G_{ii}(\omega) \sim O(d^0)$). In other words, in the leading order (nearest-neighbor site) approximation for the vertices, the internal lines give correction $\sim \frac{1}{d^{1/2}}$. Thus, taking into accunt also the fact that the nonlocal vertex correction is of order of d, Mueller-Hartmann concluded that momentum conservation can be neglected if the number of internal lines is twice as much as the number of vertices.

Neglecting the vertex momentum conservation leads to a dramatic simplification of the calculations. In particular, as it was already mentioned, it allows to neglect momentum dependence of the electron self-energy. To show this, let us consider nth order diagram for the irreducible self-energy [7]. Out of the n vertices in this diagram, the one (nonintegrated) provides equality of the total incoming and outgoing momenta. The rest of vertices, after summation over internal variables, give correction $\sim d^{n-1}$ to the diagram. On the other hand, $2n-1$ internal lines result in correction of order $\frac{1}{d^{(2n-1)\frac{1}{2}}}$ (see Eq. (3.31)). Therefore, the resulting correction to the momentum non-conserving (constant-vortex) approximation is of order $d^{n-1} \times \frac{1}{d^{(2n-1)\frac{1}{2}}} = \frac{1}{\sqrt{d}}$, i.e., it is negligible at $d \to \infty$! Thus, as Mueller-Hartmann concluded, similar to the momentum conservation in the vertices, the momentum dependence of the self-energy can be neglected if the number of internal lines is twice as much as number of vertices. Actually, the estimated above momentum-conserving correction is even smaller: it is of order $\frac{\varepsilon(\vec{k})}{d} \sim \frac{1}{d}$ since the relevant effective electron bandwidth is of order 1 (see Ref. [7]).

Next, since in infinite dimensions, finite contribution to the diagrams comes only from the local GFs, $G_{ii}(\omega) \sim O(d^0)$, one can show by using the skeleton expansion of the irreducible self-energy that self-energy has the same internal and external site arguments, i.e.

$$\Sigma_{ij\sigma}(\omega) = \delta_{ij} \Sigma_{\sigma}(\omega), \tag{3.32}$$

or, what is equivalent, it is momentum-independent in the momentum representation:

$$\Sigma_\sigma\left(\omega,\vec{k}\right)=\sum_{\left(\vec{R}_i-\vec{R}_j\right)}e^{i\vec{k}\left(\vec{R}_i-\vec{R}_j\right)}\delta_{ij}\Sigma_\sigma(\omega)=\Sigma_\sigma(\omega). \tag{3.33}$$

As it was already mentioned several times, this significantly simplifies the calculations. In particular, the electron self-energy diagrams include only same-site vertices, i.e., the diagrams are local in space and one needs to care only about the frequency conservation (an example of such diagram is shown in Fig. 3.2). Therefore, only local-in-space GFs,

$$G^{(loc)}_{\sigma,i=j}(\omega)=\int\frac{d^dk}{(2\pi)^d}G_\sigma\left(\omega,\vec{k}\right)e^{i\vec{k}\left(\vec{R}_i-\vec{R}_j\right)}=\int\frac{d^dk}{(2\pi)^d}G_\sigma\left(\omega,\vec{k}\right), \tag{3.34}$$

are relevant in infinite dimensions. Using the definition

$$G_\sigma\left(\omega,\vec{k}\right)=\frac{1}{\omega-\left(\varepsilon\left(\vec{k}\right)-\mu\right)-\Sigma_\sigma\left(\omega,\vec{k}\right)} \tag{3.35}$$

and the fact that the self-energy is momentum independent (Eq. (3.33)), one can find a simple equation that connects $G^{(loc)}_\sigma(\omega)$ and $\Sigma_\sigma(\omega)$:

$$G^{(loc)}_\sigma(\omega)=\int d\varepsilon\rho(\varepsilon)\frac{1}{\omega-(\varepsilon-\mu)-\Sigma_\sigma(\omega)}. \tag{3.36}$$

Thus, in the limit of infinite dimensions, the electron self-energy is local in space, i.e., it is frequency dependent and momentum independent. Momentum independence of the self-energy signifies that the self-energy is local in space, i.e., the spatial fluctuations of charges are neglected that corresponds to the spatial mean-field

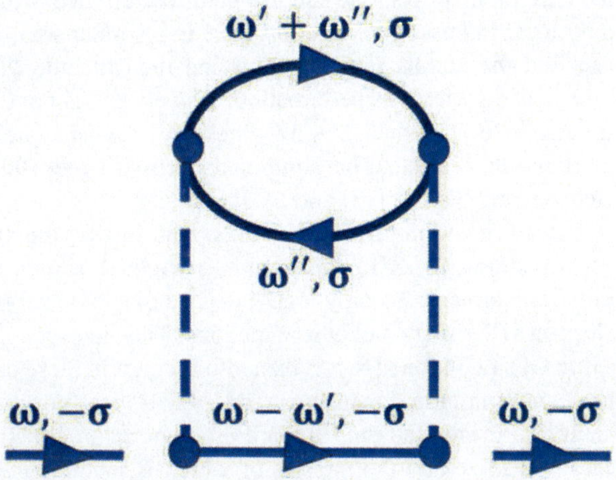

Fig. 3.2 Second-order perturbation theory diagram for the electron self-energy $\Sigma^{(2)}_{-\sigma}(\omega)$ in infinite dimensions

theory approximation. On the other hand, frequency dependence of Σ means that the time-resolved on-site electron–electron interactions (temporal charge fluctuations) are taken into account. In other words, the dynamic effects are taken into account. For the reasons above, the theory holds the name dynamical mean-field theory (DMFT).

Obviously, the main results of this subsection—momentum independence of the electron self-energy in infinite dimensions, Eq. (3.33). However, the only equation we have so far—Eq. (3.36) that connects local GF and self-energy—is not sufficient to find either GF or self-energy. Thus, the theory needs further developments, that will be described in the next subsection.

To complete this subsection, it is worth mentioning another fact about the infinite-dimensional Hubbard model. Namely, as Metzner and Vollhardt demonstrated [9], in the infinite-dimensional one-band Hubbard model the Gutzwiller approximation to the wave function gives the exact solution [9] (for the generalization of this result on the multi-orbital case, see Refs. [10, 11], and for general over-review—paper [12]).

Finally, it is important to mention that the DOS in infinite dimensions has different from Gaussian forms in the case of other lattices. One example is a parabolic DOS for the Bethe (Caley tree) lattice that will be introduced below. Such a DOS has a great advantage—it allows one to find semi-analytical solutions of the DMFT problem (due to analytical integration in Eq. (3.36)) that significantly simplifies the analysis.

3.2 Mapping the Problem on the Anderson Impurity Model

In the previous section, we have demonstrated that in infinite dimensions the Hubbard problem is significantly simplified since the electron self-energy is momentum-independent. Yet, the question of how to find $\Sigma_\sigma(\omega)$ remains open. In the case of small Us, one can use perturbation theory in U/t by taking into account only local-in-space free-electron GFs. In a similar way, in the case of large Us, one can find the atomic limit solution and then include t/U corrections by using the kinetic energy term as perturbation. However, it is not clear how to obtain $\Sigma_\sigma(\omega)$ at intermediate Us, when $U \sim W$—the regime where, probably, the most interesting physics takes place. The solution came with the 1992 breakthrough papers by Georges and Kotliar [13] and by Jarrell [14].

Let us mention their arguments that justify the theory before deriving the corresponding (DMFT) equations. As we have shown in the previous subsection, in infinite dimensions only local-in-space diagrams survive in the calculation of the electron self-energy and other functions. This, in particular, means that only local in space GFs (3.36) need to be taken into account in these diagrams. Thus, the Hubbard problem in infinite dimensions reduces to a local-in-space problem. Therefore, it is natural to try to map the Hubbard problem on an effective one-site problem that is easier to solve and the physics of which is much better understood. A single-site subsystem embedded into a bath of other sites with electrons may be regarded as such one-site, or an impurity, problem. One of the relevant model for this case is the Anderson impurity model [15]. It includes an impurity site with localized-orbital electrons that interact with each other through the Hubbard interaction, and other sites with itinerant (movable) electrons; the localized and itinerant electrons couple through hybridization terms. It was proposed [13, 14] that the lattice Hubbard problem can be mapped on the Anderson single-impurity problem with the same on-site Coulomb interaction and the same local (impurity electron) self-energies and GFs as in the lattice Hubbard model.

Let us summarize the main steps of the deriving DMFT equations by mapping the problem on the Anderson impurity problem. First, we derive the expression for the noninteracting ($U = 0$) Anderson's impurity GF by allowing hybridization between the impurity and itinerant orbital states. Namely, we show that the general expression for this GF has the form

$$\mathcal{G}(i\omega_n) = \frac{1}{i\omega_n + \mu - \Delta(i\omega_n)}, \tag{3.37}$$

where the effective hybridization field $\Delta(i\omega_n)$ describes the effect of the lattice electrons on the impurity electrons. To find $\Delta(i\omega_n)$, we solve the Anderson impurity model with the Hamiltonian:

$$\widehat{H} = \sum_{i,j,\sigma}\left(-t_{ij}c_{i\sigma}^{+}c_{j\sigma} + h.c.\right) + \sum_{i,\sigma}\left(V_i c_{i\sigma}^{+}d_\sigma + h.c.\right) + \epsilon_d\sum_{i,j,\sigma} d_\sigma^{+}d_\sigma + Un_\uparrow n_\downarrow \tag{3.38}$$

for a coupled system of the impurity-atom (d-) and itinerant (c-) electrons. The itinerant electrons hop through the lattice, and these processes are described by a similar to the Hubbard model hopping part of the Hamiltonian (the first sum in Eq. (3.38)). On the other hand, d-electrons are localized on the impurity orbital with energy ϵ_d and experience local on-site Coulomb repulsion with energy U when both spin-up and spin-down states are occupied (also, similar to the Hubbard model case). This part of the Hamiltonian consists of the last two terms in Eq. (3.38), where $n_\sigma = d_\sigma^{+}d_\sigma$. The two types of electrons hybridize, which is described by the V-terms (second sum) in Eq. (3.38).

Similar to the one-band Hubbard model (Chap. 2, Hubbard-III approximation), one can derive the equation for the impurity site time-ordered GF

$$G_{dd\sigma}(\tau - \tau') = -\left\langle T d_\sigma(\tau) d_\sigma^{+}(\tau')\right\rangle. \tag{3.39}$$

The equation has the following form:

$$\left(-\frac{\partial}{\partial\tau} - \epsilon_d + \mu\right) G_{dd\sigma}(\tau - \tau') - \sum_{i,\sigma} V_i G_{cd\sigma i}(\tau - \tau') \\ + U\langle T n_{\bar{\sigma}}(\tau) d_\sigma(\tau) d^+_{\sigma}(\tau')\rangle = \delta(\tau - \tau'), \tag{3.40}$$

where

$$G_{cd\sigma,i}(\tau - \tau') = -\langle T c_{\sigma i}(\tau) d^+_{\sigma i}(\tau')\rangle \tag{3.41}$$

is the hybridization function that satisfies the equation

$$-\frac{\partial}{\partial\tau} G_{cd\sigma,i}(\tau - \tau') - \sum_{i,\sigma} t_{ij} G_{cd\sigma j}(\tau - \tau') - V_i G_{dd\sigma}(\tau - \tau') = 0. \tag{3.42}$$

From the last equation, one can easily obtain the solution for the hybridization function. In the momentum–Matsubara frequency representation it has the following form:

$$G_{cd\sigma}\left(i\omega_n, \vec{k}\right) = \frac{V\left(\vec{k}\right)}{i\omega_n - \varepsilon\left(\vec{k}\right) + \mu} G_{dd\sigma}(i\omega_n), \tag{3.43}$$

where $V\left(\vec{k}\right)$ is the momentum Fourier transform of V_i.

Then, substituting this result in Eq. (3.40) and expressing in this equation, the Coulomb term as

$$U\langle T n_{\bar{\sigma}}(\tau) d_\sigma(\tau) d^+_{\sigma}(\tau')\rangle = \int \Sigma_{dd\sigma}(\tau - \tau'') G_{dd\sigma}(\tau'' - \tau') d\tau'', \tag{3.44}$$

one can find the following expression for the impurity GF:

$$G_{dd\sigma}(i\omega_n) = \left[i\omega_n - \epsilon_d + \mu + \sum_{\vec{k}} \frac{\left|V\left(\vec{k}\right)\right|^2}{i\omega_n - \varepsilon\left(\vec{k}\right) + \mu} - \Sigma_{dd\sigma}(i\omega_n) \right]^{-1}. \tag{3.45}$$

As it follows from the last equation, the expression for the noninteracting impurity GF has the postulated form (3.37), if one defines

$$\Delta(i\omega_n) = \sum_{\vec{k}} \frac{\left|V\left(\vec{k}\right)\right|^2}{i\omega_n - \varepsilon\left(\vec{k}\right) + \mu}. \tag{3.46}$$

In these notation, the interacting impurity GF is

$$G_{dd\sigma}(i\omega_n) = \frac{1}{i\omega_n + \mu - \Delta(i\omega_n) - \Sigma_{dd\sigma}(i\omega_n)}. \tag{3.47}$$

Thus, we found effective field $\Delta(i\omega_n)$ Eq. (3.46) of the rest of the system (bath) that acts on the impurity electrons when the interaction between the bath electrons is neglected. It is important to emphasize that the result for $\Delta(i\omega_n)$ (3.46) was derived at $U = 0$. From Eqs. (3.37) and (3.47), one can find the (Dyson) equation that connects the interacting $G_{dd\sigma}(i\omega_n)$ and noninteracting $\mathcal{G}_0(i\omega_n)$ impurity GFs and the impurity self-energy in the case of interacting impurity:

$$G^{-1}_{dd\sigma}(i\omega_n) = \mathcal{G}^{-1}_{\sigma}(i\omega_n) - \Sigma_{dd\sigma}(i\omega_n). \tag{3.48}$$

Now we are ready for the key step: we assume that the impurity and the lattice on-site GFs and self-energies are equal:

$$G_\sigma(i\omega_n) = G_{dd\sigma}(i\omega_n),$$

$$\Sigma_\sigma(i\omega_n) = \Sigma_{dd\sigma}(i\omega_n).$$

Thus, the first DMFT Eq. (3.36) for the local GF

$$G_\sigma(\omega) = \int d\varepsilon \rho(\varepsilon) \frac{1}{\omega - (\varepsilon - \mu) - \Sigma_\sigma(\omega)}, \tag{3.49}$$

is complemented by the Dyson equation emerged from the impurity problem:

$$G^{-1}_{\sigma}(i\omega_n) = \mathcal{G}^{-1}_{\sigma}(i\omega_n) - \Sigma_\sigma(i\omega_n). \tag{3.50}$$

The reader may ask: what is the advantage to have two equations (3.49) and (3.50) instead of one equation (3.49), since they still do not allow us to find the GF and the self-energy? Indeed, at this point we have two equations for three unknown functions $G_\sigma(i\omega_n)$, $\Sigma_\sigma(i\omega_n)$, and $\mathcal{G}(i\omega_n)$ (instead of one equation (3.49) for two unknown functions $G_\sigma(i\omega_n)$ and $\Sigma_\sigma(i\omega_n)$), and to solve them we need one more equation for $\mathcal{G}(i\omega_n)$ (or alternatively, we need to find the hybridization function for the impurity problem $\Delta(i\omega_n)$). Nevertheless, passing from Eq. (3.49) to the system (3.49) and (3.50) was a very important step since it allowed to obtain a closed system of DMFT equation. Namely, until this point we reduced the Anderson problem

(3.38) to the one-site problem of d-electrons in presence of the dynamical "bath" field of the itinerant electrons $\mathcal{G}(i\omega_n)$. The d-electrons interact with each other, as described by the last term in Eq. (3.38). Thus, one can write down the new, path integral, equation for the interacting impurity GF (equal to the local lattice GF) that establishes connection between $\Sigma_\sigma(i\omega_n)$, $\mathcal{G}(i\omega_n)$ and U (we use the imaginary time representation):

$$G_{dd\sigma}(\tau) \equiv G_\sigma(\tau) = \int D[\psi]D[\psi^*]\psi_\sigma(\tau)\psi_\sigma^*(0)$$
$$\times \exp\left[- \int_0^\beta d\tau' \int_0^\beta d\tau'' \sum_{\bar\sigma} \psi_{\bar\sigma}^*(\tau')\mathcal{G}_{\bar\sigma}^{R-1}(\tau' - \tau'')\psi_{\bar\sigma}(\tau'') \right.$$
$$\left. + U \int_0^\beta d\tau' \psi_\uparrow^*(\tau')\psi_\uparrow(\tau')\psi_\downarrow^*(\tau')\psi_\downarrow(\tau') \right], \quad (3.51)$$

where ψ_σ and ψ_σ^* are the anti-commuting Grassman variables and τ is the imaginary (Matsubara) time that changes in the limits $0 \le \tau \le \beta$, where $\beta = 1/T$ is the inverse temperature (once $G_\sigma(\tau)$ is found one can calculate its frequency transform $G_\sigma(i\omega_n)$).

Equation (3.51) completes the system of DMFT equations (3.49) and (3.50) for three unknown functions $G_\sigma(i\omega_n)$, $\mathcal{G}_\sigma(i\omega_n)$ and $\Sigma_\sigma(i\omega_n)$. Since the field $\mathcal{G}(i\omega_n)$ describes the dynamical effects of the itinerant electrons on the impurity electrons, it is called dynamical mean field ("mean" since it is site-independent in the case of the lattice Hubbard model). We do not give recipes here on how to calculate the path integral in the "impurity equation" (3.51), or as it is usually said "to solve the impurity problem" or to use an "impurity solver." It can be done in different ways, from using a perturbation theory (PT) or exact numerical methods to applying stochastic (e.g., Quantum Monte Carlo) approaches. Different impurity solvers are presented in Chap. 6. We intentionally do it after Chap. 5 where multi-orbital and cluster DMFT approaches are introduced, so in Chap. 6, we present approaches to solve the impurity problem in the general case of impurity that has many orbitals and/or many atoms (cluster impurity). In Chap. 4, we discuss physical results for one-band Hubbard model obtained with different impurity solvers, so before reading Chap. 4 the reader may like to make himself familiar with different solvers for the one-band case (one-atom and one-orbital impurity) reading Chap. 6.

To solve the system of DMFT equation for the one-band Hubbard model in infinite dimensions (3.49)–(3.51) one can choose different algorithms. One possibility is an iterative procedure (see, e.g., book [16]):

1. One starts with a guess self-energy function $\Sigma_\sigma^1(i\omega_n)$ (in the simplest case with $\Sigma_\sigma^1(i\omega_n) = 0$ for all frequencies) and substitutes it into Eq. (3.49) to calculate $G_\sigma(i\omega_n)$.
2. Once $G_\sigma(i\omega_n)$ has been found, one uses $\Sigma_\sigma^1(i\omega_n)$ and $G_\sigma(i\omega_n)$ to calculate the dynamical mean-field $\mathcal{G}_\sigma(i\omega_n)$ from Eq. (3.50).

3. Then, using obtained $\mathcal{G}_\sigma(i\omega_n)$ one calculates new $G_\sigma(i\omega_n)$ by solving the impurity problem Eq. (3.51).
4. Next, using $\mathcal{G}_\sigma(i\omega_n)$ and newly obtained $G_\sigma(i\omega_n)$ one calculates from the Dyson Eq. (3.50) the new self-energy $\Sigma^2_\sigma(i\omega_n)$.
5. The "old" Σ^1_σ and the "new" Σ^2_σ self-energies are compared. If the modulus of their difference for all frequencies is larger than some chosen small number, one puts $\Sigma^1_\sigma(i\omega_n) = \Sigma^2_\sigma(i\omega_n)$ and iterates steps 1–5. The procedure stops when the convergence for the self-energy has been reached.

The first DMFT solutions (with the PT [13] and exact quantum Monte Carlo [14] impurity solvers) showed that the theory is able to produce meaningful results for strongly correlated systems. In particular, as it was shown in Ref. [13] in the PM case the DMFT DOS demonstrates a pseudogap at mostly large Us. As it followed from the results of paper [14], DMFT can also describe the metal–Mott insulator transition (as parameter U/t increases and crosses the critical value) and antiferromagnetism when the system is close to half-filling regime. Details of the MIT in the PM phase were further clarified in the following after the previous two works paper [17]. Demonstrated ability of DMFT to describe pseudogap phase, MIT, and antiferromagnetism made a great impact on the community, probably especially due to a potential relevance of the results to the most popular systems in those days—high-temperature superconductors, and established DMFT as a potentially very promising approach for strongly correlated electrons.

Before proceeding with other physical properties of the one-band Hubbard model obtained with DMFT (Chap. 4) and generalization of the approach on the multi-orbital and multi-atom cases (Chap. 5), in this chapter we discuss general properties and the nature of DMFT from the point of view of many-body theory (Sect. 3.3) and mention equivalent to DMFT approaches used in the case of alloys and called coherent potential approximations (CPAs) (Sect. 3.4).

3.3 General Many-Body Theory Formulation

3.3.1 DMFT Equations

In this subsection, we will formulate DMFT based on general principles of many-body theory. We begin with discussion on how to derive the impurity equation without mapping the Hubbard problem on the Anderson impurity problem, i.e., from the "first many-body principles" (for more details, we refer the reader to review [2]). For this, let us write down the equation for the partition function of the Hubbard model:

$$Z = \int \prod_{i,\sigma} D[c_{i\sigma}] D\left[c_{i\sigma}^{+}\right] e^{-S}, \tag{3.52}$$

where

$$S = \int_0^\beta d\tau \left(\sum_{i\sigma} c_{i\sigma}^{+} \partial_\tau c_{i\sigma} - \sum_{ij\sigma} t_{ij} c_{i\sigma}^{+} c_{j\sigma} - \mu \sum_{i\sigma} c_{i\sigma}^{+} c_{i\sigma} + U \sum_i n_{i\uparrow} n_{i\downarrow} \right) \tag{3.53}$$

is the action of the system. Using the last two expressions one can also drive the equation for the effective action $S_{\text{eff}}\left[c_{0\sigma}^{+}, c_{0\sigma}\right]$ for the chosen single site $i = 0$. Namely, the single-site effective action can be found from

$$\frac{1}{Z_{\text{eff}}} e^{-S_{\text{eff}}\left[c_{0\sigma}^{+}, c_{0\sigma}\right]} = \frac{1}{Z} \int \prod_{i \neq 0, \sigma} D c_{i\sigma}^{+} D c_{i\sigma} e^{-S}, \tag{3.54}$$

where Z_{eff} is the zero-site ("effective") partition function. The total action S in Eq. (3.54) can be divided into three parts:

$$S = S_0 + S_{0-b} + S_b, \tag{3.55}$$

where

$$S_0 = \int_0^\beta d\tau \left(\sum_\sigma c_{0\sigma}^{+} \partial_\tau c_{0\sigma} - \mu \sum_\sigma c_{0\sigma}^{+} c_{0\sigma} + U n_{0\uparrow} n_{0\downarrow} \right) \tag{3.56}$$

is the zero-site action,

$$S_{0-b} = -\int_0^\beta d\tau \sum_{i\sigma} t_{i0} \left(c_{i\sigma}^{+} c_{0\sigma} + c_{0\sigma}^{+} c_{i\sigma} \right) \equiv -\int_0^\beta d\tau \sum_{i\sigma} t_{i0} \left(c_{i\sigma}^{+} \eta_{i\sigma} + \eta_{i\sigma}^{+} c_{i\sigma} \right) \tag{3.57}$$

is the part of the action that describes the hybridization between the zero site and the remaining sites (bath) electrons, where the fields $\eta_{i\sigma} = t_{i0} c_{0\sigma}$ can be regarded as sources coupled to the "bath" sites i ($i \neq 0$).

The remaining part of the action S_b is the bath action, i.e., the action of the subsystem that consists of the whole lattice minus zero site (the lattice with a cavity):

$$S_b = \int_0^\beta d\tau \left(\sum_{i \neq 0, \sigma} c_{i\sigma}^{+} \partial_\tau c_{i\sigma} - \sum_{ij \neq 0, \sigma} t_{ij} c_{i\sigma}^{+} c_{j\sigma} - \mu \sum_{i \neq 0, \sigma} c_{i\sigma}^{+} c_{i\sigma} + U \sum_{i \neq 0} n_{i\uparrow} n_{i\downarrow} \right). \tag{3.58}$$

Let us assume for a moment that the GF of the system with the cavity $G^{(0)}$ is known. Then, integration over the nonzero site states can be performed. It gives

$$S_{\rm eff} = S_0 + \sum_{n=1}^{\infty} \sum_{i_1,\ldots,i_n} \sum_{j_1,\ldots,j_n} \int d\tau \eta_{i_1}^{+}(\tau_{i_1}) \ldots \eta_{i_n}^{+}(\tau_{i_n})$$
$$G_{i_1 \ldots j_n}^{(0)}(\tau_{i_1}, \ldots, \tau_{i_n}, \tau_{j_1}, \ldots, \tau_{j_n}) \eta_{j_1}(\tau_{j_1}) \ldots \eta_{j_n}(\tau_{j_n}), \quad (3.59)$$

where $G_{i_1 \ldots j_n}^{(0)}$ are connected GFs of the system with cavity (in Eq. (3.59), the spin and site indices are incorporated into one index $(i_1, \ldots j_n)$).

As we have demonstrated in the previous subsection, the hopping parameters $t_{\vec{R}_i \vec{R}_j}$ depend on the dimensionality of the system as $d^{-|\vec{R}_i - \vec{R}_j|/2}$ (Eq. (3.30)). Then, it is possible to show that in the limit of infinite dimensions only the second-order terms in the expansion in Eq. (3.59) survive. In the proof, one can use the fact that the single-particle GFs $G_{ij}^{(0)}$ are proportional to $d^{-|\vec{R}_i - \vec{R}_j|/2}$ (Eq. (3.31)) and that higher-order (in site indices) GFs are proportional to products of these small factors $\left(d^{-|\vec{R}_i - \vec{R}_j|/2}\right)$ due to the new site indices. For example, the connected GF $G_{i_1 i_2 j_1 j_2}$ depends on the system dimension as $d^{-|\vec{R}_{i_1} - \vec{R}_{i_2}|/2} d^{-|\vec{R}_{i_1} - \vec{R}_{j_1}|/2} d^{-|\vec{R}_{i_1} - \vec{R}_{j_2}|/2}$. To complete the proof, one needs to consider possible different combinations of the sites in the above factors that give nonzero contributions to the sum in Eq. (3.59) and to show that the nth order terms in the sum in Eq. (3.59) depend on dimensionality as $d^{-(n-2)}$, thus only the second-order terms in Eq. (3.59) are finite (see, e.g., review [2]).

In this case, Eq. (3.59) becomes $S_{\rm eff} = S_0 + \sum_{i,j} \int d\tau \eta_i^{+}(\tau_i) G_{ij}^{(0)}(\tau_i, \tau_j) \eta_j(\tau_j)$, where $\eta_{i\sigma} = t_{i0} c_{0\sigma}$ (we included the spin index to be explicit), and we arrive to the same single-site effective action as in the exponent of the equation for the impurity GF (3.51):

$$S_0 = -\int_0^{\beta} d\tau \int_0^{\beta} d\tau' \sum_{\sigma} c_{0\sigma}^{+}(\tau) \mathcal{G}_{\sigma}^{-1}(\tau - \tau') c_{0\sigma}(\tau') + U \int_0^{\beta} d\tau n_{0\uparrow}(\tau) n_{0\downarrow}(\tau), \quad (3.60)$$

where the inverse dynamical mean-field function (Weiss field) $\mathcal{G}_{\sigma}^{-1}(\tau - \tau')$ has the following form in the frequency representation:

$$\mathcal{G}_{\sigma}^{-1}(i\omega_n) = i\omega_n + \mu - \sum_{ij} t_{oi} t_{oj} G_{ij\sigma}^{(0)}(i\omega_n). \quad (3.61)$$

In order to complete the system of DMFT equations, one needs to relate the cavity system GF $G_{ij}^{(0)}$ to the total lattice GFs G_{ij}. It is possible to show (for details, see Ref. [2]) that the two GFs are related as

$$G_{ij}^{(0)} = G_{ij} - \frac{G_{i0}G_{0j}}{G_{00}}. \tag{3.62}$$

This expression is rather complicated for practical use. On the other hand, it is significantly simplified in the case of the Bethe lattice and gives for the dynamical mean-field $\mathcal{G}^{-1}(i\omega_n) = i\omega_n + \mu - t^{*2}G(i\omega_n)$. This result allows one to analyze the DMFT problem analytically. In the case of other lattices, the dynamical mean-field $\mathcal{G}$ is regarded as one of the independent functions that needs to be found in order to solve the DMFT problem.

Thus, starting with the general form of the action (3.53) for the Hubbard Hamiltonian we have demonstrated how to obtain the effective impurity action (3.60) that includes the dynamical mean field (3.61). Using this result one can derive the DMFT eqs. (3.49)–(3.51) by employing similar arguments as in the previous subsection:

(a) The one-site effective action (3.60) gives the single-site GF Eq. (3.51).
(b) In the case when the electrons on zero site do not interact, the corresponding local GF is $\mathcal{G}$, as it follows from (3.60). Thus, the equation that connects the interacting and noninteracting single-site GFs is the Dyson Eq. (3.50).
(c) Using the fact that in infinite dimensions only local lattice diagrams survive, we conclude that only local GF and self-energy need to be taken into account. The GF and self-energy are connected by Eq. (3.49) that completes the system (3.49)–(3.51).

3.3.2 DMFT and Thermodynamics

In this subsection, we briefly summarize how one can calculate the thermodynamic potential and hence find the thermodynamic properties of the Hubbard model by using DMFT.

Let us begin with the equation for the thermodynamic potential

$$\Omega = \Omega_0 - T \ln \langle \mathfrak{S} \rangle, \tag{3.63}$$

where

$$\Omega_0 = -T \ln Tr\left\{ e^{-\frac{\left(\widehat{H}_0 - \mu N\right)}{T}} \right\} \tag{3.64}$$

is the thermodynamic potential of noninteracting system and the last term in Eq. (3.63) is defined by the average of the (scattering) matrix $\mathfrak{S}(\tau)$ that satisfies equation

$$\frac{\partial \mathfrak{S}(\tau)}{\partial \tau} = -H_{int}(\tau)\mathfrak{S}(\tau), \tag{3.65}$$

where $H_{int}(\tau)$ is the interacting part of the Hamiltonian (see, e.g., book [18]). The formal solution of Eq. (3.65) can be written as

$$\mathfrak{S}(\tau) = T_\tau e^{-\int_0^{1/T} H_{int}(\tau)d\tau} \tag{3.66}$$

and the average of $\mathfrak{S}(\tau)$ in Eq. (3.63) can be obtained by using PT with terms

$$\frac{(-1)^n}{n!}\int_0^{1/T} d\tau_1 \ldots \int_1^{1/T} d\tau_n \langle T_\tau H_{int}(\tau_1)\ldots H_{int}(\tau_n)\rangle. \tag{3.67}$$

Unfortunately, one cannot apply this method of the diagram summation due to presence of the factor $\frac{1}{n!}$ in front of the integrals in Eq. (3.67). Luttinger and Ward [19] and Baym and Kadanoff [20] proposed to calculate the interacting part of the thermodynamic potential by performing summation of the diagrams that include only the *exact* GFs G. The corresponding interacting part of the functional Ω' is called the Luttinger–Ward or Baym–Kadanoff functional.

Thus, the total thermodynamic potential can be decomposed as

$$\begin{aligned}\Omega = \Omega_0 - 2TV\sum_n \int \frac{d^dk}{(2\pi)^d} \\ \times\left\{\ln\left(1 - G^{(0)}\left(i\omega_n, \vec{k}\right)\Sigma\left(i\omega_n, \vec{k}\right)\right) + G\left(i\omega_n, \vec{k}\right)\Sigma\left(i\omega_n, \vec{k}\right)\right\} \\ + \Omega',\end{aligned} \tag{3.68}$$

where

$$\Omega_0 = -2TV\sum_n \int \frac{d^dk}{(2\pi)^d} \ln G^{(0)}\left(i\omega_n, \vec{k}\right) e^{i\omega_n 0^+} \tag{3.69}$$

In detail, Ω' is obtained from the skeleton expression for the self-energy (where all internal lines correspond to the exact, or full, GFs) by connecting the external vertices with fully interacting GFs and dividing each term by $2n$ (n—number of the interaction lines, $2n$—number of GFs) [21]. Thus, in this case the self-energy in Eq. (3.68) is a functional of the exact GF.

In infinite dimensions, if two internal vertices that correspond to the Hubbard interaction $Un_\uparrow n_\downarrow$ can be connected by three or more paths they must correspond to the same site, a very helpful property for the PT diagrammatic expansion that facilitates the calculations of the self-energy and other quantities. Since two internal vertices of the skeleton diagrams can be always connected by more than two paths,

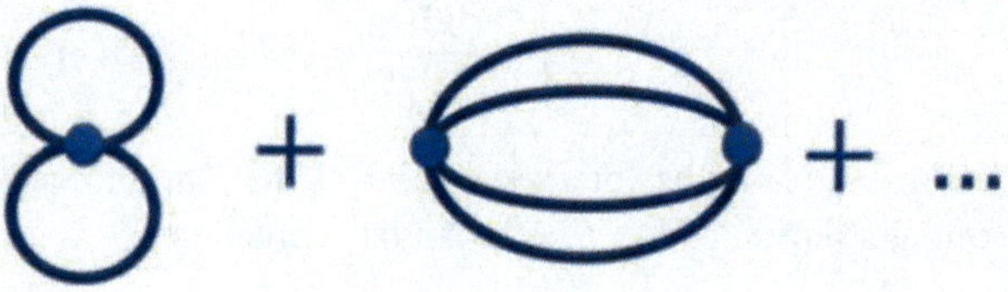

Fig. 3.3 Lowest-order terms of the skeleton-diagram expansion for the Luttinger–Ward functional Ω'

all diagrams contributing to Σ collapse to single-site diagrams. Two lowest-order skeleton graphs for the Luttinger–Ward functional are shown in Fig. 3.3.

A shorter way to show that only local-in-space diagrams contribute to the skeleton Σ in infinite dimensions is to use the equation for the differential of the Luttinger–Ward functional

$$\delta\Omega' = 2TV\sum_n \int \frac{d^3k}{(2\pi)^3} Tr\left[\Sigma\left(i\omega_n, \vec{k}\right)\delta G\left(i\omega_n, \vec{k}\right)\right] \tag{3.70}$$

that gives

$$\Sigma_{ij}(i\omega_n) = \frac{\delta\Omega'}{G_{ij}(i\omega_n)}. \tag{3.71}$$

Since as we showed above, in infinite dimensions only local-in-space GFs contribute to the skeleton Ω' the last equation gives $\Sigma_{ij}(i\omega_n) = \delta_{ij}\Sigma_{ii}(i\omega_n)$.

Next, since in infinite dimensions the Luttinger–Ward functional is defined by single-site diagrams, one can write

$$\Omega' = \sum_i \omega'(G_{ii}), \tag{3.72}$$

where $\omega'(G_{ii})$ is the contribution of the ith site to the Luttinger–Ward functional (ω' is potential per site). Equation (3.72) greatly simplificates the theory, reducing calculations of the thermodynamic potential to the single-site calculations (in particular, it gives local self-energy in Eq. (3.71)).

At the end of this subsection, we give a useful relation between the full thermodynamic potentials of the lattice and impurity models Ω and Ω_{imp} by using the equations for these two quantities:

$$\Omega = \Omega' + T\sum_{n,\sigma} \int \frac{d^dk}{(2\pi)^d} \left\{ \ln G_\sigma\left(i\omega_n, \vec{k}\right) - \Sigma_\sigma\left(i\omega_n, \vec{k}\right) G_\sigma\left(i\omega_n, \vec{k}\right) \right\}, \tag{3.73}$$

$$\Omega_{imp} = \omega' + T\sum_{n,\sigma} \{ \ln G_\sigma(i\omega_n) - \Sigma_\sigma(i\omega_n) G_\sigma(i\omega_n) \}. \tag{3.74}$$

From Eqs. (3.72)–(3.74) and from the fact that self-energy is momentum-independent one can obtain the relation

$$\frac{\Omega}{N} = \Omega_{\text{imp}} - T\sum_{n,\sigma}\left(\int d\varepsilon\rho(\varepsilon)\ln\left(i\omega_n - \varepsilon + \mu - \Sigma_\sigma(i\omega_n)\right) + \ln G_\sigma(i\omega_n)\right). \quad (3.75)$$

Finally, let us write down another useful equation - expression for the internal energy in infinite dimensions:

$$E = NT\sum_{n,\sigma}\int d\varepsilon\frac{\rho(\varepsilon)}{i\omega_n - \varepsilon + \mu - \Sigma_\sigma(i\omega_n)} + \frac{1}{2}NT\sum_{n,\sigma}\Sigma_\sigma(i\omega_n)G_\sigma(i\omega_n). \quad (3.76)$$

Thus, in DMFT, the thermodynamic properties of the system can be calculated from local GF and self-energy, which greatly simplifies the analysis.

3.4 DMFT and Equivalent CPA Approaches

As we have shown above, in DMFT the scaling of the hopping parameter with dimensionality $\left(\sim\frac{1}{\sqrt{d}}\sim\frac{1}{\sqrt{Z}}\right)$ results in a finite kinetic energy. Nonzero kinetic energy effects compete with the potential energy effects in shaping the ground state of the system. Similar ideas were also developed in 1960s by solving the alloy problems for noninteracting electrons [22–26], where it was realized that the noninteracting(!) electron self-energy is also frequency-dependent [27] and the small parameter in the theory is inversely proportional to the square root of the number of the nearest-neighbor atoms, $\frac{1}{\sqrt{Z}}$ [27]. The corresponding approaches are called coherent potential approximation (CPA). There are also generalizations of CPA on the case of interacting electrons (see below). In this subsection, we summarize the main properties of two popular interacting-electron CPA approaches used in the theory of disordered systems and equivalent to DMFT.

3.4.1 Many-Body CPA

This method is an extention of CPA from the alloys (i.e., noninteracting electrons) to the many-body (interacting electron) case by Hirooka and Shimitzu [28] who based their formulation on the finite-temperature GF approach. Namely, they considered the model of transition-metal alloy A_xB_{1-x} described by the Hubbard Hamiltonian

$$\widehat{H}=\sum_{i,j,\sigma}t_{ij}c^{+}_{i\sigma}c_{j\sigma}+\sum_{i,\sigma}\varepsilon_i n_{i\sigma}+\sum_{i}U_i n_{i\uparrow}n_{i\downarrow}, \tag{3.77}$$

where $\varepsilon_i = \varepsilon^A, \varepsilon^B$ and $U_i = U_0^A, U_0^B$ are the site energies and local Coulomb repulsions that depend on the type of the atom (*A* or *B*) that occupies given site. The inter-site hopping parameters t_{ij} do not depend on the type of atoms. This guarantees that the banstructure of two pure noninteracting systems of atoms *A* and *B* is the same. Then, the problem (3.77) was mapped on a free-fermion problem in an external time-dependent potential $\Sigma_{ij\sigma\sigma'}(t-t')$ with the Hamiltonian

$$\widehat{H}'(\tau)=\widehat{H}_0+\sum_{i,\sigma}\int_0^\beta d\tau'\sum_{j,\sigma'}c^{+}_{i\sigma}(\tau)\Sigma_{ij\sigma\sigma'}(\tau-\tau')c_{j\sigma'}(\tau'). \tag{3.78}$$

It was assumed that the potential is diagonal in space and spin indices: $\Sigma_{ij\sigma\sigma'}(\tau-\tau')=\delta_{\sigma\sigma'}\delta_{ij}\Sigma_{i\sigma}(\tau-\tau')$. Using these assumptions and also assuming that the electron self-energy is local in space and spins and is time-translational invariant (i.e., depends on one frequency variable), the authors derived a set of equations for the electron GF and self-energy, employing the concept of impurity and using the Shiba's assumption [29] that the impurity GF with local interaction embedded in "coherent potential" $\Sigma_{i\sigma}(\tau-\tau')$ is the same as the GF of the medium with the same coherent potential.

The GF $F_{ij\sigma\sigma'}(\tau-\tau')$ for the Hamiltonian (3.78) satisfies the Dyson equation

$$F_{ij\sigma\sigma}(\tau-\tau')=g_{ij\sigma\sigma}(\tau-\tau') +\sum_k\int_0^\beta d\tau_1\int_0^\beta d\tau_2 g_{ik\sigma\sigma}(\tau-\tau_1)\Sigma_{k\sigma}(\tau_1-\tau_2)F_{kj\sigma\sigma}(\tau_2-\tau'), \tag{3.79}$$

where $g_{ij\sigma\sigma}(\tau-\tau')$ is the noninteracting GF.

To determine potential $\Sigma_{i\sigma}(t-t')$ one performs the following steps. First, one considers an impurity problem on site *i*, where this coherent potential is replaced by the on-site interaction $U_i n_{i\uparrow}n_{i\downarrow}$. The corresponding impurity Hamiltonian has the form

$$\widehat{H}^{(i)}(\tau)=\widehat{H}'(\tau)-\sum_\sigma\int_0^\beta d\tau' c^{+}_{i\sigma}(\tau)\Sigma_{i\sigma}(\tau-\tau')c_{i\sigma}(\tau')+U_i n_{i\uparrow}n_{i\downarrow}. \tag{3.80}$$

Second, one writes down the Dyson equation for the impurity GF

$$G^{(i)}_{ii\sigma\sigma}(\tau-\tau') = F^{(i)}_{ii\sigma\sigma}(\tau-\tau') + \int_0^\beta d\tau_1 \int_0^\beta d\tau_2 F^{(i)}_{ii\sigma\sigma}(\tau-\tau_1)\Lambda_{i\sigma}(\tau_1-\tau_2)G^{(i)}_{ii\sigma\sigma}(\tau_2-\tau'), \quad (3.81)$$

where $\Lambda_{i\sigma}(\tau_1-\tau_2)$ is the impurity electron self-energy (coming from the interaction term $U_i n_{i\uparrow} n_{i\downarrow}$), and $F^{(i)}_{ii\sigma\sigma}(\tau-\tau')$ is the coherent GF for the system with impurity on site i. This function corresponds to the Hamiltonian

$$\widehat{H}^{(i)\prime}(\tau) = \widehat{H}'(\tau) - \sum_\sigma \int_0^\beta d\tau' c^+_{i\sigma}(\tau)\Sigma_{i\sigma}(\tau-\tau')c_{i\sigma}(\tau') \quad (3.82)$$

and satisfies the following Dyson equation:

$$F^{(i)}_{ii\sigma\sigma}(\tau-\tau') = F_{ii\sigma\sigma}(\tau-\tau') - \int_0^\beta d\tau_1 \int_0^\beta d\tau_2 F_{ii\sigma\sigma}(\tau-\tau_1)\Sigma_{i\sigma}(\tau_1-\tau_2)F^{(i)}_{ii\sigma\sigma}(\tau_2-\tau'), \quad (3.83)$$

Third, one needs to find the impurity–electron self-energy $\Lambda_{i\sigma}(\tau_1-\tau_2)$ by using some approximation, e.g., the t-matrix approximation [30], as it was done in paper [28].

The final, fourth, step is to postulate that the GFs that correspond to Hamiltonians (3.78) and (3.80) are equal:

$$G^{(i)}_{ii\sigma\sigma}(\tau-\tau') = F_{ii\sigma\sigma}(\tau-\tau'). \quad (3.84)$$

Since in the case of crystal all sites are equivalent, the self-energy (coherent potential) $\Sigma_{i\sigma}$ and coherent GF $F_{ii\sigma\sigma}$ are site-independent. Solving the system of equations for these functions and for $G^{(i)}_{ii\sigma\sigma}$ and $F^{(i)}_{ii\sigma\sigma}$ (Eqs. (3.79), (3.81), (3.83), and (3.84)), one obtains the time-dependent coherent potential $\Sigma_{i\sigma}$. Importantly, as it follows from Eq. (3.79) formal expression for the GF of the system in the external field $F_{ij\sigma\sigma}$ has the same form as the equation for the DMFT local GF:

$$F_\sigma(i\omega_n) = \int \frac{d\varepsilon\rho(\varepsilon)}{i\omega_n - \varepsilon + \mu - \Sigma_\sigma(i\omega_n)}, \quad (3.85)$$

where $\rho(\varepsilon)$ is the free-electron DOS.

3.4.2 Dynamical CPA

Single-site approaches were also developed in the theory of itinerant magnetism [31–34] by employing path-integral techniques. In these theories, the interacting Hamiltonian is transformed into a Hamiltonian of free electrons in a random time-dependent potential. The problem described by such an Hamiltonian was solved within CPA. For example, Cyrot [35] solved the problem within a static approximation approach [33] by neglecting the time dependence of the random field. A more accurate single-site theory that takes into account thermal spin fluctuations was independently proposed by Hubbard [36–38] and Hasegawa [39, 40]. Unfortunately, being a meaningful interpolating theory that connects small and large Coulomb repulsion regimes at finite temperatures, at $T = 0$ Hubbard–Hasegawa approach reduces to the ground-state HF approximation.

To improve the theory, Kakehashi and Fulde [41] formulated a variational approach that allowed to adiabatically take into account the missing in HF (in the ground state) local electron correlations. This led, in particular, to reduction of the Curie and Néel temperatures and suppression of the charge fluctuations at finite temperatures [42, 43].

As a further advancement, Kakehashi [44, 45] developed a statistically more accurate "dynamical CPA" method that fully takes into account the dynamical fluctuations of spins and charges within the single-site approximation. He demonstrated that the dynamical CPA describes the many-body energy bands and the local moment behavior in itinerant ferromagnets on the same footing [46]. Here we present basic elements of the dynamical CPA approach.

In dynamical CPA, one begins with the Hamiltonian

$$\widehat{H} = \sum_{i,j,\sigma} t_{ij} c^{+}_{i\sigma} c_{j\sigma} + \sum_{i,\sigma} (\varepsilon_i - \sigma h_i) n_{i\sigma} + \sum_{i} U n_{i\uparrow} n_{i\downarrow}, \tag{3.86}$$

where in addition to already defined quantities and operators, ε_i and h_i are the atomic energy and magnetic field on site i. Hamiltonian (3.86) can be rewritten in an equivalent form:

$$\widehat{H} = \sum_{i,j,\sigma} t_{ij} c^{+}_{i\sigma} c_{j\sigma} + \sum_{i,\sigma} (\varepsilon_i + 2\alpha U - \sigma h_i) n_{i\sigma} + \sum_{i} \frac{1}{4} \left(U_i n_i^2 - J_i m_i^2 \right), \tag{3.87}$$

where $n_i = n_{i\uparrow} + n_{i\downarrow}$, $m_i = n_{i\uparrow} - n_{i\downarrow}$, $U_i = (1 - 4\alpha U)$, $J_i = (1 + 4\alpha U)$ and α is a parameter (an arbitrary number).

The next step is calculation of the free-energy F by using the equation:

$$e^{-F/T} = Tr\left[\widehat{T} e^{-\int_0^{\beta} d\tau \widehat{H}(\tau)} \right]. \tag{3.88}$$

Using the path-integral formalism and introducing random time-dependent charge and exchange fields $\eta_i(\tau)$ and $\xi_i(\tau)$, Eq. (3.88) can be transformed to an equivalent one:

$$e^{-F/T} = \int \prod_i d\xi_i d\eta_i e^{-\beta E[\xi,\eta]}, \tag{3.89}$$

where

$$E[\xi,\eta] = -\frac{1}{\beta} \ln Z_1[\xi,\eta] + \frac{1}{4\beta} \sum_i \int_0^\beta \left[U_i \eta_i^2(\tau) + J_i \xi_i^2(\tau) \right] d\tau, \tag{3.90}$$

and

$$Z_1[\xi,\eta] = Tr\left[\widehat{T} e^{-\int_0^\beta \widehat{H}(\tau) d\tau} \right]. \tag{3.91}$$

In Eq. (3.91), the effective Hamiltonian $\widehat{H}(\tau)$ is

$$\widehat{H}(\tau) = \sum_{i,j,\sigma} t_{ij} c_{i\sigma}^+ c_{j\sigma} + \sum_{i,\sigma} v_{i\sigma} n_{i\sigma}, \tag{3.92}$$

where

$$v_{i\sigma} = \varepsilon_i - \mu - \sigma h_i + \frac{1}{2} \left[U_i i\eta_i(\tau) - J_i \xi_i(\tau) \sigma \right] \tag{3.93}$$

and μ is chemical potential.

To calculate the nontrivial part of the energy (Eqs. (3.90) and (3.91)), one can extract the free part of the Hamiltonian

$$\widehat{H}_0(\tau) = \sum_{i,j,\sigma} t_{ij} c_{i\sigma}^+ c_{j\sigma} \tag{3.94}$$

and define the free

$$g_{ij\sigma}(\tau - \tau') = -\frac{Tr\left[e^{-\beta \widehat{H}_0} \widehat{T} c_{i\sigma}(\tau) c_{j\sigma}^+(\tau') \right]}{Tr\left[e^{-\beta \widehat{H}_0} \right]}, \tag{3.95}$$

and the interacting

$$G_{ij\sigma}(\tau-\tau') = -\frac{Tr\left[\widehat{T}c_{i\sigma}(\tau)c^{+}_{j\sigma}(\tau')e^{-\int_0^\beta \widehat{H}(\tau'')d\tau''}\right]}{Tr\left[\widehat{T}e^{-\int_0^\beta \widehat{H}(\tau'')d\tau''}\right]}. \quad (3.96)$$

GFs. Then, the partition function Z_1 can be found by using the formula

$$\ln Z_1 = \ln Tr\left[e^{-\beta\widehat{H}_0}\right] + Tr\ln g - Tr\ln G^{-1}, \quad (3.97)$$

where the trace operation in the last two terms is performed over the site, time, and spin indices.

Thus, the problem now is reduced to finding the interacting GF $G_{ij\sigma}$. Since this function is connected with the free GF g by the Dyson equation

$$G^{-1} = g^{-1} - v, \quad (3.98)$$

where

$$v_{ij\sigma}(\tau,\tau') = \Delta\tau v_{i\sigma}(\tau)\delta_{ij}\delta(\tau-\tau') \quad (3.99)$$

($\Delta\tau$ is the time step), to solve the problem one needs to find the field $v_{i\sigma}(\tau)$.

This is done in the following way. First, one introduces the coherent potential $\Sigma_{i\sigma}(\tau-\tau')$ (this field is independent of $\xi_i(\tau)$ and $\eta_i(\tau)$). This gives

$$Tr\ln G^{-1} = Tr\ln \widetilde{G}^{-1} + Tr\ln\left(1-\delta v\widetilde{G}\right), \quad (3.100)$$

where

$$\widetilde{G}^{-1} = g^{-1} - \Sigma \quad (3.101)$$

is the inverse coherent GF and

$$\delta v = v - \Sigma \quad (3.102)$$

with

$$\Sigma_{ij\sigma}(\tau-\tau') = \Delta\tau\Sigma_{i\sigma}(\tau)\Delta\tau\delta_{ij}\delta. \quad (3.103)$$

Second, in order to use a single-site approximation one separates the local

$$F_{ij\sigma}(\tau-\tau')=\widetilde{G}_{ii\sigma}(\tau-\tau')\delta_{ij} \tag{3.104}$$

and nonlocal

$$F'_{ij\sigma}(\tau-\tau')=\widetilde{G}_{ij\sigma}(\tau-\tau')(1-\delta_{ij}) \tag{3.105}$$

in-site parts of GF $\widetilde{G}$, i.e., putting

$$\widetilde{G}_{ij\sigma}(\tau-\tau')=F_{ij\sigma}(\tau-\tau')+F'_{ij\sigma}(\tau-\tau'). \tag{3.106}$$

Substituting the last equation into Eq. (3.100) gives

$$\begin{aligned} Tr\ln G^{-1} &= Tr\ln \widetilde{G}^{-1}+Tr\ln(1-\delta vF) \\ &+Tr\ln\left(1-(1-\delta vF)^{-1}\delta vF'\right). \end{aligned} \tag{3.107}$$

Then, using Eqs. (3.90), (3.97), and (3.107) one can obtain the following expression for the functional of energy:

$$E[\xi,\eta]=\widetilde{F}[\Sigma]+\sum_i E_i[\xi_i,\eta_i]+\Delta E[\xi,\eta], \tag{3.108}$$

where

$$\widetilde{F}[\Sigma]=-\beta^{-1}\left(\ln Tr\left[e^{-\beta\widehat{H}_0}\right]+Tr\ln(1-\Sigma g)\right) \tag{3.109}$$

is the free energy of the effective medium,

$$E_i[\xi_i,\eta_i]=-\beta^{-1}Tr\ln(1-\delta v_iF_i)+\frac{1}{4\beta}\int_0^\beta\left[U_i\eta_i^2(\tau)+J_i\xi_i^2(\tau)\right]d\tau \tag{3.110}$$

is the single-site functional of energy and

$$\Delta E[\xi,\eta]=-\beta^{-1}Tr\ln(1-\widetilde{t}F'), \tag{3.111}$$

with

$$\widetilde{t}=(1-\delta vF)^{-1}\delta v, \tag{3.112}$$

is the term that comes from the inter-site correlations.

In the used single-site approximation the last term in Eq. (3.108) is neglected, which gives the expression for the CPA free energy

$$F_{\mathrm{CPA}} = \widetilde{F} - \beta^{-1} \sum_i \ln \int d\xi_i d\eta_i e^{-\beta E[\xi_i, \eta_i]} . \tag{3.113}$$

The coherent potential $\Sigma_{i\sigma}(\tau - \tau')$, and hence the electron GF, is found from the condition that the thermal average of the single-site t-matrix $\widetilde{t}$ is equal to zero:

$$\langle \widetilde{t}_i \rangle = - \frac{\int d\xi_i d\eta_i \, \widetilde{t}_i e^{-\beta E[\xi_i, \eta_i]}}{\int d\xi_i d\eta_i e^{-\beta E[\xi_i, \eta_i]}} = 0. \tag{3.114}$$

This equation is called dynamical CPA equation. Again, at the core of this CPA approach—the mapping of the problem on the impurity site problem—it is assumed that the local coherent GF $\widetilde{G}$ (3.101) is equal to the local GF of the impurity site obtained by using Eq. (3.95) with the single-site Hamiltonian (defined by the local part of Eq. (3.92)).

Once Σ, and hence $F_{i\sigma}$, is found, one can calculate the average charge and magnetization densities on each site:

$$\langle n_i \rangle = F_{i\uparrow}(0^-) + F_{i\downarrow}(0^-), \tag{3.115}$$

$$\langle m_i \rangle = F_{i\uparrow}(0^-) - F_{i\downarrow}(0^-). \tag{3.116}$$

In 2002, Kakehashi demonstrated [47] that many-body CPA, dynamical CPA, and DMFT are equivalent theories. We refer the reader to the original paper [47] for details of the proof.

It is important to mention that the single-site approximation for the GF in a narrow-band model with frequency-dependent electron self-energy was proposed before the CPA works by Hubbard [48]. Using an alloy–analogy approximation, he obtained the MIT in his model. Thus, both Hubbard and CPA approaches might be regarded as predecessors of DMFT (for recent developments and extensions of the CPA approaches, see, e.g., Refs. [49–56]).

References

1. Freericks, J.K.: Transport in Multilayered Nanostructures: The Dynamical Mean-field Theory Approach. Imperial College Press, London (2006)
2. Georges, A., Kotliar, G., Krauth, W., Rozenberg, M.J.: Rev. Mod. Phys. **68**, 13 (1996)
3. Brout, R.: Phys. Rev. **118**, 1009 (1960)
4. Thompson, C.J.: Commun. Math. Phys. **36**, 255 (1974)
5. Gebhard, F.: The Mott Metal-Insulator Transition. Models and Methods. Springer, Berlin (1997)
6. Metzner, W., Vollhardt, D.: Phys. Rev. Lett. **62**, 324 (1989).; Erratum Phys. Rev. Lett. **62**, 1066 (1989)
7. Muller-Hartmann, E.: Z. Phys. B. **74**, 507 (1989)

8. Kollar M.: "The Foundations of Dynamical Mean-Field Theory", in Lecture Notes of the Autumn School on Correlated Electrons 2018 "DMFT: From Infinite Dimensions to Real Materials", Eds. E. Pavarini, E. Koch, A. Lichtenstein, and D. Vollhardt (Forschungszentrum Julich, 2018)
9. Metzner, W., Vollhardt, D.: Phys. Rev. B. **37**, 7382 (1988)
10. Bünemann, J., Weber, W.: Phys. Rev. B. **55**, 4011 (1997)
11. Bünemann, J., Weber, W., Gebhard, F.: Phys. Rev. B. **57**, 6896 (1998)
12. Kotliar, G., Savrasov, S.Y., Haule, K., Oudovenko, V.S., Parcollet, O., Marianetti, C.A.: Rev. Mod. Phys. **78**, 865 (2006)
13. Georges, A., Kotliar, G.: Phys. Rev. B. **45**, 6479 (1992)
14. Jarell, M.: Phys. Rev. Lett. **69**, 168 (1992)
15. Anderson, P.W.: Phys. Rev. **124**(1), 41 (1961)
16. Anisimov, V., Izyumov, Yu.: Electronic Structure of Strongly Correlated Materials. Springer-Verlag, Berlin Heidelberg (2010)
17. Georges, A., Krauth, W.: Phys. Rev. B. **45**, 6479 (1992)
18. Abrikosov, A.A., Gorkov, L.P., Dzyaloshinski, I.E.: Methods of Quantum Field Theory in Statistical Physics. Prentice Hall Press, Upper Saddle River (1963)
19. Luttinger, J.M., Ward, J.C.: Phys. Rev. **118**, 1417 (1960)
20. Baym, G., Kadanoff, L.: Phys. Rev. **124**, 287 (1961)
21. Fabrizio, M.: Lecture Notes on Many-Body Theory. SISSA, Trieste (2013)
22. Soven, P.: Phys. Rev. **156**, 809 (1967)
23. Soven, P.: Phys. Rev. **178**, 1136 (1969)
24. Taylor, D.W.: Phys. Rev. **156**, 1017 (1967)
25. Velicky, B., Kirkpatrick, S., Ehrenreich, H.: Phys. Rev. **175**, 747 (1968)
26. Yonezawa, F.: Prog. Theor. Phys. **40**, 734 (1968)
27. Schwartz, L., Siggia, E.: Phys. Rev. B. **5**, 383 (1972)
28. Hirooka, S., Shimizu, M.: J. Phys. Soc. Jpn. **43**, 70 (1977)
29. Shiba, H.: Prog. Theor. Phys. **46**, 77 (1971)
30. Schrieffer, J.R., Mattis, D.C.: Phys. Rev. **140**, A1412 (1965)
31. Hubbard, J.: Phys. Rev. Lett. **3**, 77 (1959)
32. Stratonovich, R.L.: Sov. Phys. Dokl. **2**, 416 (1958)
33. Wang, S.Q., Evanson, W.E., Schrieffer, J.R.: Phys. Rev. Lett. **23**, 92 (1969)
34. Morandi, G., Galleani D'Agliano, E., Napoli, F., Ratto, C.F.: Adv. Phys. **23**, 867 (1974)
35. Cyrot, M.: J. Phys. (Paris). **33**, 25 (1972)
36. Hubbard, J.: Phys. Rev. B. **19**, 2626 (1979)
37. Hubbard, J.: Phys. Rev. B. **20**, 4584 (1979)
38. Hubbard, J.: Phys. Rev. B. **23**, 5974 (1981)
39. Hasegawa, H.: J. Phys. Soc. Jpn. **46**, 1504 (1979)
40. Hasegawa, H.: J. Phys. Soc. Jpn. **49**, 178 (1980)
41. Kakehashi, Y., Fulde, P.: Phys. Rev. B. **32**, 1595 (1985)
42. Kakehashi, Y., Hasegawa, H.: Phys. Rev. B. **37**, 7777 (1988)
43. Kakehashi, Y.: Phys. Rev. B. **38**, 6928 (1988)
44. Kakehashi, Y.: Phys. Rev. B. **45**, 7196 (1992)
45. Kakehashi, Y.: J. Magn. Magn. Mater. **104–107**, 677 (1992)
46. Kakehashi, Y.: Phys. Rev. B. **65**, 184420 (2002)
47. Kakehashi, Y.: Phys. Rev. B. **66**, 104428 (2002)
48. Hubbard, J.: Proc. R. Soc. London Ser. A. **281**, 401 (1964)
49. Olsson, P., Abrikosov, I.A., Vitos, L., Wallenius, J.: J. Nucl. Mater. **32**, 84 (2003)
50. Keiter, H., Baumgartner, K., Otto, D.: Generalizations of DMFT, CPA and NCA. In: Hewson, A.C., Zlatić, V. (eds.) Concepts in Electron Correlation. NATO Science Series (Series II: Mathematics, Physics and Chemistry), vol. 110. Springer, Dordrecht (2003)
51. Kakehashi, Y., Fulde, P.: Phys. Rev. B. **69**, 045101 (2004)
52. Long, N.H., Akai, H.: J. Phys. Condens. Matter. **19**, 365232 (2007)

53. Kakehashi, Y., Patoary, M.A.R., Tamashiro, T.: Phys. Rev. B. **81**, 245133 (2010)
54. Tamashiro, T., Nohara, S., Miyagi, K., Kakehashi, Y.: J. Phys. Soc. Jpn. **80**, 064702 (2011)
55. Minár, J.: J. Phys. Condens. Matter. **23**, 253201 (2011)
56. Korotin, M.A., Pchelkina, Z.V., Skorikov, N.A., Kurmaev, E.Z., Anisimov, V.I.: J. Phys. Condens. Matter. **26**, 115501 (2014)

Chapter 4
One-Band Hubbard Model: DMFT Solution

In this chapter, we present results for one atom–cell one-orbital/one-band infinite-dimensional Hubbard model obtained with DMFT. The exceptions are the cases of antiferromagnetism and superconductivity where the solutions are obtained for mostly 2D systems with two-atom unit cells. Generalization of the DMFT approach on the case of multi-atom/multi-orbital systems will be presented in the next chapter and the solutions of problems for these types of systems will be discussed in Chap. 10 on DFT + DMFT, a combined *ab initio* + many-body approach to study materials. The goal of this chapter is to give the reader a feeling about the properties of DMFT solution in the simplest but fundamental case of the one-orbital Hubbard model. Due to varieties of lattices and approximations, it is impossible to review all available literature on this topic. Therefore, we mainly focus (unless specified) on the case of infinite-dimensional Hubbard model on the hypercubic lattice.

4.1 Spectral Properties

The spectral, in general case spin-resolved, function of the Hubbard model can be calculated from the single-particle local retarded GF as:

$$A_\sigma(\omega) = -\frac{1}{\pi} ImG^R_\sigma(\omega). \tag{4.1}$$

Let us first discuss the PM (spin non-polarized) case at half-filling (the spin-polarized case will be considered in Sect. 4.4). A general idea about the dependence of the spectral function on U can be obtained from Fig. 4.1 (reproduced from early works [1, 2]). As it follows from this figure, at Us much smaller than the bandwidth, the system is in a metallic phase with slightly modified DOS as compared to the free-electron case. When U becomes of order of the bandwidth, the system develops the upper and the lower Hubbard sub-bands with shrinking quasiparticle peak at the

V. Turkowski, *Dynamical Mean-Field Theory for Strongly Correlated Materials*,
https://doi.org/10.1007/978-3-030-64904-3_4

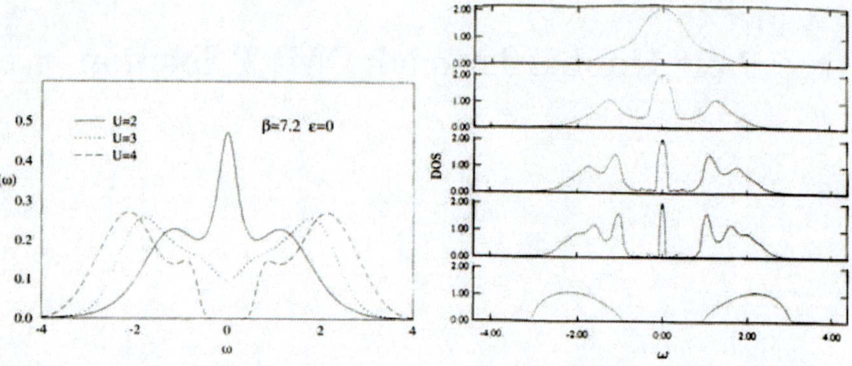

Fig. 4.1 The spectral function of the half-filled one-band Hubbard model in infinite dimensions for different values of U and different free-electron DOS. Left: the Gaussian DOS (hypercubic lattice) results at inverse temperature 7.2 (in the units of hopping). Reprinted Fig. 6 with permission from (M. Jarrell, Phys. Rev. Lett. **69**, 168 (1992)). Copyright 1992 by the American Physical Society. https://doi.org/10.1103/PhysRevLett.69.168. Right: the parabolic DOS (Bethe lattice) results for $U/D = 1, 2, 2.5, 3$, and 4 (top to bottom, D is the half-bandwidth of free electrons). Reprinted Fig. 3 with permission from (X.Y. Zhang, M.J. Rozenberg, G. Kotliar, Phys. Rev. Lett. **70**, 1666 (1993)). Copyright 1993 by the American Physical Society. https://doi.org/10.1103/PhysRevLett.70.1666

Fermi energy. Finally, when U is much larger than the bandwidth, only two Hubbard sub-bands survive. This last situation qualitatively corresponds to the atomic limit solution, with the GF

$$G_\sigma^R(\omega) = \frac{1}{2}\left(\frac{1}{\omega + \frac{U}{2} + i\delta} + \frac{1}{\omega - \frac{U}{2} + i\delta}\right). \tag{4.2}$$

The two peaks corresponding to the lower and upper Hubbard sub-bands were obtained first in Hubbard's pioneering works [3, 4].

Thus, as it follows from Fig. 4.1, as U increases the system undergoes MIT. To get more of the physical insight on the scenario of the transition (for more details, see book [5]), it is important to mention that the noninteracting bandwidth W is proportional to the bandwidth in the interacting case, i.e., to the kinetic energy of electrons, or the Fermi energy of the interacting system ϵ_F^*:

$$\frac{ZW}{2} = \epsilon_F^*, \tag{4.3}$$

where

$$Z = \left(\frac{1}{1 - \frac{\partial \Sigma(\omega)}{\partial \omega}}\right)\Bigg|_{\omega=0} \tag{4.4}$$

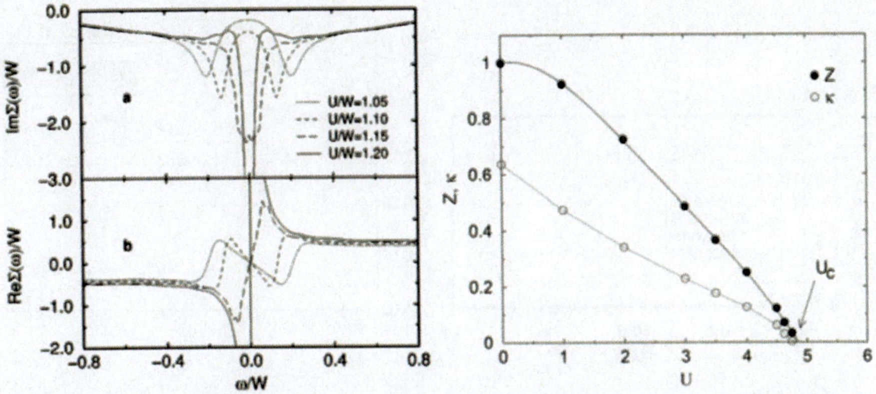

Fig. 4.2 Left: the electron self-energy at different values of U and $T/W = 0.0276$ (Bethe lattice, numerical renormalization group (NRG) solver). Reprinted Fig. 5 with permission from (R. Bulla, T. A. Costi, and D. Vollhardt, Phys. Rev. B 64, 045103 (2001)). Copyright 2001 by the American Physical Society. https://doi.org/10.1103/PhysRevB.64.045103. Right: Quasiparticle weight and compressibility as a function of U (Bethe lattice, Quantum Monte Carlo (QMC) solver) (From Ref. [7]). Reprinted Fig. 2 with permission from (J. Schlipf, M. Jarrell, P. G. J. van Dongen, N. Blümer, S. Kehrein, Th. Pruschke, and D. Vollhardt Phys. Rev. Lett. **82**, 4890 (1999)). Copyright 1999 by the American Physical Society. https://doi.org/10.1103/PhysRevLett.82.4890

is the quasiparticle weight [6]. In the case of strong correlations Z is very small, thus the kinetic energy of electrons, or ϵ_F^*, is much smaller than the free-electron bandwidth. The weight of the Hubbard sub-bands is proportional to $1 - Z$. Thus, as U increases the Hubbard sub-bands contribution to the spectrum becomes larger, while the kinetic energy contribution decreases (as the quasiparticle peak becomes smaller). Physically, in the Hubbard model, the zero-energy peak comes from scattering of electrons on the local-spin fluctuations, which gives resonance at the Fermi energy. The analogous peaks are present in the spectrum of metals with paramagnetic impurities (Abrikosov–Suhl resonance or Kondo peak) due to multiple scattering of electrons on the impurity near the Fermi surface that is accompanied by flips of the electron spins.

Disappearance of the zero-energy quasiparticle peak at critical value of U can be also tracked by analyzing the frequency dependence of the electron self-energy at different Us (left Fig. 4.2). At the transition point ($U \sim 1.2\ W$) the imaginary part of the self-energy loses its parabolic (Fermi liquid) dependence on frequency and demonstrates a peak at zero frequency, while the real part shows a discontinuity at $\omega = 0$. Such dependencies lead to a suppression of the quasiparticle peak (and, correspondingly, of the quasiparticle weight Z) at the transition (see Eq. (4.4) and right Fig. 4.2). Importantly, dependencies of the self-energy and quasiparticle peak are qualitatively and semiquantitatively the same for different free-electron DOS [8].

Temperature dependence of the zero-energy quasiparticle DOS is shown in Fig. 4.3.

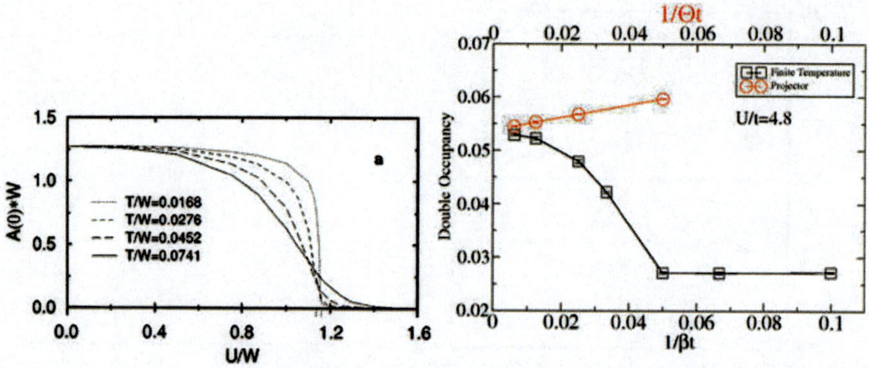

Fig. 4.3 Left: Temperature dependence of the zero-energy peak in the DOS as function of U. Reprinted Fig. 4a with permission from (R. Bulla, T. A. Costi, and D. Vollhardt, Phys. Rev. B 64, 045103 (2001)). Copyright 2001 by the American Physical Society. https://doi.org/10.1103/PhysRevB.64.045103. Right: Double occupancy as a function of temperature. The red circles correspond to different projection parameters used in the applied QMC approach. Reprinted Fig. 2 with permission from (F.F. Assaad and T.C. Lang, Phys. Rev. B **76**, 035116 (2007)). Copyright 2007 by the American Physical Society. https://doi.org/10.1103/PhysRevB.76.035116

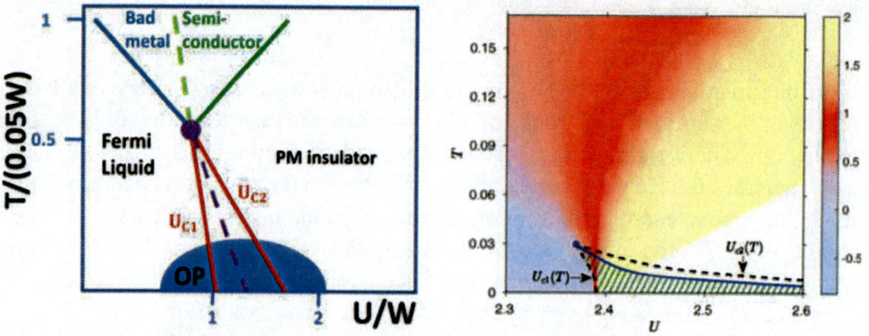

Fig. 4.4 Left: A schematic "standard" T–U phase diagram of the half-filled Hubbard model (W is the bandwidth). OP stands for the "ordered phase" and the dot corresponds to the critical point (for more discussion and a more accurate diagram, see Ref. [9]). Right: recent numerical results for the same diagram (also in units of W) obtained with the NRG approach. The red area corresponds to an exponential regime with power–law spectral function and the green dashed area is the metastable insulating-state region. Reprinted Fig. 1 with permission from (H. Eisenlohr, S.-S. B. Lee, and M. Vojta, Phys. Rev. B **100**, 155,152 (2019)). Copyright 2019 by the American Physical Society. https://doi.org/10.1103/PhysRevB.100.155152

As it follows from this figure, at finite temperatures, the value of the DOS at the energy of the quasiparticle peak drops significantly with temperature increase, but has a tail above critical U ($\approx 1.2\ W$) (see also right Fig. 4.3 for the dependence of double occupancy on temperature).

A schematic DMFT T–U phase diagram of the half-filled Hubbard model is shown in left Fig. 4.4. The diagram includes a low-temperature ordered region that

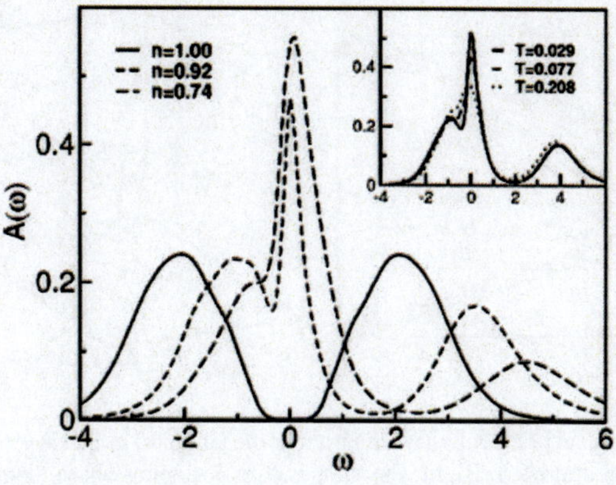

Fig. 4.5 Doping dependence of the spectral function of the Hubbard model at $U = 4.2$ and $T = 0.029$, obtained with the NRG solver (hopping parameter units). Inset: temperature dependence of the spectral function at $U = 4.2$ and filling 0.92. Reprinted Fig. 1 with permission from (J. K. Freericks, T. P. Devereaux, R. Bulla, Phys. Rev. B **67**, 155,102 (2003)). Copyright 2003 by the American Physical Society. https://doi.org/10.1103/PhysRevB.67.155102

contains also a coexistence metastable "area" of MIT spanned on the first-order transition (red) line with the second-order critical end point. The coexistence region takes place due to a hysteresis character of the MIT, with two critical values of U. Different approximations give similar values for these critical interactions at zero temperature: $U_{c1} \approx 5t$ and $U_{c2} \approx 6t$ [8, 10, 11]. There are several other crossover lines in the diagram. In particular, the Fermi liquid–bad metal line corresponds to rise of conductivity due to transformation of the wave-like electron wave function, and the insulator–semiconductor line above which the number of temperature-induced excitations significantly increases. It is important that in the coexistence region, DMFT solution gives two different values (small and large, i.e., insulating and metallic) for the double occupancy $D = \langle n_\uparrow n_\downarrow \rangle$ [8, 10]. The coexistence (hysteresis) region is probably the most intriguing part of the phase diagram which attracts a lot of attention since it is very important for proper understanding of the MIT.

As the results of recent calculations show, the T–U diagram includes a metastable insulating "exponential" regime with the quantum critical behavior of the system that extends to very low temperatures (right Fig. 4.4). The critical regime defines the quantum critical scaling above the critical end point. We discuss the consequences of such spectral features on the transport properties in the next subsection.

Now, let us consider the case of arbitrary charge density. The most interesting situation is when the doping is small, i.e., when the system is not far from half-filling (and the correlations are almost as strong as at half-filling). As it follows from DMFT calculations (see, e.g., Ref. [12] and presented here Fig. 4.5 from this paper), at small dopings the spectral function of the insulator develops a zero-energy peak. The height of the peak grows with doping increasing (see also other early works [13–16]

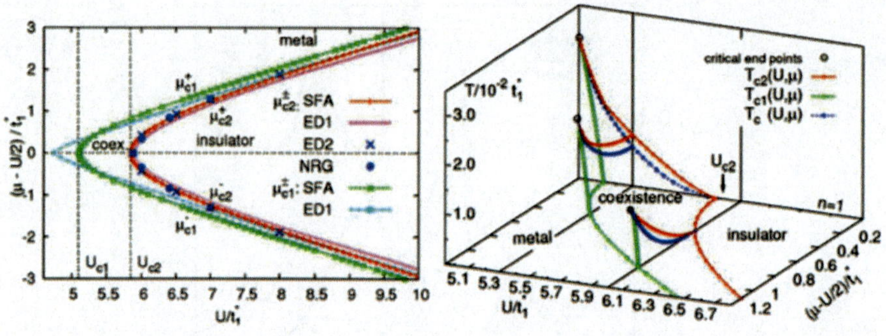

Fig. 4.6 Left: U-chemical potential phase diagram of the Hubbard model at $T = 0$ obtained with the self-energy functional approach. Right: the temperature U-doping phase diagram of the Hubbard model. Reprinted Figs. 1 and 2 with permission from (M. Eckstein, M. Kollar, M. Potthoff, and D. Vollhardt, Phys. Rev. B **75**, 125,103 (2007)). Copyright 2007 by the American Physical Society. https://doi.org/10.1103/PhysRevB.75.125103

for more details on the spectral properties of the Hubbard model away from half-filling).

Similar to the undoped system, in the doped case there is a regime with coexistent metallic and insulating phases at Coulomb repulsions $U_{c1} < U < U_{c2}$ (at large Us). Moreover, it has its doping boundaries defined by the critical values of the chemical potentials μ_{c1} and μ_{c2}, see Fig. 4.6 from Ref. [17], where the $U - \mu$ phase diagram for both electron and hole-doped system is shown. Though as results of Ref. [15] show, in the coexistence region the metallic solution has a lower energy as compared to the insulating one. It is also important that as the doping changes from small negative to positive values, the chemical potential experiences a jump (thus, the charge density is not a continuous function of the chemical potential).

Finally, let us discuss the temperature effects away from half-filling. Probably, the most important phenomenon in this case is existence of a phase separation region between the metallic and insulating phases close to half-filling at finite temperatures. In the case when the next nearest-neighbor (frustration) hopping term that breaks the electron–hole symmetry (with respect to half-filling point) is taken into account, this region significantly extends and even includes the half-filling point (see right Fig. 4.6 and the original paper [17], for details).

At this point, we conclude the discussion of the spectral properties of the Hubbard model at arbitrary doping in the PM case (for more results on the phase diagram of the doped system, see, e.g., Refs. [18, 19]). The spin-polarized case will be considered in Sect. 4.5, where we discuss magnetic properties of the Hubbard model.

4.2 Transport Properties

In this section, we summarize the most important results for the dc conductivity in the Hubbard model obtained with DMFT. Before this, we show how one can obtain this quantity with DMFT. Namely, we derive general expression for the optical (frequency-dependent) conductivity that gives formula for the DC conductivity in the limit of zero-frequency (the frequency-dependent expression will be used in the next section to analyze the optical response of the system). Details of the derivation can also be found in Refs. [5, 13, 20].

Optical conductivity tensor can be calculated from the current susceptibility as

$$\sigma^{\alpha\beta}(\omega) = \frac{1}{i\omega}\left(\chi^{\alpha\beta}(\omega + i\delta) - \chi^{\alpha\beta}(i\delta)\right), \tag{4.5}$$

where the functions $\chi^{\alpha\beta}(\omega + i\delta)$ on the r.h.s. are the real-frequency current susceptibilities obtained by analytical continuation of the corresponding functions in the Matsubara representation (most convenient for the DMFT calculations):

$$\chi^{\alpha\beta}\left(i\omega_n, \vec{q}\right) = \int_0^\beta d\tau e^{i\omega_n\tau}\sum_l e^{i\vec{q}\vec{R}_l}\left\langle T_\tau j^\alpha\left(\tau, \vec{R}_l\right) j^\beta\left(\vec{0}, 0\right)\right\rangle \tag{4.6}$$

where

$$j^\alpha\left(\tau, \vec{R}_l\right) = i\sum_\sigma \hat{c}^+_{l+x^\alpha}\left(\hat{c}_{l+x^\alpha} - \hat{c}_l\right) \tag{4.7}$$

are components of the current (x^α is the αth component of the nearest neighbor to $\vec{R}_l$ site). Eq. (4.6) is the expression for general, momentum-dependent, susceptibility (we will put $\vec{q} = 0$ when we will calculate the optical conductivity (4.5)).

In the momentum representation, $j^\alpha\left(\tau, \vec{q}\right) = i\sum_{\vec{k}\sigma} v^\alpha_{\vec{k}} \hat{c}^+_{\vec{k}\sigma} \hat{c}_{\vec{k}+\vec{q}\sigma}$ ($v^\alpha_{\vec{k}\sigma} = 2\sin\left(k^\alpha a\right)$ are the velocity components). Using this expression, Eq. (4.6) and performing the ladder expansion for the susceptibility (Fig. 4.7) one can obtain the following result:

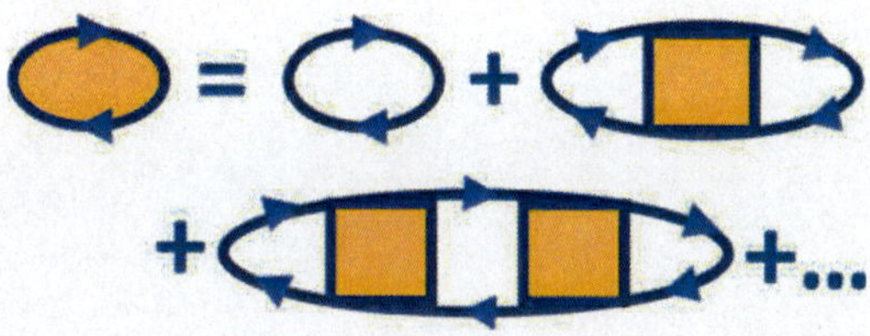

Fig. 4.7 Diagram representation of the ladder expansion for the susceptibility (Eq. (4.8))

$$\chi^{\alpha\beta}\left(i\omega_n,\vec{q}\right) = -\sum_{\vec{k},m,\sigma} v^{\alpha}_{\vec{k}\sigma} G\left(i\omega_m,\vec{k}\right) G\left(i\omega_m + i\omega_n,\vec{k}+\vec{q}\right) v^{\beta}_{\vec{k}+\vec{q}\sigma}$$
$$+ \sum_{\vec{k},\vec{p},m,l,\sigma,\sigma'} v^{\alpha}_{\vec{k}\sigma} G\left(i\omega_m,\vec{k}\right) G\left(i\omega_m + i\omega_n,\vec{k}+\vec{q}\right) \Gamma^{\sigma\sigma'}_{\vec{k},\vec{p},\vec{q}}(i\omega_m, i\omega_l, i\omega_n)$$
$$\times G\left(i\omega_l,\vec{p}\right) G\left(i\omega_l + i\omega_n,\vec{p}+\vec{q}\right) v^{\beta}_{\vec{p}+\vec{q}\sigma'} + \ldots . \tag{4.8}$$

Functions $\Gamma^{\sigma\sigma'}_{\vec{k},\vec{p},\vec{q}}(i\omega_m, i\omega_l, i\omega_n)$ in Eq. (4.8) (yellow "boxes" in Fig. 4.7) are irreducible vertices. Due to frequency conservation, they depend on three, not four, frequencies. Using the arguments similar to the ones in the previous chapter, one can show that in infinite dimensions these functions are local in space, i.e., momentum-independent (see also Sect. IV of [20] and Chap. 9). Putting in Eq. (4.8) $\vec{q} = 0$ (since we also need $\chi^{\alpha\beta}\left(i\omega_n, \vec{q} = 0\right)$ to calculate the optical conductivity (4.5)), one gets

$$\chi^{\alpha\beta}\left(i\omega_n,\vec{q}=0\right) = -\sum_{\vec{k},m,\sigma} v^{\alpha}_{\vec{k}\sigma} G\left(i\omega_m,\vec{k}\right) G\left(i\omega_m + i\omega_n,\vec{k}\right) v^{\beta}_{\vec{k}\sigma}$$
$$+ \sum_{\vec{k},\vec{p},m,l,\sigma,\sigma'} v^{\alpha}_{\vec{k}\sigma} G\left(i\omega_m,\vec{k}\right) G\left(i\omega_m + i\omega_n,\vec{k}\right) \Gamma^{\sigma\sigma'}(i\omega_m, i\omega_l, i\omega_n)$$
$$G\left(i\omega_l,\vec{p}\right) G\left(i\omega_l + i\omega_n,\vec{p}\right) v^{\beta}_{\vec{p}\sigma'} + \ldots . \tag{4.9}$$

It is easy to see from the expansion on the r.h.s. in Eq. (4.9) that all terms, except the first one, contain momentum-independent factors $\sum_{\vec{k}} v^{\alpha}_{\vec{k}\sigma} G\left(i\omega_m,\vec{k}\right) G\left(i\omega_m + i\omega_n,\vec{k}\right)$. These terms are equal zero since $v^{\alpha}_{\vec{k}\sigma}$ are odd and the GFs, which depend on $\varepsilon_{\vec{k}}$, are even functions of the momentum. Thus, only one-loop term contributes to the susceptibility. Below, for simplicity we consider the case of the hypercubic lattice where the conductivity tensor is diagonal:

$$\chi^{\alpha\beta}\left(i\omega_n,\vec{q}=0\right) = -\delta^{\alpha\beta} \sum_{\vec{k},m,\sigma} \left(v^{\alpha}_{\vec{k}\sigma}\right)^2 G\left(i\omega_m,\vec{k}\right) G\left(i\omega_m + i\omega_n,\vec{k}\right). \tag{4.10}$$

Next, the substitution

$$G\left(i\omega_m,\vec{k}\right)G\left(i\omega_m+i\omega_n,\vec{k}\right)\to\int d\varepsilon\delta\left(\varepsilon-\varepsilon\left(\vec{k}\right)\right)G(i\omega_m,\varepsilon)G(i\omega_m+i\omega_n,\varepsilon) \tag{4.11}$$

in Eq. (4.10) gives

$$\chi^{\alpha\beta}\left(i\omega_n,\vec{q}=0\right)=-\frac{2}{N}\sum_m\int d\varepsilon\left[\sum_{\vec{k}}\left(v^{\alpha}_{\vec{k}\sigma}\right)^2\delta\left(\varepsilon-\varepsilon\left(\vec{k}\right)\right)\right]G(i\omega_m,\varepsilon)G(i\omega_m+i\omega_n,\varepsilon). \tag{4.12}$$

Using a similar strategy as in deriving the hypercubic Gaussian DOS, one can show that for the hypercubic lattice in infinite dimensions

$$\sum_{\vec{k}}\left(v^{\alpha}_{\vec{k}\sigma}\right)^2\delta\left(\varepsilon-\varepsilon\left(\vec{k}\right)\right)=\frac{d}{2}A_0(\varepsilon), \tag{4.13}$$

where $A_0(\varepsilon)=-\frac{1}{\pi}ImG_0(\varepsilon)$ is the noninteracting DOS (see Appendix B in Ref. [13]).

Then, substituting Eq. (4.13) and

$$G(i\omega_m,\varepsilon)=\int d\omega'\frac{A(\omega',\varepsilon)}{i\omega_m-\omega'} \tag{4.14}$$

($A(\omega',\varepsilon)=-\frac{1}{\pi}ImG^R(\omega',\varepsilon)$ is the spectral function of interacting system) into Eq. (4.12), performing summation over ω_m, making the analytical continuation $i\omega_n\to\omega+i\delta$, and substituting the result into Eq. (4.5) one can obtain the DMFT expression for the optical conductivity:

$$\sigma(\omega)=\sigma_0\int d\omega'\int d\varepsilon A_0(\varepsilon)A(\varepsilon,\omega')A(\varepsilon,\omega'+\omega)\frac{f(\omega')-f(\omega'+\omega)}{\omega}, \tag{4.15}$$

where

$$\sigma_0=\frac{\pi e^2a^2t^{*2}}{2\hbar}\frac{N}{V} \tag{4.16}$$

is the unit of conductivity.

The static conductivity can be obtained from Eq. (4.15) by taking the limit $\omega\to 0$:

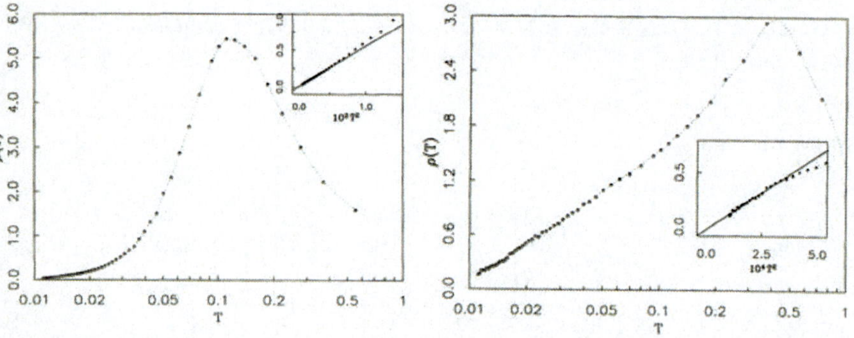

Fig. 4.8 Resistivity as a function of temperature obtained with the non-crossing approximation (NCA) solver for the half-filled Hubbard model in a metallic state close to MIT ($U = 3\,t$) (left) and in a doped-insulator state ($U = 4t$, density of electrons 0.97) (right). Insets—resistivity in the low-temperature limit. Reprinted Figs. 9 and 14 with permission from (T. Pruschke, D.L. Cox, M. Jarrell, Phys. Rev. B **47**, 3553 (1993)). Copyright 1993 by the American Physical Society. DOI: https://doi.org/10.1103/PhysRevB.47.3553

$$\sigma(\omega = 0) = \sigma_0 \int d\omega' \int d\varepsilon A_0(\varepsilon) A^2(\varepsilon, \omega') \left[-\frac{df(\omega')}{d\omega'} \right]. \tag{4.17}$$

This result is in agreement with a more general result obtained in Ref. [21] for the 3D case with the nonlocal in space GFs but local (DMFT) self-energy, where instead of $A(\varepsilon, \omega')$ there are two site-dependent GFs (and summation over the sites is performed).

Results for the DMFT conductivity as a function of temperature at different Us and dopings are shown in Fig. 4.8. In the metallic case at half-filling, the resistivity at low temperatures grows with temperature as

$$\rho(T) = \rho_o + \rho_M \left(\frac{T}{T_0} \right)^2, \tag{4.18}$$

according to the Fermi-liquid law. Then, it reaches maximum and starts to decrease at high temperatures, typically for a semimetal case.

Technically, it is very difficult to reach zero temperature limit with DMFT, however, as different calculations show, ρ_o approaches to zero as $T \to 0$ (Figs. 4.8 and 4.9, [13, 22]; a small negative ρ_o found in Ref. [13] (inset in left Fig. 4.8) is due to strongly underestimated DOS near the chemical potential calculated with NCA at $\frac{T}{t}$ of order or less than 0.01). In Eq. (4.18), constant $\rho_M \sim \frac{ha}{e^2}$ is the Mott limit of resistivity (with the mean free path of order of the lattice parameter), and temperature $T_0 = \varepsilon_F^* = ZD$ is the effective bandwidth parameter. A sharp increase of the resistivity at $T \sim \varepsilon_F^*$ indicates a bad-metal state of the system (with suppressed quasiparticle peak, i.e., a large imaginary part of the self-energy, $\mathrm{Im}\Sigma(\omega)$ (see Ref. [5, 22]). A similar temperature dependence of the resistance was found in the doped

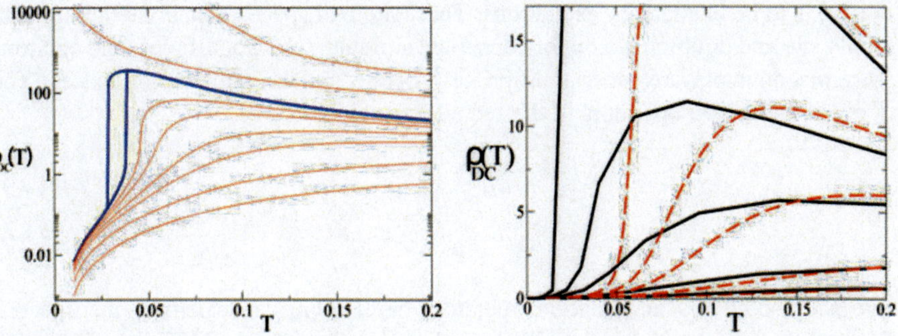

Fig. 4.9 Left: The iterative perturbation theory (IPT) solver solution for the resistivity at different values of U (below blue curves—metallic, above—insulating dependence on temperature; blue curves define the coexistence region). Right: Resistivity at different values of U obtained with the IPT (dashed lines) and NRG (solid lines) solvers. From Ref. [22]

case, right Fig. 4.8. As it was demonstrated in the original paper [13], there are no signs that ρ_o might be negative in this case (see inset in right Fig. 4.8). Interestingly, as it was discussed in Ref. [13] the values of the maximum-resistivity temperature in the undoped and doped cases are not related to T_0, contrary to the Kondo case (though in the Kondo case, the scale is a function of temperature, since the DOS at the Fermi energy is temperature-dependent).

The resistivity as a function of temperatures at different values of U (metallic, coexistence, and insulating regimes) is shown in the left Fig. 4.9 [22]. Since the resistivity is basically defined by $\mathrm{Im}\Sigma(\omega)$ and near the MIT, $\mathrm{Im}\Sigma(\omega)$ is a scaling function of $\frac{\omega}{\varepsilon_F^*}$ and $\frac{T}{\varepsilon_F^*}$ [23], and at low temperatures, $T \ll \varepsilon_F^*$, one gets near the MIT $\mathrm{Im}\Sigma(\omega) \sim W\left(\frac{T}{\varepsilon_F^*}\right)^2 \sim \frac{T^2}{Z^2 W}$. Thus, one would expect a strong enhancement of resistivity as U approaches the critical value when $T \to 0$. Though the results in left Fig. 4.9 confirm qualitatively this statement, they underestimate the growth of resistivity due to missing factor $\frac{1}{Z^2}$ in $Im\Sigma(\omega = 0)$. Indeed, at low temperatures as a more accurate NRG solver solution shows, the resistivity grows much faster with temperature increase (right Fig. 4.9 [22]).

4.3 Optical Properties

4.3.1 Excitations

A major part of the optical response of materials is defined by elementary excitations in the system. In the "standard" (weakly correlated) metals, semiconductors, and insulators, the elementary low-energy excitations are electrons and holes (one must also remember about more complex multiparticle excitations, like plasmons and excitons). In strongly correlated (Mott) insulators, one expects the holons and doublons to be elementary excitations. These are real space excitations: holon is an empty site and doublon—a doubly occupied site state (electrons, holes, and excitons in semiconductors are most appropriately defined in the momentum space). The doublon and holon operators in the second quantization are:

$$\widehat{d}_{i\sigma} = \widehat{c}_{i\sigma}\widehat{n}_{i\bar{\sigma}}, \tag{4.19}$$

$$\widehat{h}_{i\sigma} = \widehat{c}_{i\sigma}(1 - \widehat{n}_{i\bar{\sigma}}) \tag{4.20}$$

(we show only the annihilation operators here). One can calculate the spectral function for quasiparticles described by these operators with DMFT to track what are the weights of the corresponding quasiparticles in the excitation spectrum. Namely, since $\widehat{c}_{i\sigma} = \widehat{d}_{i\sigma} + \widehat{h}_{i\sigma}$ the electron spectral function is sum of the spectral functions of the operators above: $A_{cc^+}(\omega) = A_{dd^+}(\omega) + A_{dh}(\omega) + A_{h^+d^+}(\omega) + A_{h^+d}(\omega)$. The holon and doublon (diagonal) and mixed (off-diagonal) spectral functions $A_{XX}(\omega)$ and $A_{XY}(\omega)$ can be obtained from the imaginary part of the corresponding retarded GFs: $A_{XY}(\omega) = -\frac{1}{\pi} ImG_{XY}(\omega)$, where $X, Y = \widehat{h}_{i\sigma}, \widehat{d}_{i\sigma}$.

In the particle–hole symmetric case, only two of the spectral functions are independent, since $A_{dd^+}(\omega) = A_{h^+h}(-\omega)$ and $A_{h^+d^+}(\omega) = A_{dh}(\omega)$. Such an analysis in the case of different number of "flavors" of fermions was performed by Lee et al. [24]. Here, we discuss results from this paper in the relevant to this chapter case of two flavors (spin orientations). The DMFT results of paper [24] demonstrate that even in the metallic phase there are well-pronounced doublon and holon peaks (Fig. 4.10a, b; the orange curve corresponds to the doublon peak; the holon states are symmetric with respect to the $\omega = 0$ point). Thus, the lower and the upper Hubbard sub-bands consist predominantly of the holon and doublon states, correspondingly. In addition, in the metallic phase, there is a familiar zero-energy quasiparticle peak. Another important result of this work is that while the Hubbard sub-band holon and doublon peaks correspond to the charge excitations, the zero-energy peak is formed by spin excitations (see the curves for the charge (red) and spin (magenta) susceptibilities in Fig. 4.10). As it was mentioned in Ref. [24], the spin-like collective nature of the zero-energy quasiparticle is analogous to the Kondo resonance in the single-impurity Anderson model, where the spin susceptibility has a peak at the Kondo energy scale. The spin susceptibility peak (at frequency ω_s in Fig. 4.10a) is suppressed at large Us.

The Hubbard sub-band and the zero-energy quasiparticle peaks have been already observed experimentally [25, 26]. However, as results of Ref. [24] suggest, in the metallic phase one can expect extra low-energy excitations. Namely, in the metallic phase in the inner edges of the Hubbard sub-bands, there are additional subpeaks in the doublon and holon spectra (frequency ω_p in Fig. 4.10a). The position and the width of the peaks ($\delta\omega$) decrease linearly with U increasing and are suppressed in the insulating phase (see Fig. 4.10c for the insulating solution). Performing integration over the charge fluctuations at the energy scale $U/2$ (by also employing the generalized Schrieffer–Wolff transformation) and separating the holon and doublon terms in the effective Hamiltonian, the authors of Ref. [24] wrote down the low-energy

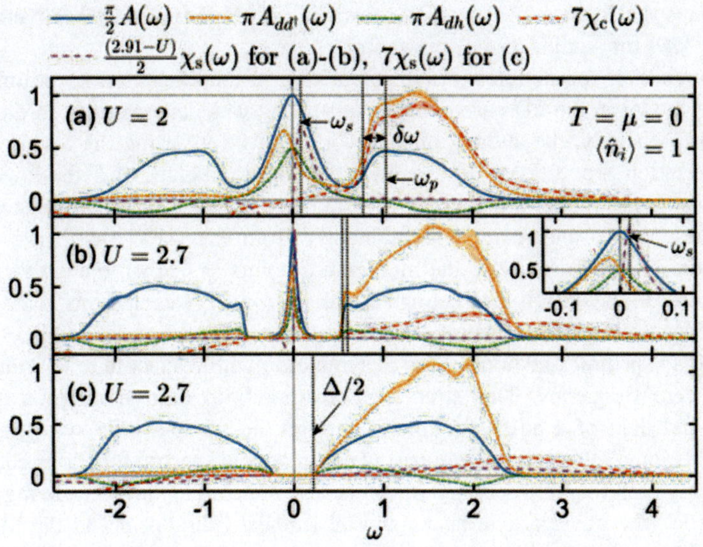

Fig. 4.10 DMFT results for the spectral function and charge and spin susceptibilities at different values of U in the metallic (**a**) and coexistence phases (**b**), (**c**). (**b**) corresponds to the metallic and (**c**) to the insulating solution in the coexistence phase (for details, see the text). Reprinted Fig. 1 with permission from (S.-S.B. Lee, J. von Delft, and A. Weichselbaum, Phys. Rev. Lett. **119**, 236,402 (2017)). Copyright 2017 by the American Physical Society. https://doi.org/10.1103/PhysRevLett.119.236402

Hamiltonian of the system. This Hamiltonian includes doublon and holon hopping terms, the Heisenberg spin-interaction term, a three-site term, and a term that describes the doublon–holon pair interaction:

$$H_{dh} = \frac{t^2}{U} \sum_{\langle i,j \rangle, \sigma, \sigma'} \left(\widehat{h}^{+}_{i\sigma} \widehat{d}^{+}_{j\sigma} + \widehat{h}^{+}_{j\sigma} \widehat{d}^{+}_{i\sigma} \right) \left(\widehat{d}_{i\sigma'} \widehat{h}_{j\sigma'} + \widehat{d}_{j\sigma'} \widehat{h}_{i\sigma'} \right). \tag{4.21}$$

Applying the mean-field theory approximation to this part of the Hamiltonian after introducing the site-independent field $\Delta_{dh} = \frac{t}{2} \sum_{\sigma} \left\langle \widehat{d}_{i\sigma} \widehat{h}_{j\sigma} + \widehat{d}_{j\sigma} \widehat{h}_{i} \right\rangle$ (an order parameter of the doublon–holon pairs), it was demonstrated that the ω_p subpeaks are present at finite Δ_{dh} (they are absent in the insulating phase when $\Delta_{dh} = 0$). Therefore, the subpeaks ω_p are generated by the doublon–holon interaction. These states have been not observed experimentally yet. They might be more easily detected in the multiband (multi-"flavor") case, where they are expected to be more pronounced. Importance of the doublon–holon interaction was confirmed in paper [27], where it was shown that in the two-band case in the metallic phase there are coupled inter-band doublon–holon states ("Hubbard excitons"). In the symmetric phase, when the inter- and intra-orbital Coulomb interactions are equal, it was shown that the quasiparticle peaks are at the Fermi level that leads to a continuous Mott transition (see also work [28] for the analysis of the properties of a three-band model and Ref. [29] for studies of the spin-polarized case).

Importance of the doublon–holon interaction in the excitation spectrum of the Hubbard model in the 2D case even at half-filling was demonstrated in the recent paper [30]. Namely, the authors of this work showed by using the density-matrix renormalization group (DMRG) approach that at large values of U the low-energy excitations in the model are coupled doublon–holon pairs ("Hubbard biexcitons"). Since these excitations come predominantly from the AFM or charge–density fluctuations, related usually to the momentum points $(\pi, \ldots, \pi)$, the authors called them π-tons. In the weakly interacting case, the low-energy excitations in the system are standard excitons, i.e., electron–hole bound states. Coming back to Fig. 4.10, the off-diagonal spectral function $A_{dh}(\omega)$ is symmetric with respect to $\omega = 0$ and has a peak at zero frequency. This structure corresponds to destruction of a doublon (holon) and then of a holon (doublon) through the particle–hole symmetric processes. Diagonal (electron and holon) and non-diagonal spectral functions contribute comparably to the zero-energy quasiparticle peak. In the insulating phase (Fig. 4.10c), the quasiparticle peaks and all subpeaks disappear and the Mott gap opens, with spreaded and suppressed charge and spin susceptibilities in the Hubbard sub-bands. It is necessary to mention that in several works the ω_p subpeaks were obtained also in the insulating phase [31–33], however, the accuracy of these solutions (e.g., the accuracy of the analytical continuation) is under question.

It is also worth to mention a recent progress in improvement of the Bethe–Salpeter equation (BSE) techniques, standard tool in the case of semiconductors, to study the electron and hole excitations within DMFT [34]. Accurate DMFT calculations of the two-particle irreducible vertex functions (see Chap. 9) are crucial in this case.

To summarize, as DMFT calculations show, Fermi-energy quasiparticles and/or doublon–holon pairs form the low-energy excitation spectrum of the Hubbard model, depending on the model parameters.

4.3.2 *Optical Conductivity*

We move to the discussion of the DMFT results for the optical response described by the optical conductivity. Another important characteristic, absorption spectrum, can be obtained from the optical conductivity. Namely, the absorption spectrum is proportional to the imaginary part of the dielectric function $\varepsilon(\omega)$ and the last quantity is connected with the optical conductivity as:

$$\varepsilon(\omega) = 1 + \frac{4\pi i \sigma(\omega)}{\omega}. \tag{4.22}$$

Similarly, one can calculate the reflectivity $r(\omega)$ from the optical conductivity by substituting the r.h.s. result for the dielectric function, Eq. (4.22), into the equation

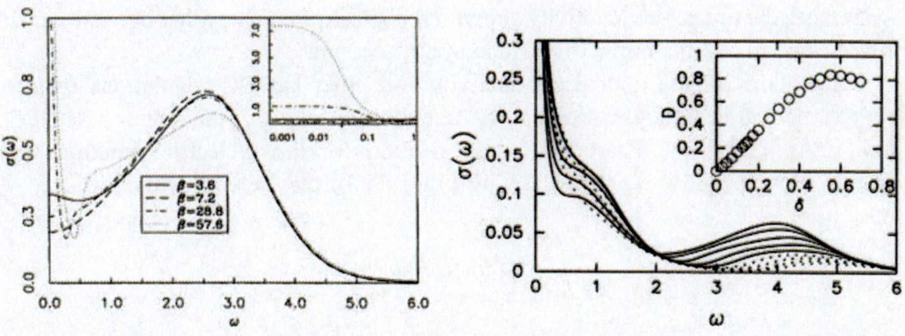

Fig. 4.11 Left: The results for the optical conductivity for the Hubbard model at half-filling and $U = 3\,t$ and different values of temperature (hypercubic lattice, QMC solver). Reprinted Fig. 10 with permission from (T. Pruschke, D.L. Cox, M. Jarrell, Phys. Rev. B **47**, 3553 (1993)). Copyright 1993 by the American Physical Society. https://doi.org/10.1103/PhysRevB.47.3553. Right: the same at different values of doping. The solid lines correspond to low dopings ($\delta < 0.25$), where the conductivity increases with doping, and the dashed ones—to higher dopings ($0.25 < \delta < 0.45$), where the conductivity decreases with doping growth (Monte-Carlo maximum entropy calculations, $U = 4t$, $\beta = 43.2$). Reprinted Fig. 2 with permission from (M. Jarrell, J.K. Freericks, and Th. Pruschke, Phys. Rev. B **51**, 11,704 (1995)). Copyright 1995 by the American Physical Society. https://doi.org/10.1103/PhysRevB.51.11704

$$r(\omega) = \left| \frac{1 - \sqrt{\varepsilon(\omega)}}{1 + \sqrt{\varepsilon(\omega)}} \right|^2. \tag{4.23}$$

DMFT results for the optical conductivity for various values of doping and temperature in infinite-dimensional Hubbard model are shown in Fig. 4.11. Looking at Fig. 4.11, one can notice a low-frequency peak coming from the intra quasiparticle band excitations due to the external field (it corresponds to delta-function Drude peak at zero temperature for metals). This peak becomes suppressed as temperature increases. The other features of the optical conductivity curves can be explained by the DMFT DOS. In the metallic phase, there is a finite-frequency peak that corresponds to transitions from the lower Hubbard sub-band to the "zero-frequency" quasiparticle band. This peak moves with doping. At higher frequencies, there is another wide peak corresponding to transitions between the Hubbard sub-bands. In the insulating phase (left figure), only the zero-frequency (Drude) and the inter-sub-band peaks survive.

The results above are for the PM case. In the case of antiferromagnetism it was shown that the raise of the AFM ordering is reflected in the following features of the optical conductivity—an enhancement of the spectral weight above the optical gap and appearance of well-separated spin-polaron peaks [35]. Recently, both PM and AFM cases were analyzed for the 2D Hubbard model by taking into account the vertex correction to the susceptibilities [36]. As it was shown in this paper, the vertex correction enhances the Drude peak and sharpens the peak that corresponds to the inter-sub-band transitions (and deepens the inter-peak valley) in the PM phase. In the

AFM case, the optical conductivity shows a dip at zero frequency that becomes even lower when the vertex correction is taken into account.

Calculations of the optical conductivity and other DMFT calculations require high numerical accuracy, which can be tested by checking if the solution satisfies sum rules. Using the Kramers–Kronig relation for the dielectric function and Eq. (4.22), one can derive the following sum rule for the dielectric function

$$\frac{2}{\pi}\int_0^\infty d\omega' Im\varepsilon(\omega')\omega' = \omega_p, \tag{4.24}$$

where the constant ω_p is the plasmon frequency that can be determined microscopically and experimentally. From this equation and Eq. (4.22), one can obtain a sum rule for the optical conductivity:

$$\int_0^\infty d\omega Re\sigma(\omega) = \frac{\omega_p^2}{8} \tag{4.25}$$

(several other sum rules can be obtained from Eq. (4.25), e.g., for reflectivity).

Since the number on the right-hand side of Eq. (4.25) is usually unknown, one needs a more useful sum–rule relation for the conductivity. In the case of DMFT conductivity (4.15), one can show that in the isotropic case (for details see Ref. [37])

$$\int_0^\infty d\omega\sigma_{xx}(\omega) = \frac{\sigma_0}{2}\int_{-\infty}^\infty d\varepsilon\widetilde{\rho}'_{xx}(\varepsilon) f_F(\varepsilon), \tag{4.26}$$

where $\widetilde{\rho}_{xx} = \frac{1}{N}\sum_{\vec{k}}\left(v^{\alpha}_{\vec{k}\sigma}\right)^2\delta\left(\varepsilon - \varepsilon\left(\vec{k}\right)\right)$ is a normalized free-electron DOS in infinite dimensions (see Eq. (4.13)). Thus, this equation can be used to test the accuracy of the calculation of the optical conductivity. Several other analytical sum rules for terms in the high-temperature expansion of the optical conductivity were obtained in paper [38] (for a review on the accuracy of recent DMFT calculations of the optical conductivity, see Ref. [37]).

Another important experimentally measure quantity is photoemission spectrum. It allows one to track, e.g., the transfer of spectral weight between the lower Hubbard sub-band and the quasiparticle peak with change of the ratio of U and W (by applying pressure or by doping the system). To obtain momentum-integrated photoemission spectrum with DMFT, one can first obtain the local spectral density and then in order to compare the results with experiment to convolute it with a Gaussian function. In the case of a half-filled Hubbard model with the Bethe-lattice DOS and Gaussian width 0.3 eV, the results for the photoemission spectrum at different values of U close to the MIT point [20] are shown in Fig. 4.12. As it follows from this figure, with U increase the emission peak weight shifts from the zero energy quasiparticle peak to the lower Hubbard sub-band. This result is expected since with increase of U the weight of the Hubbard sub-bands increases. As a matter of fact, since the

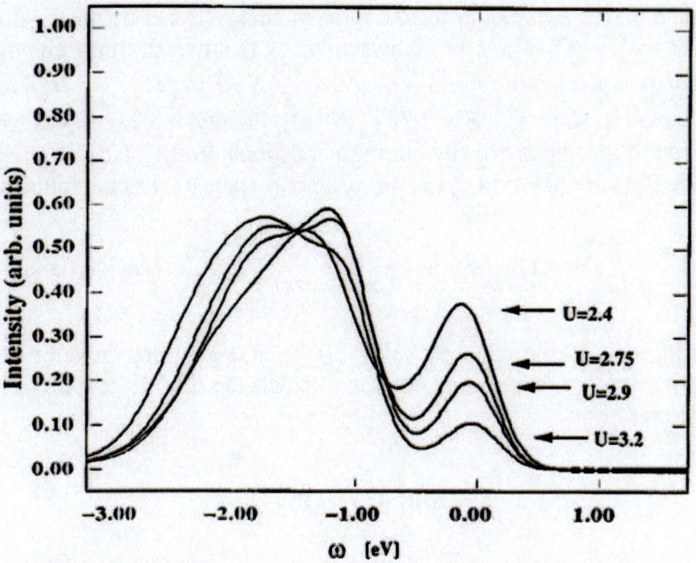

Fig. 4.12 Photoemission spectrum of the half-filled Hubbard model with the Bethe DOS and Gaussian width 0.3 eV at different values of U. Reprinted Fig. 61 with permission from (A. Georges, G. Kotliar, W. Krauth, and M.J. Rozenberg, Rev. Mod. Phys. **68**, 13 (1996)). Copyright 1996 by the American Physical Society. https://doi.org/10.1103/RevModPhys.68.13

emission spectrum strongly depends on the dispersion at the Fermi level, momentum dependence of the self-energy might also be important in this case. Thus, for realistic calculations of the photoemission spectrum one might need to go beyond DMFT.

4.4 Thermodynamics

To analyze the thermodynamic properties, one can calculate the thermodynamical potential of the system. In the DMFT approximation, the expression for this potential is [13]

$$\Omega = N\Omega_{\text{imp}} - T\sum_{\vec{k}} Trln\frac{G_{ii}}{G\left(\vec{k}\right)}, \tag{4.27}$$

where in the numerator of the logarithm argument one has the local GF and in the denumerator–momentum-dependent GF. On the other hand, it is not convenient to use the thermodynamic potential since to obtain different quantities, like specific heat, one needs to perform a double differentiation of this function with respect to temperature or another quantity, and each differentiation is numerically sensitive. Instead, one can use the internal energy, which usually requires only first-order differentiation. The expression for the internal energy E can be obtained from the relation $\Omega = E - TS - \mu N$, or equivalently, by (analytically!) differentiating the thermodynamic potential: $E = \frac{\partial(\beta\Omega)}{\partial\beta} + \mu N + \beta\frac{\partial\mu}{\partial\beta}N$. Using $\Omega = -TlnZ$ in the last equation, one can show that $E = \langle H\rangle + \mu N$ [13]. Finally, the last expression can be transformed to a computationally convenient formula in the Matsubara representation. In the PM case, the expression for the energy (per site) has the following form:

$$E = T\sum_{n,\sigma}\int_{-\infty}^{\infty} d\varepsilon\rho(\varepsilon)\frac{\varepsilon}{i\omega_n - \varepsilon + \mu - \Sigma_n(i\omega_n)} + T\sum_{n,\sigma}\Sigma_n(i\omega_n)G_n(i\omega_n) \quad (4.28)$$

(for a real-frequency formula, see Ref. [13]). Then, taking derivative of the energy (4.28) with respect to temperature one can calculate the specific heat C_v and use it to obtain the entropy

$$S = S(0) + \int_0^T dT'\frac{C_v(T')}{T'}. \quad (4.29)$$

Since the ground state of the impurity model in the insulating phase is doubly degenerate (we consider the PM case), one has for the zero-temperature paramagnetic entropy $S(0) = Nln2$, while in the metallic phase $S(0) = 0$.

To find the critical line of the first-order MIT, one equals the free energies on both sides, $F_M = F_I$, i.e.,

$$E_M - TS_M = E_I - TS_I. \quad (4.30)$$

The temperature–U-phase diagrams for the Hubbard model obtained with DMFT are shown in Fig. 4.4. We also refer the reader to Ref. [20] for more details on the thermodynamic properties of the Hubbard model obtained with DMFT. We complete this section with details of DMFT calculations and with results for two thermodynamic quantities—specific heat and entropy.

Let us begin with the temperature dependence of the specific heat in the metallic and insulating phases. In the metallic phase, it has two characteristic peaks: one at temperature of the familiar kinetic energy scale $\epsilon_F^* \sim ZW/2$ (the renormalized Fermi energy) and the other at a higher temperature, ~U. The low-temperature (low-energy) peak corresponds to local spin fluctuations (i.e., it exists at low temperatures when the zero-energy quasiparticle peak is not suppressed), while the other peak corresponds to the charge–density fluctuations. Importantly, in the metallic phase at temperatures below the first peak value, the specific heat has typical metallic dependence on temperature, $C_v = \gamma T$, where $\gamma \sim m^* \sim (U_{c2} - U)^{-1}$. Thus, the slope grows as U increases (approaches U_{c2}). With U decreasing, the slope decreases, the position of the higher energy peak moves to the lower temperatures, and both peaks merge. Doping of the system does not change the qualitative features of the specific heat curve, i.e., both spin and charge fluctuation peaks are present there as well [13]. Similarly, a two-peak structure of the specific heat curve was

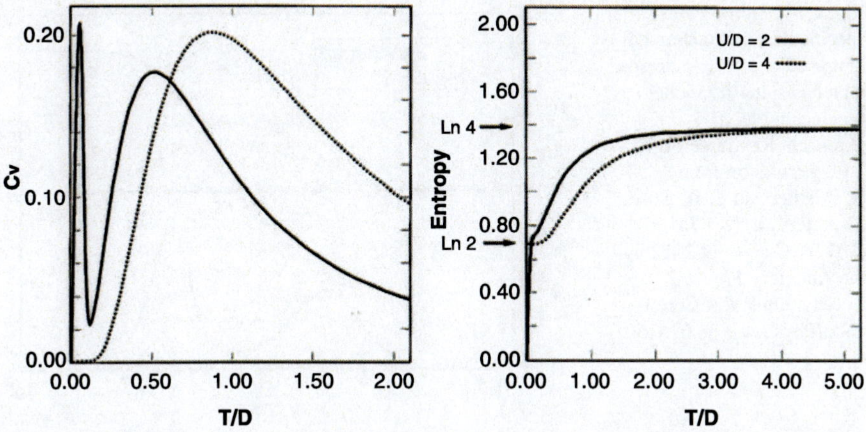

Fig. 4.13 Left: specific heat as a function of temperature for a correlated metal ($U = 2t$, solid curve) and an insulating ($U = 4t$, dashed curve cases) phases in the Hubbard model obtained with the IPT impurity solver. Right: the same for entropy. Reprinted Figs. 36 and 38 with permission from (A. Georges, G. Kotliar, W. Krauth, and M.J. Rozenberg, Rev. Mod. Phys. **68**, 13 (1996)). Copyright 1996 by the American Physical Society. https://doi.org/10.1103/RevModPhys.68.13

obtained in the 2D case with QMC [39] and VCA [40] approaches and in the case of quantum impurity model (NRG approach) [41] (For a high-temperature numerical linked cluster expansion study of the entropy and other quantities in the 3D Hubbard model, see Ref. [42]). In the insulating phase, only the charge–density fluctuations take place and the low-temperature specific heat is exponentially suppressed, $C_v \sim e^{-\frac{\Delta}{T}}$, where Δ is the gap between the Hubbard sub-bands (left Fig. 4.13).

Similar to the specific heat, the entropy in the insulating phase is also exponentially suppressed (right Fig. 4.13), with a finite zero-temperature residue value $Nln2$ coming from the state degeneracy. This (residue) result corresponds to the entropy of a system that consists of independent magnetic moments. On the first hand, this result is surprising since in DMFT the charge fluctuations and residual magnetic exchange are not neglected. As it was argued in Ref. [20], the reason for the independent magnetic moment entropy at zero temperature is as follows. In the limit of infinite dimensions, there are two exchange scales. The first is the exchange coupling between two (fixed) spins $J \sim \frac{t^2}{U} \sim O(1/d)$, that defines the splitting between the 2^N states with total zero spin and disappears at infinite dimensions as $O(1/d)$. The second is the AFM exchange energy between given spin and its $2d$ nearest neighbors that is proportional to dJ defines the Néel temperature and remains finite at $d \to \infty$. Thus, in the limit of infinite dimensions the insulating ground state remains doubly degenerated, unless the long-range AFM state is the ground state of the system. In other words, the degenerate state is valid in highly frustrated cases, relevant to infinite dimensions.

In the metallic phase, the spin entropy $ln2$ is reached at temperature of order ϵ_F^* (Fig. 4.13, right). The other $ln2$ contribution to the entropy of the metallic system comes from the charge degrees of freedom, i.e., from the integral over the high-

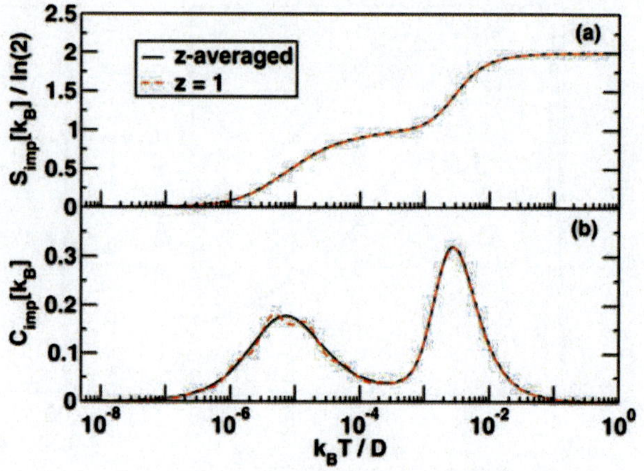

Fig. 4.14 Entropy and specific heat as functions of temperature of the Anderson quantum impurity model obtained with NRG approach. Reprinted Fig. 3 with permission from (L. Merker and T. A. Costi, Phys. Rev. B 86, 075150 (2012)). Copyright 2012 by the American Physical Society. https://doi.org/10.1103/PhysRevB.86.075150

energy peak of the specific heat, giving in total $S = \ln 4$ in the high-temperature limit (Fig. 4.13, right). Quite remarkably, in the case of a quantum impurity, the two-peak structure of the specific heat is reflected in a two-valley structure of the entropy with steps corresponding to the peak temperatures [41] (Fig. 4.14). In the last paper, the effect of the magnetic field on the thermodynamical properties of the model was also analyzed (For a study of the temperature- and doping-dependencies of the entropy with DMFT in the 3D case, see Ref. [43]).

As it was shown in Ref. [20], the internal energy changes with U much slower than its kinetic and potential components (i.e., the growth of the potential (Coulomb repulsion) energy is almost compensated by a decrease of the kinetic energy due to the electron localization).

To conclude this subsection, we mention a remarkable crossing points in the specific heat as function of temperature for the curves with different values of U, which takes place in the metallic phase at $T \approx 0.59$ (Fig. 4.15), as was first reported for the infinite-dimensional Hubbard model in Ref. [44], and later found in lower dimensions with DMFT [45, 46] and QMC simulations (see, e.g., Refs. [39, 46] for the 3D and 2D cases, correspondingly). Such an effect takes place in other thermodynamic systems, e.g., in liquid ^{3}He at different pressures and in heavy fermion systems at different pressures and magnetic fields (for references, see, e.g., Refs. [44, 47]). Thus, there is temperature at which $\frac{\partial C(T,X)}{\partial X} = 0$ (in our case $X = U$). As it was demonstrated in Ref. [47], to show that the equation $\frac{\partial C(T,X)}{\partial X} = 0$ can be satisfied, one can start with differentiating Eq. (4.29) with respect to U and taking the limit of infinite temperature. This gives

$$\frac{\partial S(T \to \infty, U)}{\partial U} = \int_0^\infty dT' \frac{1}{T'} \frac{\partial C_v(T', U)}{\partial U}. \quad (4.31)$$

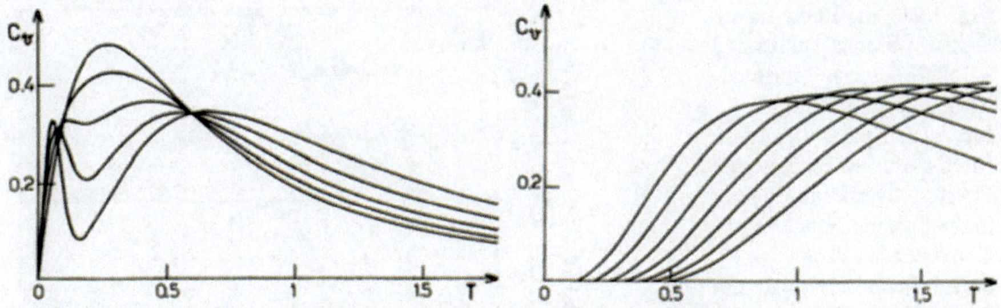

Fig. 4.15 Left: Metallic-phase specific heat as a function of temperature at different values of U (from $0.51t$ to $2t$, the low-temperature slope of C_v increases as U increases; QMC solver, infinite dimensions). Right: The same for the insulating phase ($U = 4\ t, \ldots, 10\ t$; from left to right). Reprinted Figs. 4 and 15 with permission from (A. Georges and W. Krauth, Phys. Rev. B **48**, 7167 (1993)). Copyright 1993 by the American Physical Society. https://doi.org/10.1103/PhysRevB.48.7167

Since the entropy approaches a constant as temperature goes to infinity, one gets from the last equation $\int_0^\infty dT' \frac{1}{T'} \frac{\partial C_v(T', U)}{\partial U} = 0$. Next, at high temperatures $C_v \sim \frac{U}{T}$, i.e. $\frac{\partial C_v(T', U)}{\partial U} > 0$, so at intermediate temperatures one must have $\frac{\partial C_v(T', U)}{\partial U} < 0$ to satisfy the equation in the previous sentence. Therefore, there must be at least one temperature that gives $\frac{\partial C_v(T', U)}{\partial U} = 0$. This argument is based on a *sum rule* for the specific heat,

$$\int_0^\infty dT' \frac{C_v(T', U)}{T'} = \text{constant}. \tag{4.32}$$

Thus, it is possible that there are sum rules responsible for crossing points in other systems.

As it was shown in Ref. [47] by expanding the specific heat in powers of T and U, the crossing region is small when the susceptibilities weakly depend on temperature and have linear dependence on U. Both conditions are satisfied at high temperatures. Indeed, as it is seen in left Fig. 4.15, the second (low temperature) crossing point area is much more spread comparing to the high-temperature one. Further analysis [45] has demonstrated that the difference of the crossing areas for these two points is due to the difference of the energy scales at which $C_v(T, U)$ changes: at high temperatures, the scale is the bandwidth (or hopping t), since in this case all states are excited, while at low temperatures it is the effective kinetic energy ϵ_F^*, since the low-energy excitations with this energy define the low-temperature specific heat (see Fig. 4.13 and its analysis in the text). At low Ts, the first maximum in the specific heat, and hence the first change of sign of $\frac{\partial C_v(T, U)}{\partial U}$, takes place at $T \sim \epsilon_F^*$. Therefore, the intersection point of two curves of the specific heat $C_v(T, 0)$ and $C_v(T, U)$ (at zero and some finite U) at low temperature does not take place at fixed temperature (since ϵ_F^* depends on U!). On the other hand, at high temperatures the energy scale is W, i.e., it is the same for the interacting and noninteracting system and the intersection point

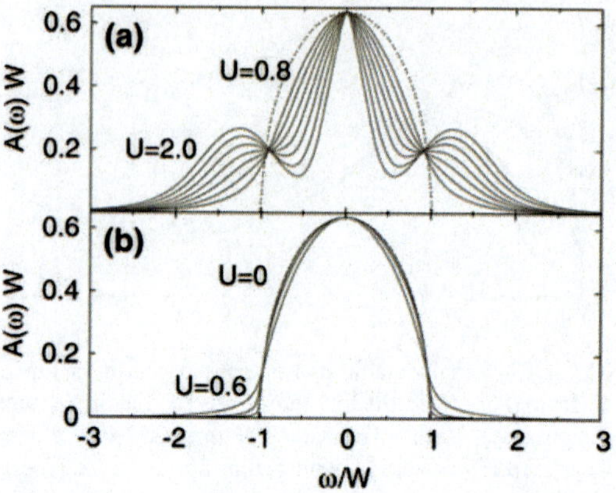

Fig. 4.16 The Bethe lattice DOS at different values of U (NRG solver). Reprinted by permission from (the Licensor): Springer Nature, Journal of Low Temperature Physics, "Isosbestic Points in the Spectral Function of Correlated Electrons," M. Eckstein, M. Kollar, and D. Vollhardt, **147**, 279 (2007). Copyright 2007

is practically U-independent. Further, it was shown in Ref. [45] by using the second-order PT in U expansion of the specific heat that the crossing point temperature depends on two parameters:

1. The first moment correction for the deviation of the used DOS from the rectangular flat one, $\delta N = \int d\varepsilon \left|\rho(\varepsilon) - \frac{1}{2}\right|$. In infinite dimensions, this parameter is small for the high-temperature crossing point and large for the low-temperature one.
2. The linear coefficient in the $1/d$ expansion (around $\varepsilon = \infty$) of the specific heat at the crossing point. It appears that even in dimensions as low as $d = 1$, this coefficient, i.e., the linear term, defines the high-temperature specific heat crossing point.

Finally, let us mention that the crossing point in the specific heat is almost the same in systems of different dimensions, i.e., systems with different free-electron DOS (see Table III in Ref. [45]). Therefore, one can put a question if such universal crossing points can also take place in the DOS curves. Indeed, as it was found in Ref. [48], for infinite-dimensional Hubbard model such a crossing point ω^* exists for the DOS $A(\omega, U)$ as a function of frequency at different values of U (Fig. 4.16). As one may guess, the reason for the existence of the crossing point is a sum rule for $A(\omega, U)$. For small Us one can show analytically that if the noninteracting DOS has a van Hove singularity, one has $\omega^* \approx W$, i.e., ω^* is U-independent. On the other hand, it was found in Ref. [48] that at larger Us the crossing point ω^* is near the edge of the free-electron DOS, i.e., the value of ω^* and $A(\omega^*, U)$ strongly depends on the form of the free-electron DOS and the crossing point does not possess the characteristic of universality of the high-temperature crossing point in the specific heat. However, as it was argued in Ref. [48], experimental finding of the DOS crossing point is still useful, since the value of the frequency at the crossing point may be used to establish the high-temperature energy scale in strongly correlated systems.

4.5 Magnetic Properties

4.5.1 *Antiferromagnetism*

In his seminal paper [1], Jarrell has demonstrated that the half-filled Hubbard model in infinite dimensions shows a PM-AFM transition as the temperature is lowered and crosses the critical value that depends on U (Fig. 4.17). In this paper, the momentum-dependent (nonlocal) susceptibility $\chi_{ij}(i\omega_n, i\omega_m)$ was calculated at different temperatures with the DMFT approximation (QMC impurity solver) as we describe below. Performing the momentum Fourier transformation of the resulting susceptibility and summing up over the frequency and site indices, one can calculate the static momentum-dependent susceptibility:

$$\chi\left(\vec{q}\right) = \frac{T}{N} \sum_{n,m,i,j} e^{-i\vec{q}\left(\vec{R}_i - \vec{R}_j\right)} \chi_{ij}(i\omega_n, i\omega_m). \tag{4.33}$$

To analyze the AFM and FM instabilities, one needs to put in Eq. (4.33) $\vec{q} = (\pi, \ldots, \pi)$ and $\vec{q} = (0, \ldots, 0)$, correspondingly, and analyze the temperature dependence of the static susceptibilities. Divergence of the susceptibility at some (critical) temperature indicates a PM-AFM or a PM-FM transition in the system. It was found in work [1] that in the AFM case the susceptibility diverges near the critical temperature as $|T - T_c|^{-\alpha}$, where $\alpha \sim 1$, establishing an AFM instability in the half-filled Hubbard model. It was also found that the AFM state extends in the region of finite doping (for densities from $n = 1$ to approximately 0.87).

Now, let us present the main details on how these results were obtained. The susceptibility $\chi_{ij}(i\omega_n, i\omega_m)$ was found by solving the equation (see also Eq. (4.8))

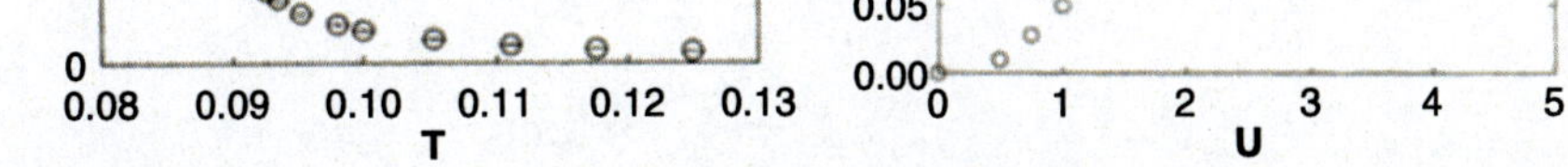

Fig. 4.17 Left: the divergence of the AFM susceptibility as temperature approaches the critical value (half-filling, $U = 1.5\ t$). Right: the average of the square of the difference of spin-up minus spin-down charge densities (i.e., of an AFM "order parameter") at critical temperature and the AFM critical temperature as a function of U at half-filling. The results were obtained with a QMC solver. Reprinted Figs. 3 and 4 with permission from (M. Jarrell, Phys. Rev. Lett. **69**, 168 (1992)). Copyright 1992 by the American Physical Society. https://doi.org/10.1103/PhysRevLett.69.168

$$\chi_{ij}(i\omega_n, i\omega_m) = \chi_{ij}^{(0)}(i\omega_n)\delta_{nm} + T\sum_{k,l}\chi_{ik}^{(0)}(i\omega_n)\Gamma(i\omega_n, i\omega_l)\chi_{kj}(i\omega_l, i\omega_m). \tag{4.34}$$

In this equation, $\chi_{ij}^{(0)}(i\omega_n)$ is the noninteracting (one-loop) part of the susceptibility, which has the following form in the momentum representation:

$$\chi^{(0)}\left(i\omega_n, \vec{q}\right) = \frac{1}{N}\sum_{\vec{k}} G\left(i\omega_n, \vec{k}\right) G\left(i\omega_n, \vec{k} + \vec{q}\right), \tag{4.35}$$

where $G\left(i\omega_n, \vec{k}\right) = \frac{1}{i\omega_n - \varepsilon\left(\vec{k}\right) + \mu - \Sigma(i\omega_n)}$ is the single-electron momentum-dependent GF with the DMFT (momentum-independent) self-energy. The noninteracting susceptibility can be found by solving the DMFT problem for the one-particle GF, while to find another unknown function in Eq. (4.34)—the local (DMFT-approximated) irreducible vertex function $\Gamma(i\omega_n, i\omega_l)$—one needs to solve a two-particle (2P) problem. Namely, $\Gamma(i\omega_n, i\omega_l)$ can be found by calculating the local DMFT susceptibility:

$$\chi_{ii}(i\omega_n, i\omega_m) = -T^2 \int_0^\beta d\tau_1 \ldots \int_0^\beta d\tau_4 e^{-i\omega_m(\tau_1-\tau_2) - i\omega_n(\tau_3-\tau_4)} \left\langle T c_{i\uparrow}(\tau_4) c_{i\downarrow}^+(\tau_3) c_{i\downarrow}(\tau_2) c_{i\uparrow}^+(\tau_1) \right\rangle, \tag{4.36}$$

substituting the result into Eq. (4.34) in the case $i = j$ and solving the resulting equation for the unknown function $\Gamma(i\omega_n, i\omega_l)$. The 2P GF in (4.36) can be found using a DMFT algorithm, similar to the one-particle case (for details, see Chap. 9).

Before proceeding with further details of the AFM solution in the Hubbard model, we present an alternative approach to solve this problem (we follow mostly the derivation steps from Anisimov and Izyumov's book [5], where the reader can find more details). The approach is based on the bipartite lattice approximation, with two nonequivalent alternating types of sites A (spin-up) and B (spin-down) and alternating staggered site-dependent magnetic field hσ. The corresponding Hubbard Hamiltonian has the following form:

$$\begin{aligned} H = &-\sum_{i,j,\sigma} t_{ij}\left(c_{Ai\sigma}^+ c_{Bj\sigma} + c_{Bj\sigma}^+ c_{Ai\sigma}\right) + U\sum_i (n_{Ai\uparrow}n_{Ai\downarrow} + n_{Bi\uparrow}n_{Bi\downarrow}) \\ &- \mu\sum_{i,\sigma}(c_{Ai\sigma}^+ c_{Ai\sigma} + c_{Bi\sigma}^+ c_{Bi\sigma}) - h\sum_{i,\sigma}\sigma(c_{Ai\sigma}^+ c_{Ai\sigma} - c_{Bi\sigma}^+ c_{Bi\sigma}), \end{aligned} \tag{4.37}$$

where the last term describes the magnetic energy of electrons with different spins. The last two sums in Eq. (4.37) can be combined into one sum

$$-\mu\sum_{i,\sigma}\left(c^{+}_{Ai\sigma}c_{Ai\sigma}+c^{+}_{Bi\sigma}c_{Bi\sigma}\right)-h\sum_{i,\sigma}\sigma\left(c^{+}_{Ai\sigma}c_{Ai\sigma}-c^{+}_{Bi\sigma}c_{Bi\sigma}\right)$$
$$=-\sum_{i,\sigma}\left(\mu_{\sigma}c^{+}_{Ai\sigma}c_{Ai\sigma}+\mu_{-\sigma}c^{+}_{Bi\sigma}c_{Bi\sigma}\right), \tag{4.38}$$

where

$$\mu_{\sigma}=\mu+h\sigma. \tag{4.39}$$

Then, one can introduce two-component annihilation and creation electron operators

$$c_{\sigma}\left(\vec{k}\right)=\begin{pmatrix} c_{A\sigma}\left(\vec{k}\right) \\ c_{B\sigma}\left(\vec{k}\right) \end{pmatrix},\quad c^{+}_{\sigma}\left(\vec{k}\right)=\left(c^{+}_{A\sigma}\left(\vec{k}\right),c^{+}_{B\sigma}\left(\vec{k}\right)\right) \tag{4.40}$$

and the time-ordered matrix GF

$$G_{\sigma}\left(\tau,\vec{k}\right)=-\left\langle T_{\tau}c_{\sigma}\left(\tau,\vec{k}\right)c^{+}_{\sigma}\left(0,\vec{k}\right)\right\rangle\equiv\begin{pmatrix} G_{AA\sigma} & G_{AB\sigma} \\ G_{BA\sigma} & G_{BB\sigma} \end{pmatrix}\left(\tau,\vec{k}\right) \tag{4.41}$$

(here we put the momentum vector variables in brackets, not as lower index variables, to make notations more clear). The last function satisfies the following matrix equation (in the Matsubara frequency representation):

$$\begin{bmatrix} i\omega_n+\mu_{\sigma}-\Sigma_{A\sigma}\left(i\omega_n,\vec{k}\right) & \varepsilon\left(\vec{k}\right) \\ \varepsilon\left(\vec{k}\right) & i\omega_n+\mu_{-\sigma}-\Sigma_{B\sigma}\left(i\omega_n,\vec{k}\right) \end{bmatrix}$$
$$\times\begin{pmatrix} G_{AA\sigma} & G_{AB\sigma} \\ G_{BA\sigma} & G_{BB\sigma} \end{pmatrix}=\begin{pmatrix} 1 & 0 \\ 0 & 1 \end{pmatrix} \tag{4.42}$$

that can be easily solved:

$$\begin{pmatrix} G_{AA\sigma} & G_{AB\sigma} \\ G_{BA\sigma} & G_{BB\sigma} \end{pmatrix}\left(i\omega_n,\vec{k}\right)$$
$$=\frac{1}{\left[i\omega_n+\mu_{\sigma}-\Sigma_{A\sigma}\left(i\omega_n,\vec{k}\right)\right]\left[i\omega_n+\mu_{-\sigma}-\Sigma_{B\sigma}\left(i\omega_n,\vec{k}\right)\right]-\varepsilon^{2}\left(\vec{k}\right)}$$
$$\times\begin{pmatrix} i\omega_n+\mu_{-\sigma}-\Sigma_{B\sigma}\left(i\omega_n,\vec{k}\right) & -\varepsilon\left(\vec{k}\right) \\ -\varepsilon\left(\vec{k}\right) & i\omega_n+\mu_{\sigma}-\Sigma_{A\sigma}\left(i\omega_n,\vec{k}\right) \end{pmatrix}. \tag{4.43}$$

Then, starting from this equation, one can use the methodology of DMFT for the problem on a bipartite lattice:

(a) Ignore momentum dependence of the self-energy: $\Sigma_{A\sigma}\left(i\omega_n, \vec{k}\right) = \Sigma_{A\sigma}(i\omega_n)$, $\Sigma_{B\sigma}\left(i\omega_n, \vec{k}\right) = \Sigma_{B\sigma}(i\omega_n)$;

(b) Introduce the local matrix GF

$$
\begin{aligned}
G_\sigma^{loc}(i\omega_n) &\equiv \begin{pmatrix} G_{AA\sigma}^{loc} & G_{AB\sigma}^{loc} \\ G_{BA\sigma}^{loc} & G_{BB\sigma}^{loc} \end{pmatrix}(i\omega_n) \\
&= \int d\varepsilon \rho(\varepsilon) \frac{1}{[i\omega_n + \mu_\sigma - \Sigma_{A\sigma}(i\omega_n)][i\omega_n + \mu_{-\sigma} - \Sigma_{B\sigma}(i\omega_n)] - \varepsilon^2} \\
&\quad \times \begin{pmatrix} i\omega_n + \mu_{-\sigma} - \Sigma_{B\sigma}(i\omega_n) & -\varepsilon \\ -\varepsilon & i\omega_n + \mu_\sigma - \Sigma_{A\sigma}(i\omega_n) \end{pmatrix};
\end{aligned} \tag{4.44}
$$

(c) Map the problem on the problem of a single impurity in the mean field $\mathcal{G}_\sigma(i\omega_n) \equiv \begin{pmatrix} \mathcal{G}_{AA\sigma} & \mathcal{G}_{AB\sigma} \\ \mathcal{G}_{BA\sigma} & \mathcal{G}_{BB\sigma} \end{pmatrix}(i\omega_n)$. Assume that the impurity GF is equal to the local GF: $G_\sigma(i\omega_n) = G_\sigma^{loc}(i\omega_{n.})$

(d) Use the Dyson equation that connects the impurity GF and the dynamical mean field:

$$
\mathcal{G}_\sigma^{-1}(i\omega_n) = G_\sigma^{-1}(i\omega_n) + \Sigma_\sigma(i\omega_n); \tag{4.45}
$$

(e) Solve the system of Eqs. (4.44) and (4.45) and the corresponding impurity problem

Then, the solution for the matrix GF gives the spin occupancies for sites A and B, which can be calculated from the spin DOS $\rho_{A\sigma}(\omega) = -\frac{1}{\pi} Im\, G_{AA\sigma}(\omega)$, $\rho_{B\sigma}(\omega) = -\frac{1}{\pi} Im\, G_{B\sigma}(\omega)$. To generate an AFM ground-state solution, one can use an infinitely small value of the alternated magnetic field h. Naturally, the antiferromagnetism can be enhanced by increasing the magnitude of h (see, e.g., Ref. [49]). Increasing h also leads to the opening of a gap in the quasiparticle spectrum. Here we discuss the case of zero alternated field.

As it was shown in Ref. [50], at half-filling and small values of U the DOS shows remnants of the square root divergence at the lower edge of the gap and has a power–law dependence on frequency on the upper edge (this result can be obtained analytically from Eq. (4.44)), where the electron self-energies can be approximated by the static Hartree result $Un_{-\sigma}$. As U increases, the DOS modifies to the familiar two Hubbard sub-band form (Fig. 4.18).

As it also follows from the results for the DOS of work [50], at $U = 3t$ the AFM state transforms into a PM one with doping change at a rather low doping ($\delta = 0.2$). Quite remarkably, at doping δ~0.13 the authors of this work obtained a strongly

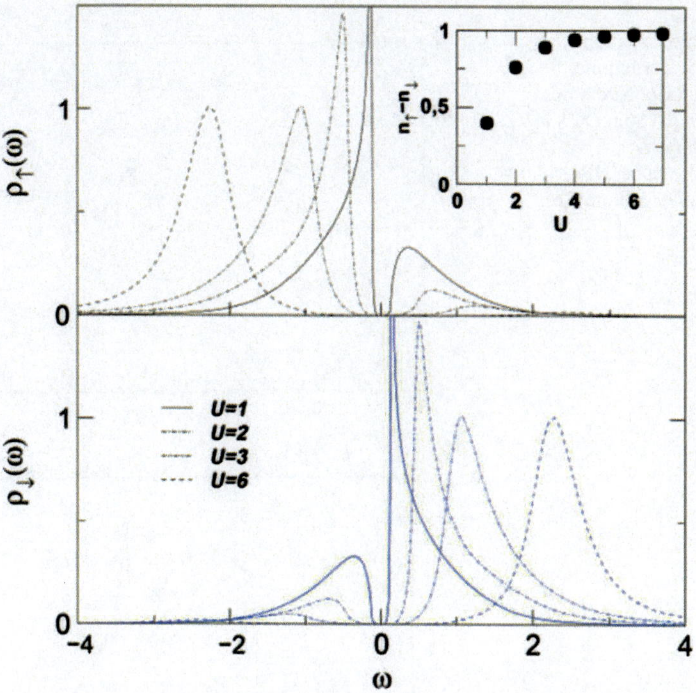

Fig. 4.18 The spin DOS of the one-band Hubbard model at half-filling and different values of U (hypercubic lattice, NRG solver). From Ref. [50]

peaked DOS, similar to the case of small Us at half-filling (the solid curves in Fig. 4.18). The AFM regions of the phase diagram of the Hubbard model are shown on a schematic diagram in Fig. 4.20. The system is in a "classic" AFM state at half-filling, while close to half-filling at small Us a phase-separated AFM state is the lowest energy configuration (from Ref. [50]). There are many other DMFT studies of the AFM phase in strongly correlated systems in the framework of both Hubbard models and real materials, e.g., in papers [35, 44, 51–56].

4.5.2 *Ferromagnetism*

The question whether DMFT solution can give an FM state in the Hubbard model was one of the key questions. The positive answer would fulfill Hubbard's dream since the main purpose of the model he introduced was to describe ferromagnetism and he was not able to obtain it with his solutions. As it was already mentioned in Chap. 1, certain optimism that such a solution can be obtained came from the Nagaoka's theorem [57] that states that in the Hubbard model with nearest-neighbor hopping and $U = \infty$ the half-filled system becomes ferromagnetic once one electron

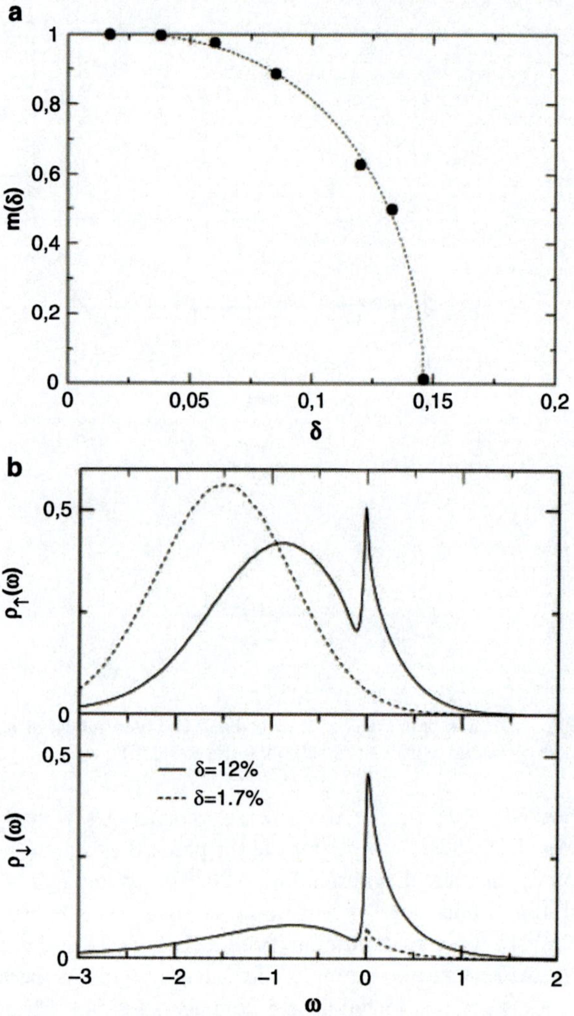

Fig. 4.19 (**a**) Magnetization as a function of doping in the infinite-dimensional Hubbard model at $U = 50t$. (**b**) Spin DOS at different dopings (hypercubic lattice, NRG impurity solver). From Ref. [50]

is removed. It was found that the saturated FM ground state is stable for several lattices, like simple cubic (for any hopping parameter t), fcp and hcp (when $t < 0$). Though the conditions of the theorem are rather artificial (nonrealistic infinite repulsion, thermodynamically irrelevant one hole—there is no extension of the proof even on the case of two holes), Nagaoka's result has the right to be regarded as the starting point to attack FM with DMFT. Obermeier et al. [58] have demonstrated with DMFT that indeed at large values of U (>20 t) the Hubbard model on the hypercubic lattice demonstrates an FM phase for a rather wide range of doping (up to ~0.3). This result was almost immediately extended on the case of generalized fcc lattice, [59, 60] more relevant to real transition–metal ferromagnets. In paper [60], it was found that large spectral weight near one of the band edges (see

Fig. 4.19) is crucial for the FM state. The enhanced weight is a result of many-body effects that renormalize the bandstructure. The requirement of the enhanced weight is a much stronger condition comparing to the Stoner criterion of FM $U\rho(\epsilon_F) > 1$ providing stability of the FM phase. It was also argued in the last paper that the obtained solution is capable to describe semiquantitatively FM phase in Ni. Later, it was found that provided one uses a realistic bandstructure in the framework of DFT + DMFT approach (see Chap. 10), DMFT can describe the FM state in many transition metals (see, e.g., Ref. [61] and Chap. 10), one of the greatest successes of the theory.

Similar to the case of AFM, possibility of an FM state can be studied by looking for a divergence of the FM susceptibility near some temperature, or by calculating spin-resolved DOS. In the last case, one needs to solve a spin-dependent DMFT problem and to calculate the spin DOS using Eq. (4.1). To solve the DMFT problem, one needs to find a solution of a system of the equations that consists of the spin-dependent local (non-matrix) GF equation

$$G_\sigma^{loc}(i\omega_n) = \int d\varepsilon \rho(\varepsilon) \frac{1}{i\omega_n + \mu_\sigma - \Sigma_\sigma(i\omega_n) - \varepsilon}, \tag{4.46}$$

Dyson Eq. (4.45) and the spin-resolved impurity problem equation.

As it was discussed above, to obtain an FM state, one needs to use rather large values of U. Typical results (at $U = 50t$) for the doping dependencies of the magnetization and spin DOS in the Hubbard model in an FM state are shown in Fig. 4.19. As it follows from this figure, the FM state persists up to high dopings

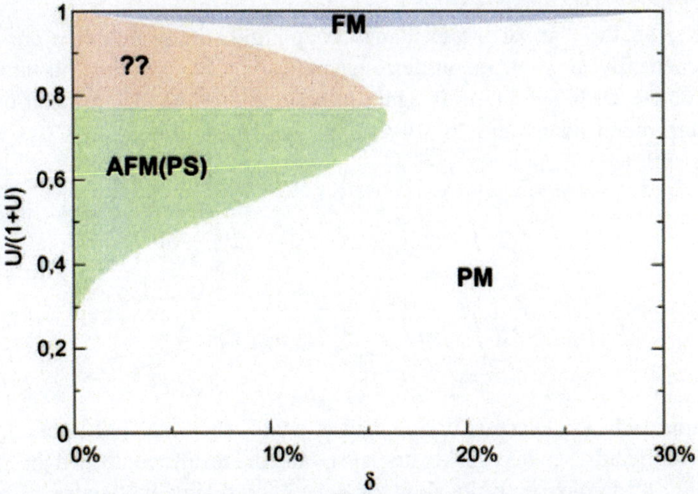

Fig. 4.20 A schematic U-doping phase diagram of the Hubbard model at zero temperature. The nature of the question-marked magnetic phase has been not clarified yet (see also the discussion of the details of the diagram in the text). From Ref. [50]

(~0.2) and the strong spectral peak in the interacting DOS mentioned above is a characteristic feature of the FM state.

A schematic (zero-temperature) U-doping phase diagram of the Hubbard model obtained with DMFT is shown in Fig. 4.20. At half-filling, the system is in an AFM state, while at low dopings and small Us a phase-separated AFM is the lowest energy configuration. For large values of $U(>25t)$ and a wide doping range (up to ~0.3) the system is in an FM state. In the region of low doping and $U > 4\,t$, the system is also in a magnetic state, though the nature of this state is not clear so far. As to the temperature-U diagram, one can get an idea about it from the left Fig. 4.4. In this figure, as one can see, the bluebell area left to the critical Us corresponds to an AFM metal state and to the right from the critical Us there is an AFM insulator state.

Recent studies of the magnetism with DMFT mostly dealed with multi-orbital systems. In particular, it was demonstrated that DMFT can give also an FM–ferroelectric transition on a hyper-perovskite lattice driven by energy splitting [62]. The orbital-resolved exchange parameters for the bcc iron were obtained by calculating the orbital susceptibilities with DMFT [63].

DMFT approximation was also successfully applied in the case of Kondo lattices—systems with local (on-site) magnetic moments in presence of itinerant electrons—described by the Hamiltonian

$$H = \sum_{\vec{k}\sigma} \varepsilon\left(\vec{k}\right) c^{+}_{\vec{k}\sigma} c_{\vec{k}\sigma} + J \sum_{i} \vec{S}_i \vec{s}_i, \tag{4.47}$$

where $\varepsilon\left(\vec{k}\right)$ is the dispersion of itinerant electrons, $\vec{S}_i$ are the local moments and $\vec{s}_i = \sum_{\alpha,\beta} c^{+}_{i\alpha}\, \vec{\sigma}_{\alpha\beta}\, c_{i\beta}$ are the itinerant magnetic moments (on sites labeled by index i). Depending on the sign of the exchange coupling J, these momenta are coupled ferromagnetically $(J < 0)$ or antiferromagnetically $(J > 0)$. For this model, the essence of the DMFT solution is again to assume that the electron self-energy is momentum-independent and to solve Eqs. (4.45) and (4.46) and the impurity equation with the action

$$S = -\sum_{\sigma} \int_0^{\beta} \int_0^{\beta} d\tau d\tau' \Psi^{+}_{\sigma}(\tau) \mathcal{G}^{-1}_{\sigma}(\tau - \tau') \Psi_{\sigma}(\tau')$$
$$- J\vec{S} \int_0^{\beta} d\tau d\tau' \sum_{\alpha,\beta} \Psi^{+}_{\alpha}(\tau)\, \vec{\sigma}_{\alpha\beta}\, \Psi_{\beta}(\tau') \tag{4.48}$$

The approach was successfully applied to study AFM and FM states in Kondo systems (for details, see Sect. 5.4 in book [5]), and the results confirmed the expected universality of dominance of local interactions in correlated materials.

Even more challenging attempt to describe the indirect magnetic exchange between magnetic atoms was performed by solving an Anderson impurity model with two impurities in Ref. [64] (see Eq. (3.38) for the one-impurity Hamiltonian).

The authors used a DMFT approach generalized on the case of site-dependent local self-energy (real-space DMFT). However, since in this approach nonlocal magnetic interactions do not affect the single-site electron self-energy defined by local interactions, DMFT has obvious limitations in this case. Thus, the relevance to nonlocal Kondo (RKKY) physics with interaction algebraically decaying with distance is not accessible with DMFT. As it was shown in the paper [64], besides RKKY limit (algebraically decaying, ~d^{-4}, oscillating with distance interactions), DMFT can successfully describe other effects of indirect magnetic exchange. Namely, as the authors showed, the DMFT results for nonlocal adatom susceptibilities (that define the exchange interactions) as functions of adatom separation are, except for the nearest-neighbor distance between the adatoms, in a good agreement with exact DMRG results.

4.6 Superconductivity

The DMFT formalism can be generalized on the superconducting case by introducing the two-component Nambu spinor $\Psi_i^+ = \left(c_{i\uparrow}^+, c_{i\downarrow}\right)$, or in momentum space $\Psi_k^+ = \left(c_{k\uparrow}^+, c_{-k\downarrow}\right)$, and the corresponding matrix GF

$$\widehat{G}\left(\tau, \vec{k}\right) = -\left\langle T\Psi_{\vec{k}}(\tau)\Psi_{\vec{k}}^+(0)\right\rangle = \begin{pmatrix} G\left(\tau, \vec{k}\right) & F\left(\tau, \vec{k}\right) \\ F^*\left(\tau, \vec{k}\right) & -G\left(-\tau, -\vec{k}\right) \end{pmatrix}, \tag{4.49}$$

where $G\left(\tau, \vec{k}\right) = -\left\langle Tc_{\vec{k}\sigma}(\tau)c_{\vec{k}\sigma}^+(0)\right\rangle$ is the familiar single-electron GF and $F\left(\tau, \vec{k}\right) = -\left\langle Tc_{\vec{k}\uparrow}(\tau)c_{-\vec{k}\downarrow}(0)\right\rangle$ is the anomalous superconducting part of the matrix GF (see Refs. [20, 65]). In infinite dimensions, $F(k, \tau)$ is a function of time and dispersion $\varepsilon(k)$, thus one can get a superconducting pairing (anomalous GF/gap function) that has the symmetry of the lattice (this can be demonstrated by absence of the vertex correction to the GF in the case of other symmetries [20, 66]). On the other hand, the time dependence of the anomalous GF, or order parameter (OP), can be highly nontrivial. In the cases of singlet and triplet (with $S_z = 0$) pairings, the anomalous GF has the following symmetries: $F\left(-\tau, -\vec{k}\right) = F\left(\tau, \vec{k}\right)$ and $F\left(-\tau, -\vec{k}\right) = -F\left(\tau, \vec{k}\right)$, correspondingly.

After introducing the Nambu GF, it is also necessary to introduce the complex self-energy

$$\widehat{\Sigma}(i\omega_n) = \begin{pmatrix} \Sigma(i\omega_n) & S(i\omega_n) \\ S(i\omega_n) & -\Sigma^*(i\omega_n) \end{pmatrix} \tag{4.50}$$

and, similarly, the dynamical mean-field $\widehat{\mathcal{G}}$ matrices. In Eq. (4.50), the diagonal and non-diagonal elements correspond to the normal and superconducting parts of the self-energy. The last component satisfies $S(i\omega_n) = S^*(-i\omega_n)$. Equations (4.49) and (4.50) give the following expression for the matrix momentum-dependent

$$\widehat{G}(k, i\omega_n) = \begin{pmatrix} i\omega_n + \mu - \varepsilon(k) - \Sigma(i\omega_n) & -S(i\omega_n) \\ -S(i\omega_n) & i\omega_n - \mu + \varepsilon(k) + \Sigma^*(i\omega_n) \end{pmatrix}^{-1} \tag{4.51}$$

and local

$$\widehat{G}(i\omega_n) = \int d\varepsilon \rho(\varepsilon) \begin{pmatrix} i\omega_n + \mu - \varepsilon - \Sigma(i\omega_n) & -S(i\omega_n) \\ -S(i\omega_n) & i\omega_n - \mu + \varepsilon + \Sigma^*(i\omega_n) \end{pmatrix}^{-1} \tag{4.52}$$

lattice GFs. In Eq. (4.51) in the diagonal elements different signs in front of $\varepsilon(k)$ are generated by σ_z matrix in the kinetic part of the Hamiltonian in Nambu representation, $-\sum_{ij} t_{ij} \Psi_i^+ \sigma_z \Psi_j$ [5].

In DMFT, the local matrix functions satisfy the matrix Dyson equation

$$\widehat{\mathcal{G}}^{-1}(i\omega_n) - \widehat{G}^{-1}(i\omega_n) = \widehat{\Sigma}(i\omega_n). \tag{4.53}$$

To complete the system of superconducting matrix DMFT Eqs. (4.52) and (4.53), one needs to write down the impurity problem equation, which has the same form as in the normal case. In particular, the expression for the impurity effective action is

$$S_{\text{eff}} = -\int_0^\beta d\tau \int_0^\beta d\tau' \Psi^+(\tau) \widehat{\mathcal{G}}^{-1}(\tau - \tau') \Psi(\tau') + U \int_0^\beta d\tau n_\uparrow(\tau) n_\downarrow(\tau). \tag{4.54}$$

The solution of the problem (4.52)–(4.54) is significantly simplified in the case of Bethe lattice, where

$$\widehat{\mathcal{G}}^{-1}(i\omega_n) = i\omega_n + \mu\sigma_z - t^2 \sigma_z \widehat{G}(i\omega_n)\, \sigma_z. \tag{4.55}$$

Unfortunately, as early calculations for the one-band Hubbard model in infinite dimensions showed, the superconducting state with neither odd- (singlet) nor even-in-frequency (triplet) order parameter is possible [66]. This would compete this section on the DMFT studies of superconductivity in one-band Hubbard model in infinite dimensions. However, we would like to mention several facts on superconducting properties of other important models obtained with DMFT. In particular, signs of possible superconductivity in the two-band case related to high-temperature cuprate superconductors were reported (still in infinite dimensions) [14, 65, 67]. Indeed, since the intra-orbital repulsive Hubbard interaction is not sufficient to form superconductive pairs, except through the Kohn–Luttinger mechanism, to get superconductivity other terms, like interorbital repulsion U' and Hund interactions J, need to be included into the Hubbard Hamiltonian [68–71]. A general Hamiltonian to study superconducting properties of the two-band Hubbard model has the following form:

$$H = -\sum_{i,j,a,\sigma} t_{ij} c^{+}_{ia\sigma} c_{ja\sigma} - \frac{1}{2}\Delta^{CF} \sum_{i,\sigma} (n_{i1\sigma} - n_{i2\sigma}) + U \sum_{i,\sigma} n_{ia\uparrow} n_{ia\downarrow} + U' \sum_{i,\sigma,\sigma'} n_{i1\sigma} n_{i2\sigma'} - \frac{1}{2} J \sum_{i} (n_{i1\uparrow} - n_{i1\downarrow})(n_{i2\uparrow} - n_{i2\downarrow}), \quad (4.56)$$

with the hopping, repulsion, and exchange parameters that are defined above and Δ^{CF}—the energy of the crystal field splitting. Signatures of the superconducting region (with a bell-like critical temperature-doping curve) in the phase diagram of such a system were found in Ref. [67] (Fig. 4.21).

Recently, it was found that superconductivity in the two-band Hubbard model may exist for a rather wide range of values of U [72] (Fig. 4.22). We refer the reader to the last paper for more details of the calculations and references on the superconducting properties of the two-band model in infinite dimensions. Similar properties were found in the case of the two-band Hubbard model in the 3D case [73, 74] (see also a review paper [75]).

We conclude this section with a brief summary of probably the most-often application of DMFT to superconductivity so far—superconductivity in the 2D Hubbard model related to high-T_c cuprate materials (for details, see book [5]). Cuprates are characterized by coexistence of the d-wave superconducting and AFM phases. To describe such a coexistence, Lichtenstein and Katsnelson [76] formulated an approach based on a cluster extension of DMFT (for further studies based on this approach, see, e.g., Refs. [77–79]; details of cluster DMFT are given in Chap. 5). Namely, it was proposed to solve the DMFT problem for a four-site cluster (Fig. 4.23).

To solve such a cluster problem, a four-component Nambu spinor $\Psi^{+}_{i}(\tau) = (\psi^{+}_{1i}, \psi^{+}_{2i}, \psi^{+}_{3i}, \psi^{+}_{4i}) \equiv \left(c^{+}_{i\uparrow}, c^{+}_{i\downarrow}, c_{i\uparrow}, c_{i\downarrow}\right)$ was introduced. Then, using this spinor the Hubbard–Stratonovich transformation in the impurity effective action was performed:

$$\exp\left[-\Delta\tau U_i n_{i\uparrow} n_{i\downarrow} + \frac{1}{2}\Delta\tau U_i \left(n_{i\uparrow} + n_{i\downarrow}\right)\right] = \frac{1}{2} \sum_{\sigma=\pm1} exp\left[\lambda_i \sigma \left(\psi^{+}_{1i}\psi_{1i} - \psi^{+}_{2i}\psi_{2i} - \psi^{+}_{3i}\psi_{3i} + \psi^{+}_{4i}\psi_{4i}\right)\right], \quad (4.57)$$

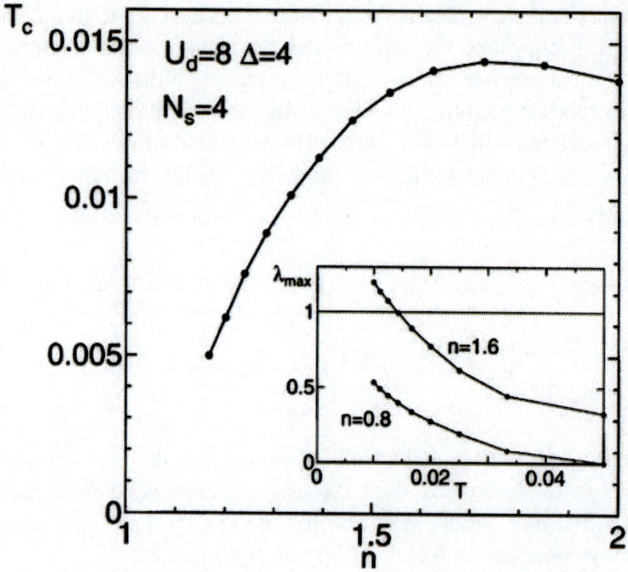

Fig. 4.21 The critical temperature as a function of doping of the infinite-dimensional two-band Hubbard model obtained with DMFT at $\Delta^{CF} = 8$ eV and $U_d = 8$ eV (Coulomb repulsion on the second (p-) orbital was assumed to be zero). The Anderson impurity problem was solved by exact diagonalization of the four-site cluster. The inset shows temperature dependence of the largest eigenvalue of the self-energy equation at different dopings. Reprinted from Journ. of Phys. and Chem. of Solids, Vol. 62, Y. Ono, K. Sano, "Mott transition and superconductivity in the two-band Hubbard model in infinite dimensions," Pages 285–288, Copyright 2001, with permission from Elsevier

where $\lambda_i = \frac{1}{2}\cosh^{-1}\left[\exp\left(\frac{1}{2}\Delta\tau U_i\right)\right]$. The form (4.57) of the effective action is convenient to introduce the AFM $\left\langle c^+_{i\uparrow} c_{j\downarrow}\right\rangle$ and superconducting $\Delta_{ij} = \langle c_{i\downarrow} c_{j\uparrow}\rangle$ order parameters. The nonzero cluster components of the order parameter are related as $\Delta = \Delta_{01} = -\Delta_{12} = \Delta_{23} = -\Delta_{30}$ (see Fig. 4.23 for the atoms numeration). Δ can be made to be real by assuming that Δ_{ij} are antisymmetric. Then, introducing the 4×4 normal and anomalous GFs, and also the self-energy and dynamical mean-field matrices, i.e., repeating the steps described by Eqs. (4.49)–(4.54), the authors analyzed the AFM and superconducting properties of the 2D Hubbard model.

Since the main goal of this chapter is the DMFT solution of the one-band Hubbard model in infinite dimensions, we refer the reader for more details on the finite-dimensional cluster equations and their simplification to Ref. [5]. Here we present a few notable DMFT results on the superconductivity in the 2D Hubbard model. Namely, as it was shown by Capone and Kotliar [79] cluster DMFT is capable to reproduce the experimentally observed AFM and superconducting parts of temperature-doping phase diagram of the cuprate superconductors at rather meaningful values of U (of order or larger than $8t$) (Fig. 4.24).

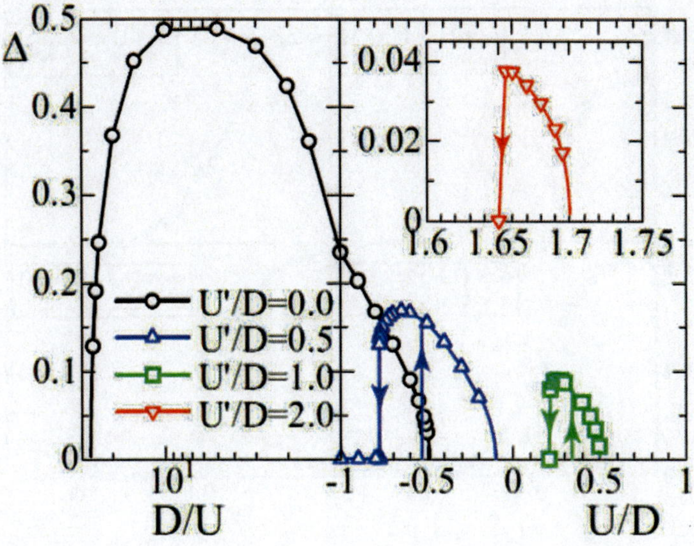

Fig. 4.22 Pair potential as a function of U at different values of U' (D is bandwidth) in the case of infinite-dimensional two-band Hubbard model at half-filling (CTQMC solver, $\Delta^{CF} = J = 0$, $\frac{T}{D} = 0.01$). Inset—results for $\frac{U'}{D} = 2$. Reprinted Fig. 1 with permission from (A. Koga and P. Werner, Phys. Rev. B **91**, 085108 (2015)). Copyright 2015 by the American Physical Society. https://doi.org/10.1103/PhysRevB.91.085108

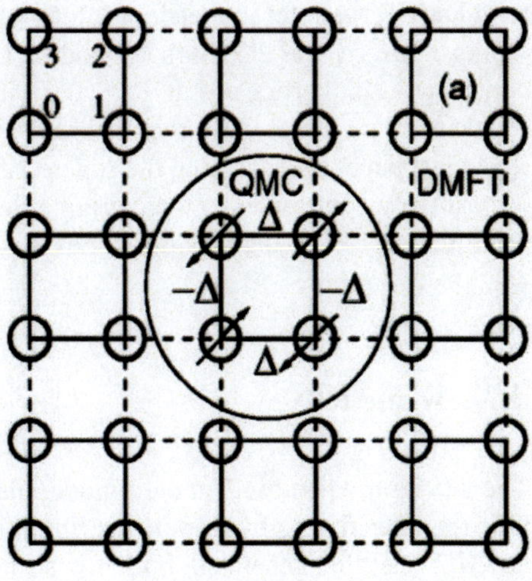

Fig. 4.23 Four-site cluster problem (one of the clusters is inside of the circle) solved with DMFT in the case of 2D Hubbard model. Numeration of the atoms in the cluster is shown in the top left corner of the lattice. Reprinted Fig. 1a with permission from (A.I. Lichtenstein, M.I. Katsnelson, Phys. Rev. B **62**, R9283 (2000)). Copyright 2000 by the American Physical Society. https://doi.org/10.1103/PhysRevB.62.R9283

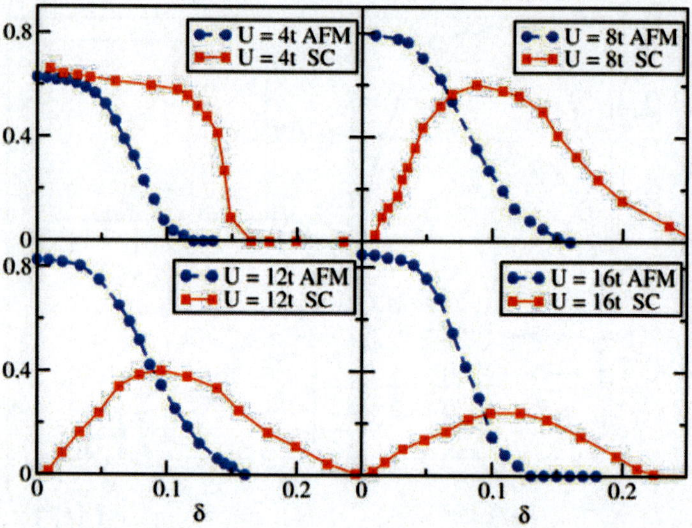

Fig. 4.24 Doping dependence of the AFM (blue) and d-wave superconducting (red, values multiplied by 10) order parameters at different values of U for the 2D Hubbard model obtained with the cellular cluster DMFT (zero temperature, exact diagonalization (ED) solver). Reprinted Fig. 1 with permission from (M. Capone, G. Kotliar, Phys. Rev. B **74**, 054513 (2006)). Copyright 2006 by the American Physical Society. https://doi.org/10.1103/PhysRevB.74.054513

For more recent interesting results for the SC solution of the 2D Hubbard model with DMFT, we refer the reader to Refs. [80–86]. In the majority of these and other works, more general 2D Hubbard models, comparing to the model in Ref. [76], were considered. In particular, in Ref. [82] it was shown that the nearest-neighbor repulsion V increases T_c in the underdoped regime (Fig. 4.25). Also, it was shown in the last paper by analyzing the superconducting spectral function ($\sim ImF(\omega)$)) that V positively contributes to the pairing at low frequencies, since this interaction can reinforce superexchange (as it satisfies $J = 4t^2/(U - V)$).

4.7 Disorder

The situation when the Hubbard model electrons move in presence of a stochastic field (generated, e.g., by random distribution of impurities) is of special interest since an MIT can be generated both by strong electron–electron correlations (Mott transition) and by disorder, more precisely by a disorder-induced Anderson localization when electrons are confined in space due to a coherent backscattering from randomly distributed impurities [87]. It is known that in the disordered case with no correlations ($U = 0$), when electrons scatter on a random field, there is a mobility edge ε_c in the continuous spectrum of the electrons: the electrons with energy $\varepsilon < \varepsilon_c$

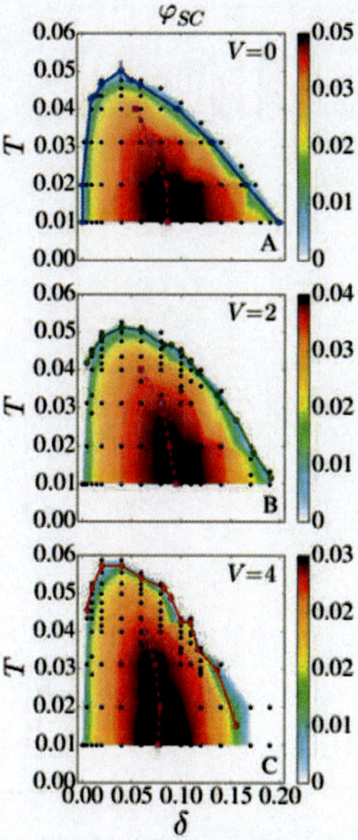

Fig. 4.25 Doping dependence of the critical temperature in the 2D Hubbard model at different values of the nearest-neighbor repulsion V/t ($U = 9\,t$, cellular DMFT, CTQMC solver; colors define the magnitude of the order parameter). Reprinted Figs. 1A, 1B, and 1C with permission from (A. Reymbaut, M. Charlebois, M. Fellous Asiani, L. Fratino, P. Sémon, G. Sordi, and A.-M. S. Tremblay Phys. Rev. B **94**, 155,146 (2016)). Copyright 2016 by the American Physical Society. https://doi.org/10.1103/PhysRevB.94.155146

are localized and with energy $\varepsilon > \varepsilon_c$ are itinerant ("movable"). Thus, the system is in the metallic state when the Fermi energy is above ε_c (for an overview, see, e.g., Ref. [88]). Obviously, the effects of local Coulomb repulsion may dramatically change the properties of disordered systems. Competition of local Coulomb repulsion and disorder can result in an interesting and exotic physics, directly related to real materials (for cartoon representations of the effects of local Coulomb interaction and disorder on the spectral properties of materials, see Fig. 4.26).

It is also known that CPA approximation, used for disordered noninteracting systems and based on an arithmetically averaged-over-disorder local DOS, does not give a solution with the Anderson localization [88, 89]. Thus, to generate the Anderson transition in the Hubbard model, effects of disorder have to be taken into account in a different way. In the case of DMFT, an appropriate approach called stochastic DMFT was proposed by Kotliar and Dobrosavljevic [90]. Below, we review this approach, as well as the second popular DMFT approach for disordered systems—DMFT+Σ [91], and summarize the main results obtained with these tools.

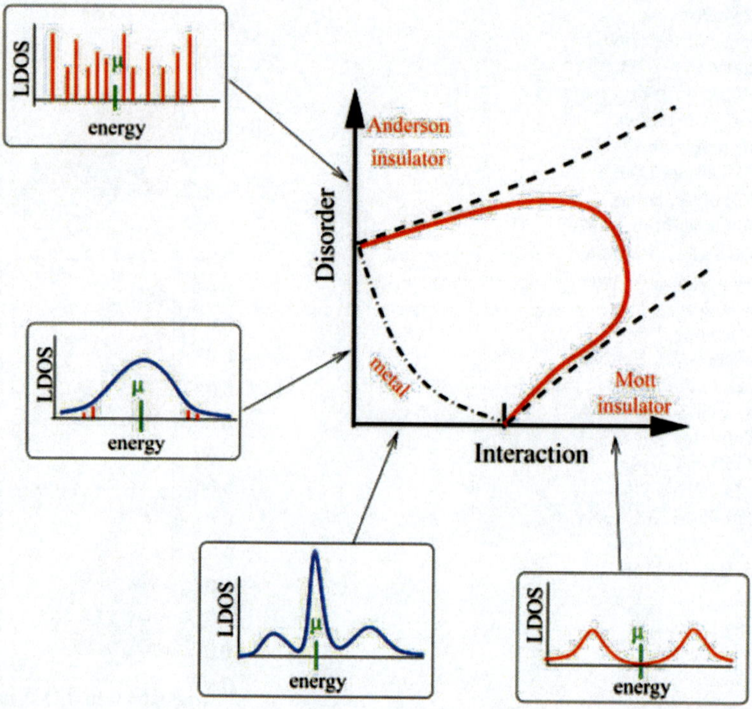

Fig. 4.26 A schematic phase diagram of systems with local repulsion (interaction) and disorder and the structure of the corresponding local DOS in different regions of the diagram. From Ref. [88]

4.7.1 Stochastic DMFT Approach

Let us consider the Hubbard Hamiltonian with randomly distributed over sites potential ε_i:

$$H = \sum_{i,j,\sigma} \left(-t_{ij} + \varepsilon_i \delta_{ij}\right) c^+_{ia\sigma} c_{ja\sigma} + U \sum_i n_{i\uparrow} n_{i\downarrow}. \tag{4.58}$$

In the case of uncorrelated (quenched) disorder, the probability distribution function (PDF) is $\mathcal{P}(\varepsilon_1, \ldots, \varepsilon_N) = \prod_{i=1}^{N} P(\varepsilon_i)$, where N is number of sites and $P(\varepsilon_i)$ is one-site distribution. One can choose continuous box PDF for the functions in the product: $P(\varepsilon_i) = \frac{1}{\Delta}\theta\left(\frac{\Delta}{2} - |\varepsilon_i|\right)$, where Δ is the disorder bandwidth and θ is the step function. For each distribution, one can solve the multisite DMFT problem, obtain the local GF for each site $G_{ii\sigma}(\omega)$, find the DOS $A_{i\sigma}(\omega) = -\frac{1}{\pi} G_{ii\sigma}(\omega)$ and average the DOS over different distributions. One of the possible averages over disorder is the arithmetic average that gives for the DOS

$$\langle A_{i\sigma}(\omega)\rangle = \int \prod_{j=1}^{N} d\varepsilon_j P(\varepsilon_i) A_{i\sigma}(\omega; \{\varepsilon_1, \ldots, \varepsilon_N\}). \tag{4.59}$$

Unfortunately, the problem with arithmetic disorder average is that it does not work well when the localization length is smaller than the size of the system. In this case, during the time evolution the particle cannot explore the full phase space (cannot probe all possible random distributions), so the system is non-self-averaging. Anderson proposed [87] instead of the arithmetically averaged DOS to find the most probable, or "typical," DOS $A_\sigma^{\mathrm{typ}}(\omega)$, such that for a given observable $P(A)$, the PDF $P\left[A_\sigma^{\mathrm{typ}}(\omega)\right]$ has the global maximum. However, usually the number of sites is very large and it is difficult to get an accurate averaging of the probability function. Thus, one should use an approximation for the typical value. Geometric mean is one of such approximations where one puts

$$A_\sigma^{\mathrm{typ}}(\omega) \approx A_\sigma^{\mathrm{geom}}(\omega) \equiv e^{\langle A_{i\sigma}(\omega)\rangle} \tag{4.60}$$

(the average in the exponent is defined in Eq. (4.59)).

Dobrosavljevic and Kotliar [90] developed the corresponding approach in the framework of DMFT and applied it to the Hubbard model at a simplified case of $U = \infty$. In the last work, the impurity problem was solved for different sites (Anderson impurity model ensemble with various ε_i). For the order parameter describing the MIT they used typical DOS (4.60) at the Fermi energy. Obtained disorder dependencies of the average and typical DOS at the Fermi energy in the interacting and noninteracting cases are shown in Fig. 4.27. As it follows from this figure, the arithmetic DOS does not disappear, i.e., does not show a transition to the insulating phase as disorder increases, while the typical DOS demonstrates such a transition at the disorder bandwidth $\Delta \sim 11\, t$ (in Ref. [90], the letter W is used instead of Δ). Quite remarkably, the typical quasiparticle weight vanishes linearly near the transition (Fig. 4.27c), similar to the case of the Mott transition. Another interesting result of work [90]—a disorder-induced transition to non-Fermi liquid regime at $W_{\mathrm{NFL}} = 7\, t$ (found from the divergence of the specific heat coefficient γ and divergence of local susceptibility at $T = 0$).

Within the same approach of stochastic DMFT, disorder bandwidth–Coulomb repulsion phase diagram was obtained by using typical DOS in work [92] (shown in the left Fig. 4.28). In addition to a "guess" diagram in Fig. 4.26, the stochastic DMFT diagram includes two more regions—a crossover region with Hubbard sub-bands in the DOS and a (hysteresis) coexistence region where the critical values of the transition U and Δ depend on the direction from which the parameters approach the critical region. The evolution of the local DOS as the system moves through the crossover and coexistence regions is shown in the right Fig. 4.28.

Extension of the analysis above on the spin-polarized case at half-filling was done by Byczuk et al. in Ref. [93]. Probably, the most important result of this paper is disorder-induced transitions from an AMF insulator to a PM metal and a PM

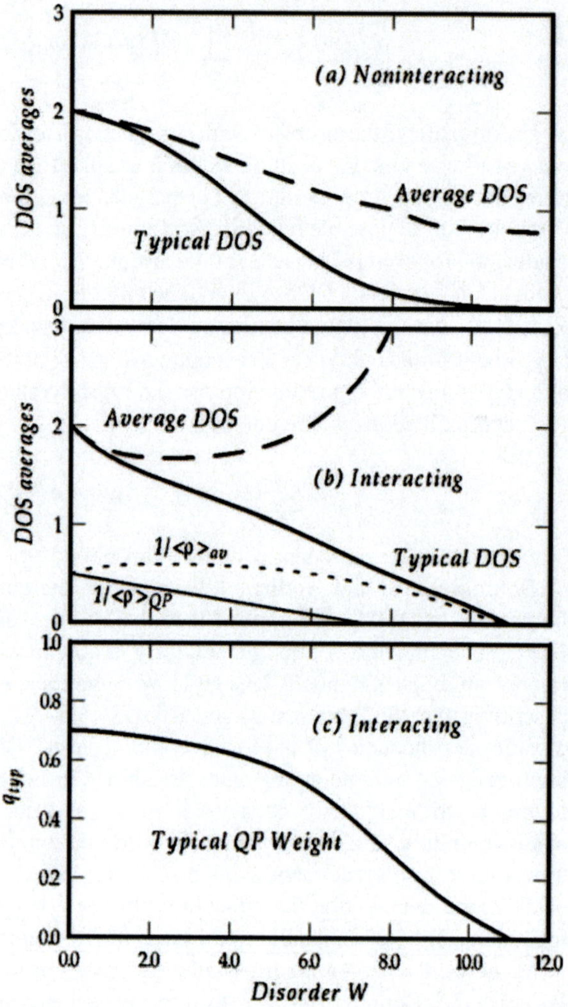

Fig. 4.27 Disorder dependencies of the average and typical DOS in the Hubbard model at the Fermi energy in the noninteracting (**a**) and interacting (**b**) cases obtained with stochastic DMFT (Bethe lattice, $U = \infty$). In (**c**), typical quasiparticle weight as a function of the disorder is shown. Reprinted Fig. 2 with permission from (V. Dobrosavljevic, G. Kotliar, Phys. Rev. Lett. **78**, 3943 (1997)). Copyright 1997 by the American Physical Society. https://doi.org/10.1103/PhysRevLett.78.3943

insulator states, i.e., destruction of the AFM order by disorder. In addition to these AFM and PM phases, an intermediate AF metal phase (with a finite DOS at the Fermi energy and a nonzero magnetic moment) was found (Fig. 4.29).

Among many other related studies, work [94] is one of the most interesting. In this paper, optical and DC conductivities in the Hubbard model at half-filling near the MIT and in presence of disorder were calculated with stochastic DMFT. It was found that the local random potential generates an effective carrier doping and makes the system more metallic, in agreement with recent experimental data on charge-transfer salts (for references, see work [94]).

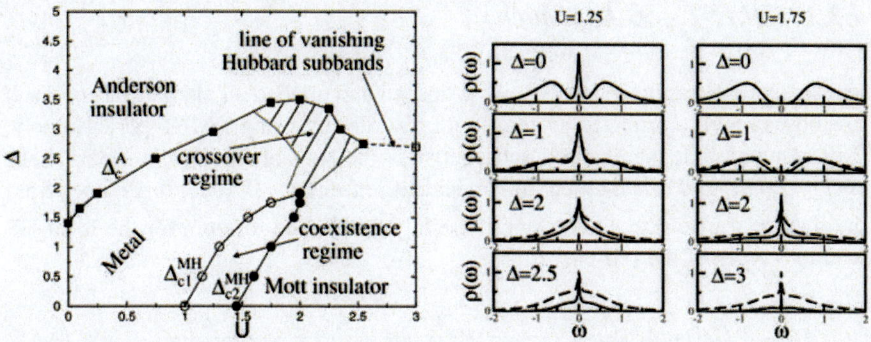

Fig. 4.28 Left: Disorder bandwidth—U phase diagram of the Hubbard model obtained with stochastic DMFT. Right: The geometrically (solid curves) and arithmetically (dashed curves) averaged local DOS for different values of U: left—small "metal" U, right—large "Mott insulator" U (Bethe lattice, mapping on the Anderson impurity model). Reprinted Figs. 1 and 4 with permission from (K. Byczuk, W. Hofstetter, and D. Vollhardt, Phys. Rev. Lett. **94**, 056404 (2005)). Copyright 2005 by the American Physical Society. https://doi.org/10.1103/PhysRevLett.94.056404

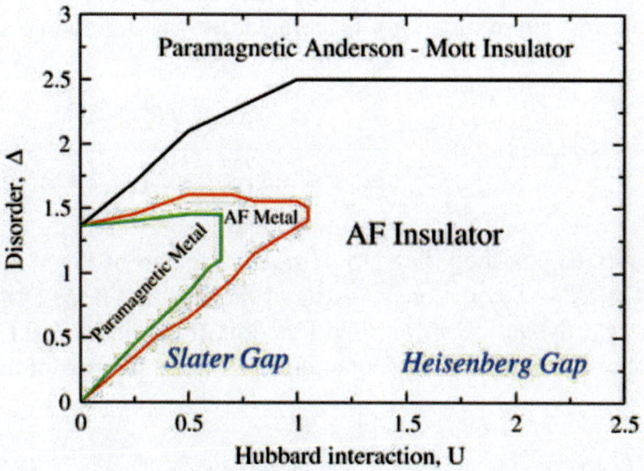

Fig. 4.29 Disorder strength—Hubbard U phase diagram of the Hubbard model at half-filling (Bethe lattice, results are obtained by mapping the problem on Andeson impurity model). The diagram includes regions with Slater (Heisenberg) gaps in the case of no disorder at small (large) with respect to bandwidth values of U. The corresponding solutions can be obtained by perturbation theory in U and $1/U$ for the Hubbard (small U) and Heisenberg (large U) models. Reprinted Fig. 1 with permission from (K. Byczuk, W. Hofstetter, and D. Vollhardt, Phys. Rev. Lett. **102**, 146,403 (2009)). Copyright 2009 by the American Physical Society. https://doi.org/10.1103/PhysRevLett.102.146403

4.7.2 DMFT + Σ Approach

Kuchinskii, Nekrasov, and Sadovskii proposed and applied an alternative approach to study strongly correlated systems with disorder by using DMFT [91]. Namely, they assumed that the electron self-energy is the sum of the DMFT (local) self-energy $\Sigma(i\omega_n)$ and of a momentum-dependent self-energy $\Sigma_{\vec{k}}(i\omega_n)$ that comes from the scattering of electrons on impurities (Fig. 4.30). The formula for the local GF with two types of the self-energy is:

$$G_{ii}(i\omega_n) = \frac{1}{N}\sum_{\vec{k}} \frac{1}{i\omega_n + \mu - \varepsilon\left(\vec{k}\right) - \Sigma(i\omega_n) - \Sigma_{\vec{k}}(i\omega_n)}. \tag{4.61}$$

The authors of Ref. [91] assumed that the PDF has the Gaussian form: $P(\varepsilon_i) = \frac{1}{\sqrt{2\pi}\Delta} e^{-\frac{\varepsilon_i^2}{2\Delta^2}}$. They also used the simplest form for the impurity-scattering (momentum-dependent) part of the self-energy, neglecting the crossing diagrams. In this case, the momentum-dependent part of the self-energy (that, as a matter of fact, is momentum-independent in this approximation) is defined by the following self-consistent equation:

$$\Sigma_{\vec{k}}(i\omega_n) = \Delta^2 G_{ii}(i\omega_n) \equiv \Delta^2 \sum_{\vec{k}} \frac{1}{i\omega_n + \mu - \varepsilon\left(\vec{k}\right) - \Sigma(i\omega_n) - \Sigma_{\vec{k}}(i\omega_n)} \tag{4.62}$$

(it corresponds to the one-loop diagram in the middle part of Fig. 4.30).

The set of DMFT + Σ equations consists of the standard three DMFT equations for the self-energy, dynamical mean-field function, (modified) local GF Eq. (4.61), and the additional defect self-energy equation (4.62). The solution of these equations

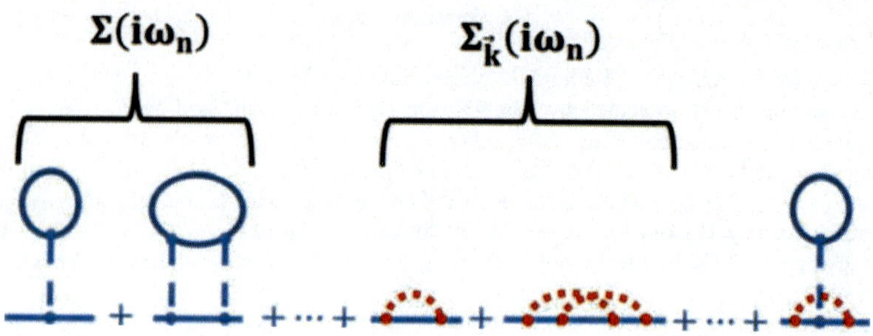

Fig. 4.30 A set of typical DMFT + Σ approach (skeleton) diagrams for the self-energy: DMFT diagrams (left), impurity scattering diagrams (center), and interference Coulomb repulsion + impurity-scattering diagrams (right). The blue dashed lines correspond to Coulomb scattering and the red-dot lines—to impurity scattering. In DMFT + Σ, only the first two types of diagrams are taken into account (see also Ref. [91])

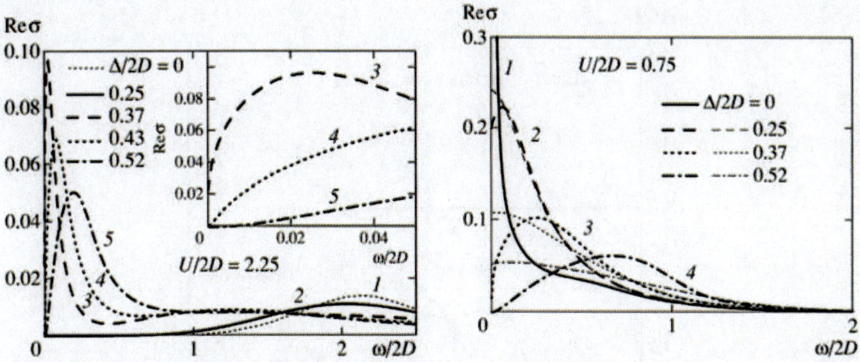

Fig. 4.31 Optical (dynamical) conductivity in the half-filled Hubbard–Anderson model as a function of disorder strength in the insulating (left) and metallic (right) phases (DMFT + Σ, semi-elliptic free-electron DOS, NRG impurity solver). From [91]

demonstrated [91] that, similar to stochastic DMFT case (right Fig. 4.28), disorder leads to the Mott insulator-to-metal transition, and in the case of strongly correlated metal it leads to a broadening of the quasiparticle peak in the spectral function (for details, see paper [91]). These features are reflected in the disorder dependence of the optical conductivity

$$\sigma(\omega) = -\lim_{\vec{q}\to 0} \frac{ie^2\omega}{\vec{q}^2} \chi^R\left(\omega, \vec{q}\right). \tag{4.63}$$

Namely, adding a disorder into a Mott insulator gives a finite conductivity within the Hubbard gap (left Fig. 4.31). This is a consequence of the zero-energy quasiparticle peak in the DOS present at finite disorder. In the metallic phase, disorder leads to a broadening of the low-frequency peak in the dynamic conductivity (right Fig. 4.31).

The disorder bandwidth—Coulomb repulsion phase diagram of the Anderson–Hubbard model obtained with DMFT + Σ is shown in Fig. 4.32. The phase diagrams obtained with the stochastic DMFT [92] (left Fig. 4.28) and DMFT + Σ [91] (Fig. 4.32) approaches have many similarities, but the last approach does not give a crossover region and the width of the coexistence region in Fig. 4.32 grows as disorder bandwidth increases. These differences can be attributed to the difference in the approaches, including their accuracy, e.g., because of a simplified one-loop approximation (4.63) for the electron self-energy used in DMFT + Σ. In general, both approaches are powerful and promising tools to study the effects of disorder in strongly correlated materials.

From recent studies, we would like to mention an analysis of the effects of disorder in a low-dimensional (2D) Hubbard model with a DMFT + CPA approach [95], analysis of the properties of the Hubbard model with a spin-selective disorder (stochastic DMFT) [96], and analytical (linearized DMFT) analysis of the DOS and

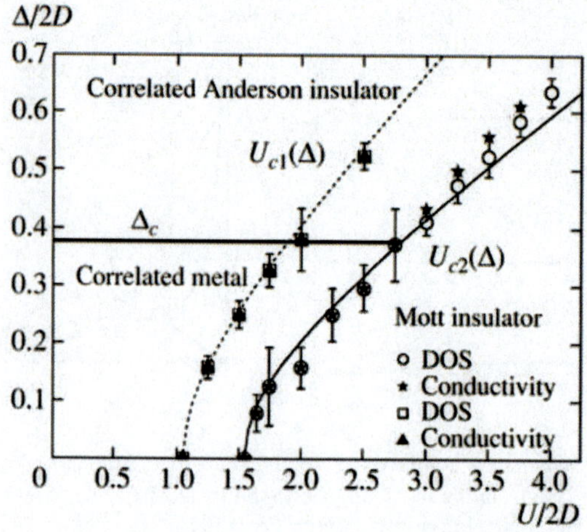

Fig. 4.32 The disorder bandwidth—U phase diagram of the Anderson–Hubbard model obtained with DMFT + Σ (semielliptic free-electron DOS, NRG impurity solver). The horizontal solid line defines the line of the Anderson transition. The non-horizontal curves define the boundaries of the coexistence region. These curves were obtained from the DOS and conductivity calculations (defined by different symbols in the figure). From Ref. [91]

the disorder–Coulomb repulsion phase diagram of the Anderson-Hubbard model on Bethe lattice [97].

4.8 Non-Fermi Liquid (Bad-Metal) Phase

Physical properties of systems of noninteracting electrons are rather simple: electrons occupy lowest possible energy states (only one fermion can occupy any state described by an individual set of quantum numbers, according to the Pauli principle). Thus, the states up to the state with some maximum energy, defined by the number of electrons in the system and called Fermi energy ε_F, are filled, and the other states are empty. Occupancy of the electronic states is described by the Fermi distribution function (left Fig. 4.33). Energies of excited electrons are above and often are close to ε_F. Similarly, momenta of the excited states $\vec{k}$ are above and often are close to the Fermi momentum $\vec{k}_F$ (we are talking about the moduli of momenta). The low-energy excitations dominate in the excitation spectrum when the temperature of the system is not high and/or the perturbation is not strong. In many physical situations these excitations define properties of the system, therefore they deserve a detailed study. The lifetime of the excited noninteracting electron in each state is infinite (due to absence of scattering processes), which is reflected in delta-function peaks in the spectral function (right Fig. 4.33, black line).

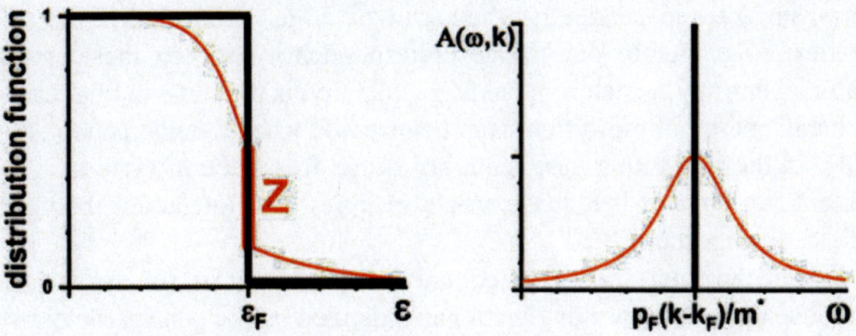

Fig. 4.33 Left: The distribution function for electrons in the noninteracting (black curve) and interacting (red curve) systems (the jump Z in the distribution function in the last case can be regarded as an order parameter of the Fermi liquid, see below). Right: the same for the spectral function of electrons

The task of finding low-energy excited states of the system becomes much more complicated when the interaction between electrons is turned on. In this case, in the Hamiltonian there is a competition between the kinetic energy term that forces occupation of low-energy states and potential energy term that often leads to a more spreaded-in-energy occupancy of states (left Fig. 4.33, red curve) and makes the lifetime of quasiparticles finite (broadened spectral function, right Fig. 4.33, red curve). In metals with typical electron densities both terms appear to be equally important. In fact, description of materials is even more difficult than that, because the uncertainty principle forbids an accurate definition of the momentum and coordinate of electrons [98].

To understand how properties of the noninteracting system, including the excited states, change when the interaction is taken into account it is convenient to turn on the interaction and slowly increase it to the real physical value. A recipe on how to track the evolution of the free-electron excitations (with fixed/labeled quantum numbers) as the interaction is turned on and grows was provided by Landau in the framework of his celebrated theory of Fermi liquids [99]. He postulated the concept of *adiabatic continuity*: the labels associated with eigenstates are more robust with respect to perturbations than the eigenstates themselves (for over-review, see Ref. [98]). In the case of electrons, it means that when the interaction is slowly turned on, there is one-to-one correspondence between the low-energy eigenenergies of the interacting and noninteracting electrons, and this correspondence holds even when the interaction reaches its maximum (real physical) value. In other words, one can assume that interacting electron and hole excitations (more accurately, *quasiparticles*) have the same quantum numbers as the corresponding noninteracting electron and hole excitations, though various characteristics of both types of the excited particles, like their mass, can be different. The concept of quasiparticle "cousins" of free excited electrons is very helpful. It can allow one to use many results from the physics of free electrons. For example, one can immediately write down the temperature dependencies of the specific heat and susceptibility of the interacting system, since the derivation of the corresponding formula only requires presence of the Fermi surface, independently whether involved fermions are free electrons or quasiparticles. Also, due to one-to-one correspondence between these states, the configurational entropy, which is an additive quantity, is the same in both cases. On the other hand, energy of the system is not conserved when the interaction is turned on: usually, in the interacting case, contrary to the free-electron systems, the total energy is not the sum of the quasiparticle energies (the interaction between the quasiparticles is important).

To construct the total energy functional for quasiparticles (in the low-energy limit), Landau suggested that this functional is defined by the quasiparticle distribution $\delta n_{\vec{k}\sigma}$, their momenta, renormalized quasiparticle mass m* (effect of the "resistance" of the disturbed ground state of the Fermi sea, and also of the lattice structure and phonons), and by the distribution function of the quasiparticles $f_{\vec{k}\sigma,\vec{k}'\sigma'}$. For the isotropic systems, the proposed energy functional is

$$E = \sum_{\vec{k},\sigma} \frac{\vec{p}_F}{m^*}\left(\hbar\vec{k} - \vec{p}_F\right)\delta n_{\vec{k}\sigma} + \frac{1}{2}\sum_{\vec{k}\vec{k}',\sigma\sigma'} f_{\vec{k}\sigma,\vec{k}'\sigma'}\delta n_{\vec{k}\sigma}\delta n_{\vec{k}'\sigma'} \tag{4.64}$$

Using Eq. (4.64) one can obtain, e.g., results for the specific heat and paramagnetic (Pauli) susceptibility of the interacting system:

$$C_V = \frac{1}{3}\frac{m^* p_F}{\hbar^3}\kappa_B^2 T, \quad \chi = \frac{m^* p_F}{\pi^2\hbar}\frac{1}{1+F_0^a}\mu_B^2. \tag{4.65}$$

These results are very similar to results in the noninteraction case, except $m \to m^*$ and a nonzero value of the Landau parameter F_0^a. This parameter is the zero-order coefficient in the Legendre expansion of the antisymmetric-in-spin distribution function $f_{\vec{k}\sigma,\vec{k}'\sigma'}$ over the angle between momenta multiplied by the DOS at the Fermi energy. Importantly, in the isotropic systems the momentum part of $f_{\vec{k}\sigma,\vec{k}'\sigma'}$ depends only on the angle between the direction of the momenta, the absolute values of the momenta are almost constants, since they are close to the isotropic Fermi energy. It must be noted that in addition to "renormalized" coefficients in the formulas for the noninteracting electrons, Landau theory also gives new, most notably collective interaction-induced excitations, like zero-sound oscillations of charges with the restoring forces described by $f_{\vec{k}\sigma,\vec{k}'\sigma'}$.

One can estimate the lifetime of quasiparticles with energy $\varepsilon > \varepsilon_F$ by using Fermi's golden rule:

$$\frac{1}{\tau_\varepsilon} = \frac{2\pi}{\hbar}\sum_f |V_{if}|^2\delta(\varepsilon - \varepsilon_F) \sim \frac{\pi}{\hbar}V^2\rho^3(\varepsilon_F)(\varepsilon - \varepsilon_F)^2. \tag{4.66}$$

To obtain the right part of this equation, the scattering matrix elements $|V_{if}|^2$ were approximated by the average value V^2. Thus, quasiparticles with energy close to the

Fermi energy (small $\varepsilon - \varepsilon_F$) have a long lifetime, i.e., they are well-defined. On the other hand, the lifetime of the high-energy states is very short and the principle of adiabatic continuity is not applicable in this case (quasiparticles decay faster than the interaction is completely turned on). At finite temperatures, the situation changes: the value of temperature defines the minimal energy scale and the low-energy quasiparticle scattering rate (and as a result, conductivity) is proportional to T^2. This means that at high temperatures, the scattering rate will be large and the quasiparticle lifetime short, i.e., the Fermi liquid theory is not valid in this case as well. To summarize, the validity of Fermi liquid theory is limited by low excitation energies and low temperatures.

Now, let us ask ourselves what could be the order parameter that characterizes properties of the Fermi liquid as the interaction is adiabatically turned on? As we already discussed, the quantum numbers like spin and angular momenta of the quasiparticles remain the same during the turning-on process. On the other hand, the wave function of electron changes and adiabaticity assumes that there is a continuous evolution of the wave function during this process, from the free-electron wave function to the quasiparticle one. The quasiparticle wave function $|\psi_{qp}\left(\sigma, \vec{k}\right)\rangle$ must contain some fraction of the "initial" free-electron wave function $|\psi_0\left(\sigma, \vec{k}\right)\rangle$ as the system adiabatically evolves:

$$|\psi_{qp}\left(\sigma, \vec{k}\right)\rangle = \sqrt{Z}|\psi_0\left(\sigma, \vec{k}\right)\rangle + \ldots, \tag{4.67}$$

where the corresponding weight coefficient Z can be considered as measurement of continuity of the adiabatic connectivity of the electron and quasiparticle states, i.e., as quantifying measure of the "Fermi-liquidness" of the system. The quasiparticle weight Z shows up in the distribution functions as a step at the Fermi energy (left Fig. 4.33). Also, it shows up in the integrated spectral function with the peak around momentum $\vec{k}$ (the integral equal Z). The integrated spectral function describes the probability that electrons can be found in the eigenstate with momentum $\vec{k}$ (see right Fig. 4.33). The width of the peak is $\sim \left(\vec{k} - \vec{k}_F\right)^2$—it defines the finite lifetime of the quasiparticle. Decrease of Z in the interacting system ($Z = 1$ in the noninteracting case) is related to the fact that electrons can be spread over many eigenstates, and the probability to find it in given (initial) state decreases.

Thus, following Ref. [98], one can summarize the main features of the Fermi Liquid:

- Equilibrium properties are similar to the noninteracting system of electrons with renormalized parameters.
- Lifetime of the low-energy excitations—fermion quasiparticles—is inversely proportional to the largest of two numbers: the square of the quasiparticle energy ε^2 and T^2.

- Overlap of the quasiparticle and the initial free-electron state at the Fermi surface Z can be considered as the order parameter of the Fermi liquid. Larger Z—the system is "more Fermi-liquid" (more similar to the free-electron system).
- New collective modes, like zero sound, generated by the interaction can be excited in the Fermi Liquid.

However, this nice and rather simple picture may be destroyed when the interaction between electrons is not weak (in general case, not only in the Hubbard model) and the behavior of the system cannot be described by the theory of Fermi Liquids anymore. Some examples are:

- Metals close to a quantum phase transition that takes place at a very low temperature (near critical point), when strong fluctuations cause also strong electron–electron scattering.
- One-dimensional systems, where due to an enhanced (result of the spatial-confinement) electron–electron interaction electrons decay on spinon and holon quasiparticles—spin and charge excitations.
- Systems with magnetic impurities, like two-channel Kondo model, where scattering of two electrons on an impurity produces a "half-of-electron" state, or Kondo model with disorder, where scattering of electrons on impurity is very strong and Fermi quasiparticles cannot be formed.

Since we are dealing with a high-dimensional non-disordered Hubbard model, we will focus mostly on the first case—the Hubbard model in a state near a zero-temperature (quantum) phase transition. Before discussing the non-Fermi liquid properties of the DMFT solutions, we would like to make several general remarks on the quantum phase transitions. When the system approaches a second-order phase transition, the fluctuations of the order parameter slow down and take place over very large wavelengths. Thus, fluctuations can be easily excited/generated by moving quasiparticles resulting in strong quasiparticle–quasiparticle scatterings. In the case of zero-temperature phase transitions, the fluctuations are quantum, since they are defined by quantum mechanics.

Possibility of the non-Fermi liquid behavior of the Hubbard model in infinite dimensions was studied with DMFT in Refs. [100–102]. As mentioned in Ref. [100], it is important that at high energy scale, which corresponds to short distances and short time scales, the local on-site excitations are important. However, in many standard materials, where the kinetic energy is dominating and the Fermi energy is of a very high scale, the local interactions can be neglected, and the Fermi-liquid picture is valid (excitation energies are small comparing to the Fermi energy). In strongly correlated materials, with narrow bands and small Fermi energy comparable or smaller than local Coulomb interaction, the last statement is not correct. Mott insulators are not in the Fermi liquid state because localized electrons are very different from itinerant-free electrons. In correlated metals, electrons are not localized, but correlation effects play a very important role. In particular, correlations result in a very low effective degeneracy temperature (below which the system is in a degenerate Fermi gas state with long-lived excitations, one of the main characteristics of the Fermi liquid). Thus, correlated metal can be in a Fermi liquid state only at very low temperatures (low energy scale), and at higher temperatures, it is a non-Fermi Liquid, a "bad metal." It means the quasiparticles are destroyed as the system is heated, which results in a transfer of the spectral weight from low-energy to intermediate- and high-energy excitations.

High-energy physics is an important part of the physics of correlated electrons (including the bad metal phase mentioned above). In order to correctly describe high-energy (on-atom) excitations, one needs to properly describe the atomic limit in the Hubbard model, including multiple excitations within the atom. In this sense, DMFT is a good technique, since it can describe both atomic (high-energy) and itinerant (low-energy) excitations, which means that it handles different timescales as well.

There is another important case where DMFT is an appropriate tool. It is known that correlations lead to a narrowing of the bandwidth of the low-energy part of the DOS that is connected with enhancement of the quasiparticle effective mass. To describe this situation, Brinkman and Rice [103], using Gutzwiller approximation (and neglecting the effect of superexchange on quasiparticles), showed that the quasiparticle weight Z disappears and the effective mass diverges as $1/Z$ when the Coulomb U approaches the critical value of MIT, i.e., when the correlations increase. However, the analysis and conclusions in Ref. [103] have a limited applicability, since the higher part of the quasiparticle spectra was not included in the analysis. On the other hand, as it was mentioned above, DMFT can describe both low-energy (zero-energy peak) excitations as well as high-energy (inter-Hubbard sub-band) excitations that become dominant with U increasing (when the zero-energy quasiparticle peak disappears).

To summarize, DMFT can handle both particle-like (on-atom, high-energy) and wave-like (itinerant, low-energy) excitations, including temperature effects through the temperature-dependent dynamical mean-field $\Delta(\omega)$. The imaginary part of this function has spectral weight mostly *at high energies* that correspond to transitions between the Hubbard sub-bands. Usually, several atomic states are involved in the transitions. For example, at half-filling opposite spin atomic states contribute to the transitions that result in magnetic moment fluctuations when charge excitations are suppressed (at $T < U$). This situation again can be described with DMFT.

At lower temperatures, the local moments, which may form in metallic phase, are screened due to the spin-flipping processes between the atom and bath electrons, i.e., due to Kondo effect. These processes become enhanced at low energies. In other words, the quasiparticles at low energies and temperatures form due to the self-consistent Kondo mechanism, and this process can be described by DMFT. Below even lower temperature T_{FL}, corresponding to the self-consistent Kondo scale, a local Fermi Liquid theory is applicable. Thus, it is important to understand below what energy scale the quasiparticles form. This scale is defined by screening of the atomic multiplet and DMFT is a proper tool for this scale as well (for further details, see Ref. [100]).

Therefore, one expects to have long-lived local Landau Fermi Liquid quasiparticles with energies at and near zero-energy quasiparticle peak in the DOS at temperatures $T < T_{FL}$ and bad metal (in the sense of transport) state at $T > T_{FL}$.

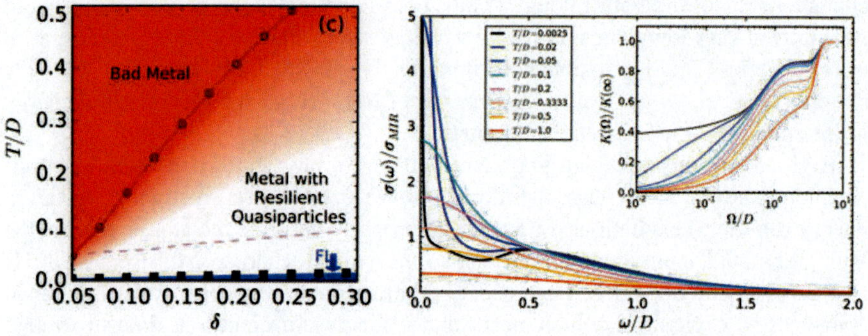

Fig. 4.34 Left: Temperature-doping phase diagram of the one-band Hubbard model with the Fermi liquid (FL), resilient quasiparticle, and bad metal regions. Right: optical conductivity as function of frequency at doping $\delta = 0.2$ and different values of temperature. The Fermi liquid T_{FL} and bad metal temperatures in this case are equal to approximately 0.05D and 0.4D, correspondingly. Inset: integrated to a given frequency spectral weight (normalized by the kinetic energy) (U = 4D, NRG and continuous-time QMC (CT-QMC) solvers). Reprinted Figs. 1c and 5 with permission from (X. Deng, J. Mravlje, R. Žitko, M. Ferrero, G. Kotliar, and A. Georges, Phys. Rev. Lett. **110**, 086401 (2013)). Copyright 2013 by the American Physical Society. https://doi.org/10.1103/PhysRevLett.110.086401

The weight Z disappears as one approaches the critical value of U, while the effective mass increases as $1/Z$. The quasiparticle decay rate (the inverse lifetime) $ZIm\Sigma(\omega)$ is proportional to T^2 and ω^2. The Fermi-liquid energy scale T_{FL} is smaller than another scale—the width of the quasiparticle peak ZD, or the (Brinkman–Rice) quasiparticle reduced bandwidth (see, e.g., Ref. [100]). As it follows from the mentioned above, there are three different phases to be approximately separated by energy scales T_{FL} and ZD: the Fermi liquid at $T < T_{FL}$, bad metal at $T > ZD$, and an intermediate, resilient quasiparticle, phase at temperatures between T_{FL} and ZD. To obtain the temperature-doping phase diagram that includes three phases, one can calculate conductivity as a function of T and doping. Then, the Fermi Liquid line can be defined as the boundary below which the resistivity has the Fermi Liquid dependence on temperature, $\rho \sim T^2$, and the bad metal phase takes place down to the temperature at which the resistivity reaches the Mott–Ioffe–Regel value $\rho \sim h/e^2 k_F$ (above this temperature, there is a significant decrease in the slope of resistivity as function of temperature). Calculations to find the temperature-doping phase diagram of the single-band Hubbard model at U = 4D were performed in Ref. [101]. The result is shown in the left Fig. 4.34. As it was found in this paper, at low temperatures (in the Fermi Liquid phase), optical conductivity demonstrates Drude and Hubbard sub-band peaks and when temperature reaches resilient quasiparticle regime, optical conductivity shows a transfer of the spectral weight that involves Drude and low-energy (infrared) regions. At even higher, bad metal, temperatures the optical conductivity is large in a much wider energy range (right Fig. 4.34).

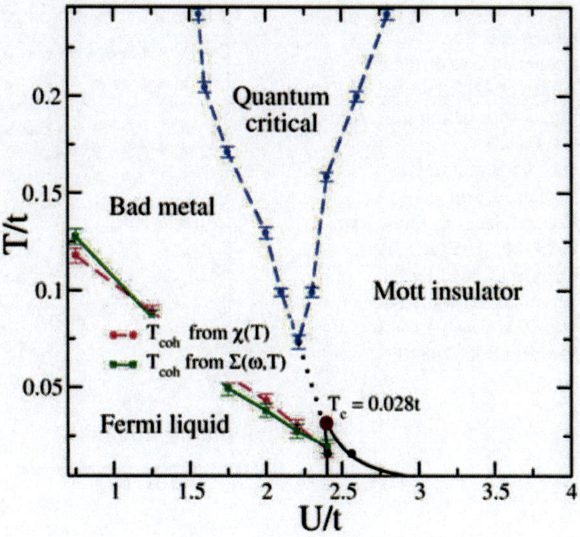

Fig. 4.35 *U–T* phase diagram of the half-filled Hubbard model obtained from the results for the local dynamical susceptibility (semicircular free-electron DOS, CT-QMC solver). Reprinted Fig. 1 with permission from (N. Dasari, N.S. Vidhyadhiraja, M. Jarrell, and R.H. McKenzie, Phys. Rev. B **95**, 165,105 (2017)). Copyright 2017 by the American Physical Society. https://doi.org/10.1103/PhysRevB.95.165105

The bad metal phase is also found in the half-filled Hubbard model at high temperatures. In particular, in Ref. [102] the local dynamic susceptibility was calculated at different values of *U* and *T* in the half-filled Hubbard model and it was found that in addition to familiar Fermi Liquid and Mott insulator phases, *U–T* diagram contains two other nontrivial phases—the bad metal and quantum critical region [102] (Fig. 4.35).

From the local dynamic susceptibility, one can also find the spin dephasing rate

$$\Gamma = \lim_{\omega \to 0} \frac{\omega Re\chi(\omega = 0)}{Im\chi(\omega)}. \tag{4.68}$$

This quantity is very useful for calculations of the phase diagram. Namely, the Fermi Liquid–bad metal transition temperature can be found from the change of the dephasing rate: at low-Ts (Fermi-liquid region) the rate is almost temperature-independent, while in the high-temperature bad-metal phase it decays with temperature (also, the Fermi liquid–bad metal transition temperature T_{FL} (T_{coh}) can be calculated [102] as the temperature above which susceptibility is no longer inversely proportional to temperature and the electron decay rate $(-Z)Im\Sigma(0)$ – to T^2). In the quantum critical region, the scattering rate was found to be linear in temperature, Γ~$0.4T$, and in the Mott-insulator phase, the rate approaches zero value as temperature goes to zero (Fig. 4.36).

In the quantum critical region, it was also found that the dynamical susceptibility shows ω/T scaling, characteristic for quantum criticality (other response functions, like optical conductivity, in the Hubbard model should demonstrate the same scaling in the critical region). Another important property of the critical region—absence of the Fermi-liquid zero-energy quasiparticle peak in the DOS (Fig. 4.37) (for more

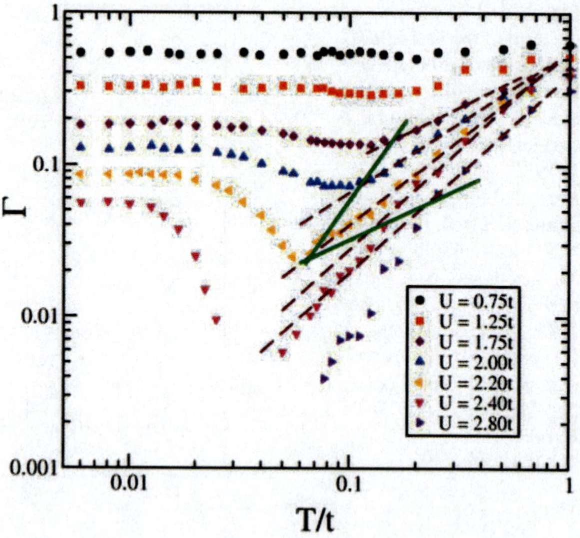

Fig. 4.36 Spin dephasing rate as function of temperature at different values of U. Reprinted Fig. 2 with permission from (N. Dasari, N.S. Vidhyadhiraja, M. Jarrell, and R.H. McKenzie, Phys. Rev. B **95**, 165,105 (2017)). Copyright 2017 by the American Physical Society. https://doi.org/10.1103/PhysRevB.95.165105

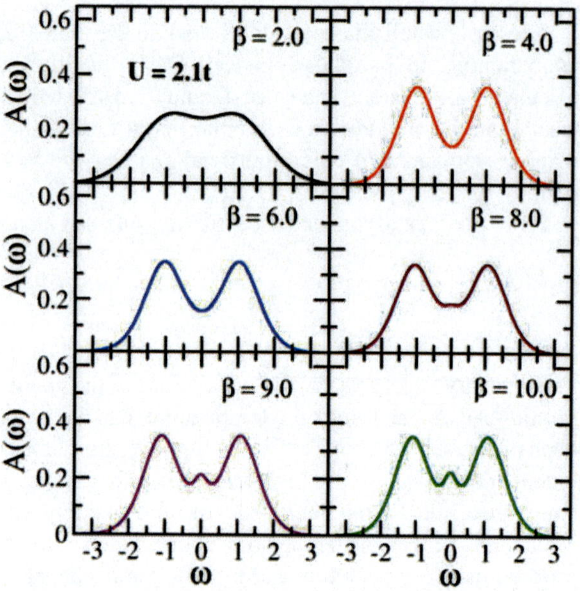

Fig. 4.37 Electron spectral function obtained at different values of temperature in the quantum critical region at $U = 2.1t$. Reprinted Fig. 6 with permission from (N. Dasari, N.S. Vidhyadhiraja, M. Jarrell, and R.H. McKenzie, Phys. Rev. B **95**, 165,105 (2017)). Copyright 2017 by the American Physical Society. https://doi.org/10.1103/PhysRevB.95.165105

DMFT studies of the properties of the system in the non-Fermi Liquid phase in the case of Hubbard model, see references in works [100, 102]).

Several interesting results on the non-Fermi Liquid properties of more complex Hubbard models were also obtained with DMFT. In particular, it was shown [111] that multi-orbital (Hubbard) systems (which is always the case for real materials) with strong Hund-rule coupling metals (e.g., 4d transition–metal oxides and iron superconductors) can be in a strong-correlation regime even when they are not close to MIT. For such systems, critical U increases with growth of the Hund's coupling and the Fermi-liquid energy scale, i.e., T_{FL}, is suppressed by this coupling since it is difficult to screen a local atomic multiplet. Another interesting result in a multi-orbital (three-orbital) case: Mott transitions at integer numbers of electrons per site that happen in frozen-momentum phase (spin–spin correlation function is nonzero as time argument goes to infinity) [105]. As it was shown in the last paper, the electron self-energy $\Sigma(\omega)$ on the critical charge–density (transition) lines demonstrates a non-Fermi liquid frequency dependence.

In the case of several nonequivalent sites (cluster DMFT discussed below), in particular in systems with strong nearest-neighbor superexchange, one needs to take into account momentum dependence of the self-energy. This leads to a rather different mechanism of evolution of the metallic system to the Mott regime as doping decreases (including the case of well-known pseudogap phase in cuprates). It must also be noted that it is much more difficult to solve the cluster DMFT equations in comparison to the one-site case since at low energies and low dopings, it requires a rather high momentum resolution (for over-review and references, see Ref. [101]).

There are also ongoing DMFT studies on possibility of non-Fermi Liquid phase and of its properties in the 2D Hubbard model. In particular, as it was shown in Ref. [106] there are signs of a non-Fermi Liquid phase in the 2D Hubbard model when the van Hove singularity in the DOS is close to the Fermi level. A hidden Fermi Liquid phase above T_{FL} with the Fermi Liquid temperature dependence (T^2) of the quasi-particle scattering rate was found in Ref. [107]. Finally, for a DMFT analysis of similar physical problems using t–J and Kondo model, see Refs. [108, 109] and Ref. [110], correspondingly.

References

1. Jarrell, M.: Phys. Rev. Lett. **69**, 168 (1992)
2. Zhang, X.Y., Rozenberg, M.J., Kotliar, G.: Phys. Rev. Lett. **70**, 1666 (1993)
3. Hubbard, J.: Proc. R. Soc. London A. **276**, 238 (1963)
4. Hubbard, J.: Proc. R. Soc. London A. **281**, 401 (1964)
5. V. Anisimov, Yu. Izyumov, Electronic structure of strongly correlated materials (Springer-Verlag, Berlin Heidelberg Series in Solids, 2010).
6. Gorkov, A.A., Dzyaloshinskii, L.P., Abrikosov, I.Y.: Quantum field theoretical methods in statistical physics. Pergamon Press, New York (1965)
7. Schlipf, J., Jarrell, M., van Dongen, P.G.J., Blümer, N., Kehrein, S., Pruschke, T., Vollhardt, D.: Phys. Rev. Lett. **82**, 4890 (1999)
8. Bulla, R.: Phys. Rev. Lett. **83**, 136 (1999)
9. Kotliar, G.: Science. **302**, 67 (2003)
10. Bulla, R., Costi, T.A., Vollhardt, D.: Phys. Rev. B. **64**, 045103 (2001)
11. Feldbacher, M., Held, K., Assaad, F.F.: Phys. Rev. Lett. **93**, 136405 (2004)
12. Freericks, J.K., Devereaux, T.P., Bulla, R., Pruschke, T.: Phys. Rev. B. **67**, 155102 (2003)
13. Pruschke, T., Cox, D.L., Jarrell, M.: Phys. Rev. B. **47**, 3553 (1993)

14. Caffarel, M., Krauth, W.: Phys. Rev. Lett. **72**, 1545 (1994)
15. Kajuter, H., Kotliar, G., Moeller, G.: Phys. Rev. B. **53**, 16214 (1996)
16. Fischer, D.S., Kotliar, G., Moeller, G.: Phys. Rev. B. **52**, 17112 (1995)
17. Eckstein, M., Kollar, M., Potthoff, M., Vollhardt, D.: Phys. Rev. B. **75**, 125103 (2007)
18. Kotliar, G., Murthy, S., Rozenberg, M.J.: Phys. Rev. Lett. **89**, 046401 (2002)
19. Potthoff, M.: Eur. Phys. J. B. **32**, 429 (2003)
20. Georges, A., Kotliar, G., Krauth, W., Rozenberg, M.J.: Rev. Mod. Phys. **68**, 13 (1996)
21. Schweitzer, H., Czycholl, G.: Phys. Rev. Lett. **67**, 3724 (1991)
22. Georges, A., Florens, S., Costi, T.A.: J. Phys. IV France. **114**, 165 (2004)
23. Moeller, G., Si, Q., Kotliar, G., Rozenberg, M., Fisher, D.S.: Phys. Rev. Lett. **74**, 2082 (1995)
24. Lee, S.-S.B., von Delft, J., Weichselbaum, A.: Phys. Rev. Lett. **119**, 236402 (2017)
25. Mo, S.-K., Denlinger, J.D., Kim, H.-D., Park, J.-H., Allen, J.W., Sekiyama, A., Yamasaki, A., Kadono, K., Suga, S., Saitoh, Y., Muro, T., Metcalf, P., Keller, G., Held, K., Eyert, V., Anisimov, V.I., Vollhardt, D.: Phys. Rev. Lett. **90**, 186403 (2003)
26. Sekiyama, A., Fujiwara, H., Imada, S., Suga, S., Eisaki, H., Uchida, S.I., Takegahara, K., Harima, H., Saitoh, Y., Nekrasov, I.A., Keller, G., Kondakov, D.E., Kozhevnikov, A.V., Pruschke, T., Held, K., Vollhardt, D., Anisimov, V.I.: Phys. Rev. Lett. **93**, 156402 (2004)
27. Nunez-Fernandez, Y., Kotliar, G., Hallberg, K.: Phys. Rev. B. **97**, 121113(R) (2018)
28. Kugler, F.B., Lee, S.-S.B., Weichselbaum, A., Kotliar, G., von Delft, J.: Phys. Rev. B. **100**, 115159 (2019)
29. Prelovsek, P., Kokalj, J., Lenarcic, Z., McKenzie, R.H.: Phys. Rev. B. **92**, 235155 (2015)
30. Kauch, A., Pudleiner, P., Astleithner, K., Ribic, T., Held, K.: preprint arXiv:1902.09342v1 (2019)
31. Nishimoto, S., Gebhard, F., Jeckelmann, E.: J. Phys. Condens. Matter. **16**, 7063 (2004)
32. Gull, E., Reichman, D.R., Millis, A.J.: Phys. Rev. B. **82**, 075109 (2010)
33. Granath, M., Schott, J.: Phys. Rev. B. **90**, 235129 (2014)
34. Tagliavini, A., Hummel, S., Wentzell, N., Andergassen, S., Toschi, A., Rohringer, G.: Phys. Rev. B. **97**, 235140 (2018)
35. Taranto, C., Sangiovanni, G., Held, K., Capone, M., Georges, A., Toschi, A.: Phys. Rev. B. **85**, 085124 (2012)
36. Sato, T., Tsunetsugu, H.: Phys. Rev. B. **94**, 085110 (2016)
37. Blumer, N.: Mott-Hubbard metal-insulator transition and optical conductivity in high dimensions. PhD Thesis, University of Augsburg (2003)
38. Perepelitsky, E., Galatas, A., Mravlje, J., Zitko, R., Khatami, E., Shastry, B.S., Georges, A.: Phys. Rev. B. **94**, 235115 (2016)
39. Paiva, T., Scalettar, R.T., Huscroft, C., McMahan, A.K.: Phys. Rev. B. **63**, 125116 (2001)
40. Seki, K., Shirakawa, T., Yunoki, S.: Phys. Rev. B. **98**, 205114 (2018)
41. Merker, L., Costi, T.A.: Phys. Rev. B. **86**, 075150 (2012)
42. Khatami, E.: Phys. Rev. B. **94**, 125114 (2016)
43. Fuchs, S., Gull, E., Pollet, L., Burovski, E., Kozik, E., Pruschke, T., Troyer, M.: Phys. Rev. Lett. **106**, 030401 (2011)
44. Georges, A., Krauth, W.: Phys. Rev. B. **48**, 7167 (1993)
45. Chandra, N., Kollar, M., Vollhardt, D.: Phys. Rev. B. **59**, 10541 (1999)
46. Staudt, R., Dzierzawa, M., Muramatsu, A.: Eur. Phys. J. B. **17**, 411 (2000)
47. Vollhardt, D.: Phys. Rev. Lett. **78**, 1307 (1997)
48. Eckstein, M., Kollar, M., Vollhardt, D.: J. Low Temp. Phys. **147**, 279 (2007)
49. Bauer, J., Hewson, A.C.: Eur. Phys. J. B. **57**, 235 (2007)
50. Zitzler, R., Pruschke, T., Bulla, R.: Eur. Phys. J. B. **27**, 473 (2002)
51. Jarrell, M., Pruschke, T.: Z. Phys. B. **90**, 182194 (1993)
52. Freerjcks, J.K., Jarrell, M.: Phys. Rev. Lett. **74**, 186 (1995)
53. Chitra, R., Kotliar Phys, G.: Rev. Lett. **83**, 2386 (1999)
54. Snoek, M., Titvinidze, I., Hofstetter, W.: Phys. Rev. B. **83**, 054419 (2011)

55. Choi, S., Kutepov, A., Haule, K., van Schilfgaarde, M., Kotliar, G.: NPJ Quantum Mater. **1**, 16001 (2016)
56. Karp, J., Botana, A.S., Norman, M.R., Park, H., Zingl M., Millis, A.: preprint arXiv:2001.06441 (2020)
57. Nagaoka, Y.: Phys. Rev. **147**, 392 (1966)
58. Obermeier, T., Pruschke, T., Keller, J.: Phys. Rev. B. **56**, R8479 (1997)
59. Vollhardt, D., Blümer, N., Held, K., Kollar, M., Schlipf, J., Ulmke, M.: Z. Phys. B. **103**, 283 (1997)
60. Ulmke, M.: Eur. Phys. J. B. **1**, 301 (1998)
61. Lichtenstein, A.: In: Pavarini, E., Koch, E., Schollwöck, U. (eds.) Lecture Notes of the Autumn School Correlated Electrons 2013, p. 148, Forschungszentrum Jülich GmbH, Institute for Advanced Simulation (2013)
62. Nguyen, H.-S., Tran, M.-T.: Phys. Rev. B. **94**, 125106 (2016)
63. Belozerov, A.S., Katanin, A.A., Anisimov, V.I.: Phys. Rev. B. **96**, 075108 (2017)
64. Titvinidze, I., Schwabe, A., Rother, N., Potthoff, M.: Phys. Rev. B. **86**, 075141 (2012)
65. Georges, A., Kotliar, G., Krauth, W.: Z. Physik. **92**, 313 (1993)
66. Jarrell, M., Pruschke, T.: Z. Physik B. **90**, 187 (1993)
67. Ono, Y., Sanod, K.: J. Phys. Chem. Solids. **62**, 285 (2001)
68. Cheung, A.K.C., Agterberg, D.F.: Phys. Rev. B. **99**, 024516 (2019)
69. Raghu, S., Kivelson, S.A., Scalapino, D.J.: Phys. Rev. B. **81**, 224505 (2010)
70. Kohn, W., Luttinger, J.M.: Phys. Rev. Lett. **15**, 524 (1965)
71. Maiti, S., Chubukov, A.V.: Lectures on the Physics of Strongly Correlated Systems XVII: Seventeenth Training Course in the Physics of Strongly Correlated Systems. In: Avella, A., Mancini, F. (ed.), AIP Conf. Proc. Vol. 1550, pp. 3–73. AIP, Melville, NY (2013)
72. Koga, A., Werner, P.: Phys. Rev. B. **91**, 085108 (2015)
73. Sakai, S., Arita, R., Aoki, H.: Physica B. **359**, 554 (2005)
74. Ishigaki, K., Nasu, J., Koga, A., Hoshino, S., Werner, P.: Phys. Rev. B. **98**, 235120 (2018)
75. Capone, M., Fabrizio, M., Castellani, C., Tosatti, E.: Rev. Mod. Phys. **81**, 943 (2009)
76. Lichtenstein, A.I., Katsnelson, M.I.: Phys. Rev. B. **62**, R9283 (2000)
77. Senechal, D., Lavertu, P.L., Marois, M.A., Tremblay, A.M.S.: Phys. Rev. Lett. **94**, 156404 (2005)
78. Aichhorn, M., Arrigoni, E.: Europhys. Lett. **72**, 117 (2005)
79. Capone, M., Kotliar, G.: Phys. Rev. B. **74**, 054513 (2006)
80. Sordi, G., Sémon, P., Haule, K., Tremblay, A.-M.S.: Phys. Rev. Lett. **108**, 216401 (2012)
81. Gull, E., Millis, A.J.: Phys. Rev. B. **91**, 085116 (2015)
82. Reymbaut, A., Charlebois, M., Fellous Asiani, M., Fratino, L., Sémon, P., Sordi, G., Tremblay, A.-M.S.: Phys. Rev. B. **94**, 155146 (2016)
83. Kitatani, M., Tsuji, N., Aoki, H.: Phys. Rev. B. **95**, 075109 (2017)
84. Vučičević, J., Ayral, T., Parcollet, O.: Phys. Rev. B. **96**, 104504 (2017)
85. Kauch, A., Hörbinger, F., Li, G., Held, K.: preprint arXiv:1901.09743v2 (2019)
86. Vilardi, D., Taranto, C., Metzner, W.: Phys. Rev. B. **99**, 104501 (2019)
87. Anderson, P.W.: Phys. Rev. **109**, 1492 (1958)
88. Byczuk, K., Hofstetter, W., Vollhardt, D.: preprint arXiv:1002.3696 (2010); published reference - Int. J. Mod. Phys. B **24**, 1727 (2010)
89. Wegner, F.: Z. Phys. B. **44**, 9 (1981)
90. Dobrosavljevic, V., Kotliar, G.: Phys. Rev. Lett. **78**, 3943 (1997)
91. Kuchinskii, E.Z., Nekrasov, I.A., Sadovskii, M.V.: JETP. **106**, 581 (2008)
92. Byczuk, K., Hofstetter, W., Vollhardt, D.: Phys. Rev. Lett. **94**, 056404 (2005)
93. Byczuk, K., Hofstetter, W., Vollhardt, D.: Phys. Rev. Lett. **102**, 146403 (2009)
94. Radonjić, M.M., Tanasković, D., Dobrosavljević, V., Haule, K.: Phys. Rev. B. **81**, 075118 (2010)
95. Lee, H., Jeschke, H.O., Valentí, R.: Phys. Rev. B. **93**, 224203 (2016)
96. Skolimowski, J., Vollhardt, D., Byczuk, K.: J. Phys. Commun. **2**, 025031 (2018)

97. Hoang, A.-T., Nguyen, T.-H.-Y., Le, D.-A.: Physica B. **570**, 320 (2019)
98. Schofield, A.J.: Contemp. Phys. **40**, 95 (1999)
99. Landau, L.D.: Soviet JETP. **3**, 920 (1957)
100. Georges, A.: In: Pavarini, E., Koch, E., Vollhardt, D., Lichtenstein, A. (eds.) DMFT at 25: infinite dimensions modeling and simulation, vol. 4. Forschungszentrum Juelich, Juelich (2014)
101. Deng, X., Mravlje, J., Zitko, R., Ferrero, M., Kotliar, G., Georges, A.: Phys. Rev. Lett. **110**, 086401 (2013)
102. Dasari, N., Vidhyadhiraja, N.S., Jarrell, M., McKenzie, R.H.: Phys. Rev. B. **95**, 165105 (2017)
103. Brinkman, W.F., Rice, T.M.: Phys. Rev. B. **2**, 4302 (1970)
104. Terletska, H., Vučičević, J., Tanasković, D., Dobrosavljević, V.: Phys. Rev. Lett. **107**, 026401 (2011)
105. Werner, P., Gull, E., Troyer, M., Millis, A.J.: Phys. Rev. Lett. **101**, 166405 (2008)
106. Schmitt, S.: Phys. Rev. B. **82**, 155126 (2010)
107. Xu, W., Haule, K., Kotliar Phys, G.: Rev. Lett. **111**, 036401 (2013)
108. Parcollet, O., Georges, A.: Phys. Rev. B. **59**, 5341 (1999)
109. Perepelitsky, E., Sriram Shastry, B.: Ann. Phys. **338**, 283 (2013)
110. Fabrizio, M.: In: Pavarini, E., Koch, E., Scalettar, R., Martin, R.M. (eds.) The physics of correlated insulators, metals, and superconductors modeling and simulation, vol. 7. Verlag des Forschungszentrum Juelich, Juelich (2017)
111. Georges, A., L. de'Medici, Mravlje, C.: Ann. Rev. Cond. Matt. Phys. **4**, 137 (2013)

Chapter 5
Multi-orbital and Cluster DMFT

5.1 Multi-orbital DMFT

In Chap. 2, we have derived the Hubbard Hamiltonian for the general multi-orbital case (Eq. (2.28))

$$\widehat{H} = -\sum_{i,j,\alpha,\beta,\sigma} t_{ij}^{\alpha\beta}\widehat{c}_{\alpha i,\sigma}^{+}\widehat{c}_{\beta j,\sigma} + \frac{1}{2}\sum_{i}\sum_{\alpha,\beta,\gamma,\delta}\sum_{\sigma,\sigma'} U_{i,\sigma\sigma'}^{\alpha\beta\gamma\delta}\widehat{c}_{\alpha i,\sigma}^{+}\widehat{c}_{\beta i,\sigma'}^{+}\widehat{c}_{\gamma i,\sigma'}\widehat{c}_{\delta i,\sigma}, \tag{5.1}$$

and approximated it by probably the most-often used Kanamori form [1] (Eq. (2.33)):

$$\widehat{H} = -\sum_{i,j,\alpha,\sigma} t_{ij}^{\alpha}\widehat{c}_{\alpha i,\sigma}^{+}\widehat{c}_{\alpha j,\sigma} - \mu\sum_{i}\sum_{\alpha,\sigma}\widehat{n}_{\alpha i,\sigma} + \frac{U}{2}\sum_{i}\sum_{\alpha,\sigma}\widehat{n}_{\alpha i,\sigma}\widehat{n}_{\alpha i,\overline{\sigma}}$$
$$+(U-2J)\frac{1}{2}\sum_{i}\sum_{\alpha\neq\beta,\sigma}\widehat{n}_{\alpha i,\sigma}\widehat{n}_{\beta i,\overline{\sigma}} + (U-3J)\frac{1}{2}\sum_{i}\sum_{\alpha\neq\beta,\sigma}\widehat{n}_{\alpha i,\sigma}\widehat{n}_{\beta i,\sigma}, \tag{5.2}$$

where U and J are the local Coulomb repulsion and exchange energies. In addition to the standard local on-site and on-orbital Hubbard interaction term in the first line (the only interaction term in the one-orbital case), Hamiltonian (5.2) includes the inter-orbital different-spin and inter-orbital same-spin interactions, or *Hund's interactions* (the first and the second terms in the last line in Eq. (5.2), correspondingly). These interactions play a very important role in the multi-orbital case, e.g., they change the Mott transition temperature. We will not analyze the solutions of the multi-orbital Hubbard model in this chapter, since the main goal here is to define the multi-orbital problem and the solutions will be discussed in Chap. 10, where we analyze properties of different materials.

One can generalize the ideology of single-site DMFT on the multi-orbital case—introduce local-in-space orbital-dependent (matrix) self-energy $\widehat{\Sigma}_{lm\sigma}(\omega)$, map the

V. Turkowski, *Dynamical Mean-Field Theory for Strongly Correlated Materials*,
https://doi.org/10.1007/978-3-030-64904-3_5

problem on a multi-orbital impurity problem with an orbital-dependent matrix dynamical mean-field $\widehat{\mathcal{G}}_{\sigma,lm}(\omega)$, and write down a generalization of the single-orbital DMFT Eqs. (3.49)–(3.51):

$$\widehat{G}^{(loc)}_{\sigma,lm}(\omega) = \sum_k \left(\left[\omega\widehat{I} - \widehat{\varepsilon}_\sigma\left(\vec{k}\right) + \mu\widehat{I} - \widehat{\Sigma}_\sigma(\omega) \right]^{-1} \right)_{lm}, \tag{5.3}$$

$$\widehat{G}^{(loc)-1}_{lm\sigma}(\omega) = \widehat{\mathcal{G}}^{-1}_{lm\sigma}(\omega) - \widehat{\Sigma}_{lm\sigma}(\omega) \tag{5.4}$$

$$G^{(loc)}_{\sigma lm}(\tau) = \int D[\psi]D[\psi^*]\psi_{\sigma l}(\tau)\psi^*_{\sigma m}(0)$$
$$\times \exp\left[-\int_0^\beta d\tau' \int_0^\beta d\tau'' \sum_{\sigma',a,b} \psi^*_{\sigma' a}(\tau') \left(\widehat{\mathcal{G}}_{\sigma'}^{-1}\right)_{ab} (\tau' - \tau'')\psi_{\sigma' b}(\tau'') \right.$$
$$\left. + \int_0^\beta d\tau' \sum_{\sigma,\sigma',abcd} U^{abcd}_{\sigma\sigma'} \psi^*_{\sigma a}(\tau')\psi^*_{\sigma' b}(\tau')\psi_{\sigma' c}(\tau')\psi_{\sigma d}(\tau') \right]. \tag{5.5}$$

This set of equations is significantly simplified in the para-orbital approximation, where $G_{\sigma lm}(\tau) = \delta_{lm} G_{\sigma l}(\tau)$ and the same for the self-energy and the dynamical mean field. Indeed, in this approximation one does not need to spend much time to invert the local GF and the dynamical mean field matrices at each step (iteration) of the calculations, since the matrices are diagonal. Next, the GF and the self-energy (Dyson) equations for different orbitals become simply disconnected:

$$G^{(loc)}_{\sigma,l}(\omega) = \sum_k \frac{1}{\omega - \varepsilon_{\sigma,l}(k) + \mu - \Sigma_{\sigma,l}(\omega)}$$
$$= \int d\varepsilon \rho_{\sigma l}(\varepsilon) \frac{1}{\omega - \varepsilon + \mu - \Sigma_{\sigma,l}(\omega)}, \tag{5.6}$$

$$G^{(loc)-1}_{\sigma,l}(\omega) = \mathcal{G}^{-1}_{\sigma,l}(\omega) - \Sigma_{\sigma,l}(\omega), \tag{5.7}$$

and the impurity Eq. (5.5) transforms into an equation where the bath part in the effective action is a simple sum of individual contributions of orbital dynamical mean fields:

$$S_{eff} = -\int_0^\beta d\tau' \int_0^\beta d\tau'' \sum_{\sigma',a} \psi^*_{\sigma' a}(\tau') \mathcal{G}_{\sigma' a}{}^{-1}(\tau' - \tau'')\psi_{\sigma' a}(\tau'')$$
$$+ \int_0^\beta d\tau' \sum_{\sigma,\sigma',abcd} U^{abcd} \psi^*_{\sigma a}(\tau')\psi^*_{\sigma' b}(\tau')\psi_{\sigma' c}(\tau')\psi_{\sigma d}(\tau'). \tag{5.8}$$

Equation (5.6) is an especially convenient approximation, since one does not need to invert the matrix under the integral for many values of momentum. The action (5.8) is also more convenient comparing to the non-diagonal matrix $\mathcal{G}$ case, since it allows a simple analytical analysis (see Sect. 6.1.1 in the next chapter).

We conclude this short section and move to another important case—DMFT for systems with several nonequivalent atoms.

5.2 Cluster DMFTs

There are different systems where the spatial (site) non-homogeneity has to be taken into account: superconductors with nonequivalent sites (discussed in previous section), antiferromagnets and other magnets, materials that contain different types of atoms, etc. In such situations, in DMFT instead of one-atom impurity problem one needs to solve an "impurity cluster" problem. Technically, the cluster problem is very similar to the many-orbital problem from the last section, where all functions are matrices in the orbital indices. In the several-atom cluster case, one again can reformulate the problem in terms of generalized orbitals, where the orbital indices are the atom number, orbital number, and spin index. Then, the GF and other DMFT equations become matrices with respect to these indices. However, there is a big difference between the cluster and the multi-orbital cases—electrons can hop within the cluster atoms, which makes the problem much more complicated—same has to deal with momentum-dependent self-energy. In this section, on cluster DMFT, we mostly follow presentation of the material in Maier et al. [2].

For simplicity, we consider the case of one-band Hubbard model with Hamiltonian

$$\widehat{H} = -\sum_{i,j,\sigma} t_{ij}\widehat{c}^{+}_{i,\sigma}\widehat{c}_{j,\sigma} - \mu\sum_{i\sigma}\widehat{n}_{i,\sigma} + \frac{1}{2}\sum_{i,j,\sigma,\sigma'} U^{\sigma\sigma'}_{i,j}\,\widehat{n}_{i,\sigma}\widehat{n}_{j,\sigma'}, \tag{5.9}$$

where the interactions $U^{\sigma\sigma'}_{i,j}$ can be nonzero for electrons on both same and different atoms. Let us formulate a DMFT approach for the case when electrons can interact inside a cluster with the number of atoms N_c. Positions of the lattice atoms x will be defined by two vectors: the cluster position vector $\widetilde{\mathrm{x}}$ and the vector of the position of each of the N_c atoms inside the cluster X, i.e., $\vec{x} = \vec{\widetilde{x}} + \vec{X}$. The corresponding $\vec{\widetilde{x}}$ and $\vec{\widetilde{x}}$ reciprocal-space momenta will be labeled $\vec{\widetilde{k}}$ and $\vec{K}$, such that the total momentum of the electron is $\vec{k} = \vec{\widetilde{k}} + \vec{K}$. One can think about coordinates $\vec{\widetilde{x}}$ as positions of sites in the superlattice formed by "clusters sites." The superlattice momentum $\vec{\widetilde{k}}$ is a smooth (technically, almost continuous) variable. For example, in the case of a hypercubic lattice with total number of atoms N one has for the momentum components in the first Brillouin zone $\widetilde{k}_i = \frac{2\pi}{(N/N_c)^{1/d}} n_i$, where $n_i = 0, \ldots, \left(\frac{N}{N_c}\right)^{\frac{1}{d}} - 1$ and

$i = 1, \ldots, d$. Since N is very large, the number of momenta $\widetilde{k}_i$ is also very large. On the other hand, the number of the "cluster" momentum points $\vec{K}$ is rather small, i.e., the variables K_α are discrete. In particular, for the d-dimensional hypercubic lattice with N_c atoms in the cluster (such that $N_c = L_c{}^d$, where L_c is the cell parameter (length)) one has

$$K_\alpha = \frac{2\pi}{L_c} n_a, \tag{5.10}$$

where $L_c = (N_c)^{\frac{1}{d}}$ and $n_a = 0, \ldots, L_c - 1$ (a = 1,..., d).

At this point, it is useful to give formulas for different Fourier transforms with respect to the space and momenta variables: $f\left(\vec{X}, \vec{\widetilde{x}}\right) = (N_c/N)\sum_{\vec{\widetilde{k}}} e^{i\vec{\widetilde{k}}\vec{\widetilde{x}}} f\left(\vec{X}, \vec{\widetilde{k}}\right)$, $f\left(\vec{X}, \vec{\widetilde{k}}\right) = \sum_{\vec{\widetilde{x}}} e^{-i\vec{\widetilde{k}}\vec{\widetilde{x}}} f\left(\vec{X}, \vec{\widetilde{x}}\right) = \frac{1}{N_c}\sum_{\vec{K}} e^{i\left(\vec{K} + \vec{\widetilde{k}}\right)\vec{X}} f\left(\vec{X}, \vec{\widetilde{k}}\right)$, $f\left(\vec{K}, \vec{\widetilde{k}}\right) = \sum_{\vec{X}} e^{-i\left(\vec{K} + \vec{\widetilde{k}}\right)\vec{X}} f\left(\vec{X}, \vec{\widetilde{k}}\right)$. These coordinate, momentum, and mixed coordinate–momentum functions will be used below.

In the space- and momentum-vector notations above, the hopping and self-energy functions can be divided into separated intra- and inter-cluster parts:

$$t\left(\vec{\widetilde{x}}_i - \vec{\widetilde{x}}_j\right) = \widehat{t}_c \delta_{\vec{\widetilde{x}}_i \vec{\widetilde{x}}_j} + \delta t\left(\vec{\widetilde{x}}_i - \vec{\widetilde{x}}_j\right), \tag{5.11}$$

$$\Sigma\left(\omega, \vec{\widetilde{x}}_i - \vec{\widetilde{x}}_j\right) = \widehat{\Sigma}_c(\omega) \delta_{\vec{\widetilde{x}}_i \vec{\widetilde{x}}_j} + \delta\Sigma\left(\omega, \vec{\widetilde{x}}_i - \vec{\widetilde{x}}_j\right), \tag{5.12}$$

where $\widehat{t}_c$ and $\widehat{\Sigma}_c(\omega)$ are the hopping and self-energy matrices (in $\vec{X}$ indices), and the last terms are the corresponding inter-cluster parts (the inter-cluster self-energy part is often neglected).

In these notations, the intra-cluster $N_c \times N_c$ matrix GF is

$$\widehat{g}(\omega) = \left[(\omega + \mu)\widehat{I} - \widehat{t}_c - \widehat{\Sigma}_c(\omega)\right]^{-1} \tag{5.13}$$

(in the last three equations and below the intra-cell functions with the hat are matrices in the intra-cell atomic indices). Using the intra-cluster GF $\widehat{g}(\omega)$, one can write down the equation for the supercell GF

$$G\left(\omega, \vec{x}_i - \vec{x}_j\right) = \hat{g}(\omega)\delta_{\vec{x}_i \vec{x}_j}$$
$$+\hat{g}(\omega)\sum_{l=1}^{N/N_c}\left[\delta t\left(\vec{x}_i - \vec{x}_l\right) + \delta\Sigma\left(\omega, \vec{x}_i - \vec{x}_l\right)\right]G\left(\omega, \vec{x}_l - \vec{x}_j\right), \quad (5.14)$$

or in the momentum representation,

$$G\left(\omega, \vec{k}\right) = \hat{g}(\omega) + \hat{g}(\omega)\left[\delta t\left(\vec{k}\right) + \delta\Sigma\left(\omega, \vec{k}\right)\right]G\left(\omega, \vec{k}\right). \quad (5.15)$$

The key approximation in the cluster DMFT is to ignore the contribution of the intercell self-energy, $\delta\Sigma\left(\omega, \vec{k}\right) = 0$, that corresponds to intra-cell approximation for the potential energy (see Ref. [2]). In this case, one can obtain the following formula for the supercell GF:

$$G\left(\omega, \vec{k}\right) = \left[\hat{g}^{-1}(\omega) - \delta t\left(\vec{k}\right)\right]^{-1}. \quad (5.16)$$

As it follows from this equation, in the cluster formalism the problem reduces to find $\hat{g}(\omega)$, or the intra-cluster self-energy $\hat{\Sigma}_c(\omega)$ (see Eq. 5.13) that depends on frequency and intra-cluster coordinates only, a great simplification of the original problem. Since $\hat{\Sigma}_c(\omega)$ depends only on the intra-cluster degrees of freedom, it must be a functional of supercell momentum average of the total GF $G\left(\omega, \vec{k}\right)$,

$$\overline{G}(\omega) = \frac{N_c}{N}\sum_{\vec{k}} G\left(\omega, \vec{k}\right), \quad (5.17)$$

that is a function restricted to the cluster ($\overline{G}(\omega)$ depends only on the intra-cluster coordinates). Thus, $\hat{\Sigma}_c(\omega) = \mathcal{F}\left[\overline{G}(\omega)\right]$.

From Eq. (5.16) it follows that $\overline{G}(\omega) = \frac{N_c}{N}\sum_{\vec{k}}\left[\hat{g}^{-1}(\omega) - \delta t\left(\vec{k}\right)\right]^{-1}$. Then, to get rid of the momentum variable (in the sum in Eq. 5.17) in the definition of the momentum-independent function $\overline{G}(\omega)$ one can introduce a (matrix) hybridization function $\hat{\Gamma}(\omega)$, that satisfies $\left[\hat{g}^{-1}(\omega) - \hat{\Gamma}(\omega)\right]^{-1} = \frac{N_c}{N}\sum_{\vec{k}}\left[\hat{g}^{-1}(\omega) - \delta t\left(\vec{k}\right)\right]^{-1}$.

This gives

$$\overline{G}(\omega) = \left[\widehat{g}^{-1}(\omega) - \widehat{\Gamma}(\omega)\right]^{-1}. \tag{5.18}$$

The intra-cluster dynamical field matrix $\widehat{\Gamma}(\omega)$ takes into account the effects of hopping of electrons in and out of the cluster, and for this reason it is called the hybridization function. Since this function plays a key role in the cluster DMFT, we present the explicit expression for $\widehat{\Gamma}(\omega)$ that can be obtained from Eqs. (5.16) to (5.18):

$$\widehat{\Gamma}(\omega) = \left[\widehat{I} + \frac{N_c}{N}\sum_{\vec{\tilde{k}}} \delta t\left(\vec{\tilde{k}}\right) G\left(\omega, \vec{\tilde{k}}\right)\right]^{-1} \frac{N_c}{N}\sum_{\vec{\tilde{k}}} \delta t\left(\vec{\tilde{k}}\right) G\left(\omega, \vec{\tilde{k}}\right) \delta t\left(\vec{\tilde{k}}\right). \tag{5.19}$$

This is a very useful expression, e.g., it can be easily used in the PT (in the intercluster hopping $\delta t\left(\vec{\tilde{k}}\right)$) approach to cluster DMFT. Equations (5.16)–(5.19) form the basis of the cluster DMFT. At this point, one can write down the DMFT equation for the local lattice GF, as well as for the self-energy (Dyson) and impurity equations generalized on the cluster case (for the equations and their solution, see below).

However, before discussion of different approximate solutions of the cluster DMFT equations, we must fix one potential problem—lack of the translation invariance inside the cluster (with respect to X vectors). Such an invariance is needed if one would like to use momentum representation for the cluster problem. The reason for absence of the translational invariance is nonequivalence of the cluster atoms: there are hopping processes between the surface cluster sites and the rest of the system ($\delta t\left(\vec{\tilde{k}}\right)$ terms), but there are no such hopping processes between the bulk (non-surface) cluster sites and the rest of the system. In detail, the intra-cluster hopping matrix elements in the momentum representation for the (coarse-grain) superlattice indices can be written as:

$$t_{\vec{X}_i\vec{X}_j}\left(\vec{\tilde{k}}\right) = \frac{1}{N_c}\sum_{\vec{K}} e^{i\left(\vec{\tilde{k}}+\vec{K}\right)\left(\vec{X}_i-\vec{X}_j\right)} \varepsilon\left(\vec{K}+\vec{\tilde{k}}\right). \tag{5.20}$$

If we had for the exponent function in Eq. (5.20) $e^{i\vec{K}\left(\vec{X}_i-\vec{X}_j\right)}$, the translational invariance would be preserved. However, the extra factor $e^{i\vec{\tilde{k}}\left(\vec{X}_i-\vec{X}_j\right)}$ breaks it. To restore the invariance a dynamical cluster approximation (DCA) for cluster DMFT was proposed [3, 4]. In this approximation, $\vec{\vec{k}}$ in the exponent in Eq. (5.20) is neglected, and the following intra-cluster hopping matrix is used:

$$t^{\text{DCA}}_{\vec{X}_i\vec{X}_j}\left(\vec{\vec{k}}\right) = \frac{1}{N_c}\sum_{\vec{K}} e^{i\vec{K}\left(\vec{X}_i-\vec{X}_j\right)}\varepsilon\left(\vec{K}+\vec{\vec{k}}\right). \tag{5.21}$$

This is a very convenient formula, since it allows one to separate the intra-(averaged over $\vec{\vec{k}}$) and inter-($\vec{\vec{k}}$ – dependent) cluster hopping parts of the matrix:

$$\begin{aligned} t^{\text{DCA}}_{\vec{X}_i\vec{X}_j}\left(\vec{\vec{k}}\right) &= \frac{1}{N_c}\sum_{\vec{K}} e^{i\vec{K}\left(\vec{X}_i-\vec{X}_j\right)}\bar{\varepsilon}\left(\vec{K}\right) \\ &+ \frac{1}{N_c}\sum_{\vec{K}} e^{i\vec{K}\left(\vec{X}_i-\vec{X}_j\right)}\left[\varepsilon\left(\vec{K}+\vec{\vec{k}}\right)-\bar{\varepsilon}\left(\vec{K}\right)\right] \\ &\equiv t^{\text{DCA}-c}_{\vec{X}_i\vec{X}_j} + \delta t^{\text{DCA}}_{\vec{X}_i\vec{X}_j}\left(\vec{\vec{k}}\right), \end{aligned} \tag{5.22}$$

where $\bar{\varepsilon}\left(\vec{K}\right) = \frac{N_c}{N}\sum_{\vec{\vec{k}}}\varepsilon\left(\vec{K}+\vec{\vec{k}}\right)$ is an effective in-cluster dispersion. At the same time, the effective inter-cluster hopping (dispersion) is given by the K-momentum Fourier transform of the function $\delta t\left(\vec{K},\vec{\vec{k}}\right) = \varepsilon\left(\vec{K}+\vec{\vec{k}}\right) - \bar{\varepsilon}\left(\vec{K}\right)$. Since the last hopping functions are intra-cluster translation-invariant, the cluster self-energy $\widehat{\Sigma}_c(\omega)$ and the hybridization $\widehat{\Gamma}(\omega)$ are also invariant matrices. Then, one can perform K-Fourier transform in Eq. (5.14) and get the diagonal in the cluster momentum GF:

$$G\left(\omega,\vec{K}+\vec{\vec{k}}\right) = \left[g^{-1}\left(\omega,\vec{K}\right) - \delta t\left(\vec{K}+\vec{\vec{k}}\right)\right]^{-1}, \tag{5.23}$$

where

$$g^{-1}\left(\omega,\vec{K}\right) = \omega - \bar{\varepsilon}\left(\vec{K}\right) + \mu - \Sigma_c\left(\omega,\vec{K}\right) \tag{5.24}$$

(compare Eq. (5.23) with Eq. (5.16)). In fact, GF in Eq. (5.24) is basically the Fourier transform of the translation-invariant version of Eq. (5.13) (with function $\bar{\varepsilon}\left(\vec{K}\right)$ used instead of Fourier transform of the matrix $\hat{t}_c$).

Using the results above, one can show [2] that in the translational-invariant case the function $\Sigma_c\left(\omega,\vec{K}\right)$ is a functional of the GF

$$\overline{G}\left(\omega,\vec{K}\right)=\frac{N_c}{N}\sum_{\vec{\tilde{k}}}G\left(\omega,\vec{K}+\vec{\tilde{k}}\right)=\frac{1}{g^{-1}\left(\omega,\vec{K}\right)-\Gamma\left(\omega,\vec{K}\right)},\tag{5.25}$$

where similar to Eq. (5.19), the cell momentum-dependent hybridization function is

$$\Gamma\left(\omega,\vec{K}\right)=\left[1+\frac{N_c}{N}\sum_{\vec{\tilde{k}}}\delta t\left(\vec{K}+\vec{\tilde{k}}\right)G\left(\omega,\vec{K}+\vec{\tilde{k}}\right)\right]^{-1}\frac{N_c}{N}\sum_{\vec{\tilde{k}}}\left[\delta t\left(\vec{K}+\vec{\tilde{k}}\right)\right]^2G\left(\omega,\vec{\tilde{k}}\right).\tag{5.26}$$

Now, we have completed general formulation of cluster DMFT. We will proceed with different approximations to solve the corresponding equations (i.e., to find $\Sigma_c\left(\omega,\vec{K}\right)$)—cluster PT, cellular DMFT, and DCA, Sects. 5.2.1–5.2.3. In each subsection, we will also give a microscopic reasoning and discuss the meaning of each approximation by considering the structure of the diagrams that contribute to the corresponding Kadanoff–Baym functional $\Phi[G, U]$ that defines the electron self-energy, $\Sigma=\frac{\delta\Phi[G,U]}{\delta G}$. $\Phi[G, U]$ is a sum of all compact and connected (skeletal) diagrams build from the GFs and interactions U. Thus, as soon as an appropriate approximation for $\Phi[G, U]$ is made, one can obtain the desired result for the self-energy.

In any approximation, the solution for Σ should be thermodynamically consistent. To show this, one can consider the grand potential Ω defined by the Kadanoff–Baym functional: $\Omega[G,U]=-k_BT[\Phi[G,U]-\mathrm{Tr}\ln(-G)-\mathrm{Tr}(\Sigma G)]$, where Tr means summation/integration over $X,\tilde{k},\omega$, and spin. From $\Sigma=\frac{\delta\Phi[G,U]}{\delta G}$ it follows that the grand potential satisfies the stationary condition $\frac{\delta\Omega[G,U]}{\delta G}=0$, which means that the solution is thermodynamically consistent. Namely, equality $\Sigma=\frac{\delta\Phi[G,U]}{\delta G}$ (and hence $\frac{\delta\Omega[G,U]}{\delta G}=0$), together with the energy and momentum conservations, guarantees that the solution for Σ satisfies the conservation laws (sum rules). Finally, it is important to mention that in the cluster DMFT approximations below in evaluating Φ only intra-cluster ("one-supercell-site") diagrams are taken into account.

5.2.1 Cluster Perturbation Theory

Probably, the simplest approximation for the Kadanoff–Baym potential is to substitute the lattice GFs in the diagrams by the cluster ones, Eq. (5.13), which gives the cluster-approximated self-energy $\Sigma_c(\omega)$. Though, such an approximation is non-self-consistent and, besides, one needs to assume that the obtained this way cluster self-energy is the same as the self-energy obtained from the isolated cluster calculations. The resulting lattice GF is

$$G\left(\omega, \vec{k}\right) = \left[\widehat{G}_0^{-1}\left(\omega, \vec{k}\right) - \widehat{\Sigma}_c(\omega)\right]^{-1} \equiv \left[\widehat{g}^{-1}(\omega) - \delta\widehat{t}\left(\vec{k}\right)\right]^{-1}, \tag{5.27}$$

where $\widehat{G}_0^{-1}\left(\omega, \vec{k}\right) = (\omega + \mu)\widehat{I} - \widehat{t}\left(\vec{k}\right)$ and the last equality in Eq. (5.27) is obtained by using Eqs. (5.11) and (5.13).

The above approximation, called the cluster PT approximation [5], consists essentially of substituting the lattice self-energy in the equation for the lattice GF (5.27) by the cluster self-energy $\widehat{\Sigma}_c(\omega)$. As it was shown by Pairault et al. [6], this approximation corresponds to the solution in the lowest order expansion in the inter-site hopping $\delta\widehat{t}\left(\vec{k}\right)$ (i.e., to the strong-coupling expansion). Therefore, since $\delta\widehat{t}\left(\vec{k}\right)$ is a perturbation, in the lowest order approximation one can calculate $\widehat{g}(\omega)$ for the isolated cluster. This can be done by using different boundary conditions (for advantages and disadvantages of different boundary conditions and for other developments of the cluster PT approach, see work [2] and references therein).

5.2.2 Cellular DMFT

The next in accuracy approximation is cellular DMFT [7], where $\widehat{g}(\omega)$ in the cluster diagrams of the Kadanoff–Baym potential are substituted by the lattice GF $G\left(\omega, \vec{k}\right)$ (the site indices in the diagrams in the thermodynamic potential are limited to the cluster!). One can get a deeper insight on the meaning of such an approximation by considering the (vertex) Laue function $\Delta = \sum_{\vec{x}} e^{i\vec{x}\left(\vec{k}_1+\vec{k}_2+\ldots-\vec{k}'_1-\vec{k}'_2-\ldots\right)}$. In one-site DMFT, one neglects the momentum conservation in each vertex, i.e., one neglects the condition of equality of total incoming and outcoming momenta $\vec{k}_1 + \vec{k}_2 + \ldots - \vec{k}'_1 - \vec{k}'_2 - \ldots$ by substituting the Laue function with 1 (see Chap. 3). This results in absence of nonlocal GFs in the diagrams in the Kadanoff–Baym potential. A similar approximation is made in the cluster PT case, where only the cluster GFs contribute to the potential. In the cellular DMFT, the Laue function is approximated by a more accurate expression:

$$\sum_{\vec{x}} e^{i\vec{x}\left(\vec{k}_1+\vec{k}_2+\ldots-\vec{k}'_1-\vec{k}'_2-\ldots\right)} \equiv \sum_{\vec{x}} e^{i\left(\vec{X}+\vec{\tilde{x}}\right)\left(\vec{K}_1+\vec{\tilde{k}}_1+\vec{K}_2+\vec{\tilde{k}}_2\ldots-\vec{K}'_1-\vec{\tilde{k}}'_1-\vec{K}'_2-\vec{\tilde{k}}'_2-\ldots\right)}$$

$$\approx \sum_{\vec{X}} e^{i\vec{X}\left(\vec{K}_1+\vec{\tilde{k}}_1+\vec{K}_2+\vec{\tilde{k}}_2\ldots-\vec{K}'_1-\vec{\tilde{k}}'_1-\vec{K}'_2-\vec{\tilde{k}}'_2-\ldots\right)} \equiv \Delta^{\mathrm{CDMFT}}, \tag{5.28}$$

where the position of clusters in the lattice (described by the omitted factor $e^{i\vec{x}\left(\vec{k}_1+\vec{k}_2...-\vec{k}'_1-\vec{k}'_2-...\right)}$) is ignored and hence the translational invariance is broken by the remaining factor $e^{i\vec{X}\left(\vec{k}_1+\vec{k}_2...-\vec{k}'_1-\vec{k}'_2-...\right)}$.

In the cellular DMFT approximation, the GFs, self-energies, and other functions that survive in the Kadanoff–Baym functional depend only on the intra-cluster indices ($\vec{X}$ or $\vec{K}$), e.g., each GF $G\left(\vec{X}_1+\vec{x}_1,\vec{X}_2+\vec{x}_2\right)$ is replaced by

$$\begin{aligned}\overline{G}\left(\vec{X}_1,\vec{X}_2\right) &= G\left(\vec{X}_1,\vec{X}_2;\vec{x}_1=\vec{x}_2=0\right)\\ &=\frac{1}{N^2}\sum_{\{K,\tilde{k}_i\}} e^{i\vec{X}_1\left(\vec{K}_1+\vec{k}_1\right)} G\left(\vec{K}_1,\vec{K}_2;\vec{k}_1,\vec{k}_2\right)e^{-i\vec{X}_2\left(\vec{K}_2+\vec{k}_2\right)}\\ &=\frac{N_c^2}{N^2}\sum_{\vec{k}_1,\vec{k}_2} G\left(\vec{X}_1,\vec{X}_2;\vec{k}_1,\vec{k}_2\right)\equiv\frac{N_c}{N}\sum_{\vec{k}}G\left(\vec{X}_1,\vec{X}_2;\vec{k}\right).\end{aligned} \tag{5.29}$$

The last equality comes from the fact that $G\left(\vec{K}_1,\vec{K}_2;\vec{k}_1,\vec{k}_2\right)$ is diagonal in $\vec{k}$ indices due to translational invariance of the superlattice.

Similarly, each interaction line in the diagrams is replaced by the coarse-grain expression,

$$\begin{aligned}U\left(\vec{X}_1,\vec{X}_2\right) &\equiv U\left(\vec{X}_1+\vec{x}_1,\vec{X}_2+\vec{x}_2\right)\to U\left(\vec{X}_1,\vec{X}_2;\vec{x}_1=\vec{x}_2=0\right)\\ &\equiv\overline{U}\left(\vec{X}_1,\vec{X}_2\right)=\frac{N_c}{N}\sum_{\vec{k}}U\left(\vec{X}_1,\vec{X}_2;\vec{k}\right).\end{aligned} \tag{5.30}$$

As a result, the summation in the diagrams is performed only over the cluster sites, and hence the self-energy $\widehat{\Sigma}=\frac{\delta\Phi\left[\widehat{G},\widehat{U}\right]}{\delta\widehat{G}}$ also depends only on the intra-cluster indices. Substituting $\Phi\left[\overline{G},\overline{U}\right]$ into the expression for the grand potential Ω, and using the equation $\frac{\delta\Omega}{\delta\widehat{G}}=0$ one can get the relation between the cluster and the lattice self-energies [2].

Cluster DMFT has been mostly applied to the 2D Hubbard model. For example, in Ref. [8] cellular DMFT (with short-range, intra-cluster electron–electron

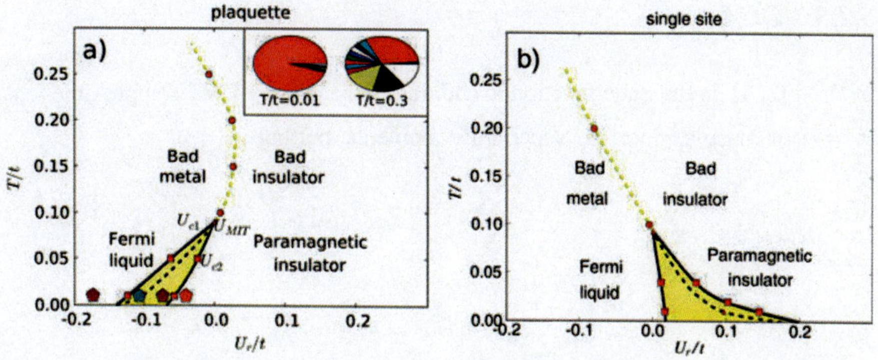

Fig. 5.1 Interaction–temperature phase diagram of the 2D half-filled Hubbard model obtained by using the CT-QMC with two solutions, cellular (2 × 2 plaquette) DMFT, and single-site DMFT. In the inset, the histogram of the two insulating states is shown (the probability for a given half-filled cluster eigenstate among the 16 eigenstates, see Ref. [8] for details). (Reprinted Fig. 1 with permission from (H. Park, K. Haule, and G. Kotliar, Phys. Rev. Lett. **101**, 186403 (2008)). Copyright 2008 by the American Physical Society. https://doi.org/10.1103/PhysRevLett.101.186403)

interactions, as discussed above) was compared with the one-site DMFT (local on-site interaction) results. The calculations were performed at half-filling. It was shown that, similar to the one-site DMFT case, cellular DMFT gives a first-order PM MIT, but the critical Us, the shapes of the transition lines, and the physics near the MIT critical point are rather different. The Coulomb interaction–temperature phase diagrams obtained with both approximations are shown in Fig. 5.1. Important differences in the spectral properties obtained with two approaches are an anomalous (low-coherence scale) metallic state on the metallic side near the transition line and sharp peaks near the band edge in the insulating state, both found with cellular DMFT [8].

Another important cellular DMFT studies were performed in Ref. [9], where low-temperature solutions for the GF and self-energy of the doped 2D Hubbard model were obtained at different sizes of clusters (from 4 to 16 atoms). Different system periodization schemes were compared in order to get results in the limit of infinite system and it was found that even four-atom cluster solution is not far from the thermodynamic limit. One of the conclusions the authors had made from this result is that four atom-cluster cellular DMFT can be appropriate to analyze properties of doped cuprates (for more applications of the approach, see references in works [2, 8, 9]).

5.2.3 DCA

In DCA [3, 4], in the Laue function in addition to the cellular DMFT approximation $\vec{\tilde{x}} = 0$, one also neglects the superlattice momenta, putting $\vec{\tilde{k}} = 0$:

$$\sum_{\vec{x}} e^{i\vec{x}\left(\vec{k}_1+\vec{k}_2+\ldots-\vec{k}'_1-\vec{k}'_2-\ldots\right)} \equiv \sum_{\vec{x}} e^{i\left(\vec{X}+\vec{\tilde{x}}\right)\left(\vec{K}_1+\vec{\tilde{k}}_1+\vec{K}_2+\vec{\tilde{k}}_2\ldots-\vec{K}'_1-\vec{\tilde{k}}'_1-\vec{K}'_2-\vec{\tilde{k}}'_2-\ldots\right)}$$
$$\approx \sum_{\vec{x}} e^{i\vec{X}\left(\vec{K}_1+\vec{K}_2+\ldots-\vec{K}'_1-\vec{K}'_2-\ldots\right)} \equiv \Delta^{\mathrm{DCA}}, \tag{5.31}$$

This approximation conserves the cluster momenta in the vertices, $\vec{K}_1 + \vec{K}_2 + \ldots = \vec{K}'_1 + \vec{K}'_2 + \ldots$. The GF legs in the DCA Kadanoff–Baym potential, similar to the cellular DMFT, are replaced by $\vec{\tilde{k}}$-independent (more accurately, by $\vec{\tilde{k}}$-averaged) GFs

$$\overline{G}\left(\omega, \vec{K}\right) = \frac{N_c}{N} \sum_{\vec{\tilde{k}}} G\left(\omega, \vec{K} + \vec{\tilde{k}}\right). \tag{5.32}$$

The interaction functions in the diagrams are also replaced by the averaged one:

$$\overline{U}\left(\vec{K}\right) = \frac{N_c}{N} \sum_{\vec{\tilde{k}}} U\left(\vec{K} + \vec{\tilde{k}}\right). \tag{5.33}$$

Thus, similar to cellular DMFT, in DCA the interactions are taken into account only within the clusters, and the diagrams in the Kadanoff–Baym functional are reduced to a set of intra-cluster diagrams. The wave vectors are approximated by the cluster wave vectors, and the lattice GFs G are approximated by the coarse-grained GFs $\overline{G}$ (Fig. 5.2).

From the above discussion, it follows that the diagrams in the Kadanoff–Baym functional depend only on $\vec{K}$. Thus, the DCA self-energy depends only on the cluster momenta and frequency: $\Sigma_c\left(\omega, \vec{K}\right) = \frac{\delta\Phi\left[\overline{G}\left(\omega,\vec{K}\right),\overline{U}\left(\vec{K}\right)\right]}{\delta\overline{G}\left(\omega,\vec{K}\right)}$. Substituting the Kadanoff–Baym functional $\Phi\left[\overline{G}\left(\omega, \vec{K}\right), \overline{U}\left(\vec{K}\right)\right]$ into the expression for the grand potential and using $\Sigma\left(\omega, \vec{K}\right) = \Sigma_c\left(\omega, \vec{K}\right)$ one can obtain the equation $\frac{\delta\Omega}{\delta G\left(\vec{K}\right)} = 0$, which means that the grand potential is stationary with respect to the lattice GF, and the solution is thermodynamically consistent.

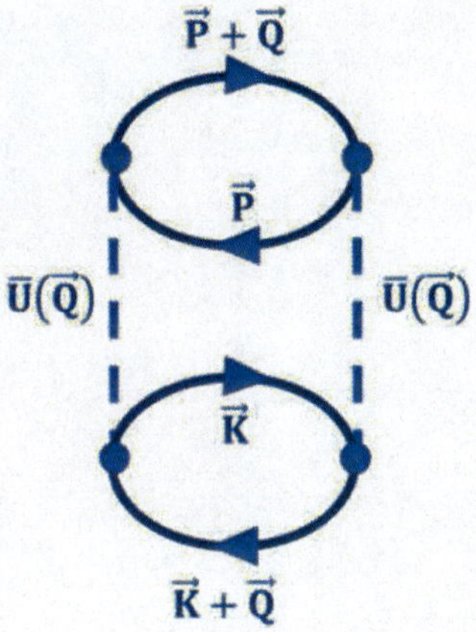

Fig. 5.2 A second-order contribution to the Kadanoff–Baym functional in the DCA approximation. The dashed lines are interactions $\overline{U}$, the solid lines are the coarse-grained GFs $\overline{G}$, and the momenta are in-cell reciprocal vectors ("cluster momenta") defined in Eq. (5.10)

As expected, in DCA the self-consistent condition $G_0^{-1}\left(\omega,\vec{K}\right) - G^{-1}\left(\omega,\vec{K}\right) = \Sigma\left(\omega,\vec{K}\right) = \frac{\delta\Phi[G,U]}{\delta G\left(\omega,\vec{K}\right)}$ gives the Dyson equation for the coarse-grained GF $\overline{G}\left(\omega,\vec{K}\right) = \sum_{\vec{\tilde{k}}}\left[\widehat{G}_0^{-1}\left(\omega,\vec{K}+\vec{\tilde{k}}\right) - \widehat{\Sigma}_c\left(\omega,\vec{K}\right)\right]^{-1}$, where $\widehat{G}_0^{-1}\left(\omega,\vec{K}+\vec{\tilde{k}}\right) = \omega - \varepsilon\left(\vec{K}+\vec{\tilde{k}}\right) + \mu$.

DCA was also mainly used in the case of 2D Hubbard model at different values of doping to study possibility of MIT, AFM, and pseudogap phases, and to describe superconducting properties of cuprates (see, e.g., Refs. [10–12] and for earlier studies—references in review [2]). In particular, in work [11] the connection between the charge susceptibility and pseudogap was studied with the eight-site DCA and it was found that the susceptibility demonstrates a well-pronounced peak at a characteristic frequency. The phases found in this work are shown in the doping–temperature phase diagram in Fig. 5.3.

In Ref. [12], the boundary line between the PM metal and the AFM insulator phases in the half-filled 2D Hubbard model was obtained with the eight-site cluster DCA at different values of temperature, U and nearest-neighbor repulsion V. Several important results were found: a hysteresis between the AFM–charge order and normal-charge order boundaries and a decrease of the AFM T_c with V increase. The obtained diagrams are shown in Figs. 5.4 and 5.5, and the V-dependence of the AFM and the charge-order order parameters in the AFM–charge-order hysteresis region are shown in Fig. 5.6.

Thus, as these results show the nearest-neighbor repulsion, though it is much weaker than the on-site repulsion, can play a very important role. At this point, it is necessary to mention that the original DMFT equations with local self-energy were derived based on the local interaction. The cluster generalization of DMFT naturally includes inter-site interactions inside the cluster, which results in a nonlocal short-

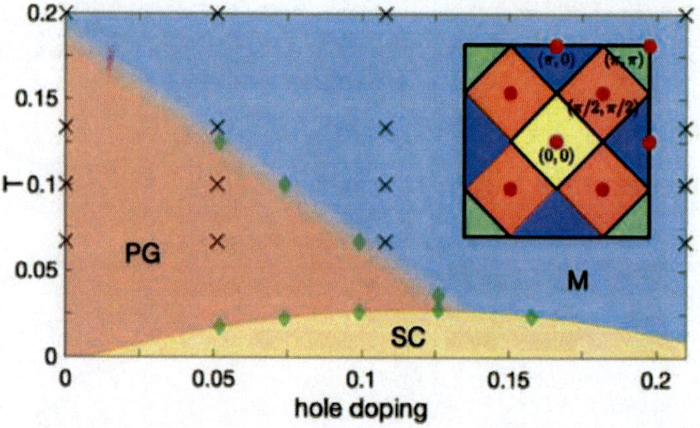

Fig. 5.3 The doping–temperature phase diagram of the 2D Hubbard model obtained with the eight-site DCA (square lattice, geometry of the cluster and the used cluster momenta are shown in the inset; $U = 7t$, $t' = -0.15t$, CT-QMC solver). Different phases are marked as: *M* metallic, *PG* pseudogap, *SC* superconducting. Cross points correspond to the values of temperature and doping at which the calculations were performed, the green diamond points were used to define the phase boundaries. (Reprinted Fig. 1 with permission from (X. Dong, X. Chen, and E. Gull, Phys. Rev. B **100**, 235107 (2019)). Copyright 2019 by the American Physical Society. https://doi.org/10.1103/PhysRevB.100.235107)

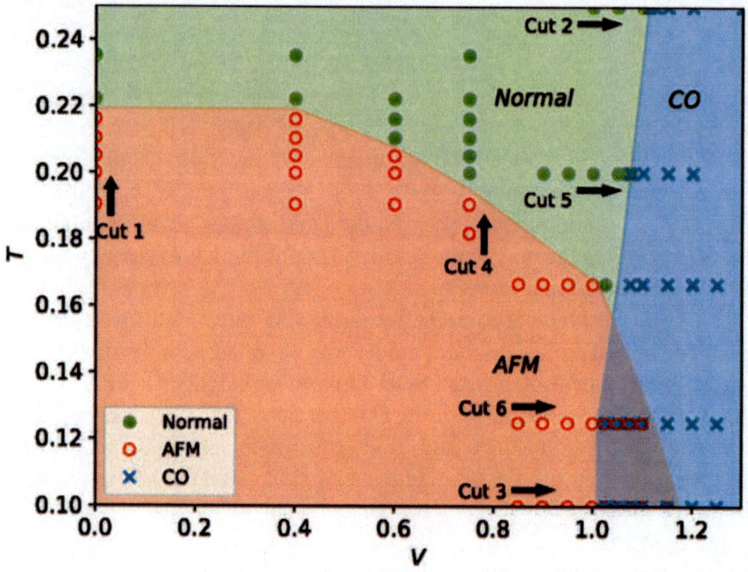

Fig. 5.4 The 8-site DCA T–V phase diagram in the half-filled 2D Hubbard model obtained with marked simulation points (square lattice, $U = 4t$, CT auxiliary field QMC solver). The simulation points are marked by different symbols and the different overlapping symbols in the (gray) hysteresis area show the first-order phase coexistence. (Reprinted Fig.1 with permission from [J. Paki, H. Terletska, S. Iskakov, and E. Gull, Phys. Rev. B **99**, 245146 (2019)). Copyright 2019 by the American Physical Society. https://doi.org/10.1103/PhysRevB.99.245146)

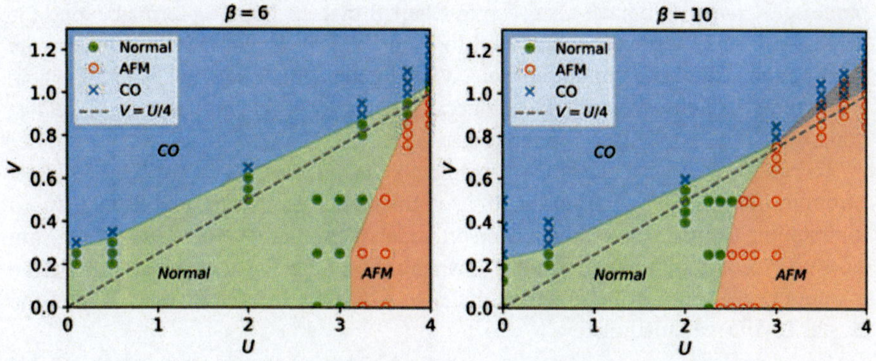

Fig. 5.5 The 8-site DCA $U - V$ phase diagrams in the case of different temperatures in the 2D half-filled Hubbard model. Details of the solution and other parameters are given in the caption to Fig. 5.4. (Reprinted Fig. 2 with permission from (J. Paki, H. Terletska, S. Iskakov, and E. Gull, Phys. Rev. B **99**, 245146 (2019)). Copyright 2019 by the American Physical Society. https://doi.org/10.1103/PhysRevB.99.245146)

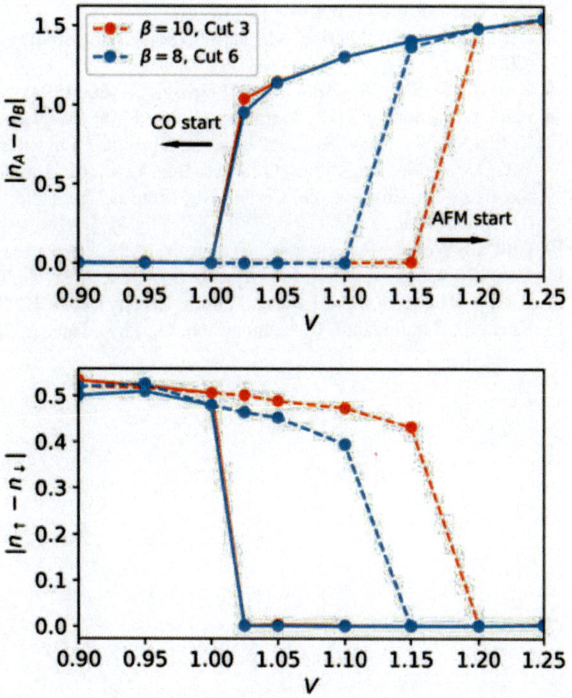

Fig. 5.6 The inter-site interaction V-dependence of the charge order and the AFM order parameters for the half-filled 2D Hubbard model at $U = 4t$ and inverse temperatures $\beta = 10/t$ and $\beta = 8/t$ obtained with 8-site DCA. Details of the solution and other parameters are given in the caption to Fig. 5.4. (Reprinted Fig. 5 with permission from (J. Paki, H. Terletska, S. Iskakov, and E. Gull, Phys. Rev. B **99**, 245146 (2019)). Copyright 2019 by the American Physical Society. https://doi.org/10.1103/PhysRevB.99.245146)

range self-energy. We have also seen in Chap. 3 that the hopping terms beyond the nearest-neighbor approximation (used to derive the Gaussian DOS in the one-site DMFT) can also play an important role. More on this will be said in Chap. 10, where the Hubbard model with nonlocal interaction and/or next- and next-next-nearest hopping will be applied to different materials.

To conclude this subsection, we would like to mention Ref. [13] where cluster sum rules for the hybridization functions in the Anderson impurity model V_{lm} (used for mapping the problem on the Anderson impurity problem) were found. These sum rules connect the hybridization functions with the connectivity of the lattice and the hopping parameters of the model and can be used, e.g., to test the accuracy of the cluster DMFT calculations.

References

1. Kanamori, J.: Prog. Theor. Phys. **30**, 275 (1963)
2. Maier, T., Jarrell, M., Pruschke, T., Hettler, M.H.: Rev. Mod. Phys. **77**, 1027 (2005)
3. Hettler, M.H., Tahvildar-Zadeh, A.N., Jarrell, M., Pruschke, T., Krishnamurthy, H.R.: Phys. Rev. B. **58**, R7485 (2000)
4. Hettler, M.H., Mukherjee, M., Jarrell, M., Krishnamurthy, H.R.: Phys. Rev. B. **61**, 12739 (2000)
5. Gros, C., Valenti, R.: Ann. Phys. (Leipzig). **3**, 460 (1994)
6. Pairault, S., Senechal, D., Tremblay, A.-M.: Phys. Rev. Lett. **80**, 5389 (1998)
7. Kotliar, G., Savrasov, S.Y., Palsson, G., Biroli, G.: Phys. Rev. Lett. **87**, 186401 (2001)
8. Park, H., Haule, K., Kotliar, G.: Phys. Rev. Lett. **101**, 186403 (2008)
9. Sakai, S., Sangiovanni, G., Civelli, M., Motome, Y., Held, K., Imada, M.: Phys. Rev. B. **85**, 035102 (2012)
10. Gull, E., Werner, P., Wang, X., Troyer, M., Millis, A.J.: Eur. Phys. Lett. **84**, 37009 (2008)
11. Dong, X., Chen, X., Gull, E.: Phys. Rev. B. **100**, 235107 (2019)
12. Paki, J., Terletska, H., Iskakov, S., Gull, E.: Phys. Rev. B. **99**, 245146 (2019)
13. Koch, E., Sangiovanni, G., Gunnarsson, O.: Phys. Rev. B. **78**, 115102 (2008)

Chapter 6
DMFT Impurity Solvers

In this chapter, we give details of solving the multi-orbital impurity problem, Eq. (5.5),

$$G_{\sigma lm}(\tau) = \int D[\psi]D[\psi^*]\psi_{\sigma l}(\tau)\psi^*_{\sigma m}(0)$$
$$\times \exp\left[-\int_0^\beta d\tau' \int_0^\beta d\tau'' \sum_{\sigma', a, b} \psi^*_{\sigma' a}(\tau') \left(\widehat{\mathcal{G}}_{\sigma'}^{-1} \right)_{ab} (\tau' - \tau'') \psi_{\sigma' b}(\tau'') \right.$$
$$\left. + \int_0^\beta d\tau' \sum_{\sigma, \sigma', abcd} U^{abcd}_{\sigma\sigma'} \psi^*_{\sigma a}(\tau') \psi^*_{\sigma' b}(\tau') \psi_{\sigma' c}(\tau') \psi_{\sigma d}(\tau') \right] \tag{6.1}$$

using three different types of approaches:

1. analytical approximate solutions (Hubbard-I and Multi-Orbital Iterative Perturbation Theory (MO-IPT))
2. numerical exact (Hamiltonian) solutions (Exact Diagonalization (ED) and Numerical Renormalization Group (NRG))
3. stochastic solutions (Hirsch–Fye Quantum Monte Carlo (HF-QMC) and Continuous-Time Quantum Monte Carlo (CT-QMC))

In most cases, we consider one-atom cell with the on-site interaction $U^{abcd}_{\sigma\sigma'} = \delta^{ad}\delta^{bc}U^{ab}_{\sigma\sigma'}$ which corresponds to taking into account only the charge–charge repulsion $U^{ab}_{\sigma\sigma'}\widehat{n}^a_\sigma\widehat{n}^b_{\sigma'}$ and ignoring all other, e.g., exchange, interactions. Generalization of the methods on the case of clusters or more complex interactions can be straightforwardly made using the algorithms below.

V. Turkowski, *Dynamical Mean-Field Theory for Strongly Correlated Materials*,
https://doi.org/10.1007/978-3-030-64904-3_6

6.1 Analytical solvers

6.1.1 Hubbard-I Solver

Probably, the simplest nontrivial solver for the impurity model (6.1) is the Hubbard-I solver, where the electron self-energy is approximated by the infinite-U result (the Hubbard-I solution of the Hubbard model discussed above), i.e., ignoring the hybridization function. In this subsection, we restrict ourselves to the Coulomb repulsion between electrons with opposite spins, $U^{ab}\widehat{n}^a_\uparrow\widehat{n}^b_\downarrow$.

The expression for the self-energy for the one-orbital case was derived in Chapter 2 Eq. (2.51):

$$\Sigma^{(at)}_\sigma(\omega) = Un_{\bar{\sigma}} + \frac{U^2 n_{\bar{\sigma}}(1-n_{\bar{\sigma}})}{\omega - U(1-n_{\bar{\sigma}})}. \tag{6.2}$$

Let us consider general, multi-orbital, case. To find the self-energy, we first write down the equation for the retarded orbital GF $G^{ab}_{ij,\sigma}(t-t') = -i\theta(t-t') \times \left\langle\left\{\widehat{c}^a_{i,\sigma}(t), \widehat{c}^{b+}_{j,\sigma}(t')\right\}\right\rangle$ (for a better visualization, we use the upper case for the orbital indices). It has the following form:

$$\left(i\frac{d}{dt}+\mu\right)G^{ab}_{ij,\sigma}(t-t') = \delta^{ab}\delta_{ij}\delta(t-t') - \sum_{l,c} t^{ac}_{il}G^{cb}_{lj,\sigma}(t-t') + \Gamma^{ab}_{ij,\sigma}(t-t'), \tag{6.3}$$

where

$$\Gamma^{ab}_{ij,\sigma}(t-t') = (-i)\theta(t-t')\sum_c U^{ac}_i\left\langle\left\{\widehat{c}^a_{i,\sigma}(t)\widehat{n}^c_{i,\bar{\sigma}}(t), \widehat{c}^{b+}_{j,\sigma}(t')\right\}\right\rangle. \tag{6.4}$$

Thus, to find the GF one needs to find the retarded function $\Gamma^{ab}_{ij,\sigma}(t-t')$. Since we consider the Hubbard-I solution, let us find this function in the atomic limit (when the hopping parameters are zero), where $\Gamma^{ab}_{ij,\sigma}(t-t')$ satisfies the following equation of motion:

$$\begin{aligned} i\frac{d}{dt}\Gamma^{ab}_{ij,\sigma}(t-t') &= \delta^{ab}\delta_{ij}\delta(t-t')\sum_c U^{ac}_i\langle\widehat{n}^c_{i,\bar{\sigma}}(t)\rangle - \mu\Gamma^{ab}_{ij,\sigma}(t-t') \\ &\quad - i\theta(t-t')\sum_{c,d} U^{ac}_i U^{ad}_i\left\langle\left\{\widehat{c}^a_{i,\sigma}\widehat{n}^c_{i,\bar{\sigma}}\widehat{n}^d_{i,\bar{\sigma}}(t), \widehat{c}^{b+}_{j,\sigma}(t')\right\}\right\rangle. \end{aligned} \tag{6.5}$$

The last term in this equation is a higher-order the same two lines below GF. The equation of motion for this function contains the next-order GF. If one writes down equation of motion for this next-order GF, a higher-order GF will appear. Thus, to solve the problem one needs to truncate the system of equations. For this reason, we use a mean-field (HF) approximation for the last term in Eq. (6.5), where we put

$$\begin{aligned}\widehat{n}^{c}_{i,\bar{\sigma}}\widehat{n}^{d}_{i,\bar{\sigma}}(t) &\equiv \delta^{cd}\widehat{n}^{c}_{i,\bar{\sigma}} + \left(1-\delta^{cd}\right)\widehat{n}^{c}_{i,\bar{\sigma}}\widehat{n}^{d}_{i,\bar{\sigma}} \\ &\approx \delta^{ab}\widehat{n}^{c}_{i,\bar{\sigma}} + \left(1-\delta^{cd}\right)\left(\langle\widehat{n}^{c}_{i,\bar{\sigma}}\rangle\widehat{n}^{d}_{i,\bar{\sigma}} + \widehat{n}^{c}_{i,\bar{\sigma}}\left\langle\widehat{n}^{d}_{i,\bar{\sigma}}\right\rangle\right).\end{aligned} \tag{6.6}$$

Then, after simple transformations, the last term in Eq. (6.5) becomes

$$-i\theta(t-t')\sum_{c}\left[U^{ac}_{i} + \sum_{d\neq c}U^{ad}_{i}\left\langle\widehat{n}^{d}_{i,\bar{\sigma}}\right\rangle\right]U^{ac}_{i}\left\langle\left\{\widehat{c}^{a}_{i,\sigma}\widehat{n}^{c}_{i,\bar{\sigma}}(t),\widehat{c}^{b+}_{j,\sigma}(t')\right\}\right\rangle. \tag{6.7}$$

Finally, to express (6.7) in terms of $\Gamma^{ab}_{ij,\sigma}(t-t')$, one can substitute $\left[U^{ac}_{i} + \sum_{d\neq c}U^{ad}_{i}\left\langle\widehat{n}^{d}_{i,\bar{\sigma}}\right\rangle\right]$ by an average over orbitals (with index c) function

$$\begin{aligned}\overline{U}^{a}_{i} &= \frac{1}{N_{orb}}\sum_{c}\left[U^{ac}_{i} + \sum_{d\neq c}U^{ad}_{i}\left\langle\widehat{n}^{d}_{i,\bar{\sigma}}\right\rangle\right] \\ &\equiv \frac{1}{N_{orb}}\sum_{c}U^{ac}_{i}\left[1-\langle\widehat{n}^{c}_{i,\bar{\sigma}}\rangle\right] + \sum_{c}U^{ac}_{i}\langle\widehat{n}^{c}_{i,\bar{\sigma}}\rangle,\end{aligned} \tag{6.8}$$

take this function out of the sum in Eq. (6.7), put the result into Eq. (6.5) instead of the last term. This will result in the required equation:

$$\begin{aligned}i\frac{d}{dt}\Gamma^{ab}_{ij,\sigma}(t-t') = {} & \delta^{ab}\delta_{ij}\delta(t-t')\sum_{c}U^{ac}_{i}\langle\widehat{n}^{c}_{i,\bar{\sigma}}(t)\rangle - \mu\Gamma^{ab}_{ij,\sigma}(t-t') \\ & + \overline{U}^{a}_{i}\Gamma^{ab}_{ij,\sigma}(t-t').\end{aligned} \tag{6.9}$$

This equation can be easily solved in the frequency representation:

$$\Gamma^{ab}_{ij,\sigma}(\omega) = \frac{\delta^{ab}\delta_{ij}\sum_{c}U^{ac}_{i}\langle\widehat{n}^{c}_{i,\bar{\sigma}}\rangle}{\omega+\mu-\overline{U}^{a}_{i}}. \tag{6.10}$$

It is advantageous that within used approximation, Γ is diagonal in the site and orbital indices, which also results in the (atomic-limit) diagonal GF, since as it follows from Eq. (6.3),

$$G_{ij,\sigma}^{ab}(\omega) = \frac{\delta^{ab}\delta_{ij} + \Gamma_{ij,\sigma}^{ab}(\omega)}{\omega + \mu}. \tag{6.11}$$

Substituting solution (6.10) into Eq. (6.11), one gets the multi-orbital GF in the atomic limit (Hubbard-I approximation):

$$G_{ii,\sigma}^{aa}(\omega) = \frac{(1/\overline{U}_i^a)\sum_c U_i^{ac}\langle\widehat{n}_{i,\bar{\sigma}}^c\rangle - 1}{(\omega+\mu)} + \frac{(1/\overline{U}_i^a)\sum_c U_i^{ac}\langle\widehat{n}_{i,\bar{\sigma}}^c\rangle}{(\omega+\mu-\overline{U}_i^a)}. \tag{6.12}$$

Comparing the equation and $G_{ii,\sigma}^{aa}(\omega) = \frac{1}{\omega+\mu-\Sigma_{ii,\sigma}^{aa}(\omega)}$, one finds the multi-orbital version of the Hubbard-I self-energy (within used approximation (6.7), (6.8)):

$$\Sigma_i^{aa}(\omega) = \sum_c U_i^{ac}\langle\widehat{n}_{i,\bar{\sigma}}^c\rangle + \left(\sum_c U_i^{ac}\langle\widehat{n}_{i,\bar{\sigma}}^c\rangle\right)\frac{\frac{1}{N_{orb}}\sum_c U_i^{ac}(1-\langle\widehat{n}_{i,\bar{\sigma}}^c\rangle)}{\omega+\mu-\sum_c U_i^{ac}(1-\langle\widehat{n}_{i,\bar{\sigma}}^c\rangle)}. \tag{6.13}$$

Solutions (6.12) and (6.13) reduce to the one-orbital solutions with $N_{\mathrm{orb}} = 1$ and $U_i^{ab} = \delta^{ab}U_i$, given by Eqs. (2.50) and (6.2), correspondingly.

As it was discussed in Sect. 2.2, the Hubbard-I approximation has several serious shortcomings: besides being valid only for large values of U, it does not give the insulator-to-metal transition with changing U/t, it gives FM phase only for particular cases of the free-electron DOS and violates the electron–hole symmetry. Thus, in the next subsection, we present a more powerful analytical approximation that improves many deficiencies of the Hubbard-I method.

6.1.2 *Iterative Perturbation Theory*

The Hubbard-I solver discussed in the previous subsection is basically applicable in the case of large Us. In this subsection, we present an analytical approximate Iterative Perturbation Theory (IPT) solution of the impurity problem valid for all values of U. IPT is probably the simplest analytical solution of the impurity problem that covers all ranges of U. It is exact in the limits of small and large Us and correctly reproduces the limit of large frequency, $\omega \to \infty$. Though it is not too accurate in the other regions of the parameters, its great advantage is that it is fast comparing to other solvers. To derive the IPT electron self-energy, let us begin with the opposite to the previous subsection case and solve the impurity problem (6.1) in the weak-coupling limit.

In the one-band PM case at half-filling, i.e., at $\langle n_\uparrow\rangle = \langle n_\downarrow\rangle = n = \frac{1}{2}$, the second order in U expansion in the impurity equation (6.1) gives the following expression for the self-energy:

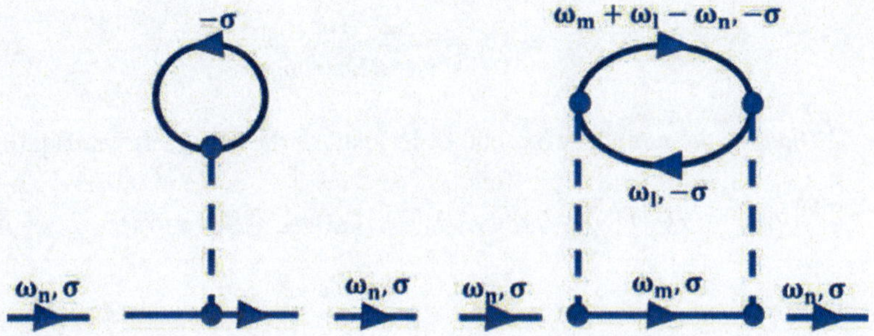

Fig. 6.1 The diagrams corresponding to Eq. (6.16) for the IPT self-energy $\Sigma_\sigma(i\omega_n)$ (in general, spin-polarized, case)

$$\Sigma(i\omega_n) = Un + U^2 \int_0^\beta d\tau e^{i\omega_n \tau} \mathcal{G}^3(\tau), \tag{6.14}$$

Similar to the atomic limit, Eq. (6.2), the self-energy (6.14) has the Hartree (static mean-field) part (the linear in U term). It does not include the Fock term that does not exist in the DMFT approximation. In the last term in Eq. (6.14), $\mathcal{G}(\tau)$ is the familiar dynamical mean-field function that has the following form in the Matsubara frequency representation:

$$\mathcal{G}(i\omega_n) \equiv \frac{1}{i\omega_n + \mu - Un - \Delta(i\omega_n)}. \tag{6.15}$$

The last function contains (in the denominator) the above-mentioned Hartree term Un that we take into account explicitly. This makes function (6.15) different from the "standard" one $\mathcal{G}(i\omega_n) = [i\omega_n + \mu - \Delta(i\omega_n)]^{-1}$.

Matsubara frequency representation on the r.h.s. of the equation (6.14) gives the expression for the IPT self-energy in terms of frequencies only:

$$\Sigma(i\omega_n) = Un - U^2T^2 \sum_{\omega_m, \omega_l} \mathcal{G}_0(i\omega_m)\mathcal{G}_0(i\omega_l)\mathcal{G}_0(i\omega_m + i\omega_l - i\omega_n) \tag{6.16}$$

(the corresponding diagrammatic representation for general (spin-dependent) case discussed below is shown in Fig. 6.1). Besides being exact at small and large Us and high frequencies, the half-filling solution (6.16) has another very important advantage with respect to the Hubbard-I result Eq. (6.2): it gives an MIT. Away from half-filling, the IPT solution does not reproduce the large-U limit. A generalized expression for the one-orbital IPT self-energy valid at arbitrary filling that gives correct large U-limit was proposed in works [1–5] (in Ref. [4] a degenerate multi-orbital case was also considered). Namely, the authors proposed the following phenomenological expression:

$$\Sigma(i\omega_n) = Un + \frac{A\Sigma^{(2)}(i\omega_n)}{1 - B\Sigma^{(2)}(i\omega_n)}, \tag{6.17}$$

where $\Sigma^{(2)}(i\omega_n)$ is the nonlinear (second-order) part of the self-energy in Eq. (6.16):

$$\Sigma^{(2)}(i\omega_n) = -U^2T^2 \sum_{\omega_m,\omega_l} \mathcal{G}_0(i\omega_m)\mathcal{G}_0(i\omega_l)\mathcal{G}_0(i\omega_m + i\omega_l - i\omega_n), \tag{6.18}$$

and

$$\mathcal{G}_0(i\omega_n) \equiv \frac{1}{i\omega_n + \tilde{\mu} - Un - \Delta(i\omega_n)} \tag{6.19}$$

is a modified dynamical mean-field function. In the last equation, $\tilde{\mu}$ is an effective chemical potential (below, we discuss how to fix it).

In addition to the parameter $\tilde{\mu}$, there two other parameters in Eq. (6.17): A and B. It was proposed in Ref. [1] to fix the parameter A from the condition that self-energy (6.17) is exact at high frequencies, and to fix B—by requiring that (6.17) correctly reproduces the atomic limit. It can be shown that in the limit of high frequencies,

$$\Sigma(\omega) = Un + \frac{U^2n(1-n)}{\omega}. \tag{6.20}$$

To obtain this result, one can use the high-frequency expansion in the lattice local GF $G(\omega) = -\frac{1}{\pi}\int d\nu \frac{ImG(\nu+i\delta)}{\omega-\nu} = \sum\limits_{m=0}^{\infty} \frac{M^{(m)}}{\omega^{m+1}}$, where the lowest moments $M^{(m)} = \int d\nu\nu^m ImG(\nu) \equiv \int d\nu\nu^m A(\nu)$ can be obtained in an analytical form by, e.g., differentiating the retarded GF m times with respect to time (see, e.g., Refs. [3, 6]). Then, inserting the expansion $G(\omega) = \sum\limits_{m=0}^{\infty} \frac{M^{(m)}}{\omega^{m+1}}$ and also similar expansion for the noninteracting GF and the expansion for the self-energy $\Sigma(\omega) = \sum\limits_{m=0}^{\infty} \frac{C^{(m)}}{\omega^m}$ into the Dyson equation, one can find analytically the first two coefficients $C^{(m)}$, and as result—to obtain Eq. (6.20).

Comparing result (6.20) with the high-frequency expression for the IPT self-energy in real-frequency representation $\Sigma(\omega) = Un + \frac{A\Sigma^{(2)}(\omega)}{1-B\Sigma^{(2)}(\omega)}$, one can find the coefficient A. Namely, to get the high-frequency result for the IPT self-energy, it is convenient to rewrite $\Sigma^{(2)}(i\omega_n)$ as:

$$\Sigma^{(2)}(i\omega_n) = -\frac{U^3}{\pi^3} \int d\omega_1 Im\mathcal{G}_0(\omega_1 + i\delta) \int d\omega_2 Im\mathcal{G}_0(\omega_2 + i\delta)$$
$$\times \int d\omega_3 Im\mathcal{G}_0(\omega_3 + i\delta)$$
$$\times \frac{f(\omega_1)f(\omega_2)(1 - f(\omega_3)) + (1 - f(\omega_1))(1 - f(\omega_2))f(\omega_3)}{i\omega_n - \omega_1 - \omega_2 + \omega_3}, \quad (6.21)$$

where the spectral representation

$$\mathcal{G}_0(z) = -\frac{1}{\pi} \int d\omega \frac{Im\mathcal{G}_0(\omega + i\delta)}{z - \omega} \quad (6.22)$$

was used.

Then, substituting the expression (6.19) for $\mathcal{G}_0$ into Eq. (6.21), performing analytical continuation $i\omega_n \rightarrow \omega + i\delta$ in this equation, taking the high-frequency limit, performing integration over other frequencies, substituting the result into Eq. (6.17), and comparing with Eq. (6.20), one gets

$$A = \frac{n(1 - n)}{n_0(1 - n_0)}, \quad (6.23)$$

where

$$n_0 = -\frac{1}{\pi} \int d\omega Im\mathcal{G}_0(\omega + i\delta) \quad (6.24)$$

is a fictitious density of electrons corresponding to the bath GF (compare with real density of electrons $n = -\frac{1}{\pi} \int d\omega ImG(\omega + i\delta)$).

To find B, let us obtain large-U limit for the self-energy (6.17) and compare it with Hubbard-I self-energy (6.2). The second-order PT part of the self-energy $\Sigma^{(2)}(\omega + i\delta)$ can be obtained from Eq. (6.21) by using the mean-field functions (6.19) at $\Delta(\omega + i\delta) \rightarrow 0$ and integrating Eq. (6.21) over frequencies. This gives

$$\Sigma^{(2)}(\omega) = \frac{U^2 n_0(1 - n_0)}{\omega + i\delta + \tilde{\mu} - U(1 - n_0)}. \quad (6.25)$$

Substituting this result and the expression for A (6.23) into Eq. (6.17) (in real-frequency representation) and comparing the resulting self-energy with the atomic-limit self-energy Eq. (6.2), one gets

$$B = \frac{(1-2n)U - \mu + \widetilde{\mu}}{n_0(1-n_0)U^2}. \tag{6.26}$$

The remaining parameter $\widetilde{\mu}$ can be found by using different reasonings and/or approximations [3] (for over-review, see, e.g., Ref. [6]). The simplest approximation is $\widetilde{\mu} = \mu$. However, in this case, the DMFT results are very different from more reliable exact diagonalization (ED) solver results [3]. Another option would be to put $n_0 = n$ and to solve the resulting equation $\int d\omega Im G_0(\omega + i\delta) = \int d\omega Im G(\omega + i\delta)$ for $\widetilde{\mu}$ (similar to the solution of the periodic Anderson model [6]). However, this approximation has no sufficient physical justification.

A more physically adjustable method to fix $\widetilde{\mu}$ by using Luttinger theorem [7] (or, equivalently, Friedel's sum rules [8]) was proposed in Ref. [1]. As it was shown in work [9], in infinite dimensions, the Luttinger theorem (the volume enclosed by the Fermi surface is proportional to the particle density and does not depend on the interaction between the particles) gives the necessary equation:

$$\mu = \mu_0 + \Sigma(\omega = 0), \tag{6.27}$$

where μ_0 is the chemical potential in the case $U = 0$ and $\Sigma(\omega = 0)$ is the $\widetilde{\mu}$-dependent IPT self-energy (6.17).

Both $n_0 = n$ and Luttinger theorem solutions are in a good agreement with the ED results [3]. However, in some cases, the first approach is preferable, since Eq. (6.27) has to be solved in the difficult $T = 0$ limit. Moreover, the Luttinger theorem in the first case is violated no more than by a few percent [5].

Thus, the one-orbital IPT self-energy in general case away from half-filling (6.17) is an interpolative solution that is exact in three limits—small and large Us and high frequencies. It is shown to be quite accurate in many cases. For example, IPT results are in a good agreement with the QMC and ED results, except very close to the Mott transition point [1, 10]. To conclude discussion of the one-band case, we would like to mention that looking for a generalized IPT solution by satisfying three limits for the self-energy above is not a unique approach. One can require to fulfill other exact properties, like the value of double occupancy (see Ref. [11] for both one- and multi-orbital case).

The IPT formalism discussed in this subsection can be straightforwardly generalized on the multi-orbital (including spin-dependent) case [4, 12–14], where one can also require fulfillment of the Luttinger theorem, or generalized Friedel's sum rules. In detail, the multi-orbital GF

$$G^a(\omega) = \frac{1}{N}\sum_{\vec{k}} \frac{1}{\omega - \varepsilon^a\left(\vec{k}\right) - \Sigma^a(\omega)} \tag{6.28}$$

depends on the orbital self-energy $\Sigma^a(\omega)$ that can have the following (generalizing the one-orbital expression, in Matsubara frequency representation) form:

$$\Sigma^a(i\omega_n) = \sum_b U^{ab} n^b + \frac{\sum_b A^{ab} \Sigma^{(2)ab}(i\omega_n)}{1 - \sum_b B^{ab} \Sigma^{(2)ab}(i\omega_n)}, \tag{6.29}$$

where the second-order PT part of the self-energy is

$$\Sigma^{(2)ab}(i\omega_n) = -\left(U^{ab}\right)^2 T^2 \sum_{\omega_m, \omega_l} \mathcal{G}_0^a(i\omega_m) \mathcal{G}_0^b(i\omega_l) \mathcal{G}_0^b(i\omega_m + i\omega_l - i\omega_n), \tag{6.30}$$

and

$$A^{ab} = \frac{n^a(1 - 2n^a) + D^{ab}[n]}{n_0^a(1 - n_0^a)}, \tag{6.31}$$

$$B^{ab} = \frac{n_a(1 - 2n_a)U^{ab} + \mu - \mu^a}{2\left(U^{ab}\right)^2 n_0^a(1 - n_0^a)}. \tag{6.32}$$

In Eq. (6.30),

$$\mathcal{G}_0^a(i\omega_n) \equiv \frac{1}{i\omega_n + \mu^a - \Delta^a(i\omega_n)} \tag{6.33}$$

is the effective bath (dynamical mean-field) function for the orbital a (in general case, the "orbital" index also includes spin variable), and $\Delta^a(i\omega_n)$ and μ^a are the orbital dynamical mean-field and orbital chemical potential. Similar to the one-orbital case, $n_0^a = -\frac{1}{\pi}\int d\omega Im \mathcal{G}_0^a(\omega + i\delta)$ is the "fictitious" and n_a is the usual orbital charge density (we consider the PM case and drop the spin indices). The last undefined object, matrix $D^{ab}[n]$, is the orbital charge correlation function that can be defined self-consistently:

$$D^{ab}[n] = \left\langle n^a n^b \right\rangle = \langle n^a \rangle \left\langle n^b \right\rangle - \frac{1}{U^{ab}\pi} \int d\omega f(\omega) Im[\Sigma^a(\omega) G^a(\omega)]. \tag{6.34}$$

Usually, the orbital chemical potentials are fixed by putting $\mu^a = \mu$.

The above approximation is able to describe properties of strongly correlated systems, including the low-T Fermi Liquid phase and MIT as functions of Coulomb repulsion. It is important to mention that there are other IPT approaches, for example, the continuous fraction expansion of the self-energy [15]:

$$\Sigma^a(i\omega_n) = \Sigma^a(\infty) + \frac{A^a}{i\omega_n - B^a - \frac{C^a}{i\omega_n - D^a - \frac{E^a}{i\omega_n - F^a - \ldots}}}, \tag{6.35}$$

where the coefficients can be fixed by requiring the self-energy to satisfy some exact limiting behaviors, like the low- and the high-frequency limits [10, 15]. Another approach that derives the coefficients for the self-energy from satisfying these limits was proposed in Ref. [16], where a similar to Eq. (6.35) expansion was used, and three-particle correlation functions in the equation were decoupled into one- and two-particle correlation functions.

6.2 Exact (Hamiltonian-Based) Solvers

6.2.1 Exact Diagonalization

The ED approach is based on discretizing the continuum of lattice states by a finite set of bath levels and solving exactly the problem of interacting impurity in the field of this finite-state bath. The idea was originally proposed in Ref. [17] and the method was further refined by implementing Lanczos algorithm [18–21].

Let us mention the main elements of the ED approach in the one-orbital case [17]. In DMFT, it is assumed that the on-site lattice GF is equal to the d-electron impurity GF for the Anderson impurity model (AIM) with the Hamiltonian:

$$H^A = \varepsilon_d \sum_\sigma d_\sigma^+ d_\sigma + \sum_{\sigma, l=2}^{n_s} \varepsilon_l a_{l\sigma}^+ a_{l\sigma} + U n_{d\uparrow} n_{d\downarrow} + \sum_{\sigma, l=2}^{n_s} \left(V_l a_{l\sigma}^+ d_\sigma + h.c. \right), \tag{6.36}$$

where in addition to the quantities defined in Sect. 3.2, n_s is the number of sites. The dynamical mean-field function (the bath GF) $\mathcal{G}(i\omega_n)$ is equal to the Anderson impurity GF $G^{(0)A}(i\omega_n)$ *at* $U = 0$. The last function can be easily calculated (we drop the spin indices to simplify notations):

$$G^{(0)A}(i\omega_n) = \frac{1}{i\omega_n - \varepsilon_d + \mu - \sum_{l=2}^{n_s} \frac{(V_l)^2}{i\omega_n - \varepsilon_l}}. \tag{6.37}$$

The goal is to find the parameters for the AIM to make the function (6.37) $G^{(0)A}(i\omega_n)$ as close to the DMFT function $\mathcal{G}(i\omega_n)$ as possible. In other words, one finds parameters ε^l and V^l from $G^{(0)A}(i\omega_n) = \mathcal{G}(i\omega_n)$, puts them into Hamiltonian (6.36), solves this problem, and finds the interacting GF. Thus, Eqs. (6.36) and (6.37) connect the local GF $G(i\omega_n)$ with the dynamical mean-field $\mathcal{G}(i\omega_n)$, similar to the path-integral impurity equation. Generally speaking, the number of the bath levels n_s is infinite, but to realize the task above in practice, one chooses a finite n_s. One can find parameters ε_l and V_l for $G^{(0)A}(i\omega_n)$ by using different methods, e.g., by minimizing the function $\sum_{n=1}^{n_{\max}} \left| \mathcal{G}^{-1}(i\omega_n) - G^{(0)A-1}(i\omega_n) \right|^2$, where $n_{\max}$ is large enough, such that $\omega_{n_{\max}}$ is larger than the other energy scales in the system.

The algorithm of the DMFT solution with the ED solver has the following steps:

1. One starts with a trial self-energy $\Sigma(i\omega_n)$.
2. Finds $G(i\omega_n)$ from Eq. (6.28).
3. Finds $\mathcal{G}(i\omega_n)$ from the DMFT Dyson equation (5.4).
4. Finds the Anderson model parameters ε_l and V_l from $G^{(0)A}(i\omega_n) = \mathcal{G}(i\omega_n)$.
5. Solves the interacting Anderson model (6.36) with these parameters and finds the impurity, i.e., local, GF $G(i\omega_n)$.
6. Use this $G(i\omega_n)$ to find the new self-energy from Eq. (5.4) and repeat the steps until the self-energy converges.

When n_s is not large (≤ 6), one can immediately find GF $G(i\omega_n)$ from Eq. (6.36) by diagonalizing the Anderson Hamiltonian. In the case of larger n_s's (up to ~10), one can still (numerically) diagonalize the Hamiltonian (6.36) at zero temperature by using the Lanczos algorithm (for details, we refer the reader to Refs. [19, 22]). The ED approach was also extended on the case of clusters [23], nearest-site interaction (extended DMFT) [24] (for over-review, see papers [25, 26]), and the multi-orbital case [26, 27]. In the multi-orbital case, one is looking for the local orbital GFs $G^a(i\omega_n)$ defined by Eq. (6.28), the self-energy $\Sigma^a(i\omega_n)$ and the dynamical mean-field function $\mathcal{G}^a(i\omega_n)$, all connected by the Dyson equation $G^{a-1}(i\omega_n) = \mathcal{G}^{a-1}(i\omega_n) - \Sigma^a(i\omega_n)$. The remaining ED impurity problem can be formulated similar to the one-orbital case. Namely, one can introduce the multi-orbital Anderson Hamiltonian

$$H^A = \sum_{\sigma,a=1}^{n_{\text{orb}}} \varepsilon_d^a d_\sigma^+ d_\sigma + \sum_{\sigma,l=2}^{n_s} \varepsilon_l a_{l\sigma}^+ a_{l\sigma} + \sum_{a,b} U^{ab} n_d^a n_d^b + \sum_{\sigma,a=1}^{n_{orb}} \sum_{\sigma,l=n_{\text{orb}}+1}^{n_s} \times \left(V_l^a a_{l\sigma}^+ d_{a\sigma} + h.c.\right), \tag{6.38}$$

with n_{orb} impurity orbitals and $n_s - n_{\text{orb}}$ bath levels. In this case, the $U = 0$ orbital impurity GF is:

$$G^{(0)Aa}(i\omega_n) = \frac{1}{i\omega_n - \varepsilon_d^a + \mu - \sum\limits_{l=n_{orb}+1}^{n_s} \frac{\left(V_l^a\right)^2}{i\omega_n - \varepsilon_l}}. \tag{6.39}$$

The problem can be further simplified by using the fact that the lattice GF is diagonal in the orbital indices. In this case, one can assume that each impurity orbital couples to its own bath, i.e., assuming that for each orbital only a certain (small) number of coefficients V_l^a are nonzero. From $G^{(0)Aa}(i\omega_n) = \mathcal{G}^a(i\omega_n)$ one can find the coefficients ε_l and V_l^a ($a = 1, \ldots, n_{\text{orb}}$), then find the impurity GF and so on, repeating the steps of the one-orbital DMFT calculations.

Let us briefly show how to also generalize the ED-DMFT approach on the case of finite temperatures. The finite-temperature mean-field has the following form:

$$\mathcal{G}^a_\sigma(i\omega_n) = \frac{1}{Z}\sum_{b,c} e^{-E^b/T}\left(\frac{\left|\langle c|d^{a+}_\sigma|b\rangle\right|^2}{i\omega_n + E^c - E^b} + \frac{\left|\langle c|d^a_\sigma|b\rangle\right|^2}{i\omega_n + E^b - E^c}\right), \tag{6.40}$$

where E^b and $|b\rangle$ are the eigenenergies and eigenvectors of the interacting Hamiltonian (6.38) and $Z = \sum_b e^{-E^b/T}$ is the partition function. Other ED-DMFT equations—local GF and the Dyson equations and $G^{(0)Aa}(i\omega_n) = \mathcal{G}^a(i\omega_n)$—have the standard Matsubara representation form (we refer the reader to the review [26] for more details).

To conclude this subsection, we mention that ED together with the Quantum Monte Carlo approaches, described in section 6.3, are the most accurate approaches in the cases of multi-band and cluster systems.

6.2.2 Numerical Renormalization Group (NRG)

Up to now, we considered the Anderson impurity Hamiltonian that consists of a finite discrete set of the energy levels of the itinerant electrons. However, in general case, the itinerant states form a wide-energy range continuum of degrees of freedom and one has to take all energies into account (from sub-meV excitations to several eVs (bandwidth)). This is not an easy task. Another complication comes from the infrared divergencies of the PT (due to coupling to small-energy excitations). Therefore, in the case of an environment with a continuous wide-range excitation spectrum, nonperturbative approaches are needed. One of such approaches is the method of renormalization group (RG) [28]. There are different types of this tool that were applied to the impurity model. Here, we summarize how RG tool (NRG) can be adopted in DMFT to solve the Anderson impurity problem [29–32] (for an over-review, see paper [33]; in this subsection, we mostly follow the presentation in this work).

Since we are interested in the Anderson impurity model for the DMFT solver, our main goal is to formulate how to solve the model for given dynamical mean-field $\Delta(\omega)$ that defines the bath of noninteracting electrons. In NRG, $\Delta(\omega)$ is the only function that is used to solve the model and other details, like dimensionality of the system, are not important. For simplicity, we assume that $\Delta(\omega)$ is a nonzero constant in the frequency interval $[-1 : 1]$ (1 is the energy scale). Before presenting details of the formalism, we mention the key steps of the RG approach:

1. To divide the frequency intervals "logarithmically": $[-\Lambda^{-n} : -\Lambda^{-(n+1)}]$ for the negative frequencies part and $[\Lambda^{-(n+1)} : \Lambda^{-n}]$ for the positive frequencies one, where $\Lambda > 1$ and $n = 0, 1, 2, \ldots$ (Fig. 6.2a).
2. To approximate the continuous spectrum $\Delta(\omega)$ in each interval by one discrete state (logarithmic discretization, Fig. 6.2b).
3. To map the discretized model on a semi-infinite chain model (Fig. 6.2c) that is iteratively diagonalized.

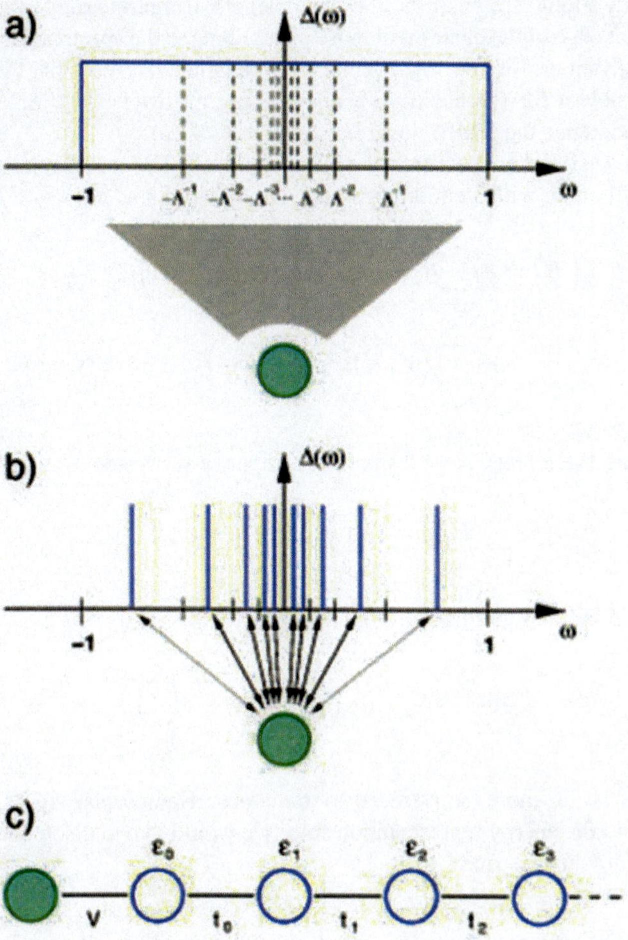

Fig. 6.2 Steps of the NRG mapping of the Anderson impurity model with a continuous bath spectrum on a semi-infinite discrete chain (see the text). Reprinted Fig. 1 with permission from (R. Bulla, Th.A. Costi, and Th. Pruschke, Rev. Mod. Phys. **80**, 395 (2008)). Copyright 2008 by the American Physical Society. DOI: https://doi.org/10.1103/RevModPhys.80.395

4. Increasing the number of intervals by one, to calculate the energy spectrum, keeping only the low-energy part of the spectrum and to continue until this part of the spectrum converges.

During the iterative diagonalization, besides the energy spectrum one can also extract other physical quantities (defined by fixed points, see below).

Logarithmic discretization is very important for the NRG analysis since after such a procedure the hopping parameters decrease fast along the chain, as $t_n \sim \Lambda^{-n/2}$ (indeed, it can be shown that adding one interval/site to the chain results in a decrease of the relevant energy scale by $\sqrt{\Lambda}$). Since the hopping parameters decrease exponentially along the chain as the chain length increases, only states near the impurity (left edge of the chain) will mostly contribute to the spectrum of the chain—the main advantage of the logarithmic discretization. By solving the Anderson impurity problem for the chain with discrete energy levels (Fig. 6.2c) one finds $G^{imp}(\omega)$, and hence the DMFT local GF $G(\omega) \equiv G^{imp}(\omega)$.

Now, let us give more details of the NRG approach. The used Anderson impurity model Hamiltonian with a continuous bath can be written as

$$H^A = \varepsilon_d \sum_\sigma d_\sigma^+ d_\sigma + U n_{d\uparrow} n_{d\downarrow} + \sum_{\sigma,\vec{k}} \varepsilon\left(\vec{k}\right) a_{\vec{k}\sigma}^+ a_{\vec{k}\sigma} + \sum_{\sigma,\vec{k}} \left(V\left(\vec{k}\right) a_{\vec{k}\sigma}^+ d_\sigma + h.c. \right). \tag{6.41}$$

In this model, the effects of the bath on the impurity subsystem

$$H_{imp} = \varepsilon_d \sum_\sigma d_\sigma^+ d_\sigma + U n_{d\uparrow} n_{d\downarrow} \tag{6.42}$$

are described by the hybridization function

$$\Delta(\omega) = \pi \sum_{\vec{k}} \left(V\left(\vec{k}\right) \right)^2 \delta\left(\omega - \varepsilon\left(\vec{k}\right) \right). \tag{6.43}$$

However, it is more convenient to transform Hamiltonian (6.41) from the momentum to the energy representation since we would like to discretize the energy domain ($\Delta(\omega)$ changes from -1 to 1):

$$H^A = \varepsilon_d \sum_\sigma d_\sigma^+ d_\sigma + U n_{d\uparrow} n_{d\downarrow} + \sum_\sigma \int_{-1}^{1} d\varepsilon g(\varepsilon) a_{\varepsilon\sigma}^+ a_{\varepsilon\sigma} + \sum_\sigma \int_{-1}^{1} d\varepsilon h(\varepsilon) \times \left(a_{\varepsilon\sigma}^+ d_\sigma + h.c. \right), \tag{6.44}$$

where the spectrum $g(\varepsilon)$ and the hybridization $h(\varepsilon)$ functions are connected to $\Delta(\omega)$ as

$$\Delta(\omega) = \pi \frac{d\varepsilon(\omega)}{d\omega} h[\varepsilon(\omega)]^2 \tag{6.45}$$

and $\varepsilon(\omega)$ is the inverse function of $g(\varepsilon)$ $[g(\varepsilon(\omega)) = \omega]$ (Ref. [33]). In the case of a constant dynamical mean field, $\Delta(\omega) = \Delta_0$, one can choose $\varepsilon(\omega) = \omega$, or $g(\varepsilon) = \varepsilon$, and $h^2(\varepsilon) = \Delta_0/\pi$.

For the discrete energy points in the Hamiltonian (6.44), $x_n = \pm\Lambda^{-n}$, with the energy interval lengths $d_n = \Lambda^{-n}(1 - \Lambda^{-1})$ ($n = 0, 1, 2, \ldots$; see Fig. 6.2) one can introduce a complete set of orthonormal functions for each interval ($x_{n+1} < \pm\varepsilon < x_n$):

$$\psi_{np}^{\pm}(\varepsilon) = \frac{1}{\sqrt{d_n}} e^{\pm i\omega_n p\varepsilon}, \tag{6.46}$$

where $\omega_n = 2\pi/d_n$ is the fundamental frequency for nth interval and p is the integer number that changes from $-\infty$ to $+\infty$ (each $\psi_{np}^{-}(\varepsilon)$ is zero beyond its energy interval $x_{n+1} < \pm\varepsilon < x_n$). Then, expanding the bath fermion operators in the basis (6.46), $a_{\varepsilon\sigma} = \sum_{n,p}\left[a_{np\sigma}\psi_{np}^{+}(\varepsilon) + b_{np\sigma}\psi_{np}^{-}(\varepsilon)\right]$, and putting the result into the Hamiltonian (6.44), one obtains the following Hamiltonian for the discrete variables:

$$H^A = \varepsilon^d \sum_{\sigma} d_{\sigma}^{+} d_{\sigma} + U n_{d\uparrow} n_{d\downarrow} + \sum_{n,\sigma}\left[\xi_n^{+} a_{n\sigma}^{+} a_{n\sigma} + \xi_n^{-} b_{n\sigma}^{+} b_{n\sigma}\right]$$
$$+ \frac{1}{\sqrt{\pi}} \sum_{n,\sigma}\left(d_{\sigma}^{+}\left[\gamma_n^{+} a_{n\sigma} + \gamma_n^{-} b_{n\sigma}\right] + h.c.\right), \tag{6.47}$$

where

$$\xi_n^{+} = \int_{x_{n+1}}^{x_n} \varepsilon\Delta(\varepsilon)d\varepsilon \Big/ \int_{x_{n+1}}^{x_n} \Delta(\varepsilon)d\varepsilon, \quad \xi_n^{-} = \frac{\int_{-x_n}^{-x_{n+1}} \varepsilon\Delta(\varepsilon)d\varepsilon}{\int_{-x_n}^{-x_{n+1}} \Delta(\varepsilon)d\varepsilon}, \tag{6.48}$$

$$\gamma_n^{+} = \sqrt{\int_{x_{n+1}}^{x_n} \Delta(\varepsilon)d\varepsilon}, \quad \gamma_n^{-} = \sqrt{\int_{-x_n}^{-x_{n+1}} \Delta(\varepsilon)d\varepsilon}. \tag{6.49}$$

In deriving the Hamiltonian Eq. (6.47), the following assumptions were made. It was assumed that $h(\varepsilon)$ is a constant in each interval, $h(\varepsilon) = h_n^{\pm}$ for $x_{n+1} < \pm\varepsilon < x_n$, where $h_n^{\pm} = 1/\sqrt{\pi d_n}\gamma_n^{\pm}$ are the averaged in each interval hybridizations (the square roots in Eq. (6.49) are used because the expressions under the root have the unit of $[\text{energy}]^2$, which gives for the hybridization/hopping the required unit of energy). In the kinetic (bath) part of the Hamiltonian (6.47), the mixed terms with different states p and p', like $f_n^{+}(p,p')a_{np\sigma}^{+}a_{np'\sigma}$ (and similar for the $b_{np\sigma}^{+}b_{np'\sigma}$ part), were neglected. Moreover, all the terms with $p \neq 0$ were neglected too (the reason for this is that these states couple to the impurity only indirectly (through the $p = 0$ state) and the coupling between $p = 0$ and $p \neq 0$ states vanishes as $1 - \Lambda^{-1}$ when $\Lambda \to 1$ (Λ has to be close to 1 to get an accurate solution of the impurity problem)). As the NRG calculations show, this approximation in the kinetic part of the Hamiltonian gives very good results in a majority of cases.

The Hamiltonian (6.47) has to be mapped on the semi-infinite chain. In the case $\Delta(\omega) = \Delta_0$, the coefficients in this Hamiltonian have an extremely simple form: $\xi_n^{\pm} = \frac{1}{2}\Lambda^{-n}(1 + \Lambda^{-1})$, $\gamma_n^{\pm} = \sqrt{\Delta_0\Lambda^{-n}(1 - \Lambda^{-1})}$. From Eq. (6.47) one can see that the impurity electrons in this discrete model interact with the electrons described by the operator

$$c_{0\sigma} = \frac{1}{\sqrt{\xi_0}} \sum_{n,\sigma} \left[\gamma_n^+ a_{n\sigma} + \gamma_n^- b_{n\sigma} \right], \tag{6.50}$$

where the multiplier

$$\xi_0 = \left(\gamma_n^+\right)^2 + \left(\gamma_n^-\right)^2 \tag{6.51}$$

is introduced for normalization. Thus, we assume that the site 0 with states described by electron state operator (6.50) is next to the impurity ("d-operator") site. Now, let us show that all other operators (the last sum in the first line of Eq. (6.47)) can be identified with electron operators $c_{n>0,\sigma}$ (n is the site number) with states on the other atoms in the semi-infinite chain (Fig. 6.2). This situation is described by the Hamiltonian

$$\begin{aligned} H^A = {} & \varepsilon_d \sum_\sigma d_\sigma^+ d_\sigma + U n_{d\uparrow} n_{d\downarrow} + \sqrt{\frac{\xi_0}{\pi}} \frac{1}{\sqrt{\pi}} \sum_\sigma \left(d_\sigma^+ c_{0\sigma} + c_{0\sigma}^+ d_{0\sigma} \right) \\ & + \sum_{\sigma,n=0}^{\infty} \left[\varepsilon_n c_{n\sigma}^+ c_{n\sigma} + t_n c_{n\sigma}^+ c_{n+1\sigma} + t_n c_{n\sigma}^+ c_{n+1\sigma} \right]. \end{aligned} \tag{6.52}$$

To prove this statement, one can express $a_{n\sigma}$ and $b_{n\sigma}$ (Eq. 6.47) in terms of new $c_{n\sigma}$ operators:

$$a_{n\sigma} = \sum_{m=0}^{\infty} u_{nm} c_{m\sigma}, \quad b_{n\sigma} = \sum_{m=0}^{\infty} v_{nm} c_{m\sigma}, \quad c_{n\sigma} = \sum_{m=0}^{\infty} \left(u_{nm} a_{m\sigma} + v_{nm} b_{m\sigma} \right). \tag{6.53}$$

From Eq. (6.50) one finds

$$u_{0m} = \frac{\gamma_m^+}{\sqrt{\xi_0}}, \qquad v_{0m} = \frac{\gamma_m^-}{\sqrt{\xi_0}}. \tag{6.54}$$

The remaining coefficients u_{nm} and v_{nm} can be found from:

$$\sum_{n,\sigma} \left[\xi_n^+ a_{n\sigma}^+ a_{n\sigma} + \xi_n^- b_{n\sigma}^+ b_{n\sigma} \right] = \sum_{\sigma,n=0}^{\infty} \left[\varepsilon_n c_{n+1\sigma}^+ c_{n\sigma} + t_n c_{n\sigma}^+ c_{n+1\sigma} + t_n c_{n+1\sigma}^+ c_{n\sigma} \right] \tag{6.55}$$

with the "starting" relations

$$\varepsilon_0 = \frac{1}{\xi_0} \int_{-1}^{1} \varepsilon \Delta(\varepsilon) d\varepsilon, \quad t_0^2 = \sum_m \left[\left(\xi_n^+ - \varepsilon_0\right)^2 \left(\gamma_m^+\right)^2 + \left(\xi_n^- - \varepsilon_0\right)^2 \left(\gamma_m^-\right)^2 \right], \tag{6.56}$$

$$u_{1m} = \frac{1}{t_0}\left(\xi_n^+ - \varepsilon_0\right)u_{0m}, \quad u_{1m} = \frac{1}{t_0}\left(\xi_n^+ - \varepsilon_0\right)u_{0m}. \tag{6.57}$$

For $n > 1$ one has

$$\varepsilon_n = \sum_m [\xi_m^+ u_{nm}^2 + \xi_m^- v_{nm}^2],$$
$$t_n^2 = \sum_m \left[\left(\xi_n^+\right)^2 u_{nm}^2 + \left(\xi_n^-\right)^2 v_{nm}^2\right] - t_{n-1}^2 - \varepsilon_n, \tag{6.58}$$

$$u_{n+1m} = \frac{1}{t_n}\left(\xi_m^+ - \varepsilon_n\right)u_{nm} - t_{n-1}u_{n-1,m},$$
$$v_{n+1m} = \frac{1}{t_n}\left(\xi_m^- - \varepsilon_n\right)v_{nm} - t_{n-1}v_{n-1,m}. \tag{6.59}$$

Importantly, for an even function, $\Delta(\omega) = \Delta(-\omega)$, all ε_n are equal zero, a significant simplification for the model. For one of such an even dynamical mean-field functions, $\Delta(\omega) = \Delta_0 =$ constant, one can also find a simple result for the hoppings:

$$t_n = \frac{\left(1 + \Lambda^{-1}\right)\left(1 - \Lambda^{-n-1}\right)}{2\sqrt{1 - \Lambda^{-2n-1}}\sqrt{1 - \Lambda^{-2n-3}}}\Lambda^{-n/2} \to \frac{1}{2}\left(1 + \Lambda^{-1}\right)\Lambda^{-n/2} \quad \text{at} \quad n \to \infty. \tag{6.60}$$

This result was originally obtained by Wilson for the Kondo impurity model with constant DOS (see Ref. [28]). It is very important for the next steps that the hopping parameters (6.60) fall exponentially with distance.

Thus, we obtained a discrete-energy semi-infinite chain representation of the Anderson impurity model with continuous bath states. The mapped problem is defined by Eqs. (6.52)–(6.54) and (6.56)–(6.59). Now, we can move to the iterative diagonalization of the Hamiltonian (6.52).

The chain of the iterative Hamiltonians approaches the original continuous-bath Hamiltonian H as number of sites N increases to infinity:

$$H = \lim_{N\to\infty} \Lambda^{-(N-1)/2} H_N, \tag{6.61}$$

where

$$H_N = \Lambda^{(N-1)/2}\left(\varepsilon_d \sum_\sigma d_\sigma^+ d_\sigma + U n_{d\uparrow} n_{d\downarrow} + \sqrt{\frac{\xi_0}{\pi}}\frac{1}{\sqrt{\pi}}\sum_\sigma \left(d_\sigma^+ c_{0\sigma} + c_{0\sigma}^+ d_{0\sigma}\right)\right.$$
$$\left. + \sum_{\sigma,n=0}^{N} \varepsilon_n c_{n+1\sigma}^+ c_{n\sigma} + \sum_{\sigma,n=0}^{N-1} \left[t_n c_{n\sigma}^+ c_{n+1\sigma} + t_n c_{n+1\sigma}^+ c_{n\sigma}\right]\right). \tag{6.62}$$

In the last equation, the factor $\Lambda^{(N-1)/2}$ was introduced to cancel Λ-dependence of the hopping parameter between the last two sites, since for these sites $t_{N-1} \sim \Lambda^{-(N-1)/2}$ (see Eq. 6.60). This scaling is convenient for the analysis of fixed points (see below). To compensate $\Lambda^{(N-1)/2}$ in Eq. (6.62), the inverse pre-factor was added in the exact Hamiltonian (6.61).

From Eq. (6.62), one can get the starting two-site Hamiltonian:

$$H_0 = \Lambda^{-1/2}\left(\varepsilon_d \sum_\sigma d_\sigma^+ d_\sigma + U n_{d\uparrow} n_{d\downarrow} + \sqrt{\frac{\xi_0}{\pi}} \frac{1}{\sqrt{\pi}} \sum_\sigma \left(d_\sigma^+ c_{0\sigma} + c_{0\sigma}^+ d_{0\sigma}\right) + \sum_\sigma \varepsilon_0 c_{0\sigma}^+ c_{0\sigma}\right), \tag{6.63}$$

and the recursive relation:

$$\begin{aligned} H_{N+1} = \sqrt{\Lambda} H_N &+ \Lambda^{N/2} \sum_\sigma \varepsilon_{N+1} c_{N+1\sigma}^+ c_{N+1\sigma} \\ &+ \Lambda^{N/2} \sum_\sigma \left[t_N c_{N\sigma}^+ c_{N+1\sigma} + t_N c_{N+1\sigma}^+ c_{N\sigma}\right]. \end{aligned} \tag{6.64}$$

The last relation is the RG transformation

$$H_{N+1} = R[H_N]. \tag{6.65}$$

Usually, in the RG approaches such a transformation allows one to map a Hamiltonian with a set of parameters, or coupling constants, P into a formally the same Hamiltonian with modified parameters P', renormalized coupling constants. Analyzing the flow of the coupling constants allows one to describe different properties of the system. Since in our case the Hamiltonians before and after the NRG iteration have different forms, it is impossible to renormalize the coupling constants, and we characterize H_N by the energy spectrum (set of eigenenergies) $E_N(r)$:

$$H_N |r\rangle_N = E_N(r) |r\rangle_N, \tag{6.66}$$

where $|r\rangle_N$ are the corresponding eigenvectors ($r = 1, \ldots, N_s$; N_s is the dimension of the Hamiltonian H_N). Once the problem (6.66) is solved, one can obtain the flow of $E_N(r)$ to analyze properties of the system (see below). To solve the problem iteratively, the new basis set can be built by using the previous one as

$$|s; r\rangle_{N+1} = |r\rangle_N \otimes |s(N+1)\rangle. \tag{6.67}$$

The basis set $|s; r\rangle_{N+1}$ is the product of the basis set $|r\rangle_N$ build of eigenvectors of H_N and of a new site basis of states $|s(N+1)\rangle$. Using (6.67), one can construct a new Hamiltonian: $H_{N+1}(r, s; r', s') = {}_{N+1}\langle r, s| H_{N+1} |r', s'\rangle_{N+1}$ and obtain a new eigen-

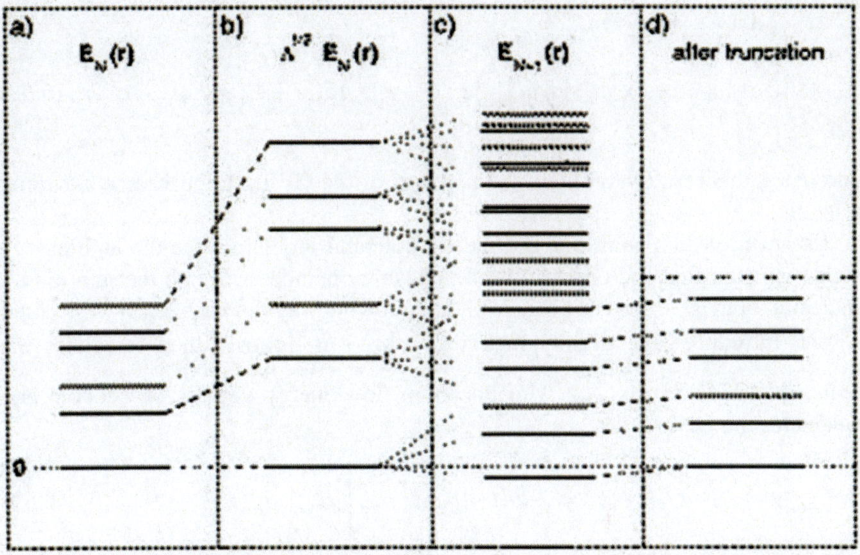

Fig. 6.3 The NRG iterative steps of generating the spectrum by diagonalizing the Hamiltonian matrix (see the text). Reprinted Fig. 3 with permission from (R. Bulla, Th.A. Costi, and Th. Pruschke, Rev. Mod. Phys. **80**, 395 (2008)). Copyright 2008 by the American Physical Society. DOI: https://doi.org/10.1103/RevModPhys.80.395

spectrum $E_{N+1}(s, r)$, etc. We will not discuss details of the construction of the new eigenvectors and refer the reader to review [33]. Instead, we demonstrate how one can solve the problem by tracing the spectrum as the number of sites increases.

The algorithm is illustrated in Fig. 6.3. First, one obtains the set of eigenenergies for the N-site Hamiltonian (Fig. 6.3), then one gets the eigenenergies for the N + 1 case (Fig. 6.3c) (scaled by $\sqrt{\Lambda}$, see Eq. (6.64); the lowest eigenenergy is always fixed to 0). Since the number of states significantly increases with increasing the number of sites even by 1, one keeps the number of eigenenergies, and hence the dimension of the Hilbert space during the iterative diagonalizations fixed (equal N_s, Fig. 6.3d). Such a truncation scheme appears to work well since usually high-energy states weakly affect the low-energy physics (in particular, because addition of a new site is a weak perturbation, $t_N \sim \Lambda^{-N/2}$ (the accuracy of the truncation approximation decreases when Λ decreases and approaches 1)).

The iteration procedure (increasing number of sites) is stopped when a converged truncated spectrum is obtained, i.e., when the energies have flown to a fixed point. Once the spectrum E_a and eigenvectors $|a\rangle$ are obtained, one can calculate the on-site d-electron GF. Namely, one can obtain the imaginary part of the GF (the spectral function) from the Lehmann representation

$$
\begin{aligned}
A_\sigma(\omega) &= -\frac{1}{\pi} Im G_\sigma(\omega) \\
&= \frac{1}{Z}\sum_{a,b} \left| \left\langle b \middle| \widehat{d}_\sigma \middle| a \right\rangle \right|^2 \left(e^{-\frac{E_a}{T}} + e^{-\frac{E_b}{T}} \right) \delta(\omega - E_b + E_a),
\end{aligned}
\tag{6.68}
$$

and using this result—to obtain the real part of the GF from the Kramers–Kronig relation.

However, at this point one can meet a potential problem since the high-energy states are truncated and one may need to have to include them into the sum (6.68). Inclusion of these states can be done in the following way: for each chain with length N one introduces the density matrix $\widehat{\rho} = \sum_a e^{-\frac{E_a}{T}} |a\rangle \langle a|$, then calculates the reduced density matrix $\widehat{\rho}_{\text{red}}$ with traced-out low-energy degrees of freedom and calculates the GF as:

$$
G_\sigma(t - t') = -i\theta(t - t')\text{Tr}\left(\widehat{\rho}_{\text{red}} \left\{ \widehat{d}_\sigma(t), \widehat{d}_\sigma^{+}(t') \right\} \right). \tag{6.69}
$$

Another potential difficulty is calculation of the self-energy by subtracting the inverse free and interacting impurity GFs, since the inverse differences are a potential source of errors. An alternative way to calculate the self-energy is: $\Sigma_\sigma(\omega) = U\frac{F_\sigma(\omega)}{G_\sigma(\omega)}$, were $F_\sigma(\omega)$ is the Fourier transform of the correlation function $F_\sigma(t - t') = -i\theta(t - t')Tr\left(\widehat{\rho}_{\text{red}} \left\{ \widehat{d}_\sigma(t) \widehat{d}_{\bar{\sigma}}^{+}(t) \widehat{d}_{\bar{\sigma}}(t), \widehat{d}_\sigma^{+}(t') \right\} \right)$ (for details, see Refs. [6, 29, 33]. We conclude here the one-orbital case, referring the reader to two important papers: [34], where accuracy of the NRG results was tested, and to [35], where an alternative to Eq. (6.68) approach to obtain the finite-temperature spectral function was proposed.

The first generalization of the NRG approach on the multi-band case was proposed by Pruschke and Bulla in work [36], and later this approach was refined in works [37–39]. In Ref. [36], a two-orbital model with focus on the role of exchange energy J in the MIT was studied by using two different algorithms. In the first approach, the authors included explicitly the orbital degrees into the iterative construction of the basis states, though this formalism is not applicable when the Hamiltonian contains terms that break the orbital symmetry. In the second approach, an asymmetric truncation scheme was used. In this scheme, the Hilbert space is truncated after adding each orbital separately (not simultaneously). Similar to the first approach, it led to a significant reduction of the typical matrix size. Though the second approach violates the orbital symmetry (which is not a problem in the case of a crystal-field splitting). Study of the Mott transition with these methods in Ref. [36] showed that the transition qualitatively and quantitatively depends on the value of J and the type of the interaction.

One can refine the scheme by introducing different types of chains for each orbital (flavor) [37–39]. One possibility is to map the multi-orbital bath on Wilson's semi-infinite chain consisting of m identical sub-chains for each orbital (m—number of orbitals) with discretization energy points [38]

$$\varepsilon_{n\nu}^{\pm} = \begin{cases} \pm X_n, & n = 0 \\ \pm X_n \Lambda^{-n+z_\nu}, & n = 1, 2, \ldots \end{cases}, \tag{6.70}$$

where n is the atom index, ν—orbital index and z_ν—twist (any continuous number equal or larger than 0 and smaller than 1) needed to shift the discretization points for different sub-chains.

One can discretize the charge density by a single pole in each interval with equal weight:

$$\rho_\nu^{\text{disc}}(\omega, z) = \sum_{n=0}^{\infty} \sum\nolimits_{\lambda=\pm} \gamma_{n\nu}^{\lambda}(z) \delta\left(\omega - \xi_{n\nu}^{\lambda}(z)\right), \tag{6.71}$$

where $\gamma_{n\nu}^{\lambda}(z) = \int_{\varepsilon_{n\nu}^{\lambda}(z)}^{\varepsilon_{n\nu}^{\lambda}(z)} d\omega \rho_\nu(\omega)$ are the pole weights. The pole positions $\xi_{n\nu}^{\lambda}(z)$ are defined by a differential equation obtained from the condition that when the number of separate runs N_z goes to infinity, one recovers continuous bath spectrum after averaging over z, $\rho_\nu(\omega) = \int_0^1 dz \rho_\nu^{disc}(\omega, z)$ [34].

For a constant DOS, one has

$$\gamma_{n\nu}^{\lambda}(z) = X_n \begin{cases} 1 - \Lambda^{z-n}, & n = 0 \\ \left(1 - \dfrac{1}{\Lambda}\right) \Lambda^{-n+z_\nu}, & n = 1, 2, \ldots \end{cases}, \tag{6.72}$$

$$\xi_{n\nu}^{\lambda}(z) = \lambda \frac{\gamma_{n\nu}^{\lambda}(z)}{\Lambda} \begin{cases} z, & n = 0 \\ 1, & n = 1, 2, \ldots \end{cases}. \tag{6.73}$$

Each sub-chain ν is characterized by a DOS $\rho_\nu(\omega)$ and all sub-chains form the total chain with discretized Hamiltonian:

$$\widehat{H}_{\text{bath}}^{\text{disc}}(z) = \sum_{\nu=1}^{m} \sum_{n=0}^{\infty} \left(\varepsilon_{n\nu} d_{n\nu}^{+} d_{n\nu} + t_{n\nu} d_{n+1\nu}^{+} d_{n\nu} + t_{n\nu} d_{n\nu}^{+} d_{n+1\nu} \right), \tag{6.74}$$

where the coefficients $\varepsilon_{n\nu}$ and $t_{n\nu}$ can be found similarly to the one-orbital case (technically, by using the Lanczos tridiagonalization). At large n's, the hopping parameters decay exponentially, $\frac{t_{n\nu}}{D_\nu} \sim \Lambda^{z_\nu - n/2}$ (D_ν is the orbital bandwidth). For equal bandwidths $D_\nu = D$ and twists $z_\nu = z$, one has $t_{n+1\nu}/t_{n\nu} \sim \Lambda^{-1/2}$. In the last case, all sub-chains are identical and there is no scale separation between different flavors for each site n, they all have energy at scale $\sim\Lambda^{-n/2}$, thus all subsites (sites in different chains) should be treated equivalently at each NRG step. Similar to the

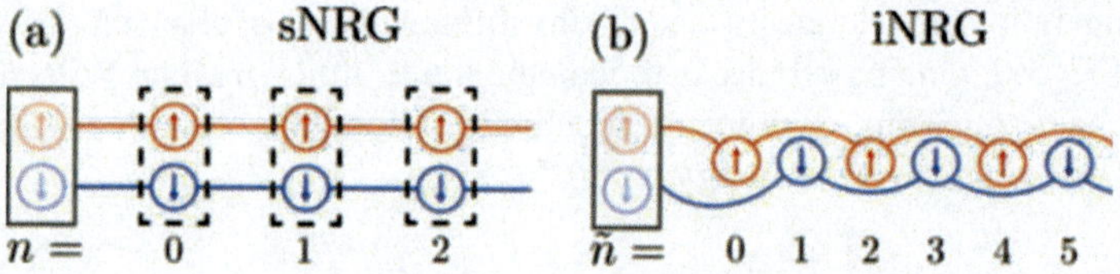

Fig. 6.4 A schematic illustration of the structure of the NRG standard (**a**) and interleaved (**b**) chains in the case of two-flavor (one-orbital, two-spin) system. Reprinted Figs. 1(a) and 1(b) with permission from (K. M. Stadler, A. K. Mitchell, J. von Delft, and A. Weichselbaum, Phys. Rev. B **93**, 235101 (2016)). Copyright 2016 by the American Physical Society. DOI: https://doi.org/10.1103/PhysRevB.93.235101

one-band case, here one has to diagonalize the Hamiltonian by truncating high-energy levels.

An alternative sub-chain structure was proposed in Ref. [38]. It allowed to separate the energy scales between each subsite of supersite and between different supersites. The idea used in Ref. [38] is to interleave each subsite from different subchains in linear sequence, creating m times longer linear chain (see Fig. 6.4). The total-chain atoms are labeled as $\tilde{n} \equiv (n, \nu) = mn + \nu - 1$, and the corresponding hopping parameters $\tilde{t}_{\tilde{n}} = \tilde{t}_{n,\nu}$ describe the hopping between the subsites ν and between nearest sites for n. Contrary to the previous standard-chain order, in the case of interleaved chain order, there is no nearest-neighbor hopping. $\tilde{t}_{\tilde{n}}$ decreases as $\tilde{n}$ increases, and the ratio $\tilde{t}_{\tilde{n}+m}/\tilde{t}_{\tilde{n}} \sim \Lambda^{-1/2}$ is consistent with the ratio $t_{n+1\nu}/t_{n\nu} \sim \Lambda^{-1/2}$ in the "standard" case, which insures a net rate of decrease of the hopping along the sub-chains. The decrease corresponds to a uniform energy separation along the chain when one goes from one to the next subsite with $\tilde{t}_{\tilde{n}+1}/\tilde{t}_{\tilde{n}} \sim \Lambda^{-1/m}$ (in the standard case, $\tilde{t}_{\tilde{n}}$ is the same for given n and different ν's). Such energy separation can be achieved by choosing different z_ν for each orbital ν as $z_{\nu+1} = z_\nu - \frac{1}{2m}$. It can be shown that using this scheme one can get a separation of energies for different subsites $\Lambda^{-\frac{(\tilde{n}-1)}{2} - \frac{\nu}{2m}}$, resulting in a more fine-grained, and hence more accurate, RG scheme (we refer the reader for more details to Ref. [38]).

It is important to mention another RG approach often used to study 1D systems [40–42]—the density matrix renormalization group (DMRG). Similar to NRG, in DMRG one also increases the system size step-by-step and uses an iterative approach, but the states are chosen differently, using the density matrix. Contrary to NRG, DMRG is applicable to systems where the parameters do not fall off exponentially along the chain. Since NRG works only in the last case, the areas where these two methods can be applied do not completely coincide. Though DMRG has been also applied to the DMFT impurity models (see Ref. [43] and other references in the Table in Sect. 6.4). In general, NRG approach is more suitable, comparing to DMRG, for analyzing the fixed points and other low-energy properties of impurity systems.

6.3 Quantum Monte Carlo (QMC) Solvers

Stochastic QMC methods to solve the impurity problem are very powerful and, probably, the most popular tools used for this purpose in DMFT (for general concepts of the QMC approach, see review [44]). Here, we present two such methods—early Hirsch–Fye QMC and more recent, state-of-art, Continuous-Time QMC (CT-QMC) approaches.

6.3.1 *Hirsch–Fye QMC*

Before presenting details of the Hirsch–Fye QMC solution of the DMFT impurity problem, let us summarize the main idea of the Hirsch–Fye transformation used in this method. As the first step, it is convenient to rewrite the Hubbard Hamiltonian in the following form:

$$\widehat{H} = \widehat{H}_0 + \widehat{H}_{\text{int}}, \tag{6.75}$$

where

$$\widehat{H}_0 = -\sum_{i,j,\sigma} t_{ij} c_{i\sigma}^{\dagger} c_{j\sigma} - \left(\mu - \frac{U}{2}\right) \sum_i (n_{i\uparrow} + n_{i\downarrow}) \tag{6.76}$$

is the free electron and

$$\widehat{H}_{\text{int}} = U\left(n_\uparrow n_\downarrow - \frac{n_\uparrow + n_\downarrow}{2}\right) \tag{6.77}$$

is the interacting part.

Then, dividing the imaginary time interval on L equal pieces with time step $\Delta\tau = \beta/L$ the equation for the partition function of the system

$$Z = \text{Tr}\, e^{-\beta\left[\widehat{H}_0 + \widehat{H}_{\text{int}}\right]} \tag{6.78}$$

can be approximated by the following one:

$$
\begin{aligned}
Z &= \mathrm{Tr}\, e^{-\Delta\tau\left[\widehat{H}_0(\tau_1)+\widehat{H}_{\rm int}(\tau_1)\right]} \times \ldots \times e^{-\Delta\tau\left[\widehat{H}_0(\tau_L)+\widehat{H}_{\rm int}(\tau_L)\right]} \\
&\equiv \mathrm{Tr} \prod e^{-\Delta\tau\left[\widehat{H}_0+\widehat{H}_{\rm int}\right]} \\
&\approx \mathrm{Tr}\, e^{-\Delta\tau\left[\widehat{H}_0(\tau_1)+\widehat{H}_{\rm int}(\tau_1)+\ldots+\widehat{H}_0(\tau_L)+\widehat{H}_{\rm int}(\tau_L)\right]} + O(\Delta\tau^2) \\
&\equiv \mathrm{Tr}\, T_\tau e^{-\int_0^\beta d\tau\left[\widehat{H}_0(\tau)+\widehat{H}_{\rm int}(\tau)\right]} + O(\Delta\tau^2)
\end{aligned} \tag{6.79}
$$

Next, using Trotter formula, one gets

$$
Z = \mathrm{Tr} \prod e^{-\Delta\tau\left[\widehat{H}_0+\widehat{H}_{\rm int}\right]} \equiv \mathrm{Tr} \prod e^{-\Delta\tau \widehat{H}_0} e^{-\Delta\tau \widehat{H}_{\rm int}}. \tag{6.80}
$$

In order to get rid of the product of four fermion operators ($\sim n_\uparrow n_\downarrow$) in the interaction part of the Hamiltonian one can make the Hirsch–Fye transformation [45]:

$$
e^{-\Delta\tau U n_\uparrow n_\downarrow + \Delta\tau\frac{U}{2}(n_\uparrow+n_\downarrow)} = \frac{1}{2}\sum_{s=\pm1} e^{\lambda s(n_\uparrow - n_\downarrow)} \equiv \frac{1}{2}\sum_{s=\pm1} e^{\lambda s \sum_\sigma \sigma\cdot n_\sigma}, \tag{6.81}
$$

where s are "spin" variables and

$$
cosh\, \lambda = e^{\Delta\tau\frac{U}{2}} \tag{6.82}
$$

(Equation (6.81) can be easily verified by inserting all possible values of the spin densities, $n_\uparrow = 0, 1$ and $n_\downarrow = 0, 1$, into its left and right part.)

In the case of the DMFT partition function

$$
Z = \int D[c]D[c^+]e^{-\int d\tau_1 \int d\tau_2 \sum_\sigma c_\sigma^+(\tau_1)\mathcal{G}_\sigma^{-1}(\tau_1,\tau_2)c_\sigma(\tau_2)+U\int d\tau_1 n_\uparrow(\tau_1)n_\downarrow(\tau_1)} \tag{6.83}
$$

the transformation (6.81) gives

$$
\begin{aligned}
Z &= \sum_{s_l=\pm1}\int D[c]D[c^+]e^{-\Delta\tau^2\sum_{m,n,\sigma} c_\sigma^+(\tau_m)\mathcal{G}_\sigma^{-1}(\tau_m,\tau_n)c_\sigma(\tau_n)+\Delta\tau\lambda\sum_{l,\sigma} s_l\cdot\sigma\cdot n_\sigma(\tau_l)} \\
&= \sum_{s_l=\pm1}\int D[c]D[c^+]e^{-\Delta\tau^2\sum_{l,m,\sigma} c_\sigma^+(\tau_l)\left[\mathcal{G}_\sigma^{-1}(\tau_l,\tau_m)+\sigma\lambda s_l\frac{\delta_{l,m+1}}{\Delta\tau}\right]c_\sigma(\tau_m)},
\end{aligned} \tag{6.84}
$$

where

$$\delta_{l,m+1} = \begin{cases} 1 \text{ if } \ l = m+1, l = 2, \ldots, L-1 \\ -1, \text{if } \ l = 1, m = L \\ 0, \quad \text{otherwise} \end{cases} . \tag{6.85}$$

To shorten notations, let us use the following functions equal to the functions in the square brackets in the exponent in the last integral in Eq. (6.84):

$$G_{\sigma}^{-1}{}_{\{s_1,\ldots,s_L\}}(\tau_l, \tau_m) = \ \mathcal{G}_{\sigma}^{-1}(\tau_l, \tau_m) + \sigma\lambda s_l \frac{\delta_{l,m+1}}{\Delta\tau} \tag{6.86}$$

($\{s_1, \ldots, s_L\}$ is a set of the "spin" values for different time points). In the matrix form, Eq. (6.86) is:

$$\begin{bmatrix} G_\sigma(\tau_1,\tau_1) & G_\sigma(\tau_1,\tau_2) & \ldots & G_\sigma(\tau_1,\tau_{L-2}) & G_\sigma(\tau_1,\tau_{L-1}) & G_\sigma(\tau_1,\tau_L) \\ G_\sigma(\tau_2,\tau_1) & G_\sigma(\tau_2,\tau_2) & \ldots & G_\sigma(\tau_2,\tau_{L-2}) & G_\sigma(\tau_2,\tau_{L-1}) & G_\sigma(\tau_2,\tau_L) \\ G_\sigma(\tau_3,\tau_1) & G_\sigma(\tau_3,\tau_2) & \ldots & G_\sigma(\tau_3,\tau_{L-2}) & G_\sigma(\tau_3,\tau_{L-1}) & G_\sigma(\tau_3,\tau_L) \\ \ldots & \ldots & \ldots & \ldots & \ldots & \ldots \\ G_\sigma(\tau_{L-1},\tau_1) & G_\sigma(\tau_{L-1},\tau_2) & \ldots & G_\sigma(\tau_{L-1},\tau_{L-2}) & G_\sigma(\tau_{L-1},\tau_{L-1}) & G_\sigma(\tau_{L-1},\tau_L) \\ G_\sigma(\tau_L,\tau_1) & G_\sigma(\tau_L,\tau_2) & \ldots & G_\sigma(\tau_L,\tau_{L-2}) & G_\sigma(\tau_L,\tau_{L-1}) & G_\sigma(\tau_L,\tau_L) \end{bmatrix}^{-1}_{\{s_1,\ldots,s_L\}}$$

$$= \begin{bmatrix} \mathcal{G}_\sigma(\tau_1,\tau_1) & \mathcal{G}_\sigma(\tau_1,\tau_2) & \ldots & \mathcal{G}_\sigma(\tau_1,\tau_{L-2}) & \mathcal{G}_\sigma(\tau_1,\tau_{L-1}) & \mathcal{G}_\sigma(\tau_1,\tau_L) \\ \mathcal{G}_\sigma(\tau_2,\tau_1) & \mathcal{G}_\sigma(\tau_2,\tau_2) & \ldots & \mathcal{G}_\sigma(\tau_2,\tau_{L-2}) & \mathcal{G}_\sigma(\tau_2,\tau_{L-1}) & \mathcal{G}_\sigma(\tau_2,\tau_L) \\ \mathcal{G}_\sigma(\tau_3,\tau_1) & \mathcal{G}_\sigma(\tau_3,\tau_2) & \ldots & \mathcal{G}_\sigma(\tau_3,\tau_{L-2}) & \mathcal{G}_\sigma(\tau_3,\tau_{L-1}) & \mathcal{G}_\sigma(\tau_3,\tau_L) \\ \ldots & \ldots & \ldots & \ldots & \ldots & \ldots \\ \mathcal{G}_\sigma(\tau_{L-1},\tau_1) & \mathcal{G}_\sigma(\tau_{L-1},\tau_2) & \ldots & \mathcal{G}_\sigma(\tau_{L-1},\tau_{L-2}) & \mathcal{G}_\sigma(\tau_{L-1},\tau_{L-1}) & \mathcal{G}_\sigma(\tau_{L-1},\tau_L) \\ \mathcal{G}_\sigma(\tau_L,\tau_1) & \mathcal{G}_\sigma(\tau_L,\tau_2) & \ldots & \mathcal{G}_\sigma(\tau_L,\tau_{L-2}) & \mathcal{G}_\sigma(\tau_L,\tau_{L-1}) & \mathcal{G}_\sigma(\tau_L,\tau_L) \end{bmatrix}^{-1}$$

$$+\frac{\lambda\sigma}{\Delta\tau} \begin{bmatrix} 0 & 0 & \ldots & 0 & 0 & -s_1 \\ s_2 & 0 & \ldots & 0 & 0 & 0 \\ 0 & s_3 & \ldots & 0 & 0 & 0 \\ \ldots & \ldots & \ldots & \ldots & \ldots & \ldots \\ 0 & 0 & \ldots & s_{L-2} & 0 & 0 \\ 0 & 0 & \ldots & 0 & s_{L-1} & 0 \end{bmatrix} . \tag{6.87}$$

Since the action in the exponent in Eq. (6.84) is bilinear in operators, the path integration in this equation can be performed analytically by using the formula

$$\int D[c]D[c^+]e^{-\int d\tau d\tau' c^+(\tau)A(\tau,\tau')c(\tau')} = \det A. \tag{6.88}$$

The result is

$$Z = \sum_{\{s_1=\pm 1,\ldots,s_L=\pm 1\}} \det\left[G_{\uparrow}^{-1}{}_{\{s_1,\ldots,s_L\}}\right] \cdot \det\left[G_{\downarrow}^{-1}{}_{\{s_1,\ldots,s_L\}}\right]. \tag{6.89}$$

Similarly, by using the path-integral formula

$$\int D[c]D[c^+]c(\tau_l)c^+(\tau_m)e^{-\int d\tau d\tau' c^+(\tau)A(\tau,\tau')c(\tau')} = \det A \cdot A(\tau_m,\tau_l), \tag{6.90}$$

one can get a formula for the local GF

$$G_\sigma(\tau_l,\tau_m) = \int D[c]D[c^+]c_\sigma(\tau_l)c_\sigma^+(\tau_m)e^{-\int d\tau_1 \int d\tau_2 \sum_{\sigma'} c_{\sigma'}^+(\tau_1)\mathcal{G}_{\sigma'}^{-1}(\tau_1,\tau_2)c_{\sigma'}(\tau_2)+U\int d\tau_1 n_\uparrow(\tau_1)n_\downarrow(\tau_1)}. \tag{6.91}$$

Namely, the Hirsch–Fye QMC formula for the local GF is

$$G_\sigma(\tau_l,\tau_m) = \sum_{\{s_1=\pm 1,\ldots,s_L=\pm 1\}} \det\left[G_{\uparrow}^{-1}{}_{\{s_1,\ldots,s_L\}}\right] \cdot \det\left[G_{\downarrow}^{-1}{}_{\{s_1,\ldots,s_L\}}\right] G_{\sigma\{s_1,\ldots,s_L\}}(\tau_l,\tau_m). \tag{6.92}$$

Thus, we have obtained the Hirsch–Fye QMC expressions for the impurity partition function and impurity GF. Summation over all "Ising spin" indices $s_1 = \pm 1, \ldots, s_L = \pm 1$ in Eqs. (6.89) and (6.92) gives the exact results for the partition function and GF. Unfortunately, such a summation would take an extremely long time since the number of the terms in the sums is very large. For example, even for a modest number of time points $L = 100$ one has to sum up $2^{100} \sim 10^{30}$ terms. For this reason, usually a stochastic method is used to calculate Z and GF from Eqs. (6.89) to (6.92). In detail, one can begin with the GF (6.86) with some spin configuration, e.g., $s = \{s_1, s_2, \ldots, s_l, \ldots, s_{L-1}, s_L\} = \{1, \ldots, 1\}$ and generates new Ising configurations, or "moves," by flipping one of the "spins" at a time:

$$s = \{s_1, s_2, \ldots, s_l, \ldots, s_{L-1}, s_L\} \rightarrow s' = \{s_1, s_2, \ldots, -s_l, \ldots, s_{L-1}, s_L\}. \tag{6.93}$$

Then, the corresponding terms are included into the sums in Eqs. (6.89) and (6.92) in the case the move is "good" and it is accepted. According to the Metropolis algorithm, acceptance of the move is defined by the ratio

$$P(s \to s') = \min\left(1, \frac{\det\left[G_{\uparrow}^{-1}{}_{\{s'\}}\right]\det\left[G_{\uparrow}^{-1}{}_{\{s'\}}\right]}{\det\left[G_{\uparrow}^{-1}{}_{\{s\}}\right]\det\left[G_{\uparrow}^{-1}{}_{\{s\}}\right]}\right). \tag{6.94}$$

The move is accepted if $P(s \to s')$ is larger than a randomly selected number between 0 and 1.

Thus, equations (6.86), (6.89), (6.92)–(6.94) are the recipe how to perform Hirsch–Fye QMC simulations. Unfortunately, inversion of matrices $G_{\sigma}^{-1}{}_{\{s_1,\ldots,s_L\}}(\tau_l, \tau_m)$ (to calculate GF Eq. (6.92)) is extremely time-consuming. For this reason, another set of equivalent equations are used in practical implementation of the Hirsch–Fye QMC approach [15, 46, 47]. We describe first this scheme for one-orbital case, and later generalize it on multi-orbitals.

To derive the equations, let us begin with a discrete-approximated equation for the partition function for the lattice Hubbard model:

$$\begin{aligned} Z &\approx Tr\prod e^{-\Delta\tau\widehat{H}_0}e^{-\Delta\tau\widehat{H}_{int}} \\ &\approx \sum_{m_L,\ldots,m_1}\left\langle m_L|e^{-\Delta\tau\widehat{H}_0}e^{-\Delta\tau\widehat{H}_{int}}|m_{L-1}\right\rangle \\ &\times\left\langle m_{L-1}|e^{-\Delta\tau\widehat{H}_0}e^{-\Delta\tau\widehat{H}_{int}}|m_{L-2}\right\rangle\ldots\left\langle m_2|e^{-\Delta\tau\widehat{H}_0}e^{-\Delta\tau\widehat{H}_{int}}|m_1\right\rangle, \end{aligned} \tag{6.95}$$

where in the last line of this equation we have inserted total sums of operators in the occupation basis $\sum_m |m\rangle\langle m|$ (equal to identity operator). Then, in each term in Eq. (6.95), one can perform the Hirsch–Fye transformation (6.81). At this point, it is convenient to introduce the matrix (inverse of $G_{\sigma\{s\}}$, see below):

$$O_{\sigma\{s\}} = \begin{pmatrix} I & 0 & 0 & \cdots & 0 & e^{-\Delta\tau K V^L_{\sigma\{s\}}} \\ e^{-\Delta\tau K V^1_{\sigma\{s\}}} & I & 0 & \cdots & 0 & 0 \\ 0 & e^{-\Delta\tau K V^2_{\sigma\{s\}}} & I & \cdots & 0 & 0 \\ 0 & 0 & e^{-\Delta\tau K V^3_{\sigma\{s\}}} & \cdots & 0 & 0 \\ \cdots & \cdots & \cdots & \cdots & \cdots & \cdots \\ 0 & 0 & 0 & \cdots & e^{-\Delta\tau K V^{L-1}_{\sigma\{s\}}} & I \end{pmatrix}, \tag{6.96}$$

where the kinetic matrix K_{ij} is defined by

$$\widehat{\mathrm{H}}_0 = \sum_{\mathrm{i,j},\sigma} \mathrm{c}_{\mathrm{i}\sigma}^{+} \mathrm{K}_{\mathrm{ij}} \mathrm{c}_{\mathrm{j}\sigma} \tag{6.97}$$

(each element in Eq. (6.96) is $N \times N$ matrix, where N is number of sites) and

$$V^{l}_{\sigma,ij\{s\}} = \begin{cases} \lambda\sigma s_{il}\delta_{ij}, & i \text{ in cluster} \\ 0, & \text{otherwise} \end{cases} \tag{6.98}$$

(i, j are site and l is time index). Then, the matrix $e^{V_{\sigma\{s\}}}$ is

$$e^{V_{\sigma\{s\}}} = \begin{pmatrix} e^{\lambda\sigma s} & 0 & 0 & \ldots & 0 & 0 \\ 0 & 1 & 0 & \ldots & 0 & 0 \\ 0 & 0 & 1 & \ldots & 0 & 0 \\ 0 & 0 & 0 & \ldots & 0 & 0 \\ \ldots & \ldots & \ldots & \ldots & \ldots & \ldots \\ 0 & 0 & 0 & \ldots & 0 & 1 \end{pmatrix}. \tag{6.99}$$

In this formalism, Eq. (6.89) can be written as:

$$Z \approx \mathrm{Tr}_{\{s\}}\left[\det O_{\uparrow\{s\}} \cdot \det O_{\downarrow\{s\}}\right], \tag{6.100}$$

where the matrix $O_{\sigma\{s\}}$, as it is possible to show, is the inverse of $G_{\sigma\{s\}}$:

$$G^{-1}_{\sigma\{s\}} = O_{\sigma\{s\}}. \tag{6.101}$$

Then, Eqs. (6.96) and (6.101) give

$$G^{-1}_{\sigma\{s\}} = I - Te^{-\Delta\tau K} e^{V_{\sigma\{s\}}} \tag{6.102}$$

where

$$T_{il,jl'} = \delta_{l-1,l'}\delta_{ij}, \tag{6.103}$$

and

$$V^{l}_{\sigma,il,jl'\{s\}} = \begin{cases} \lambda\sigma s_{il}\delta_{ll'}\delta_{ij}, & i \text{ in cluster} \\ 0, & i \text{ or } j \text{ are not in cluster} \end{cases} \tag{6.104}$$

In order to get relation between $G^{-1}_{\sigma\{s\}}$ and $G^{-1}_{\sigma\{s'\}}$ for two different iso-spin configurations $\{s\}$ and $\{s'\}$, let us multiply Eq. (6.102) by $\mathrm{e}^{-V_{\sigma\{s\}}}$:

$$G^{-1}_{\sigma\{s\}} e^{-V_{\sigma\{s\}}} = e^{-V_{\sigma\{s\}}} - \mathrm{T} e^{-\Delta\tau K}. \tag{6.105}$$

One can write a similar equation for the $\{s'\}$ configuration:

$$G'^{-1}_{\sigma\{s'\}} e^{-V'_{\sigma\{s'\}}} = e^{-V'_{\sigma\{s'\}}} - \mathrm{T} e^{-\Delta\tau K}. \tag{6.106}$$

Subtracting Eq. (6.105) from Eq. (6.106) and performing simple transformations one gets the equation that relates the "spin-flipped" and the "old" functions:

$$G'_{\sigma\{s'\}} = G_{\sigma\{s\}} + \left(G_{\sigma\{s\}} - I\right)\left(e^{\left(V'_{\sigma\{s'\}} - V_{\sigma\{s\}}\right)} - I\right) G'_{\sigma\{s'\}}. \tag{6.107}$$

Besides the fact that expression (6.107) allows one to calculate the new-move function directly without inversion, it is also very convenient, since it does not depend on the kinetic energy and can be directly applied to the impurity model.

The acceptance rate Eq. (6.94) can be calculated by using the formula that follows from (6.107):

$$G_{\sigma\{s\}} G'^{-1}_{\sigma\{s'\}} = I - \left(G_{\sigma\{s\}} - I\right)\left(e^{\left(V'_{\sigma\{s'\}} - V_{\sigma\{s\}}\right)} - I\right) \tag{6.108}$$

(indeed, the expression on the right-hand side of Eq. (6.94) is the ratio of products of determinants of the spin-up an spin-down functions given by Eq. (6.108)).

Let us provide more technical details on how Eqs. (6.107) and (6.108) are used in practical calculations. Equation (6.107) can be rewritten in components as

$$G'_{pn} = G_{pn} + \sum_m \left(G_{pm} - \delta_{pm}\right)\left(e^{\sigma\lambda\left(s'_m - s_m\right)} - 1\right) G'_{mn}. \tag{6.109}$$

Next, using last the equation one can get $G'_{mn} = G_{mn} + (G_{mm} - 1) \times \left(e^{\sigma\lambda\left(s'_m - s_m\right)} - 1\right) G'_{mn}$ and then

$$G'_{mn} = G_{mn}\left[1 - (G_{mm} - 1)\left(e^{\sigma\lambda\left(s'_m - s_m\right)} - 1\right)\right]^{-1}. \tag{6.110}$$

Substitution of the last expression into the sum in Eq. (6.109) gives

$$G'_{pn} = G_{pn} + \sum_m \frac{\left(G_{pm} - \delta_{pm}\right)\left(e^{\sigma\lambda\left(s'_m - s_m\right)} - 1\right)}{1 - \left(G_{mm} - 1\right)\left(e^{\sigma\lambda\left(s'_m - s_m\right)} - 1\right)} G_{mn}. \tag{6.111}$$

This is the equation for updating the components of configuration functions used in the practical implementation.

Finally, using Eq. (6.106), one can rewrite the rate

$$R(s \to s') = \frac{\det\left[G_{\uparrow}^{-1}{}_{\{s'\}}\right]\det\left[G_{\uparrow}^{-1}{}_{\{s'\}}\right]}{\left(\det\left[G_{\uparrow}^{-1}{}_{\{s\}}\right]\det\left[G_{\uparrow}^{-1}{}_{\{s\}}\right]\right)} \equiv \prod_{\sigma}\det\left[G_{\sigma}^{-1}{}_{\{s'\}}\right]\det\left[G_{\sigma\{s\}}\right] \tag{6.112}$$

in the acceptance rate formula (6.94) as

$$R(s \to s') = \prod_{\sigma}\left[1 - \left(G_{\sigma mm} - 1\right)\left(e^{\sigma\lambda\left(s'_m - s_m\right)} - 1\right)\right]. \tag{6.113}$$

Equations (6.110) and (6.113) are very convenient, since they don't include matrix inversion (only a simple function inversion in Eq. (6.110)).

For newer developments of QMC methods for one-band systems, like an auxiliary-field QMC method, where different Hubbard–Stratonovich transformations depending on the auxiliary field of interest are applied, or a dual fermion (DF) approach based on a diagrammatic expansion around DMFT, see Refs. [48, 49], correspondingly. Works on other extensions of the QMC approach include Refs. [50–53] (static case) and Ref. [54] (the case out-of-equilibrium—the real-time QMC (see also Chapter 9)).

The one-band QMC approach above can be straightforwardly extended on the multi-orbital case [55] with the interaction part of the Hamiltonian

$$\widehat{H}_{int} = \sum_{m,m'} U^{mm'}\left(n^m n^{m'} - \frac{n^m + n^{m'}}{2}\right) \tag{6.114}$$

(the "flavor" indices m and m' are combined spin-orbital indices) by applying the following Hubbard–Stratonovich transformation:

$$e^{-\Delta\tau U^{mm'}\left(n^m n^{m'} - \frac{n^m + n^{m'}}{2}\right)} = \frac{1}{2}\sum_{S^{mm'}=\pm 1} e^{\lambda^{mm'} S^{mm'}\left(n^m - n^{m'}\right)}, \tag{6.115}$$

where $S^{mm'}$ are multi-orbital (matrix) auxiliary fields (for given time point) and $\lambda^{mm'}$ are defined by

$$cosh\lambda^{mm'} = e^{\Delta\tau\frac{U^{mm'}}{2}}. \tag{6.116}$$

Similar to the one-orbital case, the partition function can be calculated as trace over the product of all flavors (orbitals + spins instead of spins):

$$Z \approx \mathrm{Tr}_{\{s^{mm'}(\tau_l)\}} \prod_m \det O^m_{\{s^{mm'}(\tau_l)\}}, \tag{6.117}$$

where $O^m_{\{s^{mm'}(\tau_l)\}}$ is defined by Eq. (6.96) with the kinetic energy matrix K^m of rank NL (N is the number of bath states (momenta) plus 1 (impurity state)), and

$$V_l^m = \sum_{m' \neq m} \lambda^{mm'} Q^{mm'}(\tau_l)|m\rangle\langle m|, \tag{6.118}$$

$$Q^{mm'}(\tau_l) = \begin{cases} s^{mm'}(\tau_l) \text{ if } m > m' \\ -s^{mm'}(\tau_l) \text{ if } m < m' \end{cases}. \tag{6.119}$$

The remaining one-band QMC equations (6.101), (6.107), (6.110) and (6.113) can be also straightforwardly generalized on the multi-flavor case (they include a new flavor index m):

$$G^{m-1}_{\{S_{mm'}\}} = O^m_{\{s^{mm'}\}}, \tag{6.120}$$

$$G^{m\prime} = G^m + \sum_m (G^m - I)\left(e^{V^{m'} - V^m} - 1\right) G^{m\prime}, \tag{6.121}$$

$$G^{m\prime} = \left[I - \sum_m (G^m - I)\left(e^{V^{m'} - V^m} - 1\right)\right]^{-1} G^m, \tag{6.122}$$

$$\begin{aligned} R(s \to s') &= \prod_m \det\left[O^{m-1}_{\{s^{mm'}\}}\right] \det\left[O^{m'}_{\{s^{mm'}\}}\right] \\ &= \prod_m \left[I - (G^m - I)\left(e^{V^{m'} - V^m} - I\right)\right]. \end{aligned} \tag{6.123}$$

Since one makes only one spin flip at a time, the determinants in Eq. (6.123) can be easily evaluated and one gets

$$R = R^m R^{m'}, \tag{6.124}$$

where

$$R^m = 1 - \left(G^m_{ll} - 1\right)\left(e^{-2\lambda^{mm'} s^{mm'}} - 1\right), \tag{6.125}$$

$$R^{m'} = 1 - \left(G^{m'}_{ll} - 1\right)\left(e^{2\lambda^{mm'} s^{mm'}} - 1\right), \tag{6.126}$$

and G^m_{ll} are matrices in the time indices. If the flip is accepted, one obtains the new spin-configuration functions by using

$$G^{m\ \prime}_{l_1 l_2} = G^m_{l_1 l_2} + \sum_l \left(G^m_{l_1 l} - \delta_{l_1 l}\right)\left(e^{-2\lambda^{mm'} s^{mm'}} - 1\right)\frac{1}{R^m} G^m_{l l_2}, \tag{6.127}$$

$$G^{m'\ \prime}_{l_1 l_2} = G^{m'}_{l_1 l_2} + \sum_l \left(G^{m'}_{l_1 l} - \delta_{l_1 l}\right)\left(e^{-2\lambda^{mm'} s^{mm'}} - 1\right)\frac{1}{R^{m'}} G^{m'}_{l l_2}. \tag{6.128}$$

Similar to the one-orbital case, where one starts with GF Eq. (6.86) with some initial set of spins, in the multi-flavor case one can start from $G^{m-1}_{\{S_{mm'}\}} = O^m_{\{s^{mm'}\}}$, where $O^m_{\{s^{mm'}\}}$ is defined by multi-orbital Eq.(6.96) with K and V matrices defined by Eqs. (6.97) and (6.98).

It is important to mention that the acceptance rate R can be negative—the so-called sign problem. While in the one-band case, this problem is absent, it can emerge in the multi-orbital case at low temperatures [46]. The simplest way to avoid this problem is to re-weight the determinants by multiplying them by a number with the appropriate sign or by a function (calculated with the statistical average) (see, e.g., Section II.D in Ref. [57]). We will discuss this situation in more detail in the next subsection.

Before concluding the discussion of the Hirsch–Fye QMC solver, we mention that the multi-orbital QMC approach was recently generalized on the case out-of-equilibrium [56].

6.3.2 *Continuous-Time Quantum Monte Carlo (CT-QMC)*

The Hirsch–Fye QMC was the state-of-art approach for DMFT during 1990s and 2000s and in the last decade it was replaced by the CT-QMC method. There were several reasons for this replacement, the most important—a high computational cost of the Hirsch–Fye QMC in the multi-orbital case. Other reasons include problems to get equilibrations at low temperatures and large interactions and need to use equal-step time discretization (for details, see Ref. [57]).

The discretization of the imaginary time interval is a really big problem in the Hirsch–Fye QMC, the time step should be small enough to properly catch the fast change of the GF and also to accurately take into account its discontinuous derivatives at $\tau = 0$ and β. Thus, one needs to have a large number of points in the used homogeneous grid, especially at low temperatures. CT-QMC is an approach that is not based on explicit discretization of time. In this method, first proposed and elaborated for boson systems [58–61] and then generalized on the case of fermions [62], the terms are sampled in a diagrammatic expansion instead of sampling the configurations in a complete set of states. Also, the fermionic systems have much less issues with the sign problem in the case of impurity models [63, 64] (see also Ref. [57] for further results). In this subsection, we follow mainly the presentation in work [57].

The goal of the CT-QMC approach is to solve the impurity problem described by a rather general Hamiltonian:

$$
\begin{aligned}
H = & \sum_{a,b} E^{ab} d_a^+ d_b + \sum_{a,b} I^{pqrs} d_p^+ d_q^+ d_r d_s + \sum_{\vec{k},a} \varepsilon^a\left(\vec{k}\right) c_{\vec{k}a}^+ c_{\vec{k}a} \\
& + \sum_{\vec{k},a}\left(V^{ab}\left(\vec{k}\right) c_{\vec{k}a}^+ d_b + h.c.\right) + \sum_{abcd,\, \vec{k}_1 \vec{k}_2} J^{abcd}_{\vec{k}_1 \vec{k}_2} c_{\vec{k}_1 a}^+ c_{\vec{k}_2 b} d_c^+ d_d \\
\equiv & H_0^{\text{loc}} + H_I^{\text{loc}} + H_{\text{bath}} + H_{\text{hyb}} + H_{\text{hyb}}^{\text{exchange}}.
\end{aligned}
\tag{6.129}
$$

The corresponding DMFT action for the impurity problem is:

$$
\begin{aligned}
Z = & \int D[d] D[d^+] e^{-S} \\
= & \int D[d] D[d^+] e^{-\int d\tau_1 d\tau_2 d_a^+(\tau_1)\left[\left(\partial_{\tau_1} + E^{ab}\right)\delta(\tau_1 - \tau_2) + \Delta^{ab}(\tau_1 - \tau_2)\right] d_b(\tau_2) + \int d\tau_1 H_I^{loc}(\tau_1)} \\
\equiv & \int D[d] D[d^+] e^{-\int d\tau_1 d\tau_2 d_a^+(\tau_1)\left(\mathcal{G}^{-1}\right)^{ab}(\tau_1 - \tau_2) d_b(\tau_2) + \int d\tau_1 H_I^{loc}(\tau_1)}
\end{aligned}
\tag{6.130}
$$

where

$$
\Delta^{ab}(i\omega_n) = \sum_{\vec{k},a} V^{aa*}\left(\vec{k}\right)\left(i\omega_n - \varepsilon^a\left(\vec{k}\right)\right)^{-1} V^{ab}\left(\vec{k}\right). \tag{6.131}
$$

The idea used in CT-QMC to solve the problem with partition function (6.130) might be explained by splitting the Hamiltonian in the exponent of the partition function $Z = \text{Tr}e^{-\beta H}$ on two parts: H_a, for which the solution is known (e.g., for the noninteracting or infinite-U Hubbard models) and the remaining ("difficult") part H_b. Then, one can write down Z in the continuous path-integral form in the interaction representation with respect to H_a using expansion in powers of H_b:

$$Z = TrT_\tau e^{-\beta H_a} e^{-\int_0^\beta H_b(\tau)d\tau}$$
$$= \sum_k (-1)^k \int_0^\beta d\tau_1 \ldots \int_{\tau_{k-2}}^\beta d\tau_{k-1} \int_{\tau_{k-1}}^\beta d\tau_k Tr\left[e^{-\beta H_a} H_b(\tau_k) H_b(\tau_{k-1}) \ldots H_b(\tau_1)\right]. \tag{6.132}$$

Despite the fact that the summation in Eq. (6.132) is performed to infinite order, in the case of a finite-range spectrum of H_b the contributions of the high-order terms to the sum are exponentially suppressed due to $1/k!$ factor in the exponent expansion. Thus, one does not need to truncate the diagrams, since the result of summation is finite in any case. Depending on the choice of H_a and H_b, different expansions are possible. Below, we give details of two, probably most popular and complementary, expansions: CT-INT [63] (H_b—the interaction term) and CT-HYB [65, 66] (H_b—the hybridization term $\sum_{\sigma,l=2}^{n_s} \left(V_l a_{l\sigma}^+ d_\sigma + h.c.\right)$ in the impurity model). Two other popular types of expansions, which we do not discuss here are CT-AUX [62, 67] and infinite-U CT-J [68], which are in some sense "cousins" of CT-INT and CT-HYB, correspondingly. CT-AUX is based on the same methodology of expansion in powers of interaction as CT-INT, but the interaction term there is decomposed by using an auxiliary field, similar to the Hirsch–Fye algorithm. CT-AUX tool is the most efficient in the case of many orbitals and/or many atoms in the cluster. CT-J, similar to CT-HYB, is based on the expansion in term of the (exchange) hybridization part of the Hamiltonian that has structure $JScc^+$, where J is the exchange coupling, S is the spin matrix and the last two operators are the fermion bath operators. This method was applied to Kondo-like problems where the interaction on the impurity (spin) site is infinitely large. For details of the CT-AUX and CT-J approaches, see review [57].

6.3.2.1 Small-U Expansion (CT-INT)

In CT-INT [63], one chooses the interaction part $S_U = \int d\tau_1 H_I^{loc}(\tau_1) = U \int d\tau_1 n_\uparrow(\tau_1) n_\downarrow(\tau_1)$ in the action in (6.132) as the expansion unit. This gives

$$\frac{Z}{Z_0} = 1 + \frac{(-U)}{1!} \int_0^\beta d\tau_1 \langle n_\uparrow(\tau_1) n_\downarrow(\tau_1)\rangle_0$$
$$+ \frac{(-U)^2}{2!} \int_0^\beta d\tau_1 \int_0^\beta d\tau_2 \langle n_\uparrow(\tau_1) n_\downarrow(\tau_1) n_\uparrow(\tau_2) n_\downarrow(\tau_2)\rangle_0 \ldots, \tag{6.133}$$

where $Z_0 = \int D[d]D[d^+]e^{-S_0}$ and the expectation values are $\langle\ldots\rangle_0 = \frac{1}{Z_0} \times \int D[d^+]D[d][\ldots]e^{-S_0}$. Using the Wick theorem, one can obtain for each (kth order) term in Eq. (6.133)

$$\langle n_\uparrow(\tau_1)n_\downarrow(\tau_1)\dots n_\uparrow(\tau_k)n_\downarrow(\tau_k)\rangle_0 = \det D_\uparrow^k \cdot \det D_\downarrow^k, \tag{6.134}$$

where D_σ^k are $k \times k$ matrices with the dynamical mean-field functions as matrix elements:

$$\left(D_\sigma^k\right)_{ij} = \mathcal{G}_\sigma\left(\tau_i - \tau_j\right) = -\left\langle T d_\sigma(\tau_1) d_\sigma^+(\tau_2)\right\rangle_0. \tag{6.135}$$

This gives

$$\frac{Z}{Z_0} = \sum_k \frac{(-U)^k}{k!}\int_0^\beta d\tau_1 \dots \int_0^\beta d\tau_k \prod_\sigma \det D_\sigma^k. \tag{6.136}$$

The terms in the last expression can be negative because $\det D_\sigma^k < 0$ or because for odd k's, $(-U)^k < 0$. It was shown in Ref. [69] that the first situation is impossible in the case of one-band model. To get rid of potentially negative terms due to negative pre-factors $(-U)^k$, it was proposed [70] to perform a particle–hole transformation for spin-down electrons, $d_\downarrow \to \widetilde{d}_\downarrow^+$, $d_\downarrow^+ \to \widetilde{d}_\downarrow$, which gives $(-U) \to U$ and each pre-factor $(-U)^k$ in Eq. (6.136) becomes positive. Other parameters and functions also change after such a transformation, most importantly: $\Delta_\downarrow(\tau) \to -\Delta_\downarrow(-\tau)$, $\varepsilon_{0\uparrow} \to \varepsilon_{0\uparrow} + U$, $\varepsilon_{0\downarrow} \to -\varepsilon_{0\downarrow}$. As result, the interaction part of the Hamiltonian changes to $Un_\uparrow(n_\downarrow - 1)$ and in $D_\downarrow^k$ the diagonal terms are shifted by (−1): $\mathcal{G}_\downarrow(0) \to \mathcal{G}_\downarrow(0) - 1$.

Though these changes lead to spin-up–spin-down asymmetry in the equations, the total sum in the partition function remains spin-invariant. However, it is difficult to preserve this symmetry dynamically during the Monte Carlo sampling. For this reason, a more symmetric transformation to get rid of the sign problem, which consists of shifting both spin-up and spin-down densities in the interaction part, was proposed [70]:

$$S_U = U\int d\tau\left(n_\uparrow(\tau) - \alpha_\uparrow\right)\left(n_\downarrow(\tau) - \alpha_\downarrow\right). \tag{6.137}$$

It gives $\varepsilon_{0\sigma} \to \varepsilon_{0\sigma} - \alpha_{-\sigma} + U$, $\mathcal{G}_\sigma \to \widetilde{\mathcal{G}}_\sigma$ (new dynamical mean-field function) and

$$\begin{aligned}\det D_\sigma^k &= \langle (n_\sigma(\tau_1) - \alpha_\sigma)\dots(n_\sigma(\tau_k) - \alpha_\sigma)\rangle_0 \\ &= \det\begin{bmatrix} \widetilde{\mathcal{G}}_\sigma(0) - \alpha_\sigma & \widetilde{\mathcal{G}}_\sigma(\tau_1 - \tau_2) & \dots & \widetilde{\mathcal{G}}_\sigma(\tau_1 - \tau_k) \\ \widetilde{\mathcal{G}}_\sigma(\tau_2 - \tau_1) & \widetilde{\mathcal{G}}_\sigma(0) - \alpha_\sigma & \dots & \widetilde{\mathcal{G}}_\sigma(\tau_2 - \tau_k) \\ \dots & \dots & \dots & \dots \\ \widetilde{\mathcal{G}}_\sigma(\tau_k - \tau_1) & \widetilde{\mathcal{G}}_\sigma(\tau_k - \tau_2) & \dots & \widetilde{\mathcal{G}}_\sigma(0) - \alpha_\sigma \end{bmatrix}.\end{aligned} \tag{6.138}$$

It was shown in Ref. [70] that to avoid the sign problem, it is sufficient to use $\alpha_\uparrow$ and $\alpha_\downarrow$ that satisfy $\alpha_\uparrow + \alpha_\downarrow = 1$ and $\alpha_\uparrow\alpha_\downarrow \le 0$, i.e., for one spin α has to be positive and for the other—negative. To remove the spin asymmetry caused by non-equal $\alpha_\uparrow$ and $\alpha_\downarrow$, one can introduce random auxiliary fields $s_i = \pm 1$, and substitute α_σ in Eqs. (6.137) and (6.138) by time-dependent $\alpha_{s_\tau\sigma}$. In particular, in this case the interaction action becomes

$$S_U = \frac{U}{2} \int d\tau \left(n_\uparrow(\tau) - \alpha_{s_\tau\uparrow}\right)\left(n_\downarrow(\tau) - \alpha_{s_\tau\downarrow}\right) \tag{6.139}$$

(the factor ½ takes into account two possible values of spin $s_\tau = \pm 1$). In practice, since the matrices (6.138) are almost singular, to avoid numerical instabilities it is convenient instead of $\alpha_{s_\tau\sigma}$ to use $\alpha_{s_\tau\sigma} = \delta_{s_\tau\sigma}\left(\frac{1}{2} + \alpha\right) - \delta_{s_\tau\bar{\sigma}}\alpha$, where α is a small positive number (~0.01) [57]. Now, the remaining task is to calculate the sum (6.136) with the functions under the integrals defined by Eq. (6.138). To define the CT-QMC scheme, let us rewrite (6.136) as

$$Z = \sum_{k=0}^{\infty} \sum_{\gamma \in \Gamma_k} \int_0^\beta d\tau_1 \ldots \int_{\tau_{k-2}}^\beta d\tau_{k-1} \int_{\tau_{k-1}}^\beta d\tau_k w(k, \gamma, \tau_1, \ldots, \tau_k), \tag{6.140}$$

where $\gamma \in \Gamma_k$ includes all parameters of the k-order diagrams (spins, pseudospins, orbital indices, and the diagram topology). The configuration $w(k, \gamma, \tau_1, \ldots, \tau_k)$, that can be represented as

$$x = (k, \gamma, (\tau_1, \ldots, \tau_k)), \tag{6.141}$$

have the weight

$$p(x) = w(k, \gamma, \tau_1, \ldots, \tau_k) d\tau_1 \ldots d\tau_k \equiv w(k) \prod_{i=1}^{k} d\tau_i, \tag{6.142}$$

where due to time ordering, the time points are arranged as $\tau_1 < \tau_2 \ldots < \tau_k$. Equation (6.142) is the distribution describing probability of the k − order diagram with vertices at times $(\tau_1, \ldots, \tau_k)$. Below, we omit index γ.

The CT-QMC evaluation of Z in Eq. (6.140) is based on updating configurations in the form of increase of order k by adding a vertex at time τ and of decrease of order k by removing a vertex at some time point (Fig. 6.5). Starting from an arbitrary configuration, any configuration can be reached with this procedure. The updates with change of k are necessary and usually are sufficient to satisfy requirement of the ergodicity (of exploring all configuration space) [57]. The acceptance ratio for the proposing vertex insertion update with probability $\frac{\Delta\tau}{2\beta}$ and a removal update with probability $\frac{1}{(k+1)}$ is

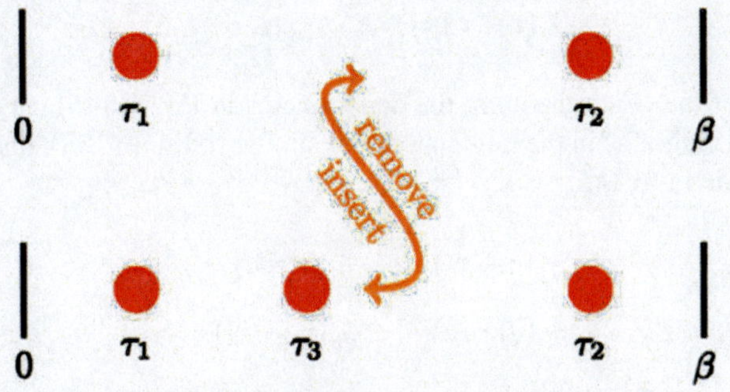

Fig. 6.5 A schematic representation of the insert (top-to-bottom) and removal (bottom-to-top) update in the CT-INT simulations. Reprinted Fig. 2 with permission from (E. Gull, A.J. Millis, A.I. Lichtenstein, A.N. Rubtsov, M. Troyer and P. Werner, Rev. Mod. Phys. **83**, 349 (2011)). Copyright 2016 by the American Physical Society. DOI: https://doi.org/10.1103/RevModPhys.83.349

$$R = \frac{\beta U}{(k+1)} \prod_{\sigma} \frac{\det D_{\sigma}^{k+1}}{\det D_{\sigma}^{k}}. \tag{6.143}$$

Next, when the insertion (removal) is approved, one adds (removes) the corresponding term in the GF

$$G_{\sigma}(\tau - \tau') = \frac{Z_0}{Z} \sum_{k} \frac{(-U)^k}{k!} \int_0^{\beta} d\tau_1 \ldots \int_0^{\beta} d\tau_k \langle d_{\sigma}(\tau) d_{\sigma}^{+}(\tau')\, n_{\uparrow}(\tau_1) n_{\downarrow}(\tau_1) \ldots n_{\uparrow}(\tau_k) n_{\downarrow}(\tau_k) \rangle_0 + \ldots . \tag{6.144}$$

$G_{\sigma}(\tau - \tau')$ is estimated by quantities

$$G_{\sigma\tau_1,\ldots,\tau_k}(\tau, \tau') = -\frac{\langle T_{\tau} d_{\sigma}(\tau) d_{\sigma}^{+}(\tau')\, n_{\uparrow}(\tau_1) n_{\downarrow}(\tau_1) \ldots n_{\uparrow}(\tau_k) n_{\downarrow}(\tau_k) \rangle_0}{\langle\, n_{\uparrow}(\tau_1) n_{\downarrow}(\tau_1) \ldots n_{\uparrow}(\tau_k) n_{\downarrow}(\tau_k) \rangle_0} \tag{6.145}$$

by performing the Monte Carlo averaging. Namely, the expectation value of this or another quantity is obtained by integrating this quantity with the corresponding weight over the configuration space $\langle A \rangle_p = \frac{1}{Z} \int_C dx A(x) p(x)$. In QMC, the integral can be approximated by average over the values for chosen M configurations x_i, $\langle A \rangle_p \approx \langle A \rangle_{\mathrm{MC}} = \frac{1}{M} \sum_{i=1}^{M} A(x_i)$, where the configurations are chosen with probability $p(x)/Z$. Thus,

$$G_\sigma(\tau - \tau') = \langle G_{\sigma\tau_1,\ldots,\tau_k}(\tau, \tau') \rangle_{\text{MC}}. \tag{6.146}$$

Because of the Wick theorem, the denominator in Eq. (6.145) cancels with the disconnected diagrams in the numerator, and similar to Eq. (6.138), one gets for the right-hand side in (6.145)

$$-\det \begin{bmatrix} \mathcal{G}_\sigma(\tau - \tau') & \mathcal{G}_\sigma(\tau - \tau_1) & \mathcal{G}_\sigma(\tau - \tau_2) & \ldots & \mathcal{G}_\sigma(\tau - \tau_k) \\ \mathcal{G}_\sigma(\tau_1 - \tau') & \mathcal{G}_\sigma(0) - \alpha_\sigma & \mathcal{G}_\sigma(\tau_1 - \tau_2) & \ldots & \mathcal{G}_\sigma(\tau_1 - \tau_k) \\ \mathcal{G}_\sigma(\tau_2 - \tau') & \mathcal{G}_\sigma(\tau_2 - \tau_1) & \mathcal{G}_\sigma(0) - \alpha_\sigma & \ldots & \mathcal{G}_\sigma(\tau_2 - \tau_k) \\ \ldots & \ldots & \ldots & \ldots & \ldots \\ \mathcal{G}_\sigma(\tau_k - \tau') & \mathcal{G}_\sigma(\tau_k - \tau_1) & \mathcal{G}_\sigma(\tau_k - \tau_2) & \ldots & \mathcal{G}_\sigma(0) - \alpha_\sigma \end{bmatrix}, \tag{6.147}$$

with new first raw and first column. It is important that even though $G_{\sigma\tau_1,\ldots,\tau_k}(\tau, \tau')$ is not time-translation invariant, the invariance is restored after the MC average.

Generalization of the approach on the multi-orbital case with number of orbitals N_{orb} is straightforward: one expands the partition function in powers of the multi-orbital interaction action $S_{\text{loc}} = U \int d\tau_1 \int d\tau_2 I^{\text{pqrs}} \left(d_p^+ d_s - \alpha_{ps}\right)\left(d_q^+ d_r - \alpha_{qr}\right)$, where α_{ps} is a two-index field introduced to avoid the sign problem. Then, the partition function is calculated as

$$Z = \sum_{k=0}^{\infty} \sum_{\xi_1,\ldots,\xi_k=1}^{N_\xi^k} \int_0^\beta d\tau_1 \ldots \int_0^\beta d\tau_k \frac{(-1)^k I^{\xi_1} \ldots I^{\xi_k}}{k!} \langle v_{\xi_1} \ldots v_{\xi_k} \rangle_0, \tag{6.148}$$

where $\xi_i \equiv (p_i q_i r_i s_i)$ is collective orbital index in the interaction matrices I, the functions in the averages are $v_{\xi_i} = \left\langle \left(d^+_{p_{\xi_i}} d_{s_{\xi_i}} - \alpha_{p_{\xi_i} s_{\xi_i}}\right)\left(d^+_{q_{\xi_i}} d_{r_{\xi_i}} - \alpha_{q_{\xi_i} r_{\xi_i}}\right)\right\rangle_0$, and N_ξ^k is total number of orbital combinations in the k-order term in Eq. (6.140).

The multi-orbital problem (6.148) also reduces to evaluation of determinants for the partition function and GF of type (6.138) and (6.147), where each matrix element is an $N_{\text{orb}} \times N_{\text{orb}}$ orbital matrix. In the acceptance ratio formula Eq. (6.143), in the multi-orbital case one puts $\frac{\beta U}{(k+1)} \to \frac{\beta I N_\xi^k}{(k+1)}$, where I is the average over orbitals I^{ξ_k}.

6.3.2.2 Large-U Expansion (CT-HYB)

For large Us, a meaningful approach is to expand the partition function in powers of the hybridization function $H_{\text{hyb}} = \sum_{p,j}\left[V_p^j c_p^+ d_j + V_p^{j*} d_j^+ c_p\right]$ [65, 66]. Since H_{hyb} consists of two conjugated parts $\widetilde{H}_{\text{hyb}} = \sum_{p,j} V_p^j c_p^+ d_j$ and $\widetilde{H}_{\text{hyb}}^+$, only terms that consist of equal number of $\widetilde{H}_{\text{hyb}}$ and $\widetilde{H}_{\text{hyb}}^+$, and hence only the even-power terms, will give a nonzero contribution to the sum. Therefore, the expansion of the CT-HYB partition function has the following form:

$$Z = \sum_{k=0}^{\infty} \int_0^\beta d\tau_1 \ldots \int_{\tau_{k-1}}^\beta d\tau_k \int_0^\beta d\tau_1' \ldots$$

$$\int_{\tau_{k-1}'}^\beta d\tau_k' Tr\left[T_\tau e^{-\beta H_a} \widetilde{H}_{\text{hyb}}(\tau_k)\widetilde{H}_{\text{hyb}}^+(\tau_k') \ldots \widetilde{H}_{\text{hyb}}(\tau_1)\widetilde{H}_{\text{hyb}}^+(\tau_1')\right], \tag{6.149}$$

or, using the expressions for $\widetilde{H}_{\text{hyb}}$ and $\widetilde{H}_{\text{hyb}}^+$ and separating the bath- and the impurity-electron operators,

$$Z = \sum_{k=0}^{\infty} \int_0^\beta d\tau_1 \ldots \int_{\tau_{k-1}}^\beta d\tau_k \int_0^\beta d\tau_1' \ldots \int_{\tau_{k-1}'}^\beta d\tau_k' \sum_{\substack{j_1,\ldots,j_k \\ j_1',\ldots,j_k'}} \sum_{\substack{p_1,\ldots,p_k \\ p_1',\ldots,p_k'}} V_{p_1}^{j_1} V_{p_1'}^{j_1'*} \ldots V_{p_k}^{j_k} V_{p_k'}^{j_k'*}$$

$$\times \text{Tr}_d\left[T_\tau e^{-\beta H_{\text{loc}}} d_{j_k}(\tau_k) d_{j_k'}^+(\tau_k') \ldots d_{j_1}(\tau_1) d_{j_1'}^+(\tau_1')\right]$$

$$\times \text{Tr}_c\left[T_\tau e^{-\beta H_{\text{bath}}} c_{j_k}(\tau_k) c_{j_k'}^+(\tau_k') \ldots c_{j_1}(\tau_1) c_{j_1'}^+(\tau_1')\right]. \tag{6.150}$$

Integration over the bath fields in Eq. (6.150) can be easily performed, since these fields are noninteracting. Then, using the result for the bath partition function

$$Z_{\text{bath}} = \text{Tr}\left[e^{-\beta H_{\text{bath}}}\right] \prod_p \prod_\sigma \left(1 + e^{-\varepsilon_p \beta}\right) \tag{6.151}$$

and the (anti-periodic) time-dependent hybridization function that follows from Eq. (6.131)

$$\Delta^{lm}(\tau) = \sum_p \frac{V_p^{l*} V_p^m}{e^{\varepsilon_p \beta} + 1} \times \begin{cases} -e^{-\varepsilon_p(\tau-\beta)}, & 0 < \tau < \beta \\ e^{-\varepsilon_p \tau}, & -\beta < \tau < 0 \end{cases} \tag{6.152}$$

one gets

$$\frac{1}{Z_{\text{bath}}} Tr_c\left[T_\tau e^{-\beta H_{\text{bath}}} c_{j_k}(\tau_k) c^+_{j'_k}(\tau'_k) \ldots c_{j_1}(\tau_1) c^+_{j'_1}(\tau'_1)\right] = \det\Delta, \tag{6.153}$$

where the elements of the $k \times k$ matrix Δ are $\Delta^{\ j_l j_m}(\tau_l - \tau_m)$. Using Eqs. (6.151) and (6.153) one obtains the formula for the partition function

$$Z = \sum_{k=0}^{\infty} \int_0^\beta d\tau_1 \ldots \int_{\tau_{k-1}}^\beta d\tau_k \int_0^\beta d\tau'_1 \ldots \int_{\tau'_{k-1}}^\beta d\tau'_k$$
$$\times \sum_{\substack{j_1, \ldots, j_k \\ j'_1, \ldots, j'_k}} \left[T_\tau e^{-\beta H_{\text{loc}}} d_{j_k}(\tau_k) d^+_{j'_k}(\tau'_k) \ldots d_{j_1}(\tau_1) d^+_{j'_1}(\tau'_1)\right] det\Delta_k. \tag{6.154}$$

This is the fundamental equation in the CT-HYB approach. The expression under the integrals in Eq. (6.154)

$$\left[T_\tau e^{-\beta H_{\text{loc}}} d_{j_k}(\tau_k) d^+_{j'_k}(\tau'_k) \ldots d_{j_1}(\tau_1) d^+_{j'_1}(\tau'_1)\right]$$
$$\equiv P_{j_1, j'_1, \ldots, j_k j'_k}\left(\tau_1 \tau'_1, \ldots, \tau_k \tau'_k\right) \tag{6.155}$$

has meaning of the probability distribution of the k-order function at times $\left(\tau_k \tau'_k, \ldots, \tau_k \tau'_k\right)$. Equation (6.154) and the equation for the GF written in terms of $P_{j_1, j'_1, \ldots, j_k j'_k}\left(\tau_1 \tau'_1, \ldots, \tau_k \tau'_k\right)$ are very similar to the corresponding Hirsch–Fye QMC equations written in terms of the probability distributions.

Similar to the CT-INT case, the integration in CT-HYB (6.154) is also performed by stochastic Monte Carlo sampling adding or removing one of the time values τ and τ' that gives a new time set with the new $(k+1)$- or $(k-1)$-order term. The difference from the CT-INT case, however, is that in present case the time set consists of two subsets $\{\tau\}$ and $\{\tau'\}$. Comparison of the new probability function with the old one gives the answer whether the update is accepted or not, and summation over all accepted configurations gives the partition function.

Once the update is accepted (the acceptance ratio equations are given below), it contributes to the sum in

$$G_{lm}(\tau_l - \tau_m) = -Z_{bath} \sum_{k=0}^{\infty} \int_0^\beta d\tau_1 \ldots \int_{\tau_{k-1}}^\beta d\tau_k \int_0^\beta d\tau'_1 \ldots \int_{\tau'_{k-1}}^\beta d\tau'_k$$
$$\times \sum_{\substack{j_1, \ldots, j_k \\ j'_1, \ldots, j'_k}} \left[T_\tau e^{-\beta H_{\text{loc}}} d_l(\tau_l) d^+_m(\tau_m) d_{j_k}(\tau_k) d^+_{j'_k}(\tau'_k) \ldots d_{j_1}(\tau_1) d^+_{j'_1}(\tau'_1)\right] \det\Delta_k. \tag{6.156}$$

Sometimes, it might be more convenient to use the frequency representation for calculation of $G_{lm}(\tau_l - \tau_m)$ and rewrite the last equation as [71]

$$G(i\omega_n) = \sum_{k=0}^{\infty} \int_0^{\beta} d\tau_1 \ldots \int_{\tau_{k-1}}^{\beta} d\tau_k \int_0^{\beta} d\tau_1' \ldots \int_{\tau_{k-1}'}^{\beta} d\tau_k'$$
$$\times \sum_{\substack{j_1, \ldots, j_k \\ j_1', \ldots, j_k'}} P_{j_1 j_1', \ldots, j_k j_k'}(\tau_1\tau_1', \ldots, \tau_k\tau_k') G_{j_1 j_1', \ldots, j_k j_k'}(i\omega_n, \tau_k\tau_k', \ldots, \tau_1\tau_1'), \tag{6.157}$$

where

$$G_{j_1, j_1', \ldots, j_k j_k'}(i\omega_n, \tau_1\tau_1', \ldots, \tau_k\tau_k') = T \sum_{i,j} e^{i\omega_n\tau_i} M_{ij}(i\omega_n, \tau_1\tau_1', \ldots, \tau_k\tau_k') e^{-i\omega_n\tau_j} \tag{6.158}$$

and matrix M is the inverse of matrix (6.152), $M = \Delta^{-1}$.

The update process in the case of adding (removing) of two points $\tau_i, \tau_{i'}'$ consists of adding (removing) the i*th* raw and i'*th* column in the matrix M. Similar to Hirsch–Fye QMC, it is convenient to get the new GF by using the result for the old one [65, 72]. Namely, when the new raw i and column i' are added the new GF is connected with the old one as:

$$G^{\text{new}}(\tau - \tau') = G^{\text{old}}(\tau - \tau') - T \cdot p \cdot \left(\sum_{j=1}^{k} e^{i\omega_n\tau_j} L_j \right) \left(\sum_{j'=1}^{k} e^{-i\omega_n\tau_{j'}'} R_{j'} \right). \tag{6.159}$$

In the last equation,

$$p = \left[\Delta(\tau - \tau') - \sum_{i,j} \Delta(\tau - \tau_j) M_{ji}^{\text{old}} \Delta(\tau_i - \tau') \right]^{-1} \tag{6.160}$$

and

$$L_j = (\widetilde{L}_1, \ldots, \widetilde{L}_{j-1}, -1, \widetilde{L}_{j+1}, \ldots, \widetilde{L}_k), \tag{6.161}$$
$$R_{j'} = (\widetilde{R}_1, \ldots, \widetilde{R}_{j'-1}, -1, \widetilde{R}_{j'+1}, \ldots, \widetilde{R}_k), \tag{6.162}$$

where

$$\widetilde{L}_j = \sum_{i,j} M_{ij}^{\text{old}} \Delta(\tau_i - \tau'), \tag{6.163}$$

$$\widetilde{R}_j = \sum_{i,j} \Delta(\tau - \tau_i)\, M_{ij}^{\text{old}}. \tag{6.164}$$

When the raw i and the column i' are removed, one has

$$G^{\text{new}}(\tau - \tau') = G^{\text{old}}(\tau - \tau') + \frac{T}{M_{ii'}} \left(\sum_{j=1}^{k} e^{i\omega_n \tau_j} M_{ji}^{\text{old}} \right) \times \left(\sum_{j'=1}^{k} M_{i'j'}^{\text{old}}\, e^{-i\omega_n \tau'_{j'}} \right). \tag{6.165}$$

The acceptance ratio in the case a raw and a column are added is

$$\frac{\det M^{\text{old}}}{\det M^{\text{new}}} = \frac{1}{p}, \tag{6.166}$$

and in the case of removal of a raw and a column:

$$\frac{\det M^{\text{old}}}{\det M^{\text{new}}} = M_{kk}. \tag{6.167}$$

We refer the reader for an information about alternative implementations of CT-HYB to Ref. [57], where they are discussed. The generalization of the CT-HYB approach on the multi-orbital case is rather straightforward [66, 72].

We end the discussion of details of the CT-HYB solver by mentioning that even though the partition function includes all-orders terms in the hybridization expansion, similar to CT-INT, the sum converges for any value of the hybridization (see, e.g., book [71], Figs. 3.6 and 3.7).

6.4 Other Solvers

In this section, we present for convenience a table with a list of (not all!) one- and multi-orbital solvers including the solvers above (in alphabetical order) and the corresponding references. The references include the original and/or review paper (s), thus their number is rather limited but hopefully is sufficient to give enough information for the reader to make her/him able to use these tools. The solvers are grouped by three types, as the subsections in this section: analytical, Hamiltonian-form, and stochastic solvers. The multi-orbital solver papers are marked with superscript "M." In the second group, "ED" and "RG" in brackets are put to define the type of the solver, where it is not obvious.

Analytical	
Equations-of-motion decoupling schemes	[46, 85]
Equations-of-motion for local GFs solver	[100]M
Hierarchical equations-of-motion solvers	[101, 102]M
Hubbard-I	[2] (Chapter 2), [15], [90], [91]
Iterative Perturbation Theory (IPT)	[1–3, 5, 6, 9, 46], [4]M, [12–14]M
Local moment approach (LMA) solvers	[103], [104], [105]M
Non-crossing approximation (NCA)	[81–84]
One-crossing approximation (OCA)	[81], [94–97]
Rational interpolation for the self-energy	[15], [93]
Slave-boson solvers	[15, 46, 87–89], [92]M
Slave quantum rotor solver	[92]M
Hamiltonian-form	
Adaptively truncated Hilbert space solver (RG)	[116]
Augmented hybrid ED solver	[106]M
Bath optimization and cluster PT solver (ED)	[117]
Chebyshev expansions represented with matrix product states (CheMPS, RG)	[109]
Configuration Interaction approximation to ED	[98]
Coupled-cluster impurity solvers (ED-type, multisite)	[118, 119]
DMRG	[33, 40–43]
Distributional ED	[114]
Exact diagonalization (ED)	[17–21, 23–25], [46], [26]M, [27]M
ED-based real-frequency solver	[99]
Fork Tensor–Product States (RG)	[115]M
Functional Renormalization Group (fRG)	[107], [108]M
Imaginary-time evolution of matrix product states (ITMPS, RG)	[113]M
Multi-scale ED + PT solver	[111]M
Multi-scale standard basis operator equation of motion impurity solver (ED + EoM)	[112]
NRG	[29–33, 35], [36–39]M
Projective self-consistent technique (RG)	[46, 86]
Time evolving block decimation (TEBD) for matrix product states solver (RG)	[110]
Stochastic	
CT-QMC	[58–64, 68–70], [57, 67, 80]M
Double-expansion (large-U, small-J) CT-QMC	[120]M
Hirsch–Fye QMC	[15, 45–47], [55]M

6.5 Analytical Continuation

It is technically much simpler to solve DMFT equations for a set of imaginary time points or discrete Matsubara frequencies than using real time or frequencies. On the other hand, physical information is mostly contained in real frequency- and time dependencies of different functions (self-energies, GFs, etc.). Thus, once the Matsubara-frequency solution has been found, one needs to obtain the real-frequency dependencies of the functions. It can be done by the analytical continuation procedure

$$i\omega_n \to \omega + i\delta. \tag{6.168}$$

To get connection between the real-frequency and imaginary-time dependencies of functions that can be also used, one may take advantage from the fact that the imaginary part of the real-frequency (retarded) GF, $A(\omega) = -\frac{1}{\pi}G^R(\omega)$, is connected with the Matsubara-frequency GF by means of the spectral representation formula

$$G(i\omega_n) = \int_{-\infty}^{\infty} \frac{A(\omega)}{i\omega_n - \omega} d\omega. \tag{6.169}$$

Substituting this result into the Fourier transform formula

$$G(\tau) = T \sum_n e^{i\omega_n \tau} G(i\omega_n) \tag{6.170}$$

one can obtain a relation that connects the Matsubara-time GF and real-frequency retarded GF (its imaginary part):

$$G(\tau) = -\int_{-\infty}^{\infty} \frac{e^{-\tau\omega}}{1 + e^{-\beta\omega}} A(\omega) d\omega \tag{6.171}$$

Thus, one can find $A(\omega)$ from $G(\tau)$ for a given set of values of imaginary time $\{\tau_i\}$.

Below, we describe two approaches for the analytical continuation: a fast algorithm based on Eq. (6.168) (the Padé approximation method [73]) and a more accurate, method-of-choice, based on Eq. (6.171) (the maximum entropy (MaxEnt) approach [74–77]).

6.5.1 *Padé Approximation*

In this subsection, we show how to obtain the real frequency dependence of the GF from the dependence of this function on the Matsubara frequencies by using the Padé approximation method [73] developed by Vidberg and Serene who also first applied it to solve the Eliashberg equations [78].

Let us assume that we have solved the DMFT problem in the Matsubara frequency representation and found numerically the GF $G_N(i\omega_n)$ for N Matsubara frequencies ($i = 1,\ldots,N$). The strategy is to find an analytical formula that gives this dependence on the Matsubara frequencies for discrete set $\{\omega_n\}$, and then to make substitution $i\omega_n \to \omega + i\delta$ in this formula. The formalism is absolutely the same for the analytical continuation of the self-energy.

The analytical approximate formula for the GF can be obtained by using Padé approximation [78]. To write down this expression one uses the fact that GF at large frequencies has the asymptotic $G(i\omega_n) \sim 1/\omega_n$ (see, e.g., Ref. [79]). Thus, the analytical expression for this function can be a ratio of two polynomials (we use complex variable z instead of $i\omega_n$):

$$G_N(z) = \frac{A_0 + A_1 z + \ldots + A_{N/2-1} z^{(N/2-1)}}{1 + B_1 z + \ldots + B_{N/2-1} z^{(N/2-1)} + B_{N/2} z^{N/2}} \equiv \frac{A_N}{B_N} \tag{6.172}$$

for even Ns and

$$G_N(z) = \frac{A_0 + A_1 z + \ldots + A_{(N-1)/2} z^{(N-1)/2}}{1 + B_1 z + \ldots + B_{(N-1)/2} z^{(N-1)/2} + B_{(N+1)/2} z^{(N+1)/2}} \equiv \frac{A_N}{B_N} \tag{6.173}$$

for odd Ns. We put the frequency-independent coefficient B_0 in these expressions equal one, since it can be done by dividing the numerator and denominator by B_0 (a simple analysis shows that this coefficient is nonzero).

Then, substituting the values of frequencies z_i used in the DMFT calculations and the obtained values of GFs at these frequencies $G_N(z_i)$ ($i = 1, \ldots, N$) into Eqs. (6.172) and (6.173), and multiplying the right and the left parts of the equations by the denominator polynomials one gets N algebraic equations for the unknown coefficients A and B. In particular, for even Ns,

$$\begin{aligned} &A_0 + z_i A_1 + \ldots + z_i^{\left(\frac{N}{2}-1\right)} A_{\frac{N}{2}-1} - z_i G_N(z_i) B_1 - \ldots - z_i^{\frac{N}{2}} G_N(z_i) B_{\frac{N}{2}} \\ &= G_N(z_i), \quad (i = 1, \ldots, N) \end{aligned} \tag{6.174}$$

and similar for odd Ns. It must be noted that in the case of an odd N one of the coefficients in the numerator or denominator in Eq. (6.173), except the first and the last ones, has to be fixed (e.g., $B_1 = 1$ (see Eq. (6.182)), otherwise the number of the unknown coefficients will be larger than the number of equations.

Once the system of Eqs. (6.174) is solved, one can use obtained coefficients $A_0, \ldots, A_{\frac{N}{2}-1}, B_1, \ldots, B_{\frac{N}{2}}$ to construct the analytical formula for real-frequency self-energy:

$$G_N(\omega) = \frac{A_0 + A_1(\omega + i\delta) + \ldots + A_{N/2-1}(\omega + i\delta)^{(N/2-1)}}{1 + B_1(\omega + i\delta) + \ldots + B_{N/2-1}(\omega + i\delta)^{(N/2-1)} + B_{N/2}(\omega + i\delta)^{N/2}}. \quad (6.175)$$

In practice, however, one can use a faster algorithm to calculate these coefficients for $G_N(\omega)$ than to solve the system of equations (6.174) [78]. Namely, instead of the ratio of polynomials, one can use a continued fraction expansion expression for the GF

$$G_N(z) = \cfrac{a_1}{1 + \cfrac{a(z-z_1)}{1 + \cfrac{a_3(z-z_2)}{\cdots \cfrac{\cdots}{\cfrac{\cdots}{1 + \cfrac{a(z - z_{N-1})}{1}}}}}}, \quad (6.176)$$

where $z_1, \ldots, z_{N-1}$ are the first $N - 1$ Matsubara frequency points used in the calculations. Then, the coefficients a_i $(i = 1, \ldots N)$ can be found from the system of N equations (6.176), where on the right part one puts $z = z_1, \ldots, z_{N-1}$ and on the left—corresponding values $G_N(z_i)$, for which we use notation u_i. In other words, the system of equations for the coefficients $a_1, \ldots, a_N$ is

$$G_N(z_i) = u_i, \quad (i = 1, \ldots, N). \quad (6.177)$$

These coefficients can be found from the following recursion relations:

$$a_i = g_i(z_i), \quad g_1(z_i) = u_i, \quad (i = 1, \ldots, N) \quad (6.178)$$

$$g_j(z) = \frac{g_{j-1}(z_{j-1}) - g_{j-1}(z)}{(z - z_{j-1})g_{j-1}(z)}. \quad (j = 2, \ldots, N) \quad (6.179)$$

Namely, to solve the problem one makes the following steps:

(1) from (6.178) one gets $g_1(z_1) = u_1$, and then $a_1 = g_1(z_1)$
(2) from (6.178) and (6.179): $a_2 = g_2(z_2) = \frac{g_1(z_1)-g_1(z_2)}{(z_2-z_1)g_1(z_2)} \equiv \frac{u_1-u_2}{(z_2-z_1)u_2}$
(3) then, one finds $g_2(z_3) = \frac{g_1(z_1)-g_1(z_3)}{(z_3-z_1)g_1(z_3)} = \frac{u_1-u_3}{(z_3-z_1)u_3(z_3)}$
(4) next, $a_3 = g_3(z_3) = \frac{g_2(z_2)-g_2(z_3)}{(z_3-z_2)g_2(z_3)} = \frac{a_2-g_2(z_3)}{(z_3-z_2)g_2(z_3)}$, where a_2 and $g_2(z_3)$ are given in the steps (2) and (3), and so on (find $g_3(z_4)$ and then $a_4 = g_4(z_4)$, both from Eq. (6.179), i.e., one repeats consequently steps (3), (4) by increasing the function and the argument index by 1).

The found coefficients a_i have to be substituted into Eq. (6.176), and after the analytical continuation $z \to \omega + i\delta$, one finds analytical expression for $G_N(\omega)$.

In many cases, ratio-of-polynomials results are more convenient than representation Eq. (6.176). We present here a fast algorithm how to find $A_N(z)$ and $B_N(z)$ polynomials (here A and B are the frequency-dependent total polynomials, not coefficients!). They can be found from a coefficients by using the following recursive relations:

$$A_{i+1}(z) = A_i(z) + \left(z - z_j\right)a_{\,j+1}A_{i-1}(z), \tag{6.180}$$

$$B_{i+1}(z) = B_i(z) + \left(z - z_j\right)a_{\,j+1}B_{i-1}(z) \tag{6.181}$$

(a_{j+1} are defined in Eq. (6.178), beginning with $i = 0$ and using

$$A_0 = 0, B_0 = B_1 = 1. \tag{6.182}$$

Equations (6.180)–(6.182) can be solved for a needed set of real frequencies (by putting there $z = \omega + i\delta$) that gives

$$G_N(z) = \frac{A_N(z)}{B_N(z)}. \tag{6.183}$$

Equations (6.180)–(6.183) complete our discussion of the analytical continuation by using Padé approximation.

6.5.2 *Maximum Entropy Approach*

The results for GF can be sensitive to the number of used Matsubara time points (this number has to be rather large at low temperatures, which makes DMFT calculations very slow). At the same time, in the currently most popular CT-QMC approaches $G(\tau)$ is calculated for a sufficiently small number of time points. Therefore, the frequency-dependent GF $G(i\omega_n)$ (obtained by using the inverse to (6.170) transformation) might be not accurate enough to be used in the Pade approximation. Therefore, in such cases, one may try to use Eq. (6.171) to get the real-frequency GF (its imaginary part) from the Matsubara-time GF. However, then one meets with another problem: besides a small number of time points, the CT-QMC (or another QMC) result for the GF is obtained with a stochastic error. Moreover, the kernel in Eq. (6.171) $\frac{e^{-\tau\omega}}{1+e^{-\beta\omega}}$ exponentially decays at high frequencies, thus $G(\tau)$ is not sensitive to the values of $A(\omega)$ at high frequencies when $|\omega| > T$. This makes calculation of $A(\omega)$ from Eq. (6.171) not a well-defined problem [71, 77]. An alternative approach for the analytical continuation of "stochastically obtained" functions is the maximum entropy (MaxEnt) method developed in Refs. [74–76]. Below we summarize the main steps of this approach following review [77] and book [71].

Since $A(\omega) > 0$ and $\int_{-\infty}^{\infty} A(\omega) = 1$, one can regard $A(\omega)$ as a probability distribution function (in general case, the last integral is just finite, and equal 1 when $A(\omega)$ is normalized). The MaxEnt approach is based on Bayesian statistics and gives a recipe how to get the most probable function $A(\omega)$ for given values of $G(\tau)$ by using Eq. (6.171). In detail, let us consider two events: G—obtained from (CT-) QMC values $G(\tau)$, and A —values $A(\omega)$. According to the Bayes theorem, for two events G and A, their joint probability is

$$\Pr[G, A] = \Pr[G|A]\Pr[A] = \Pr[A|G]\Pr[G], \tag{6.184}$$

where $\Pr[G]$ and $\Pr[A]$ are probabilities of events G and A, respectively, $\Pr[G, A]$ is the conditional probability of event G given event A, and vice versa, $\Pr[A, G]$ is the conditional probability of event A given event G. The probabilities are normalized:

$$\int \Pr[G]dG = 1, \tag{6.185}$$

$$\Pr[G] = \int \Pr[G, A]dA, \tag{6.186}$$

$$\int \Pr[G, A]dG = 1. \tag{6.187}$$

Determination of $A(\omega)$ is not a well-defined problem since there exists infinite number of solutions $A(\omega)$ for given $G(\tau)$ and given error bars, so MaxEnt method chooses the most probable solution $A(\omega)$. As it follows from Eq. (6.184), the conditional probability of given spectrum $A(\omega)$ for given GF $G(\tau)$ is

$$\Pr[A|G] = \frac{\Pr[G|A]\Pr[A]}{\Pr[G]}. \tag{6.188}$$

Before analyzing and using the parts of this equation, we mention that $\Pr[G|A]$ in Eq. (6.188) is called the likelihood function (conditional probability of $G(\tau)$ given $A(\omega)$), probability $\Pr[A]$ is prior probability that contains prior information on $A(\omega)$, and the "normalization"—probability $\Pr[G]$—is called the evidence.

The probability $\Pr[A|G]$ is maximal when $\Pr[G|A]$ Pr $[A]$ is maximal, i.e., we have to find maximum $\Pr[A]$ and $\Pr[G|A]$. Thus, we need to find the function $A(\omega)$ that gives for these two probabilities maximum values. The principle of maximum entropy says that $A(\omega)$ should be such that it maximizes the entropy

$$S = -\int d\omega A(\omega) \log \frac{A(\omega)}{m(\omega)}, \tag{6.189}$$

where $m(\omega)$ is called "default model." In the absence of data one has $\Pr[A|G] = \Pr[A]$, and

$$\Pr[A] = e^{\alpha S}, \tag{6.190}$$

where α is a positive number (it will be found below by a maximization procedure).

The usually used form of entropy

$$S = \int d\omega \left(A(\omega) - m(\omega) - A(\omega) \log \frac{A(\omega)}{m(\omega)} \right). \tag{6.191}$$

is little different from Eq. (6.189). It reduces to (6.189) when $A(\omega)$ and $m(\omega)$ have the same normalization, the condition not difficult to satisfy. The entropy S has its maximum value, zero, when $A(\omega) = m(\omega)$, otherwise $S < 0$.

The form of entropy function depends on different assumptions on $A(\omega)$, like the system independence, coordinate invariance, fulfilling sum rules, correct asymptotic behavior, etc. (see Refs. [71, 77]). The entropy is defined relative to the default model, thus knowledge of the prior probability $P[A]$ that contains previous information on $A(\omega)$ is very helpful. Otherwise, in the absence of data, the resultant spectrum will be identical to the model one.

To find the next quantity, $P[G|A]$, one uses the central limit theorem that says that the probability distribution for a random variable χ is always Gaussian when its measurements are completely independent of each other, i.e.

$$P(\chi) = \frac{1}{2\pi\sigma} e^{-\frac{\chi^2}{2}}, \tag{6.192}$$

where σ is dispersion. Our strategy is the following—first, to express χ in terms of $G(\tau)$ ($G(\tau)$ is a result of average of independent QMC measurements (accepted configurations) $G_{s_1,\ldots,s_L}(\tau_l, \tau_{l'})$) Then, since $G(\tau)$ can be expressed in terms of $A(\omega)$ (Eq. (6.171)), - to get functional $\chi[A]$. Then, using Eqs. (6.188), (6.190), and (6.192) it follows that to get maximum $\Pr[A|G]$ one has to maximize the functional

$$Q_\alpha[A] = \alpha S[A] - \frac{\chi^2[A]}{2} \tag{6.193}$$

with respect to α and $A(\omega)$, which will give the maximum conditional probability

$$Pr[A|G] \sim e^{Q_\alpha[A]}. \tag{6.194}$$

In other words, one can find α and, most importantly, $A(\omega)$ by solving the system of equations

$$\frac{\delta Q_\alpha[A]}{\delta A(\omega)} = 0, \tag{6.195}$$

$$\frac{\partial Q_\alpha[A]}{\partial \alpha} = 0. \tag{6.196}$$

Our remaining task is to find $\chi[G]$. As we mentioned, $G(\tau)$ is a result of average of independent QMC measurements (accepted configurations) $G_{s_1,\ldots,s_L}(\tau_l, \tau_{l'})$. Let us divide the accepted configurations on N_d subsets (bins) and make averages in each subset that give average subsets $\{G_l^i\}$, where l is the time point and i is the subset index. Then, one can find an error, or random deviation from average value, for each bin $\Delta G_l^i = G_l^i - \overline{G}_l$, where $\overline{G}_l = \frac{1}{N_d}\sum_{i=1}^{N_d} G_l^i$ is the average over all bins. Since all ΔG_l^i are independent, G_l^i should be normally- (Gaussian-) distributed, i.e., with the probability

$$P\left(G_l^i\right) = \frac{1}{2\pi\sigma} e^{-\frac{\chi_i^2}{2}}, \tag{6.197}$$

where

$$\chi_i^2 = \frac{1}{\sigma^2}\left(\Delta G_l^i\right)^2 = \frac{1}{\sigma^2}\left(\overline{G}_l - G_l^i\right)^2, \tag{6.198}$$

and the dispersion σ is

$$\sigma = \sqrt{\frac{1}{N_d(N_d-1)}\sum_{i=1}^{N_d}\left(\Delta G_l^i\right)^2}. \tag{6.199}$$

Thus, now one has the expression for χ_i^2 which is a functional of G_l^i and hence a functional of $A(\omega)$ (by means of Eq. (6.171)). $\chi_i^2[A]$ can be substituted into Eq. (6.192) and the last expression can be used in the maximizing equations (6.195) and (6.196) that define maximal entropy α and $A(\omega)$.

However, in practice, QMC results for G_l^i for different bins are not really independent, i.e., are they are correlated, and $P\left(G_l^i\right)$ deviates from the normal one. In this case, one uses

$$\chi_{\mathrm{i}}^2 = \sum_{1} \Delta G_l^i \left(C^{-1}\right)_{ll'} \Delta G_{l'}^i, \tag{6.200}$$

where

$$C_{ll'} = \frac{1}{N_d(N_d-1)}\sum_{i=1}^{N_d} \Delta G_l^i \left(C^{-1}\right)_{\mathrm{ll'}} \Delta G_{l'}^i \tag{6.201}$$

is the covariation matrix.

Then, to get a Gaussian form for χ_i^2, i.e., a sum of squares of deviations of GFs on the right-hand side in Eq. (6.200), one needs to diagonalize $C_{ll'}$: $C' = U^{-1}CU$, and simultaneously to transform GFs: $G' = U^{-1}G$. This gives

$$\chi_i^2 = \sum_l \left(\frac{\Delta G_l'^i}{\sigma_l} \right)^2, \tag{6.202}$$

where σ_l are the eigenvalues of matrix C. Because of these transformations, in Eq. (6.200) one has to use G' instead of G.

There are also other ways to normalize the distribution of G_l^i (to eliminate the correlations of bins), e.g., by using a more coarse mesh of bins (re-binning), combining them in blocks and averaging G_l^i for all the data in the block. We refer the reader for details of this procedure and other details to Ref. [77].

References

1. Kajueter, H., Kotliar, G.: Phys. Rev. Lett. **77**, 131 (1996)
2. Kajueter, H., Kotliar, G., Moeller, G.: Phys. Rev. B. **53**, 16214 (1996)
3. Potthoff, M., Wegner, T., Nolting, W.: Phys. Rev. B. **55**, 16132 (1997)
4. Anisimov, V.I., Poteryaev, A.I., Korotin, M.A., Anokhin, A.O., Kotliar, G.: J. Phys.: Condens. Matter. **9**, 7359 (1997)
5. Martin-Rodero, A., Louis, E., Flores, F., Tejedor, C.: Phys. Rev. B. **33**, 1814 (1986)
6. Helmes, R.: Dynamical Mean Field Theory of inhomogeneous correlated systems. Cologne University, Germany (2008) PhD Thesis
7. Luttinger, J.M., Ward, J.C.: Phys. Rev. **118**, 1417 (1960)
8. Langreth, D.C.: Phys. Rev. **150**, 516 (1966)
9. Muller-Hartmann, E.: Z. Phys. B. **76**, 211 (1989)
10. Oudovenko, V.S., Haule, K., Savrasov, S.Y., Villani, D., Kotliar, G.: preprint cond-mat/0401539 (2004).
11. Arsenault, L.-F., Semon, P., Tremblay, A.-M.S.: Phys. Rev. B. **86**, 085133 (2012)
12. Craco, L., Laad, M.S., Mueller-Hartmann, E.: Phys. Rev. Lett. **90**, 237203 (2003)
13. Laad, M.S., Craco, L., Mueller-Hartmann, E.: Phys. Rev. B. **67**, 033105 (2003)
14. Laad, M.S., Craco, L., Mueller-Hartmann, E.: Phys. Rev. B. **73**, 195120 (2006)
15. Kotliar, G., Savrasov, S.Y., Haule, K., Oudovenko, V.S., Parcollet, O., Marianetti, C.A.: Rev. Mod. Phys. **78**, 865 (2006)
16. Dasari, N., Mondal, W.R., Zhang, P., Moreno, J., Jarrell, M., Vidhyadhiraja, N.S.: Eur. Phys. J. B. **89**, 202 (2016)
17. Caffarrel, M., Krauth, W.: Phys. Rev. Lett. **72**, 1545 (1994)
18. Jaklic, J., Prelovsek, P.: Phys. Rev. B. **49**, 5065 (1994)
19. Jaklic, J., Prelovsek, P.: Adv. Phys. **49**, 1 (2000)
20. Aichhorn, M., Daghofer, M., Evertz, H.G., von der Linden, W.: Phys. Rev. B. **67**, 161103 (R) (2003)
21. Capone, M., de' Medici, L., Georges, A.: Phys. Rev. B. **76**, 245116 (2007)
22. Haydock, R., Heine, V., Kelly, M.J.: J. Phys. C. **8**, 2591 (1975)
23. Liebsch, A., Tong, N.-H.: Phys. Rev. B. **80**, 165126 (2009)
24. Medvedeva, D., Iskakov, S., Krien, F., Mazurenko, V.V., Lichtenstein, A.I.: Phys. Rev. B. **96**, 235149 (2017)

25. Lu, Y., Haverkort, M.W.: Eur. Phys. J. Special Top. **226**, 2549 (2017)
26. Liebsch, A., Ishida, H.: J. Phys.: Condens. Matter. **24**, 053201 (2012)
27. Perroni, C.A., Ishida, H., Liebsch, A.: Phys. Rev. B. **75**, 045125 (2007)
28. Wilson, K.G.: Rev. Mod. Phys. **47**, 773 (1975)
29. Bulla, R., Hewson, A.C., Pruschke, T.: J. Phys.: Condens. Matter. **10**, 8365 (1998)
30. Sakai, O., Kuramoto, Y.: Solid State Commun. **89**, 307 (1994)
31. Bulla, R.: Phys. Rev. Lett. **83**, 136 (1999)
32. Bulla, R., Costi, T.A., Vollhardt, D.: Phys. Rev. B. **64**, 045103 (2001)
33. Bulla, R., Costi, T.A., Pruschke, T.: Rev. Mod. Phys. **80**, 395 (2008)
34. Žitko, R., Pruschke, T.: Phys. Rev. B. **79**, 085106 (2009)
35. Osolin, Z., Zitko, R.: Phys. Rev. B. **87**, 245135 (2013)
36. Pruschke, T., Bulla, R.: Eur. Phys. J. B. **44**, 217 (2005)
37. Stadler, K.M., Yin, Z.P., von Delft, J., Kotliar, G., Weichselbaum, A.: Phys. Rev. Lett. **115**, 136401 (2015)
38. Stadler, K.M., Mitchell, A.K., von Delft, J., Weichselbaum, A.: Phys. Rev. B. **93**, 235101 (2016)
39. Mitchell, A.K., Galpin, M.R., Wilson-Fletcher, S., Logan, D.E., Bulla, R.: Phys. Rev. B. **89**, 121105(R) (2014)
40. Noack, R.M., Manmana, S.R.: AIP Conf. Proc. **789**, 93 (2005)
41. Schollwöck, U.: Rev. Mod. Phys. **77**, 259 (2005)
42. Hallberg, K.A.: Adv. Phys. **55**, 477 (2006)
43. Nishimoto, S., Jeckelmann, E.: J. Phys.: Condens. Matter. **16**, 613 (2004)
44. Blankenbecler, R., Scalapino, D.J., Sugar, R.L.: Phys. Rev. B. **24**, 2278 (1981)
45. Hirsch, J.E., Fye, R.M.: Phys. Rev. Lett. **56**, 2521 (1986)
46. Georges, A., Kotliar, G., Krauth, W., Rozenberg, M.J.: Rev. Mod. Phys. **68**, 13 (1996)
47. A. Macridin, M. Jarrell, and K. Mikelsons, Quantum Monte Carlo algorithms for the quantum cluster problem. In Eds. A. Avella and F. Mancini (Eds.), Lectures on the physics of strongly correlated systems XII: Twelfth Training Course in the Physics of Strongly Correlated Systems, AIP Conference Proceedings/Materials Physics and Applications) (No. 12), (2008).
48. Vitali, E., Shi, H., Qin, M., Zhang, S.: Phys. Rev. B. **94**, 085140 (2016)
49. Gukelberger, J., Kozik, E., Hafermann, H.: Phys. Rev. B. **96**, 035152 (2017)
50. N. Blümer, preprint arXiv:0801.1222 (2008).
51. Gorelik, E.V., Blumer, N.: Phys. Rev. A. **80**, 051602 (2009)
52. Alvarez, G., et al.: Proceedings of the 2008 ACM/IEEE Conference on Supercomputing, SC '08, pp. 61:1–61:10. IEEE Press, Piscataway, NJ (2008)
53. Nukala, P.K.V.V., Maier, T.A., Summers, M.S., Alvarez, G., Schulthess, T.C.: Phys. Rev. B. **80**, 195111 (2009)
54. Dong, Q., Krivenko, I., Kleinhenz, J., Antipov, A.E., Cohen, G., Gull, E.: Phys. Rev. B. **96**, 155126 (2017)
55. Takegahara, K.: J. Phys. Soc. Jpn. **62**, 1736 (1993)
56. Shinaoka, H., Assaad, F., Blumer, N., Werner, P.: Eur. Phys. J. Special Top. **226**, 2499 (2017)
57. Gull, E., Millis, A.J., Lichtenstein, A.I., Rubtsov, A.N., Troyer, M., Werner, P.: Rev. Mod. Phys. **83**, 349 (2011)
58. Prokof'ev, N.V., Svistunov, N.B.V., Tupitsyn, I.S.: JETP Lett. **64**, 911 (1996)
59. Beard, B.B., Wiese, U.-J.: Phys. Rev. Lett. **77**, 5130 (1996)
60. Prokof'ev, N.V., Svistunov, B.V.: Phys. Rev. Lett. **81** (1998)
61. Prokof'ev, N.V., Svistunov, B.V., Tupitsyn, I.S.: Sov. Phys. JETP. **87**, 310 (1998)
62. Rombouts, S.M.A., Heyde, K., Jachowicz, N.: Phys. Rev. Lett. **82**, 4155 (1999)
63. Rubtsov, A.N., Lichtenstein, A.I.: JETP Lett. **80**, 61 (2004)
64. Rubtsov, A.N., Savkin, V.V., Lichtenstein, A.I.: Phys. Rev. B. **72**, 035122 (2005)
65. Werner, P., Comanac, A., de' Medici, L., Troyer, M., Millis, A.J.: Phys. Rev. Lett. **97**, 076405 (2006)
66. Werner, P., Millis, A.J.: Phys. Rev. B. **74**, 155107 (2006)

67. E. Gull, Ph.D. Thesis (ETH Zurich, 2008).
68. Otsuki, J., Kusunose, H., Werner, P., Kuramoto, Y.: J. Phys. Soc. Jpn. **76**, 114707 (2007)
69. Yoo, J.S., Chandrasekharan, R.K., Kaul, D.U., Baranger, H.U.: J. Phys. A. **38**, 10307 (2005)
70. A.N. Rubtsov, preprint arXiv:cond-mat/0302228 (2003).
71. V. Anisimov, Yu. Izyumov, Electronic structure of strongly correlated materials (Springer Series in Solids, 2010).
72. Haule, K.: Phys. Rev. B. **75**, 155113 (2007)
73. Baker Jr., G.A.: Essentials of Padé Approximants. Academic Press, New York (1975) Chapter 8
74. Silver, R.N., Sivia, D.S., Gubernatis, J.E.: Phys. Rev. B. **41**, 2380 (1990)
75. Silver, R.N., Sivia, D.S., Gubernatis, J.E., Jarrell, M.: Phys. Rev. Lett. **65**, 496 (1990)
76. Gubernatis, J.E., Jarrell, M., Silver, R.N., Sivia, D.: Phys. Rev. B. **44**, 6011 (1991)
77. Jarrell, M., Gubernatis, J.E.: Phys. Rep. **269**, 133 (1996)
78. Vidberg, H.J., Serene, J.W.: J. Low Temp. Phys. **29**, 179 (1977)
79. Turkowski, V., Freericks, J.K.: Phys. Rev. B. **77**, 205102 (2008)
80. E. Gorelov, Ph.D. Thesis (University of Hamburg, 2007).
81. Pruschke, T., Cox, D.L., Jarrell, M.: Phys. Rev. B. **47**, 3553 (1993)
82. Jarrell, M., Pruschke, T.: Phys. Rev. B. **49**, 1458 (1993)
83. Jarrell, M., Pruschke, T.: Z. Phys. B. **90**, 187 (1993)
84. Pruschke, T., Cox, D.L., Jarrell, M.: Europhys. Lett. **21**, 593 (1993)
85. Gros, C.: Phys. Rev. B. **50**, 7295 (1994)
86. G. Moeller, V. Dobrosavljevic, and A. Ruckenstein, unpublished (1995).
87. Kotliar, G., Ruckenstein, A.: Phys. Rev. Lett. **57**, 1362 (1986)
88. Kotliar, G.: In: Emery, V.J. (ed.) Correlated electron systems. World Scientific, Singapore (1993)
89. Kotliar, G.: In: Doucot, B., Zinn-Justin, J. (eds.) Strongly interacting fermions and high Tc superconductivity. Elsevier Science, New York (1994)
90. Lichtenstein, A.I., Katsnelson, M.I.: Phys. Rev. B. **57**, 6884 (1998)
91. Svane, A., Kanchana, V., Vaitheeswaran, G., Santi, G., Temmerman, W.M., Szotek, Z., Strange, P., Petit, L.: Phys. Rev. B. **71**, 045119 (2005)
92. Florens, S., Georges, A.: Phys. Rev. B. **66**, 165111 (2002)
93. Savrasov, S.Y., Oudovenko, V.S., Haule, K., Villani, D., Kotliar, G.: Phys. Rev. B. **71**, 115117 (2005)
94. Pruschke, T., Grewe, N.: Z. Phys. B. **74**, 439 (1989)
95. Haule, K., Kirchner, S., Kroha, J., Wolfle, P.: Phys. Rev. B. **64**, 155111 (2001)
96. Tosi, L., Roura-Bas, P., Llois, A.M., Manuel, L.O.: Phys. Rev. B. **83**, 073301 (2011)
97. Vildosola, V., Pourovskii, L.V., Manuel, L.O., Roura-Bas, P.: J. Phys.: Condens. Matter. **27**, 485602 (2015)
98. Zgid, D., Gull, E., Chan, G.K.-L.: Phys. Rev. B. **86**, 165128 (2012)
99. Lu, Y., Hoppner, M., Gunnarsson, O., Haverkort, M.W.: Phys. Rev. B. **90**, 085102 (2014)
100. Feng, Q., Oppeneer, P.M.: J. Phys.: Condens. Matter. **23**, 425601 (2011)
101. Hou, D., Wang, R., Zheng, X., Tong, N.H., Wei, J.H., Yan, Y.J.: Phys. Rev. B. **90**, 045141 (2014)
102. Li, Z.H., Tong, N.H., Zheng, X., Hou, D., Wei, J.H., Hu, J., Yan, Y.J.: Phys. Rev. Lett. **109**, 266403 (2012)
103. Logan, D.E., Eastwood, M.P., Tusch, M.A.: J. Phys.: Condens. Matter. **10**, 2673 (1998)
104. Barman, H.: Phys. Rev. B. **94**, 045106 (2016)
105. Kauch, A., Byczuk, R.: Physica B. **407**, 209 (2012)
106. Weber, C., Amaricci, A., Capone, M., Littlewood, P.B.: Phys. Rev. B. **86**, 115136 (2012)
107. Kinza, M., Ortloff, J., Bauer, J., Honerkamp, C.: Phys. Rev. B. **87**, 035111 (2013)
108. Kinza, M., Honerkamp, C.: Phys. Rev. B. **88**, 195136 (2013)
109. Holzner, A., Weichselbaum, A., McCulloch, I.P., Schollwock, U., von Delft, J.: Phys. Rev. B. **83**, 195115 (2011)

110. Ganahl, M., Aichhorn, M., Evertz, H.G., Thunstrom, P., Held, K., Verstraete, F.: Phys. Rev. B. **92**, 155132 (2015)
111. Kananenka, A.A., Gull, E., Zgid, D.: Phys. Rev. B. **121111**(R), 91 (2015)
112. Li, H., Tong, N.-H.: Eur. Phys. J. B. **88**, 319 (2015)
113. Wolf, F.A., Go, A., McCulloch, I.P., Millis, A.J., Schollwöck, U.: Phys. Rev. X. **5**, 041032 (2015)
114. Motahari, S., Requist, R., Jacob, D.: Phys. Rev. B. **94**, 235133 (2016)
115. Bauernfeind, D., Zingl, M., Triebl, R., Aichhorn, M., Evertz, H.G.: Phys. Rev. X. **7**, 031013 (2017)
116. Go, A., Millis, A.J.: Phys. Rev. B. **96**, 085139 (2017)
117. Zingl, M., Nuss, M., Bauernfeind, D., Aichhorn, M.: Physica B. **536**, 254 (2018)
118. Shee, A., Zgid, D.: J. Chem. Theory Comput. **15**, 6010 (2019)
119. Zhu, T., Jiménez-Hoyos, C.A., McClain, J., Berkelbach, T.C., Chan, G.K.-L.: Phys. Rev. B. **100**, 115154 (2019)
120. Steiner, K., Nomura, Y., Werner, P.: Phys. Rev. B. **92**, 115123 (2015)

Chapter 7
Surfaces, Interfaces, Films, and Nanosystems

DMFT is regarded to be a successful theory in the 3D case where the local self-energy is a rather good approximation. Thus, different bulk systems were studied with DMFT approximation (see Chap. 10) assuming that the electron self-energy is local in space, and by using a realistic "noninteracting" electron DOS (instead of the infinite-dimensional DOS, e.g., hypercubic lattice and Bethe lattice DOS). In this chapter, we generalize the DMFT approach and apply it to "smaller" correlated systems, i.e., systems with restricted geometry: surfaces (semi-infinite systems), interfaces, films or heterostructures (all three constrained in one direction), and nanosystems (systems constrained in three directions). These structures have a great potential to be used in technological applications, such as electronic and magnetic recording devices, sensors and photovoltaics, since due to restricted geometry the correlation effects are enhanced in these systems and they can be further tuned by changing the system size and geometry. In most cases, such a tuning also leads to a modification of the properties of materials. Thus, the role of correlation effects in low-dimensional systems needs to be understood.

The main philosophy behind applicability of DMFT to the system above is based on the assumption of high *average* atomic coordination number that makes the local self-energy a good approximation. We do not discuss extended 2D (monolayer) systems, considered in a great detail in Chap. 5. We also do not discuss 1D atomic chains and rods. The reason for this is that DMFT is not a proper approximation for atomic chains due to a small coordination number in these systems, and the approach for a few atom or thicker rods (including nano-rods) is technically very similar to the one for films or heterostructures, constrained in two instead of one dimension. In other words, in the case of rods, instead of coupled monolayers, one can define the problem as a problem of coupled "flakes" parallelly "strung" on one of the coordinate axes. Thus, the rod problem is equivalent to a problem of coupled 2D (finite) nanosystems (discussed in the last subsection).

V. Turkowski, *Dynamical Mean-Field Theory for Strongly Correlated Materials*,
https://doi.org/10.1007/978-3-030-64904-3_7

7.1 Surfaces, Interfaces, and Films

In their pioneering works [1, 2], Potthoff and Nolting proposed to study surface properties of strongly correlated materials by solving a semi-infinite Hubbard model that consists of coupled layers. The following Hamiltonian was proposed [2]

$$H = \sum_{i,j,a,b,\sigma} t_{ia,jb} c^{+}_{ia,\sigma} c_{jb,\sigma} + \sum_{i,a} U_a n_{ia,\uparrow} n_{ia,\downarrow} \tag{7.1}$$

where a and b are the layer indices (they run from 1 to ∞), i, j are site indices within the layer, such that the atom index is defined by the center of layer R_a and the in-layer r_i vectors: $R = R_a + r_i$. The on-site interaction U_a is in general layer-dependent. In Ref. [2], it was assumed that it has two different values: U_1 for the first (surface) layer and U—for all others. The hopping parameters $t_{ia,\ jb}$ where assumed to be nonzero for the nearest-neighbor (intra- and interlayer) atoms and equal $-$t for all layers except the hoppings involving the first-layer atoms. The zero-energy (on-site) energy $t_{ia,\ ia}$ was assumed to be finite and equal t_0 in the first layer, for the other layers it was put $t_{ia,\ ia} = 0$.

In this model, the translational symmetry is preserved within layers, thus one can introduce the 2D matrix spectrum:

$$\varepsilon_{ab}\left(\vec{k}\right) = \frac{1}{N_{\parallel}} \sum_{i,j} e^{-i\vec{k}\left(\vec{R}_i - \vec{R}_j\right)} t_{ia,jb}, \tag{7.2}$$

where $\vec{k}$ is the 2D wave vector in the first Brillouin zone and $N_{\parallel}(\rightarrow\infty)$ is the number of sites in each layer. From Eq. (7.2) and the structure of the hopping matrix, one can find two types of dispersions: the "in-plane" $\varepsilon_{aa}\left(\vec{k}\right) = t\varepsilon_{\parallel}\left(\vec{k}\right)$ and the "out-of-plane" $\varepsilon_{a,a+1}\left(\vec{k}\right) = t\varepsilon_{\perp}\left(\vec{k}\right)$ (for a $\geq$ 2) one. These dispersions have a rather different form for different types of surfaces in solids. For example, in the case of simple cubic lattice and (100) surface $\varepsilon_{\parallel}\left(\vec{k}\right) = -2t\left(\cos k_x + \cos k_y\right)$ and $\varepsilon_{\perp}\left(\vec{k}\right) = \pm 1$.

Using these notations, one can write the expression for the total free-electron energy matrix (in the layer indices):

$$\hat{\varepsilon}\left(\vec{k}\right) = \begin{pmatrix} t_{11}\varepsilon_{\|}\left(\vec{k}\right) + t_0 & t_{12}\varepsilon_{\perp}\left(\vec{k}\right) & 0 & 0 & 0 & \cdots \\ t_{21}\varepsilon_{\perp}\left(\vec{k}\right) & t\varepsilon_{\|}\left(\vec{k}\right) & t\varepsilon_{\perp}\left(\vec{k}\right) & 0 & 0 & \cdots \\ 0 & t\varepsilon_{\perp}\left(\vec{k}\right) & t\varepsilon_{\|}\left(\vec{k}\right) & t\varepsilon_{\perp}\left(\vec{k}\right) & 0 & \cdots \\ 0 & 0 & t\varepsilon_{\perp}\left(\vec{k}\right) & t\varepsilon_{\|}\left(\vec{k}\right) & t\varepsilon_{\perp}\left(\vec{k}\right) & \cdots \\ 0 & 0 & 0 & t\varepsilon_{\perp}\left(\vec{k}\right) & t\varepsilon_{\|}\left(\vec{k}\right) & \cdots \\ \cdots & \cdots & \cdots & \cdots & \cdots & \cdots \end{pmatrix}. \tag{7.3}$$

In the DMFT approximation it is assumed that the corresponding self-energy matrix is momentum-independent and diagonal in the layer indices: $\Sigma_{ab}(\omega) = \delta_{ab}\Sigma_a(\omega)$. Importantly, the value of the matrix element of the self-energy depends on the layer index. Using the matrices $\hat{\varepsilon}\left(\vec{k}\right)$ and $\hat{\Sigma}(\omega)$ one can define the local layer GF with components

$$G_a(\omega) = \frac{1}{N_{\|}}\sum_{\vec{k}}\left(\left[(\omega+\mu)\hat{I} - \hat{\varepsilon}\left(\vec{k}\right) - \hat{\Sigma}(\omega)\right]^{-1}\right)_{aa} \tag{7.4}$$

In the case of DMFT with no translational-invariance, one still assumes that the functional Φ depends on the local, but layer-dependent, GFs $G_a(\omega)$. Then, since $\Sigma_a(\omega) = \frac{\Phi[\{G_i\}]}{G_a(\omega)} = S[G_a]$, each orbital component of the self-energy is a functional of the corresponding local GF $G_a(\omega)$: $\Sigma_a(\omega) = S[G_a]$. Therefore, for each layer one can formulate an independent single-impurity problem with the Anderson impurity model Hamiltonian

$$H_{\text{imp}}^{(a)} = \sum_{\sigma}\varepsilon_d^a d_\sigma^+ d_\sigma + U^a n_{d,\uparrow} n_{d,\downarrow} + \sum_{\vec{k},\sigma}\varepsilon^a\left(\vec{k}\right) c_{\vec{k}\sigma}^+ c_{\vec{k}\sigma}$$
$$+ \sum_{\vec{k},\sigma}\left(V^a\left(\vec{k}\right) c_{\vec{k}\sigma}^+ d_\sigma + h.c.\right) \tag{7.5}$$

with the impurity atom energy level ε_d^a, the itinerant electron dispersion $\varepsilon^a\left(\vec{k}\right)$ and the hybridization strength $V^a\left(\vec{k}\right)$.

Similar to the previous chapters, one can derive (from Eq. (7.5)) the expression for the layer hybridization fields $\Delta^a(\omega)$ connected with the impurity self-energies and GFs as

$$\Delta^a(\omega) = \omega + \mu - \varepsilon_d^a - \Sigma_{\text{imp}}^a(\omega) - G_{\text{imp}}^{a-1}(\omega). \tag{7.6}$$

The result is

$$\Delta^a(\omega) = \sum_{\vec{k},\sigma} \frac{\left|V^a\left(\vec{k}\right)\right|^2}{\omega - \varepsilon^a\left(\vec{k}\right)}. \tag{7.7}$$

Then, one can find the impurity self-energy and GF for each a (using the effective Hamiltonian that follows from Eqs. (7.5) and (7.7), see Chap. 3), postulate equality of the local lattice and impurity GFs and of self-energies, and find the local GF from Eq.(7.4).

Potthoff and Nolting proposed the following algorithm to solve the surface problem: (1) start from a chosen initial self-energy functions for each layer $\Sigma_a(\omega)$; (2) calculate local GFs from Eq. (7.4); (3) using resulting $G_a(\omega)$ and $\Sigma_a(\omega) = \Sigma_{\text{imp}}^a(\omega)$ calculate the hybridization field $\Delta^a(\omega)$ from Eq. (7.6) (we use the upper symbol for the orbital index of the impurity self-energy, to make the notations more transparent); (4) solve the impurity problem for each layer and find $\Sigma_{\text{imp}}^a(\omega) = \Sigma_a(\omega)$. The procedure continues until the self-energies converge. Naturally, one can use an equivalent scheme by solving Eq. (7.4), Dyson equation $G^{a-1}(\omega) = \mathcal{G}^{a-1}(\omega) - \Sigma_a(\omega)$ that connects the local GF and dynamical mean-field $\mathcal{G}^a(\omega) = [\omega + \mu - \Delta^a(\omega)]^{-1}$, and solving the impurity problem for the GF defined in terms of the path integral as described in the previous chapter. It is important to note that despite the fact that each impurity problem is solved independently, the layers are coupled since the local GF for each layer calculated from Eq. (7.4) depends on the self-energies of all other layers.

As it was argued in Ref. [2] (see also Refs. [3, 4]), since near the critical U the quasi-particle peak is well-separated from the Hubbard peaks at $\omega = \pm U/2$, one can use a single-pole approximation for the hybridization function: $\Delta^a(\omega) \to \frac{\Delta_N^a}{\omega}$, where Δ_N^a are layer- and iteration-dependent coefficients (N is the iteration number). Such an approximation (together with assumption that the electron spectrum has also delta peaks at $\pm U/2$) allows one to connect the $(N+1)$th and Nth iteration coefficients by means of the linear equations $\Delta_{N+1}^a = \sum_b K_{ab}\Delta_N^a$ and to use these equations for an analytical, "linearized DMFT," analysis of the properties of the system near MIT. In particular, this approximation allows one to find the critical value for the Coulomb repulsion U_c for different types of surfaces [2]. For example, in the case of (100) surface in the system with a finite number of layers, the following analytical result was found: $U_c = 6t\sqrt{4 + 2\cos\frac{\pi}{N+1}}$, where N is the number of layers. We refer the reader to work [2] for more details.

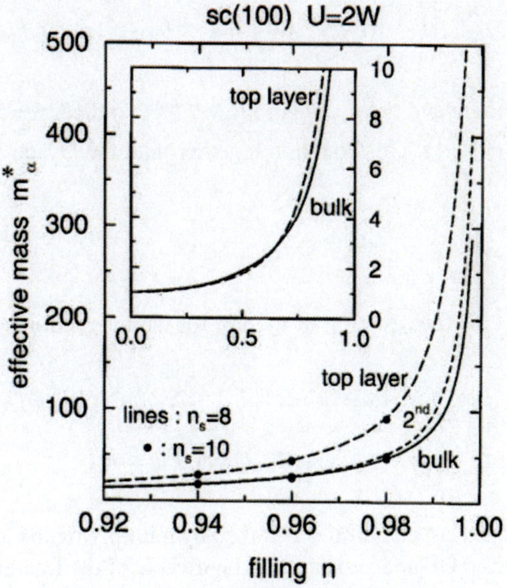

Fig. 7.1 Effective mass of electron in the top- and subsurface layers and in the bulk as a function of doping (15-layer system, ED solver). (Reprinted from Physica B, Vols. 259–261, M. Potthoff, M. and W. Nolting, "Effective mass at the surface of a Fermi liquid," Pages 760–761, Copyright 1999, with permission from Elsevier)

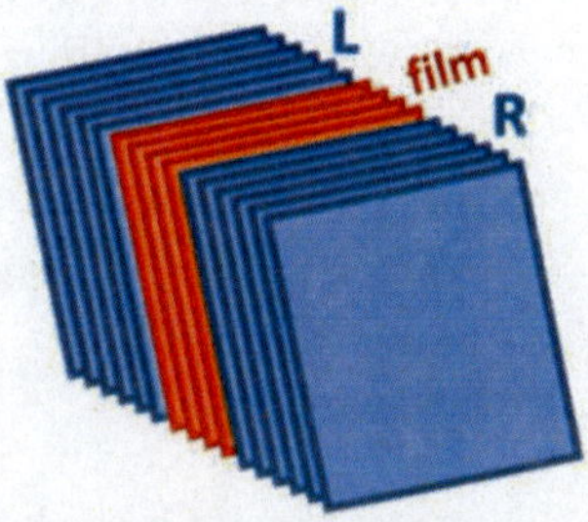

Fig. 7.2 A cartoon representation of a film (multilayer) structure embedded between the left (L) and right (R) infinite substrates

Later the method developed in Ref. [2] was used by its authors to analyze the problem of MIT on the surface and in thin films (see Refs. [3–6]). In particular, in Ref. [5] a 15-layer simple-cubic lattice system with (100) surface was studied and it was found that when the doping approaches half-filling, correlation effects lead to a much larger increase of the value of the effective mass of electrons in the surface layer as compared to the others layers, suggesting existence of correlation-generated surface quasiparticles (Fig. 7.1).

Another approach to solve the DMFT problem for multilayer system sandwiched between substrate layers is *Quantum Zipper Algorithm* [7] (a cartoon of the system is shown in Fig. 7.2). In this method, one begins with obtaining the solution for the $\pm\infty$ substrate layers (in the considered case–metallic, with no electron interaction), and then using recurrence relations between the neighbor-layer GFs, one step-by-step obtains the solution for the GFs for the central layer (film) GFs.

To derive the recurrence relations for the GFs, let us first write down the equation for the noninteracting layer GFs in the momentum–frequency representation:

$$\sum_c \left[\left(\omega + \mu - \varepsilon^a\left(\vec{k}\right)\right)\delta_{ac} + t_{a,a+1}\delta_{a+1,c} + t_{a,a-1}\delta_{a-1,c} \right] G^{(0)}_{cb}\left(\omega, \vec{k}\right) = \delta_{ab} \quad (7.8)$$

(here and below, we use lower layer indices for all functions, except the dispersions $\varepsilon^a\left(\vec{k}\right)$). This equation together with the Dyson equation

$$G_{ab}\left(\omega, \vec{k}\right) = G^{(0)}_{cb}\left(\omega, \vec{k}\right) + \sum_c G^{(0)}_{ac}\left(\omega, \vec{k}\right)\Sigma_{cc}(\omega)G_{cb}\left(\omega, \vec{k}\right) \quad (7.9)$$

give the equation of motion for the interacting layer GFs:

$$\left[\omega + \mu - \varepsilon^a\left(\vec{k}\right) - \Sigma_{aa}(\omega)\right] G_{ab}\left(\omega, \vec{k}\right) + t_{a,a+1}G_{a+1,b}\left(\omega, \vec{k}\right) + t_{a,a-1}G_{a-1,b}\left(\omega, \vec{k}\right) = \delta_{ab} \quad (7.10)$$

(the last equation is obtained by multiplying the left part of Eq.(7.9) by inverse of the free GF and substituting the inverse of the free-electron GF (that can be found from Eq.(7.8)) into the modified Eq. (7.9)).

Eq. (7.10) (at $a = b$) gives the following formula that connects the diagonal and the nearest-neighbor non-diagonal GFs:

$$G_{aa}\left(\omega, \vec{k}\right) = \frac{1}{\omega + \mu - \varepsilon^a\left(\vec{k}\right) - \Sigma_{aa}(\omega) + \frac{G_{a,a-1}\left(\omega, \vec{k}\right)}{G_{aa}\left(\omega, \vec{k}\right)} t_{a-1,a} + \frac{G_{a,a+1}\left(\omega, \vec{k}\right)}{G_{aa}\left(\omega, \vec{k}\right)} t_{a+1,a}}. \quad (7.11)$$

To define a recursive procedure, one needs an equation that connects only diagonal, or only non-diagonal, functions. If, for example, the non-diagonal GFs are found, one can get the diagonal ones from Eq. (7.11).

One can also write down equations for the non-diagonal GFs by using Eq. (7.10). Namely, Eq. (7.10) for the non-diagonal GF can be transformed to the following form:

$$L_{a-n}\left(\omega, \vec{k}\right) = \omega + \mu - \varepsilon^{a-n}\left(\vec{k}\right) - \Sigma_{a-n}(\omega) - \frac{t_{a-n,a-n-1}t_{a-n-1,a-n}}{L_{a-n-1}\left(\omega, \vec{k}\right)}, \quad (7.12)$$

where

$$L_{a-n}\left(\omega,\vec{k}\right)=-\frac{G_{a,a-n+1}\left(\omega,\vec{k}\right)t_{a-n+1,a-n}}{G_{a,a-n}\left(\omega,\vec{k}\right)} \tag{7.13}$$

is defined by the ratio of two non-diagonal GFs. Equations (7.12) and (7.13) are valid for $n>0$. They can be used recursively starting from the left edge of the system (formally, with $n=-\infty$).

A similar set of equations can be written for the right functions:

$$R_{a+n}\left(\omega,\vec{k}\right)=\omega+\mu-\varepsilon^{a+n}\left(\vec{k}\right)-\Sigma_{a+n}(\omega)-\frac{t_{a+n,a+n+1}t_{a+n+1,a+n}}{R_{a+n+1}\left(\omega,\vec{k}\right)}, \tag{7.14}$$

$$R_{a+n}\left(\omega,\vec{k}\right)=-\frac{G_{a,a+n-1}\left(\omega,\vec{k}\right)t_{a+n-1,a+n}}{G_{a,a-n}\left(\omega,\vec{k}\right)}. \tag{7.15}$$

Equation (7.14) can also be solved recursively starting from the right layer of the system (with $n=\infty$).

To start the recursive procedure from left or/and right, one needs $L_{-\infty}$ or/and R_{∞}. One can find $L_{-\infty}$ by putting $n=\infty$ in Eq. (7.12) and neglecting a finite number a in the indices in all functions. This gives

$$L_{-\infty}\left(\omega,\vec{k}\right)=\frac{1}{2}\left(\omega+\mu-\varepsilon^{-\infty}\left(\vec{k}\right)-\Sigma_{-\infty}(\omega)\right)$$
$$\pm\frac{1}{2}\sqrt{\left(\omega+\mu-\varepsilon^{-\infty}\left(\vec{k}\right)-\Sigma_{-\infty}(\omega)\right)^{2}-4t_{-\infty}^{2}}, \tag{7.16}$$

where $t_{-\infty}^2=t_{-\infty,\infty-1}t_{\infty-1,-\infty}$ (in practice, one uses a finite integer number instead of $-\infty$). As it is mentioned in book [7], the sign in front of the square root in Eq. (7.16) is chosen (for the reasons of causality) to yield the imaginary part smaller (larger) than 0 when ω omega is in the upper (lower) half-plane. For a real $L_{-\infty}\left(\omega,\vec{k}\right)$, one chooses the sign that gives $\left|L_{-\infty}\left(\omega,\vec{k}\right)\right|>t_{-\infty}^2$. Then, using the bulk result for the edge layer self-energy $\Sigma_{-\infty}(\omega)$ and substituting it into Eq. (7.16), one obtains $L_{-\infty}(\omega)$ that is used in Eq. (7.12) to start the recursive solution.

Similarly, one can get the function for the right edge by using Eq. (7.14):

$$R_{\infty}\left(\omega,\vec{k}\right)=\frac{1}{2}\left(\omega+\mu-\varepsilon^{\infty}\left(\vec{k}\right)-\Sigma_{\infty}(\omega)\right)$$
$$\pm\frac{1}{2}\sqrt{\left(\omega+\mu-\varepsilon^{\infty}\left(\vec{k}\right)-\Sigma_{\infty}(\omega)\right)^{2}-4t_{\infty}^{2}}, \tag{7.17}$$

where $t_{\infty}^2 = t_{\infty,\infty+1} t_{\infty+1,\infty}$, with the same choice of the sign in front of the root as in Eq. (7.16). Again, similarly to the case of $L_{-\infty}\left(\omega, \vec{k}\right)$, one can use the bulk self-energy for $\Sigma_{\infty}(\omega)$ to calculate $R_{\infty}\left(\omega, \vec{k}\right)$ and use the last function to solve recursively Eq. (7.14) starting from the right edge ($n = \infty$) and moving to the central layer ($n = 0$).

The obtained solutions $L_a(\omega, k)$ and $R_a\left(\omega, \vec{k}\right)$ for all layers give the solution for the layer GFs, since as it follows from Eq. (7.11),

$$G_{aa}\left(\omega, \vec{k}\right) \equiv G_a\left(\omega, \vec{k}\right) = \frac{1}{L_a\left(\omega, \vec{k}\right) + R_a\left(\omega, \vec{k}\right) - \left(\omega + \mu - \varepsilon^{\infty}\left(\vec{k}\right) - \Sigma_a(\omega)\right)}. \tag{7.18}$$

One can find the local layer GFs by performing the momentum summation in Eq. (7.18).

It was found that the obtained solution is accurate if it gives bulk results for ~30 edge layers on each side [7]. We refer the reader for more details of the approach and for a summary of some applications of the method (mostly in the framework of the Falicov–Kimball model) to book [7] and Refs. [8–11].

Finally, we mention another, a more recent *embedding method* [12, 13] used to solve the DMFT problem for multilayers stacked (embedded) between semi-infinite substrates, like leads. In this approach, the effect of the substrates on the surface and other parts of the multilayers is described by a frequency-dependent embedding potential. This potential is defined by the DMFT self-energies of the substrates obtained from the bulk calculations. Once the embedding potential is found, the rest of the calculations, to find the GFs of the multilayer system, are performed with the standard DMFT procedure.

In detail, in the embedding method, one starts from the following multilayer Hamiltonian:

$$\widehat{H} = \widehat{H}^{(0)} + \widehat{U}, \tag{7.19}$$

where

$$\widehat{H}^{(0)} = \widehat{H}^{(0)}_{\mathrm{LL}} + \widehat{H}^{(0)}_{\Omega\Omega} + \widehat{H}^{(0)}_{\mathrm{RR}} + \left(\widehat{H}^{(0)}_{\Omega\mathrm{L}} + \widehat{H}^{(0)}_{\mathrm{L}\Omega} + \widehat{H}^{(0)}_{\Omega\mathrm{R}} + \widehat{H}^{(0)}_{\mathrm{R}\Omega}\right) \tag{7.20}$$

is the kinetic part and

$$\widehat{U} = \sum_{i,abcd} U_i^{abcd} \widehat{c}_{ia\uparrow}^{+} \widehat{c}_{ib\downarrow}^{+} c_{ic,\downarrow} \widehat{c}_{id,\uparrow} \tag{7.21}$$

is the standard on-site intra- and interorbital interaction part of the Hamiltonian (it includes the repulsion and the exchange terms). The kinetic part of the Hamiltonian (7.20) consists of the film $\widehat{H}^{(0)}_{\Omega\Omega}$, left and right substrate $\widehat{H}^{(0)}_{\mathrm{LL}}$ and $\widehat{H}^{(0)}_{\mathrm{RR}}$, and the film–substrate interaction terms (in brackets). All noninteracting terms have a simple free-electron form, e.g.,

$$\widehat{H}^{(0)}_{\Omega\Omega} = \sum_{ij,ab,\sigma} t_{ij}^{ab} \widehat{c}_{ia\sigma}^{+} \widehat{c}_{jb\sigma}, \tag{7.22}$$

where the indices a, b include the site, orbital and spin indices are t_{ij}^{ab} are the hopping parameters, nonzero within the film.

Using the Hamiltonian Eq. (7.19), one can find the GFs of the noninteracting system (that consists of three subsystems coupled through the interfaces!) from

$$\widehat{G}^{(0)-1}_{ab}(\omega) = \left[(\omega + \mu)\widehat{I} - \widehat{H}^{(0)}\right]_{ab}. \tag{7.23}$$

It is possible to show [12, 13] that the effect of the substrates on the film GF $\widehat{G}^{(0)}_{\Omega}$ can be taken into account by the embedding potentials

$$\widehat{S}^{L(0)}_{aa'}(\omega) = \sum_b t_{ab} \widehat{G}^{L(0)}_{ab}(\omega) t_{ba'}, \tag{7.24}$$

$$\widehat{S}^{R(0)}_{aa'}(\omega) = \sum_b t_{ab} \widehat{G}^{R(0)}_{ab}(\omega) t_{ba'}, \tag{7.25}$$

where the indices a and a' belong to the film (Ω) and $\widehat{G}^{(0)L-1}_{ab}(\omega) = \left[(\omega + \mu)\widehat{I} - \widehat{H}^{(0)}_{\mathrm{LL}}\right]_{ab}$, $\widehat{G}^{(0)R-1}_{ab}(\omega) = \left[(\omega + \mu)\widehat{I} - \widehat{H}^{(0)}_{\mathrm{RR}}\right]_{ab}$.

In detail, the noninteracting film GF can be expressed in terms of the embedding potentials as

$$\widehat{G}^{(0)-1}_{ab}(\omega) = \left[(\omega + \mu)\widehat{I} - \widehat{H}^{(0)}_{\mathrm{em}}\right]_{\mathrm{ab}} \tag{7.26}$$

where a and b belong to the film and

$$\widehat{H}^{(0)}_{\mathrm{em}} = \widehat{H}^{(0)}_{\Omega\Omega} + \widehat{S}^{L(0)}(\omega) + \widehat{S}^{R(0)}(\omega) \tag{7.27}$$

is the effective Hamiltonian in the embedded region.

In principle, in embedded DMFT the electron–electron interaction can be taken into account also in the substrates. The generalization of the approach described above on this case is straightforward: in the equations for each GF one adds the region-dependent local-in-space electron self-energy. In this case, the total-system GF and the film GF become

$$\widehat{G}_{ab}^{-1}(\omega) = \left[(\omega + \mu)\widehat{I} - \widehat{H}^{(0)} - \widehat{\Sigma}(\omega)\right]_{ab} \tag{7.28}$$

and

$$\widehat{G}_{ab}^{-1}(\omega) = \left[(\omega + \mu)\widehat{I} - \widehat{H}_{\rm em}\right]_{ab}, \tag{7.29}$$

correspondingly. In the last equation,

$$\widehat{H}_{\rm em} = \widehat{H}_{\Omega\Omega}^{(0)} + \widehat{S}^{L}(\omega) + \widehat{S}^{R}(\omega) + \widehat{\Sigma}_{\Omega\Omega}(\omega) \tag{7.30}$$

is the embedding effective Hamiltonian matrix. Similar to the case of "non-correlated" substrates (Eq. (7.27)) this matrix includes the parts corresponding to the interaction of the film with the substrates (the second and the third terms). In addition, it includes the electron self-energy part (the last term) that comes from the electron–electron correlations in the film. The embedding potential matrices in Eq. (7.30) are

$$\widehat{S}_{aa'}^{L}(\omega) = \sum_{b} t_{ab}\widehat{G}_{ab}^{L}(\omega)t_{ba'}, \tag{7.31}$$

$$\widehat{S}_{aa'}^{R}(\omega) = \sum_{b} t_{ab}\widehat{G}_{ab}^{R}(\omega)t_{ba'}, \tag{7.32}$$

where

$$\widehat{G}_{ab}^{L-1}(\omega) = \left[(\omega + \mu)\widehat{I} - \widehat{H}_{\rm LL}^{(0)} - \widehat{\Sigma}_{\rm LL}(\omega)\right]_{ab}. \tag{7.33}$$

$$\widehat{G}_{ab}^{R-1}(\omega) = \left[(\omega + \mu)\widehat{I} - \widehat{H}_{\rm RR}^{(0)} - \widehat{\Sigma}_{\rm RR}(\omega)\right]_{ab}, \tag{7.34}$$

The last functions include the substrate self-energy matrices $\widehat{\Sigma}_{\rm LL}(\omega)$ and $\widehat{\Sigma}_{\rm RR}(\omega)$.

Now we are ready to solve the DMFT problem with the embedding method defined by Eqs. (7.28)–(7.34). This can be done by using the following steps: (1) solve the DMFT bulk problems for the substrates to get $\widehat{\Sigma}_{\rm RR}(\omega)$ and $\widehat{\Sigma}_{\rm LL}(\omega)$. (2) Use Eqs. (7.33) and (7.34) to get the substrate GFs $\widehat{G}_{ab}^{L}(\omega)$ and $\widehat{G}_{ab}^{R}(\omega)$, and then, use Eqs. (7.31), (7.32) to find the embedding fields $\widehat{S}_{aa'}^{L}(\omega)$ and $\widehat{S}_{aa'}^{R}(\omega)$. (3) Substitute

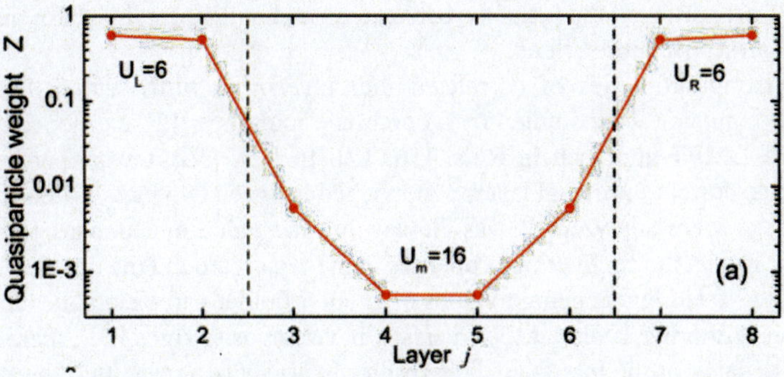

Fig. 7.3 Layer dependence of the quasiparticle weight Z in the case of four-layer Mott insulator slab (layers 3–6) sandwiched between two metal substrates (simple cubic lattice, $T = 0.02t$, ED solver). (Reprinted Fig. 4a with permission from (H. Ishida and A. Liebsch, Phys. Rev. B 79, 045130 (2009)). Copyright 2009 by the American Physical Society. https://doi.org/10.1103/PhysRevB.79.045130)

the sum of obtained fields, $\widehat{S}^{L}_{aa'}(\omega)+\widehat{S}^{R}_{aa'}(\omega)$, into the film Hamiltonian (7.30) and solve the film problem with DMFT with additional embedding potentials in the standard way. The advantage of the effective embedded potential approach is that one can significantly reduce the number of the substrate layers to solve the problem (the substrate layers contribute to the film problem only through the one shot-calculated embedding field $\widehat{S}^{L}_{aa'}(\omega)+\widehat{S}^{R}_{aa'}(\omega)$).

The authors of Refs. [12, 13] applied the approach to several embedded systems of different thicknesses, in particular they studied the layer-dependence of the quasiparticle weight Z. It was found [12] that in the case of four-layer Mott-insulator slab sandwiched between two metal substrates the metallic states penetrate into the slab, resulting in relatively large values of Z in the boundary layers of insulators. Quite remarkably, even for such a thin film, the metallic states penetration length is very short (showing stability of the correlation-induced insulating phase) (Fig. 7.3).

Thorough studies by numerous groups allowed to establish the main physical properties of the surfaces, films and interfaces in coupled and/or sandwiched structures with strong electron–electron correlations. Here we mention some of them beginning with the electronic properties. In Ref. [14], the embedding DMFT approach was used to analyze the surface (or, solid-vacuum interface) properties of a doped Mott insulator. It was shown that the surface spectral function is modified comparing to the bulk one due to both correlation and charge-transfer effects. These two factors also strongly modify the hopping amplitudes at the surface that results in forming of a "dead" surface layer with exponentially suppressed quasiparticle amplitudes. The authors of Ref. [15] used a Potthoff–Nolting type of approach to analyze the electronic properties of Mott-insulator multilayers embedded into an infinitely-extended band insulator. The results of the calculations showed that in this system there exists a few unit–cell crossover region where the bulk insulator can sustain a metallic behavior. These results demonstrate an important difference in the electronic properties of the interface between a band and a Mott insulator and of the vacuum–Mott insulator surface.

Conducting properties of correlated multilayers/thin films with a long-range Coulomb repulsion were studied by a combined multilayer [2] and nonequilibrium real-space DMFT approach in Refs. [16, 17]. In Ref. [16], the properties of the steady-state current in several layers sandwiched between two metallic leads under a bias voltage were analyzed. It was shown that the value of Hubbard U plays an important role in the conducting properties in the region around the interfaces. It was also demonstrated that at certain values of U and of other parameters the charges on the two neighboring layers at the interface have opposite signs, i.e., there exists a dipole-like layer at the interface. This result qualitatively agrees with the results in Ref. [14], where the authors found a charge transfer through the interface. Conducting properties of thin films under applied transverse electric field were studied in Ref. [17]. Several interesting properties were found, most notably existence of conducting regions near the surfaces at half-filling when a strong field is applied. Other interesting results of this work—an enhancement of conductivity on one side and a transition to the insulating state on the other side of doped system when the applied field strength increases and, similar to a bulk, there was found a hysteresis behavior of the doping-induced Mott transition.

Magnetic properties of a Mott-insulator heterostructure sandwiched into a band insulator was studied within a semiclassical DMFT approach in Ref. [18] by using the Potthoff–Nolting formalism. It was found that at large Us, along with a bulk AFM phase, there exists an interface-stabilized FM phase (Fig. 7.4). The charge density in the induced by surface effects FM region was found to be small comparing to the charge density in the AFM region. Such an AFM "core"—FM "skin" magnetic order, more relevant to thick films, might be of an interest in magnetic technologies.

In Ref. [19], an embedding DMFT approach was generalized and used to analyze the superconducting properties of a multilayer system with attractive ($U < 0$) Hubbard interaction. The system was sandwiched between two correlated (with a repulsive on-site interaction $U > 0$) substrates. It was demonstrated that the first two layers of the nonsuperconducting substrates can become superconducting due to proximity effects, independently of the rate of the charge redistribution.

Thus, as we have seen, the correlation effects are enhanced on the film–vacuum surface. On the other hand, the properties of the interface of the sandwiched structure can be tuned to different limits and to exotic states by proximity effects choosing a proper slab. In general, as it was mentioned in Ref. [20], due to a lower dimensionality of surfaces and interfaces caused by the system anisotropy, the ignored-above momentum dependence of the self-energy in these subsystems can be also very important. To take the momentum dependence of Σ into account, the authors of Ref. [20] proposed to use a generalized "space-time separated" self-energy that consists of the local (DMFT) part $\Sigma^{\mathrm{DMFT}}(\omega)$ and of momentum-dependent part $\Sigma^{\mathrm{D\Gamma A}}\left(\vec{k}\right)$, calculated by using Dynamical Vertex Approximation (DΓA), the approach that is a topic of Chap. 9.

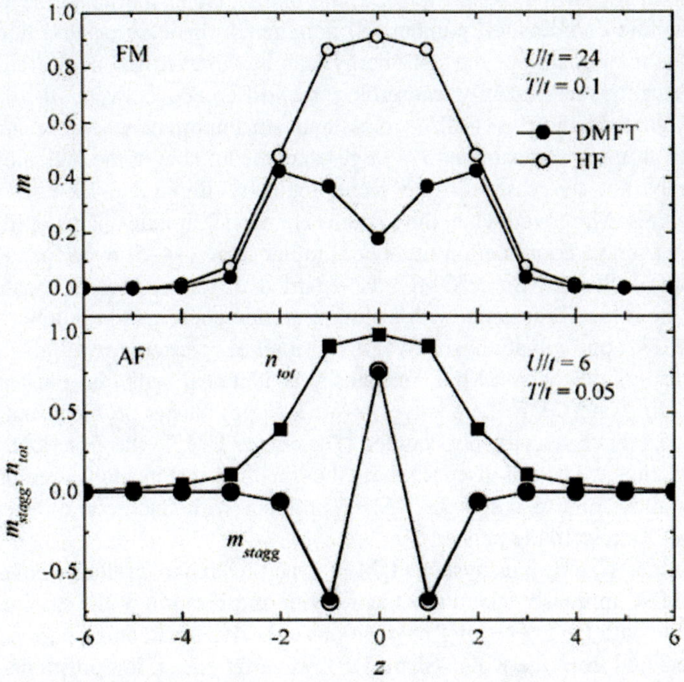

Fig. 7.4 Magnetization densities (for the FM and AFM states) in different layers of a four-layer heterostructure (centered at 0) at half-filling and different (shown) other parameters obtained with the Hartree Fock and DMFT (ED two-site solver) approximations. (Reprinted Fig. 2 with permission from (S. Okamoto and A.J. Millis, Phys. Rev. B 72, 235108 (2005)). Copyright 2005 by the American Physical Society. https://doi.org/10.1103/PhysRevB.72.235108)

7.2 Nanoparticles

In the case of "small" systems, like nanoparticles and molecules, the role of electron–electron correlations can be even more important than in films and heterostructures due to more constrained geometry that forces electrons "to meet more often" (we have already seen that on surfaces (systems with constrained geometry) correlation effects are strongly enhanced as compared to the bulk systems). Besides fundamental and modern-technology interests, understanding the role of correlation effects in small systems will help to understand other phenomena, like the dynamics of growth of nanoparticles (where electron interaction plays an important role) and different quantum properties of computationally non-accessible large nanoparticles. Furthermore, such an understanding will help to analyze usually not high enough-resolved experimental results.

On the first sight, it might not be obvious why DMFT can be applied to finite systems where there is no periodicity, i.e., there are no equivalent atoms (clusters) in the cell (supercell) that produce a thermodynamic bath. However, as it was pointed by Florens, [21] DMFT is still a good approximation in the nanocase when the average atomic coordination number (Z) is large. In this case, spatial fluctuations (i.e., spatial non-locality of the self-energy) can be "in average" neglected. Indeed, these fluctuations are mainly caused by the difference between the "surface" (low-coordinated) and the "bulk" atoms, and are important when the number of the surface atoms is comparable to or larger than the number of the bulk atoms. This is definitely not the case for the technologically important 10 nm or larger nanostructures. Moreover, when the systems are not 1D atomic chains or 2D atomic layers, the average coordination number is rather large (~4–5) even for very small (~5–10 atom) clusters. Since DMFT is regarded in many cases as a satisfactory approximation for 2D systems with a similar atomic coordination number, one can expect DMFT approximation to be valid for small 3D clusters as well.

Technically, the nano-DMFT formalism is identical with the cluster DMFT approach—in nano-DMFT one solves a (often large) cluster problem, but with no superlattice, i.e., clusters are not coupled. The cluster DMFT was described in detail in Chap. 5, thus we do not discuss it here, but mention that available computational resources allow one to solve the DMFT problem with hundreds or even a few thousands of atoms in the cluster.

Most often, to study nanosystems DMFT is applied in the framework of so-called DFT + DMFT approach, where one begins with optimization of the structure of the nanoparticle with DFT. The DFT calculations also give the hopping parameters that can be obtained from the Kohn–Sham DFT wave functions, ion positions, and ion potential as

$$t_{ij}^{ab} = \int dr \psi^{a*}\left(\vec{r} + \vec{R}_i\right)\left(-\frac{\vec{\nabla}^2}{2m} + V_{ion}\left(\vec{r}\right)\right)\psi^{b}\left(\vec{r} + \vec{R}_j\right), \tag{7.35}$$

where $\vec{R}_i$ and $\vec{R}_j$ are the ion positions and a and b are the orbital indices. Other Hubbard model parameters—the matrix elements of the Coulomb interaction—are either obtained by using *ab initio* calculations (see Chap. 10), or one chooses its values by using some reasoning, like fitting the results for some quantity (quantities) to experimental data. Then, the obtained Hubbard model is solved with DMFT (these are the main steps of a simple version of the DFT + DMFT approach for bulk and other systems; for more refined implementations of DFT + DMFT, including self-consistent algorithms, see Chap. 10).

From the very initial studies (2008–2012), nano-DMFT showed itself to be a meaningful approximation for a few- to hundred-atom systems [22–27]. In particular, in their pioneering work Boukhvalov et al. [22] have demonstrated that the DOS of the molecular magnet $(pyH)_3[Mn_4O_3Cl_3(OAc)_3] \cdot 2MeCN$ (short name Mn_4 molecule) obtained with DMFT is in a better agreement with experimental data than the results obtained with alternative DFT and DFT + U tools. Namely, DMFT gives the experimental HOMO-LUMO gap 1.8 eV (Fig. 7.5).

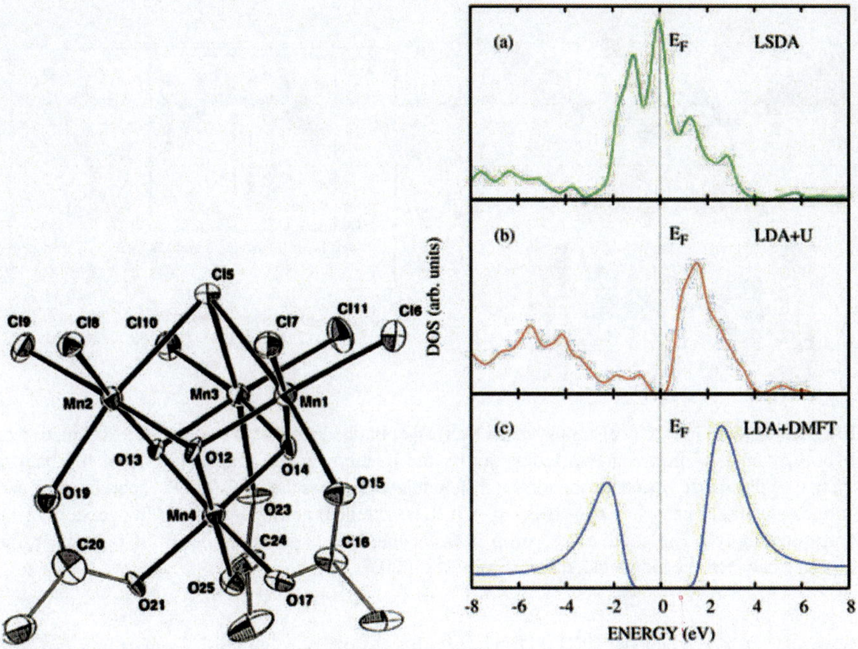

Fig. 7.5 Left: graphic representation of the Mn_4 molecule that consists of four Mn atoms and a "non-correlated" ligand (Reprinted top Fig. 4 with permission from S. Wang, H.-L. Tsai, K. Folting, W. E. Streib, D. N. Hendrickson, and G. Christou, Inorg. Chem. **35**, 7578 (1996). Copyright 1996 American Chemical Society). Right: d-orbital DOS for the Mn atoms in the Mn_4 molecule obtained with spin LDA, LDA + U, and LDA + DMFT. In the last two cases, the following parameters were used: $U = 4$ eV, $J = 0.9$ eV, $T = 0.125$ eV. In the DMFT calculations, the multi-orbital QMC impurity solver was used. Reprinted Fig. 1 with permission from (D. W. Boukhvalov, L. I. Vergara, V. V. Dobrovitski, M. I. Katsnelson, A. I. Lichtenstein, P. Kögerler, J. L. Musfeldt, and B. N. Harmon (2008)). Copyright 2008 by the American Physical Society. https://doi.org/10.1103/PhysRevB.77.180402

Similarly, it was shown in Ref. [27] that the DMFT spectrum of hydrogen molecules is more accurate (with respect to the exact spectrum) as compared to the results of another approach—the unrestricted HF.

Early studies of the transport properties of nanosystems also showed a great potential of DMFT for the nanocase. Namely, DMFT analysis of the transport properties of 110-atom quantum point contact model [24] showed that the conductivity in the contact drops as the distance between the tip and contact atoms increases, or equivalently, as the tip-atom hybridization t_{QPC} decreases. Surprisingly, the drop in the conductivity happens faster with t_{QPC} decrease than the local "Mott–Hubbard transition" (crossover) on the tip atoms (that become a "quasi-isolated" subsystem as t_{QPC} decreases), Fig. 7.6. In other words, the conductance drops faster than the exponentially decaying with distance hybridization t_{QPC}, in a qualitative agreement with similar experimental data in transition-metal point contacts with

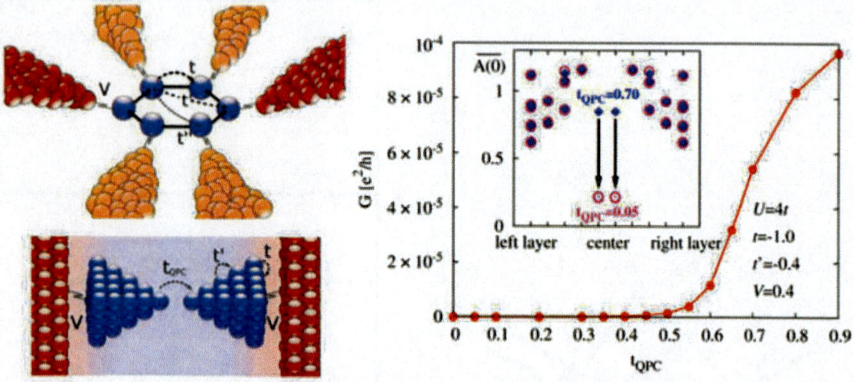

Fig. 7.6 Studied in Ref. [24] nanosystems (left) and obtained results for conductance as function of hybridization t_{QPC} between two atoms across the quantum point contact (QPC) of the bottom system (right). Inset: Spectral function for different layers across the QPC for two, "conducting" and "nonconducting," values of t_{QPC} ($t_{QPC} = 0.7t$ (blue rhombs) and $t_{QPC} = 0.05t$ (magenta circles), correspondingly). The conductance drop is faster than the exponential decay of t_{QPC} when the distance between the atoms increases. (From Ref. [24])

partially filled d-shells [28]. These results confirm a special importance of the correlation effects at the system edge.

In Ref. [23], transport properties of small Ni clusters between Cu nanowires were analyzed with DFT + DMFT. It was demonstrated that correlation effects give quasiparticle resonances at the Fermi level responsible for Fano line shapes in the conductance of nanocontacts (Fig. 7.7). Such Fano line shapes were observed experimentally (for details and references, see paper [23]).

Finally, in Refs. [25, 26] DFT + DMFT approach was applied to study the magnetic properties of small (2–5 atom) Fe and FePt clusters and it was found that DMFT gives a correct range of values for the cluster magnetic moments as compared to experimental data, despite very small cluster sizes.

Now, we move to a more recent developments and applications of the nano DMFT. In the last decade, a lot of progress has been made in improving the nano-DMFT tools and in understanding the properties of correlated nanostructures. A method to subtract the double counting of correlations in the DFT + DMFT scheme (since the correlation effects were included in both DFT and DMFT calculations, see Chap. 10) was proposed in Ref. [29]. The method was tested on the case of H_2 molecule and it was shown that the double counting-corrected approach gives very accurate ground-state energy and excitation spectrum of the system. This demonstrated that DMFT is valid even in the case of the smallest nontrivial system! In Refs. [30, 31], DMFT was applied to rings with taken into account spatial correlations through the Dynamical Vertex Approximation (Chap. 9). In Ref. [30], it was shown that in the rings the spatial correlations are important only when the coupling between the ring molecule and the metallic leads is weak and when the interatomic connectivity is low. In all other cases, standard DMFT (with local self-energy) was

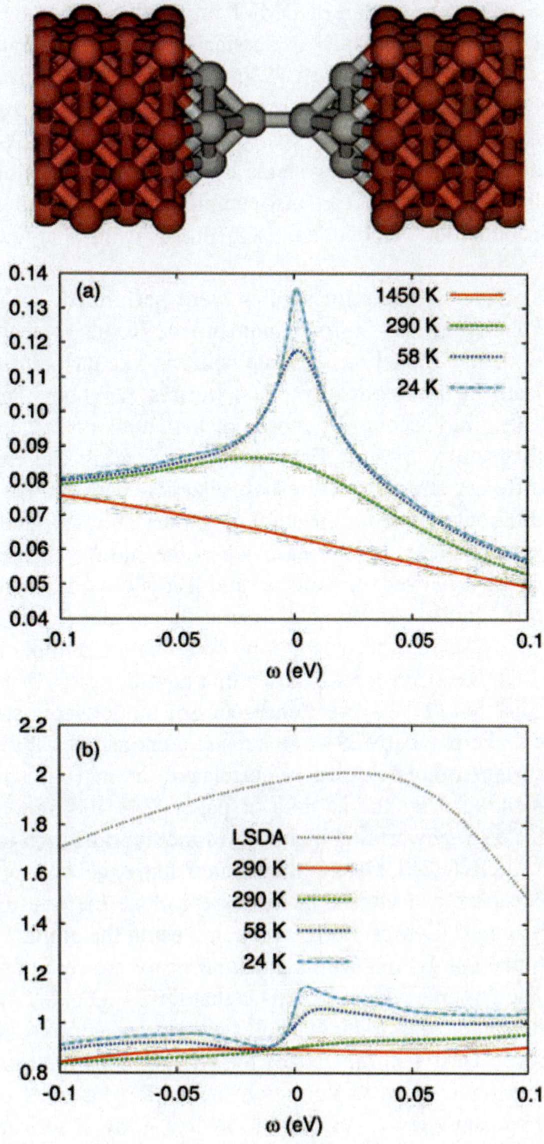

Fig. 7.7 Top: the structure of the Ni nanocontact between (001) Cu wires. Bottom: the spectral function of $3d_{xz}$ and $3d_{xz}$ orbitals of Ni tip atoms near the Fermi level (**a**) and the corresponding transmission function (**b**) at different temperatures (OCA impurity solver). The local spin-density approximation (LSDA) result for the transmission is also shown. (Reprinted Figs. 2b and 7 with permission from (D. Jacob, K. Haule, and G. Kotliar, Phys. Rev. B **82**, 195115 (2010)). Copyright 2010 by the American Physical Society. https://doi.org/10.1103/PhysRevB.82.195115)

found to be a good approximation. In Ref. [31], the spatial correlations were taken into account through the full parquet-based version of the Dynamical Vertex Approximation (see the next chapter). As the authors demonstrated in the case of semiconducting nanostructures, the last approach is capable to reproduce the exact many-body solution, contrary to the local-in-space DMFT. Thus, with the Dynamical Vertex Approximation extension, DMFT is capable to describe even 1D systems, probably the most challenging systems for this theory.

A generalization of DMFT on another important case—periodic quantum "particles" embedded into a "classical" surrounding (real-space embedding approach)—was formulated in Ref. [32]. In this approach one treats the embedded part with DMFT or other accurate approaches, while the surrounding part is treated with computationally non-expensive tools, like DFT. Such separation allows one to analyze rather large systems by using advantages of the periodicity of the system. Importantly, one can also employ this approach to study the limit of isolated nanoparticles embedded in a media, by using a large unit cell (large distance between particles).

Several important studies were performed in the case of magnetism. To test validity of DMFT for nanomagnets, Rotter et al. [33] solved exactly the double exchange model of a six-site benzene-like nanocluster with each atom coupled to a lead. In this model, nonlocal (nearest-neighbor) spin–spin correlations were also taken into account by means of the Hund's rule coupling that favors a short-range magnetic (AFM or FM) order. Comparing the exact and the DMFT results at different strength of the hybridization between the cluster and leads atoms (i.e., different number of the atom neighbors), it was found that DMFT is a rather accurate approximation in the case when the number of neighboring sites (hybridization strength between the cluster and lead atoms) is large. In other words, it was shown that, similar to the bulk case, the nonlocal correlations are not important in nanosystems when the atomic coordination number is large.

In Ref. [34] it was shown that contrary to DFT and DFT + U, DFT + DMFT is capable to give correct energetics of the low-spin state in the spin-crossover molecule $Fe(phen)_2(NCS)_2$. This result demonstrates ability of DMFT to describe even a system of one "strongly-correlated" atom (Fe) coupled to a ligand. As it was mentioned by the authors of paper [34], the important part of the approach that makes it powerful is inclusion of the hybridization in a correlated environment.

In Ref. [35] it was demonstrated that, again contrary to DFT + U, DFT + DMFT is capable, to a large degree, to reproduce the non-monotonous dependence of the magnetization of 10–100 Fe particles on the number of atoms in the system. It was found that DMFT works much better for the case of small (up to ~50 atom) clusters. The possible reason for this is that DFT + U gives overestimated frozen magnetization on the important in small systems surface atoms (charges pushed to the borders), while DMFT homogenizes the charge redistribution throughout the system due to dynamical effects. Very rich magnetic properties of zigzag graphene nanoflakes were reported by Valli et al. in Ref. [36]. It was found that, similar to extended systems, at half-filling the system is in an AFM state. It was also demonstrated that doping results in a nonzero total magnetic moment in the system at low temperatures (a ferrimagnetic (FI) state), i.e., the doping enhances FM correlations between the localized moments. These correlations are generated by delocalization of the doped holes near the edges (Fig. 7.8). It was also argued that such a system has a potential to be used in building switchable spintronic devices, since it demonstrates a hysteresis for the magnetic transition with coexistence of different magnetic orders.

A potential of hexagonal graphene nanoflakes for spintronics was also demonstrated in Ref. [37], where the spin and the charge transport properties of these

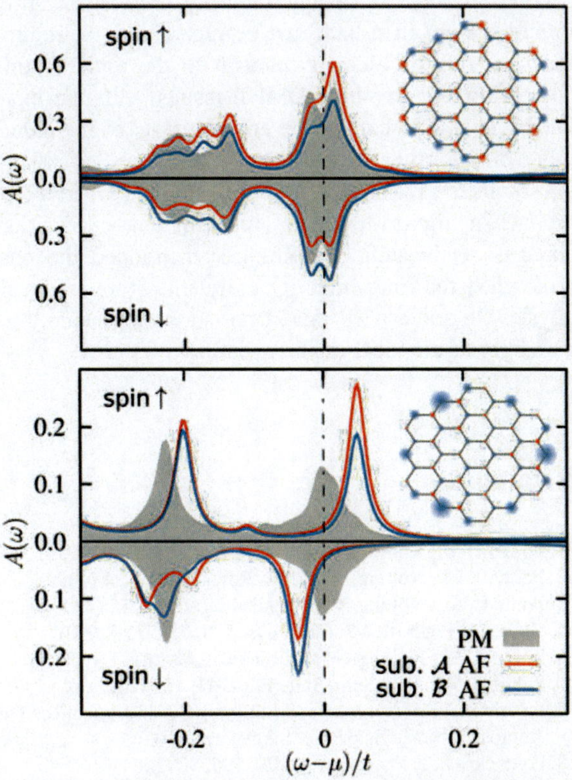

Fig. 7.8 The DOS for the three left-edge ("*y*-axis") atoms in the graphene flakes at $U = 3.75t$, number of holes $\delta = 2$, in the cases of AFM (top Figure, $T = 0.01t$) and FI (bottom Figure, $T = 0.005t$) order. The results correspond to different sub-lattices shown in the inset figures. The gray DOS were obtained with nonmagnetic calculations (from [36], see this paper for more details). (Reprinted Fig. 6 with permission from (A. Valli, A. Amaricci, A. Toschi, T. Saha-Dasgupta, K. Held, and M. Capone Phys. Rev. B **94**, 245146 (2016)). Copyright 2016 by the American Physical Society. https://doi.org/10.1103/PhysRevB.94.245146)

systems were analyzed with combined static DMFT + Landauer (nonequilibrium GF) approach at different temperatures of the junctions. In particular, it was shown that one can get a large temperature-gradient (thermoelectric effect) spin current in the case of flakes with FM zigzag edges. The spin current can be significantly increased in small flakes by applied voltage. As the size of the flake increases, the critical U that gives FM edges, and hence the spin current, decreases. There was also found the maximum value of U above which the current is destroyed by the effects of Coulomb repulsion. Thus, for the maximum-size 600-atom flakes studied in work [37] it was shown with DMFT that the spin current is nonzero at U's from ~t to ~$3t$, with the minimum and maximum values of the current a few times larger compared to the static mean-field theory solution. The difference in the results obtained by two approaches can be due to ignored in the last approach charge fluctuations (dynamical effects). We also refer the reader to work [38] for a review of different nano-DMFT approaches and their applications in the case of Hubbard-type systems.

Finally, we would like to mention paper [39], where DMFT was applied to a different type of models—multiple Kondo impurities coupled to an electronic bath. The properties of the system were analyzed with so-called impurity DMFT, where the impurity self-energy for each impurity atom was found self-consistently by solving an Anderson single-impurity problem in an effective bath. Since Kondo impurity atom in a bath can be regarded as a limiting case of nanoparticle in a medium, this problem is relevant to the topic of this section. It was found by comparing the impurity DMFT results with the exact NRG data that impurity DMFT is able to capture several elements of the Kondo physics, like the onset of dilute limit for two separated magnetic impurities and underscreened Kondo physics, but it fails to give an inter-impurity singlet (generated by RKKY interaction) and to produce a high-to-low spin quantum phase transition as the inter-dot coupling changes. As the authors of Ref. [39] mentioned, the local self-energy approximation fails when the inter-impurity correlations are responsible for suppressing Kondo physics. In some way, this is not surprising since this case is beyond the DMFT paradigm of the local-impurity bath.

References

1. Potthoff, M., Nolting, W.: Phys. Rev. B. **59**, 2549 (1999)
2. Potthoff, M., Nolting, W.: Phys. Rev. B. **60**, 7834 (1999)
3. Potthoff, M., Nolting, W.: Eur. Phys. J. B. **8**, 555 (1999)
4. Bulla, R., Potthoff, M.: Eur. Phys. J. B. **13**, 257 (2000)
5. Potthoff, M., Nolting, W.: Physica B. **259–261**, 760 (1999)
6. Potthoff, M.: Adv. Solid State Phys. **42**, 121 (2002)
7. Freericks, J.K.: Transport in Multilayered Nanostructures. The Dynamical Mean-Field Theory Approach. World Scientific, Singapore (2006)
8. Freericks, J.K.: Phys. Rev. B. **70**, 195342 (2004)
9. Chen, L., Freericks, J.K.: Phys. Rev. B. **75**, 125114 (2007)
10. Freericks, J.K., Zlatić, V., Shvaika, A.M.: Phys. Rev. B. **75**, 035133 (2007)
11. Zlatic, V., Freericks, J.K.: Phys. Rev. B. **96**, 235146 (2017)
12. Ishida, H., Liebsch, A.: Phys. Rev. B. **79**, 045130 (2009)
13. Nourafkan, R., Marsiglio, F., Capone, M.: Phys. Rev. B. **82**, 115127 (2010)
14. Nourafkan, R., Marsiglio, F.: Phys. Rev. B. **83**, 155116 (2011)
15. Okamoto, S., Millis, A.J.: Phys. Rev. B. **70**, 241104 (2004)
16. Titvinidze, I., Sorantin, M.E., Dorda, A., von der Linden, W., Arrigoni, E.: Phys. Rev. B. **98**, 035146 (2018)
17. Bakalov, P., Esfahani, D.N., Covaci, L., Peeters, F.M., Tempere, J., Locquet, J.-P.: Phys. Rev. B. **93**, 165112 (2016)
18. Okamoto, S., Millis, A.J.: Phys. Rev. B. **72**, 235108 (2005)
19. Petocchi, F., Capone, M.: Phys. Rev. B. **93**, 235125 (2016)
20. Klebel, B., Schaefer, T., Toschi, A., Tomczak, J.M.: preprint arXiv 2005, 01369 (2020)
21. Florens, S.: Phys. Rev. Lett. **99**, 046402 (2007)
22. Boukhvalov, D.W., Vergara, L.I., Dobrovitski, V.V., Katsnelson, M.I., Lichtenstein, A.I., Kögerler, P., Musfeldt, J.L., Harmon, B.N.: Phys. Rev. B. **77**, 180402 (2008)
23. Jacob, D., Haule, K., Kotliar, G.: Phys. Rev. B. **82**, 195115 (2010)
24. Valli, A., Sangiovanni, G., Gunnarsson, O., Toschi, A., Held, K.: Phys. Rev. Lett. **104**, 246402 (2010)
25. Turkowski, V., Kabir, A., Nayyar, N., Rahman, T.S.: J. Phys.: Condens. Matter. **22**, 462202 (2010)
26. Turkowski, V., Kabir, A., Nayyar, N., Rahman, T.S.: J. Chem. Phys. **136**, 114108 (2012)
27. Lin, N., Marianetti, C.A., Millis, A.J., Reichman, D.R.: Phys. Rev. Lett. **106**, 096402 (2011)

28. Krans, J.M., Muller, C.J., Yanson, I.K., Govaert, T.C.M., Hesper, R., van Ruitenbeek, J.M.: Phys. Rev. B. **48**, 14721 (1993)
29. Lee, J., Haule, K.: Phys. Rev. B. **91**, 155144 (2015)
30. Valli, A., Sangiovanni, G., Toschi, A., Held, K.: Phys. Rev. B. **86**, 115418 (2012)
31. Valli, A., Schaefer, T., Thunstroem, P., Rohringer, G., Andergassen, S., Sangiovanni, G., Held, K., Toschi, A.: Phys. Rev. B. **91**, 115115 (2015)
32. Chibani, W., Ren, X., Scheffler, M., Rinke, P.: Phys. Rev. B. **93**, 165106 (2016)
33. Rotter, D., Valli, A., Sangiovanni, G., Held, K.: Eur. Phys. J. B. **86**, 68 (2013)
34. Chen, J., Millis, A.J., Marianetti, C.A.: Phys. Rev. B. **91**, 241111 (2015)
35. Kabir, A., Turkowski, V., Rahman, T.S.: J. Phys.: Condens. Matter. **27**, 125601 (2015)
36. Valli, A., Amaricci, A., Toschi, A., Saha-Dasgupta, T., Held, K., Capone, M.: Phys. Rev. B. **94**, 245146 (2016)
37. Phung, T.T., Peters, R., Honecker, A., de Laissardiere, G.T., Vahedi, J.: preprint arXiv:2003.01139v1 (2020)
38. Schueler, M., Barthel, S., Wehling, T., Karolak, M., Valli, A., Sangiovanni, G.: Eur. Phys. J. Special Topics. **226**, 2615 (2017)
39. Mitchell, A.K., Bulla, R.: Phys. Rev. B. **92**, 155101 (2015)

Chapter 8
Nonequilibrium DMFT

Nonequilibrium properties of strongly correlated systems are another hot topic of research. Besides discovery of new fascinating phenomena, research in this field may lead to many practical ultrafast applications that include switches [1, 2], microelectromechanical system (MEMS) devices [3], smart window applications [4], biosensors [5], and ultrafast lithium storage batteries [6]. Another reason to understand nonequilibrium properties of strongly correlated systems is the need for interpretation of numerous new experimental data, especially response of the materials to pulse- and other external perturbations (see, e.g., Refs. [7–10]).

Generalization of DMFT on the nonequilibrium case [11–13] (for over-review, see Ref. [14]) was an important step in the field of nonequilibrium properties of strongly correlated systems. In this chapter, we present details of nonequilibrium DMFT and summarize important results obtained with this theory.

8.1 Basics of the Nonequilibrium Many-Body Formalism

Since for many researchers nonequilibrium many-body approach is less familiar tool than the equilibrium one, before moving to nonequilibrium DMFT, we give a brief summary of general features of the nonequilibrium many-body theory in the second quantization and path-integral formulations.

8.1.1 Second Quantization Approach

Let us assume that initially the system is in equilibrium and is perturbed at time t_0. Before this time, the system is characterized by some density operator $\widehat{\rho} = \sum_n \rho_n |n\rangle\langle n|$, where $\rho_n = e^{-E_n/T}$ are its eigenvalues. We begin with developing a PT

V. Turkowski, *Dynamical Mean-Field Theory for Strongly Correlated Materials*,
https://doi.org/10.1007/978-3-030-64904-3_8

for the time evolution of operators of a particle after a perturbation is applied. The corresponding time-dependent many-body PT, known as a closed-time path GF formalism, was developed by Keldysh [15] (before, similar relativistic quantum-field theory formulation was done by Schwinger [16]). The main goal of the developed approach is to find the time evolution of thermodynamic averages of products of Heisenberg time-dependent operators $\widehat{c}_H(x) = e^{i(t-t_0)\widehat{H}}\widehat{c}(x)e^{-i(t-t_0)\widehat{H}}$ (that satisfy the Heisenberg equation of motion $\frac{i\partial\widehat{c}_H(x)}{\partial t} = [H, \widehat{c}_H(x)]$) and time – independent wave functions. In particular, for one and two operators, the thermal averages, or averages with respect to the initial density operator, are defined as:

$$\langle\widehat{c}_H(x)\rangle_\rho = Tr[\widehat{\rho}\widehat{c}_H(x)], \quad \langle\widehat{c}_H(x)\widehat{c}_H^+(x')\rangle_\rho = Tr[\widehat{\rho}\widehat{c}_H(x)\widehat{c}_H^+(x')], \tag{8.1}$$

where $x = \left(\vec{r}, t\right)$ (here $\widehat{c}_H(x)$ is not necessarily an electron annihilation operator; thus $\langle\widehat{c}_H(x)\rangle$ can be finite). Since in most cases of interest, the fields $\widehat{c}_H(x)$ interact with each other and other (external) fields, the Hamiltonian of the system $\widehat{H}$ can be divided on a sum of the noninteracting $\widehat{H}_0$ and interacting $\widehat{H}_{\text{int}}$ parts. In PT, one expands the partition and other functions in powers of $\widehat{H}_{\text{int}}$. At this point, it is convenient to introduce the interaction, or Dirac, representation for the operators $\widehat{\varphi}_I(x) = e^{i(t-t_0)\widehat{H}_0}\widehat{c}_H(x)e^{-i(t-t_0)\widehat{H}_0}$, including the interaction part of the Hamiltonian operator, $\widehat{H}_{\text{int},I}(x) = e^{i(t-t_0)\widehat{H}_0}\widehat{H}_{int,H}(x)e^{-i(t-t_0)\widehat{H}_0}$. The advantage of the interaction representation is that in this representation the time evolution of the operator is defined only by the interaction part of the Hamiltonian, $i\frac{\partial\widehat{c}_I(x)}{\partial t} = [H_{\text{int},I}, \widehat{c}_I(x)]$. Thus, this representation is more convenient for the PT that is also defined by $H_{\text{int},I}$. To write down the equation that connects the Heisenberg and the interaction operators, one can introduce the evolution operator

$$\widehat{U}(t, t_0) = \widehat{T}e^{i\int_{t_0}^{t} dt'\widehat{H}_{int,I}(t')}. \tag{8.2}$$

Then, the connection can be written as

$$\widehat{c}_H(x) = \widehat{U}(t_0, t)\widehat{c}_I(x)\widehat{U}(t, t_0), \tag{8.3}$$

and for the averages of the Heisenberg operators, one gets

$$\langle\widehat{c}_H(x)\rangle_\rho = Tr\left[\widehat{\rho}\widehat{U}(t_0, t)\,\widehat{c}_H(x)\widehat{U}(t_0, t)\right], \tag{8.4}$$

$$\langle\widehat{c}_H(x)\widehat{c}_H^+(x')\rangle_\rho = \begin{cases} Tr\left[\widehat{\rho}\widehat{U}(t_0, t)\widehat{c}_H(x)\widehat{U}(t, t')\widehat{c}_H^+(x')\widehat{U}(t', t_0)\right], & t > t' \\ Tr\left[\widehat{\rho}\widehat{U}(t_0, t')\widehat{c}_H^+(x')\widehat{U}(t', t)\widehat{c}_H(x)\widehat{U}(t, t_0)\right], & t < t' \end{cases}. \tag{8.5}$$

Assuming that $t_0 = -\infty$, introducing the scattering operator $\widehat{S} = \widehat{U}(\infty, -\infty)$ and using the time-ordering operator $\widehat{T}$ (defined and used in previous chapters), one can show (see, e.g., Ref. [17]) that the last two equations are equivalent to

$$\langle \widehat{c}_H(x) \rangle_\rho = Tr\left[\widehat{\rho}\widehat{S}^+\widehat{T}\widehat{S}\widehat{c}_I(x)\right], \tag{8.6}$$

$$\langle \widehat{c}_H(x)\widehat{c}_H^+(x') \rangle_\rho = Tr\left[\widehat{\rho}\,\widehat{S}^+\widehat{T}\widehat{S}\widehat{c}_I(x)\widehat{c}_I^+(x')\right]. \tag{8.7}$$

In particular, to prove Eq. (8.6) one writes: $\widehat{S}^+\left[\widehat{T}\left(\widehat{S}\widehat{c}_I(x)\right)\right] = \widehat{U}(-\infty, \infty) \times \left[\widehat{U}(\infty, t)\widehat{c}_I(x)\widehat{U}(t, -\infty)\right] = \widehat{U}(-\infty, t)\widehat{c}_I(x)\widehat{U}(t, -\infty)$. From this equality, it follows that Eqs. (8.4) and (8.6) are equivalent at $t_0 = -\infty$. Similar steps can be taken to prove Eq. (8.7).

As it follows from Eqs. (8.6) and (8.7), the averages are calculated by using a nontrivial time contour with time points running from $-\infty$ to ∞ $\left(\text{forced by } \widehat{S}\right.$ operator) and backwards (forced by $\left.\widehat{S}^+\right)$. For convenience, the contour can be represented as running from $-\infty$ to ∞ above and from ∞ to $-\infty$ below the real-time axis (top Fig. 8.1).

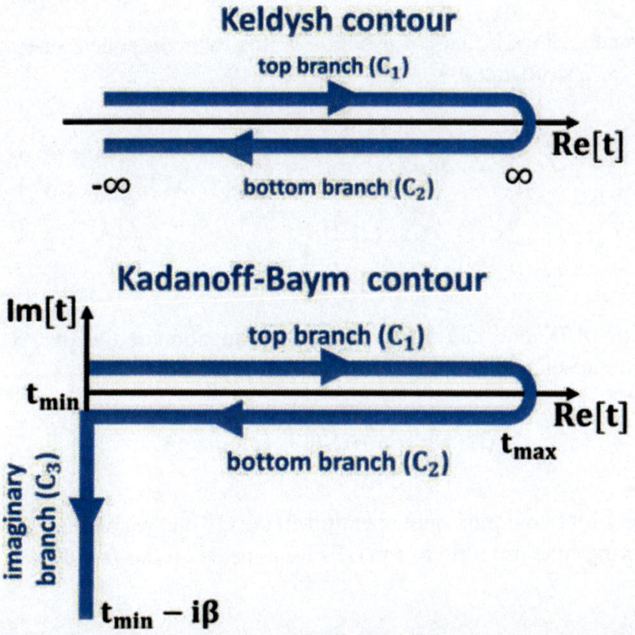

Fig. 8.1 Schematic representation of the two-branch Keldysh [15] (top) and three-branch Kadanoff–Baym [18] (bottom) time contours with the time ordering (from the initial to the final time) shown by arrows

To distinguish the operator time arguments on different branches, for the arguments on the top branch we use x_+ and on the bottom - x_-. In these notations, one can rewrite the time-dependent part on the right-hand side in Eq. (8.6) as $\widehat{S}^+\left[\widehat{T}\left(\widehat{S}\widehat{c}_I(x)\right)\right] = \widehat{T}_P\left[\widehat{S}^+\widehat{S}\widehat{c}_I(x_+)\right]$, where the path-ordering operator $\widehat{T}_P$ is the time-ordering operator along the contour P in top Fig. 8.1, with the increasing time direction shown by arrows. Using this result and the fact that $\widehat{S}^+\widehat{S} = \widehat{T}_P e^{-i\int_P \widehat{H}_{\text{int},I}(t)dt}$, one can rewrite Eq.(8.6) as

$$\langle\widehat{c}_H(x)\rangle_\rho = Tr\left[\widehat{\rho}\widehat{T}_P\widehat{S}^+\widehat{S}\widehat{c}_I(x)\right] \equiv Tr\left[\widehat{\rho}\ \widehat{T}_P e^{-i\int_P \widehat{H}_{\text{int},I}(t)dt}\widehat{c}_I(x)\right]. \tag{8.8}$$

Similarly, one can obtain

$$\begin{aligned}\langle\widehat{c}_H(x)\widehat{c}_H^+(x')\rangle_\rho &= Tr\left[\widehat{\rho}\widehat{T}_P\widehat{S}^+\widehat{S}\widehat{c}_I(x)\widehat{c}_I^+(x')\right] \\ &\equiv Tr\left[\widehat{\rho}\ \widehat{T}_P e^{-i\int_P \widehat{H}_{\text{int},I}(t)dt}\widehat{c}_I(x)\widehat{c}_I^+(x')\right].\end{aligned} \tag{8.9}$$

Then, one can introduce the fields defined on the contour in terms of two-component operator $c_{I,P}(x) = \begin{pmatrix}\widehat{c}_I(x_+) \\ \widehat{c}_I(x_-)\end{pmatrix}$, where the top (bottom) component acts on the first (second) branch. The GF that corresponds to this two-component operators has the following 2×2 structure:

$$\begin{aligned}G_P(x,x') &= -iTr\left[\widehat{\rho}\widehat{T}_P\widehat{S}^+\widehat{S}\widehat{c}_I(x)\widehat{c}_I^+(x')\right] = \begin{pmatrix}G(x_+,x'_+) & G(x_+,x'_-) \\ G(x_-,x'_+) & G(x_-,x'_-)\end{pmatrix} \\ &\equiv \begin{pmatrix}G_{++}(x,x') & G_{+-}(x,x') \\ G_{-+}(x,x') & G_{--}(x,x')\end{pmatrix}.\end{aligned} \tag{8.10}$$

Using Eq. (8.7), one can write down the equation for the matrix GF in the Heisenberg representation:

$$G_P(x,x') = -i\left\langle\widehat{T}_P\widehat{c}_H(x)\widehat{c}_H^+(x')\right\rangle_\rho. \tag{8.11}$$

Time ordering for two-component operators in the GF on the contour $\widehat{T}_P$ corresponds to the following time ordering of the GF components on the first time-branch:

$$\begin{pmatrix} G_{++}(x,x') & G_{+-}(x,x') \\ G_{-+}(x,x') & G_{--}(x,x') \end{pmatrix} = -i \begin{pmatrix} \left\langle \widehat{T}\widehat{c}_H(x)\widehat{c}_H^+(x')\right\rangle_\rho & -\langle \widehat{c}_H^+(x')\,\widehat{c}_H(x)\rangle_\rho \\ \langle \widehat{c}_H(x)\widehat{c}_H^+(x')\rangle_\rho & \left\langle \widehat{\widehat{T}}\widehat{c}_H(x)\widehat{c}_H^+(x')\right\rangle_\rho \end{pmatrix}, \tag{8.12}$$

where on the right-hand-side, x and x' are on the first ("physical") time branch. In Eq. (8.12), $\widehat{T}$ is the usual time-ordering operator (we consider the case of fermions):

$$\widehat{T}\widehat{c}_H(x)\widehat{c}_H^+(x') = \theta(t-t')\widehat{c}_H(x)\widehat{c}_H^+(x') - \theta(t'-t)\widehat{c}_H^+(x')\widehat{c}_H(x), \tag{8.13}$$

and $\widehat{\widehat{T}}$ is anti-time ordering operator:

$$\widehat{\widehat{T}}\widehat{c}_H(x)\widehat{c}_H^+(x') = -\theta(t-t')\widehat{c}_H^+(x')\widehat{c}_H(x) + \theta(t'-t)\widehat{c}_H(x)\widehat{c}_H^+(x'). \tag{8.14}$$

For the $G_{-+}(x,x')$ the time x (on the contour) is always after x', thus there is no need to include the time-ordering operator in the corresponding matrix element in Eq. (8.12). Similarly, the time-ordering is fixed for $G_{+-}(x,x')$, where x' is always later than x.

Since the components of GF satisfy

$$G_{++}(x,x') + G_{--}(x,x') = G_{+-}(x,x') + G_{-+}(x,x'), \tag{8.15}$$

there are only three independent components in the GF matrix (8.12). Four components of the matrix can be expressed in terms of independent familiar retarded G^R and advanced G^A GFs, and perhaps less familiar Keldysh GF G^K. The relations between the two sets of functions are:

$$G^R(x,x') = -i\theta(t-t')\langle\{\widehat{c}_H(x),\widehat{c}_H^+(x')\}\rangle_\rho = G_{++}(x,x') - G_{+-}(x,x'), \tag{8.16}$$

$$G^A(x,x') = i\theta(t'-t)\langle\{\widehat{c}_H(x),\widehat{c}_H^+(x')\}\rangle_\rho = G_{++}(x,x') - G_{-+}(x,x'), \tag{8.17}$$

$$G^K(x,x') = -i\langle[\widehat{c}_H(x),\widehat{c}_H^+(x')]\rangle_\rho = G_{-+}(x,x') + G_{+-}(x,x'). \tag{8.18}$$

For some purposes, it is also convenient to introduce a fourth function:

$$C(x,x') = -i\langle\{\widehat{c}_H(x),\widehat{c}_H^+(x')\}\rangle_\rho = G_{-+}(x,x') - G_{+-}(x,x'). \tag{8.19}$$

Using Eqs. (8.15)–(8.18), one can express the components of the matrix $G_P(x,x')$ in terms of G^R, G^A, and G^K:

$$G_P(x,x') = \begin{pmatrix} G_{++}(x,x') & G_{+-}(x,x') \\ G_{-+}(x,x') & G_{--}(x,x') \end{pmatrix}$$
$$= \frac{1}{2}\begin{pmatrix} G^K + G^R + G^A & G^K - G^R + G^A \\ G^K + G^R - G^A & G^K - G^R - G^A \end{pmatrix}(x,x'). \quad (8.20)$$

To make things even simpler, one can apply a (Keldysh) transformation to the matrix GF $G_P(x,x')$ to fix one of the components to be equal zero [17]:

$$\widetilde{G}(x,x') = QG_P(x,x')Q^{-1} = \begin{pmatrix} 0 & G^A \\ G^R & G^K \end{pmatrix}(x,x'), \quad (8.21)$$

where

$$Q = \frac{1}{\sqrt{2}}\begin{pmatrix} 1 & -1 \\ 1 & 1 \end{pmatrix}. \quad (8.22)$$

In this transformation, the components of the operator also change:

$$\widehat{\widetilde{\varphi}}_H(x) = Q\begin{pmatrix} \widehat{c}_H(x_+) \\ \widehat{c}_H(x_-) \end{pmatrix} = \frac{1}{\sqrt{2}}\begin{pmatrix} \widehat{c}_H(x_+) - \widehat{c}_H(x_+) \\ \widehat{c}_H(x_+) + \widehat{c}_H(x_+) \end{pmatrix}. \quad (8.23)$$

Due to the structure of the matrix GF (8.21) the product of GFs is especially simple. It always gives the 11-component equal zero, and the R (21) and A (12) elements equal to products of the corresponding GFs:

$$\widetilde{G}_1\widetilde{G}_2 = QG_{P1}(x,x')Q^{-1}QG_{P2}(x,x')Q^{-1} = \begin{pmatrix} 0 & G_1^A G_2^A \\ G_1^R G_2^R & G_1^R G_2^K + G_1^K G_2^A \end{pmatrix}. \quad (8.24)$$

This is especially convenient when one is interested only in the retarded or advanced functions: then one does not need to use the matrix formulation, and can work with the corresponding (R or A) scalar equations.

The formulated approach can be used for PT calculations, similar to the equilibrium case, by employing the Wick theorem and other powerful tools of the equilibrium many-body theory. The difference from the equilibrium case is that now instead of one-branch time-ordered GFs $G(x,x') = -i\left\langle \widehat{T}\widehat{c}_H(x)c_H^+(x')\right\rangle_\rho$ one has matrix GFs (8.20) or (8.21). This is the price one pays for the ability to study a more general, nonequilibrium, case.

One adjustment to the formalism above when it is used in the many-body case (with an arbitrary number of particles) should be made. There, the initial density matrix has the form

$$\widehat{\rho} \equiv \rho(0) = \frac{1}{Z} e^{-\mathcal{H}(0)/T}, \tag{8.25}$$

where

$$\mathcal{H}(t) = H(t) - \mu N(t), \tag{8.26}$$

is the many-body (in general, time-dependent) Hamiltonian that includes an additional number of particles term $-\mu N(t)$, where $N(t)$ is the operator of the number of particles, μ – chemical potential), and

$$Z = Tr\left[e^{-\mathcal{H}(0)/T}\right] \tag{8.27}$$

is the partition function.

Then, both "small" and many-body (solid) systems are described by the same set of equations shown above. In particular, the equation for the average of two operators is

$$\left\langle \widehat{c}_H(x)\widehat{c}_H^+(x')\right\rangle_\rho = \frac{Tr\left[\widehat{\rho}\,\widehat{T}_P e^{-i\int_P \widehat{H}_{\mathrm{int},I}(t)dt}\widehat{c}_H(x)\widehat{c}_H^+(x')\right]}{Tr\left[\widehat{\rho}\,\widehat{T}_P e^{-i\int_P \widehat{H}_{\mathrm{int},I}(t)dt}\right]}, \tag{8.28}$$

where the function in the denominator is the partition function.

The Keldysh two-branch approach is convenient to study nonequilibrium steady states of open systems, where the energy supplied by external source (perturbation) is compensated by dissipation to the environment (see Ref. [14] and the text below). Let us also present a more general, and in some way a more universal, three-branch, method that can be formulated by using Eq. (8.28). Namely, one can formally represent the density matrix (8.25) in Eq. (8.28) by integrating the Hamiltonian over the imaginary time: $\rho(0) = \frac{1}{Z}e^{-\mathcal{H}(0)/T} \equiv \frac{1}{Z}e^{-\int_0^\beta \mathcal{H}d\tau}$ (the last equality is correct since $\mathcal{H}(0)$ is time-independent). Since $\widehat{\rho}$ is at the very left end in the time-ordered product in Eq. (8.28), one can formally assume that in addition to the two Keldysh branches, there is an extra, imaginary time, branch $[t_0 : t_0 - i\beta]$ at the end of the Keldysh contour with times after the times on the two horizontal branches (bottom Fig. 8.1). Such a contour is called Kadanoff–Baym contour [18] and to identify it we will use letter C (complex, see below). Thus, the density operator can be formally incorporated into the integral part in Eq. (8.28) and the time-ordered GF can be also defined and calculated on the three-branch contour as:

$$G(x,x') \equiv -i\left\langle \widehat{T}_C \widehat{c}_H(x)\widehat{c}_H^+(x')\right\rangle = -i\frac{Tr\left[\widehat{T}_C e^{-i\int_C \mathcal{H}(t)dt}\widehat{c}_H(x)\widehat{c}_H^+(x')\right]}{Tr\left[\widehat{T}_C e^{-i\int_P \mathcal{H}dt}\right]}, \tag{8.29}$$

where we have put $\mathcal{H}(t)$ instead of $\widehat{H}_{\text{int},I}(t)$ (since the noninteracting parts of the Hamiltonian cancel after the integration over the real branches). All concepts of the two-branch formalism, including the definitions of the time-ordering (8.13), (8.14) in the GFs, remain valid in the three-branch case as well. The only difference is that in the three-branch case the time-ordering should be taken along the complex-time path (bottom Fig. 8.1).

The time-ordered GF (8.29) has the following matrix structure with respect to the time branches:

$$\widehat{G} = \begin{pmatrix} G_{11} & G_{12} & G_{13} \\ G_{21} & G_{22} & G_{23} \\ G_{31} & G_{32} & G_{33} \end{pmatrix}. \tag{8.30}$$

The components of this matrix with arguments 1 and 2 correspond to + and − branches in the Keldysh case, respectively. In addition to these components, there are five components with one or two time arguments on the third branch. Out of these five components, only three are of different type. Namely, components G_{13} and G_{23} are similar, since their first (real-branch) arguments lay before the second argument (imaginary time one) on the contour, and these components correspond to the physically the same correlator; the same is correct for G_{31} and G_{32}. In the Kadanoff–Baym representation, in addition to three-independent real-time functions (8.16)–(8.18), the following three correlators that include imaginary time are often used:

$$G^{\urcorner}(t,\tau') = \frac{1}{2}(G_{13} + G_{23})(t,\tau') = i\langle \widehat{c}_H^+(\tau')\widehat{c}_H(t)\rangle, \tag{8.31}$$

$$G^{\ulcorner}(\tau,t') = \frac{1}{2}(G_{31} + G_{32})(\tau,t') = -i\langle \widehat{c}_H(\tau)\widehat{c}_H^+(t')\rangle, \tag{8.32}$$

$$G^{M}(\tau,\tau') = -iG_{33}(\tau,\tau') = -\langle T_\tau \widehat{c}_H(\tau)\widehat{c}_H^+(\tau')\rangle, \tag{8.33}$$

where $G^{\urcorner}$ ($G^{\ulcorner}$) is the left (right) mixing GF and G^M is the familiar imaginary-time Matsubara GF.

Two other popular (real-time!) GFs are the lesser

$$G^{<}(t,t') = G_{12} = \frac{1}{2}\left(G^K - G^R + G^A\right)(t,t') = i\langle \widehat{c}_H^+(t')\widehat{c}_H(t)\rangle \tag{8.34}$$

and the greater

$$G^{>}(t,t') = G_{21} = \frac{1}{2}\left(G^K + G^R - G^A\right)(t,t') = -i\langle \widehat{c}_H(t)\widehat{c}_H^{+}(t')\rangle \tag{8.35}$$

GFs.

To find a nonequilibrium GF, similar to the equilibrium case one can write down the Heisenberg equation of motion for the GF and solve it. For example, for the retarded GF one has $i\frac{\partial G^R(t,t')}{\partial t} = \delta_C(t,t')\delta(r-r') - i\theta_C(t,t')\langle\{[\widehat{c}_H(t),\mathcal{H}],\widehat{c}_H^{+}(t')\}\rangle$. Let us consider first the noninteracting case. An external perturbation can be introduced in the noninteracting Hamiltonian in the form of a time-dependent dispersion. For example, in presence of an external vector potential $\vec{A}(t)$, the free electron dispersion becomes $\varepsilon\left(\vec{k} - e\vec{A}(t)\right)$, where $\varepsilon\left(\vec{k}\right)$ is the free-electron dispersion of the nonperturbed system. Then, the equation for the free-electron GF

$$\left[i\frac{\partial}{\partial t} - \varepsilon\left(\vec{k},t\right) + \mu\right] G^R(t,t') = \delta_C(t,t'). \tag{8.36}$$

has the following solution:

$$G_k^R(t,t') = -i\theta_C(t,t')e^{-i\int_{t'}^{t}\left(\varepsilon\left(\vec{k},\bar{t}\right)-\mu\right)d\bar{t}}, \tag{8.37}$$

where $\theta_C(t,t')$ is the theta function on the complex time contour (it is equal 1 when $t > t'$ on the contour, and zero otherwise).

In the interacting case, the procedure of finding GFs is also very similar to the equilibrium approach. The starting point could be the matrix Dyson equation:

$$\widehat{G}(t,t') = \widehat{G}^{(0)}(t,t') + \int_C dt'' \int_C dt''' \widehat{G}^{(0)}(t,t'')\widehat{\Sigma}(t'',t''')\widehat{G}(t''',t'), \tag{8.38}$$

where $\widehat{G}^{(0)}(t,t')$ is the noninteracting matrix GF that usually can be easily found (numerically). One can solve Eq. (8.38) by using different approximations, e.g., PT, similar to the equilibrium case.

Once the solution of the equations for the GFs is found, one can calculate different quantities, for example, the time-dependent DOS and the charge density, by introducing the relative $t_{\text{rel}} = t - t'$ and the average $t_{\text{ave}} = \frac{t+t'}{2}$ time arguments:

$$A(t_{\text{ave}},\omega) = -\frac{1}{\pi}Im\sum_{\vec{k}} G^R\left(t_{\text{ave}},\omega,\vec{k}\right), \quad n(t_{\text{ave}}) = -i\sum_{\vec{k}} G_k^{<}\left(t_{\text{ave}},t_{\text{ave}}+\delta,\vec{k}\right), \tag{8.39}$$

where $G^R\left(t_{\text{ave}},\omega,\vec{k}\right) = \int e^{i\omega t_{\text{rel}}} G^R\left(t_{\text{ave}},t_{\text{rel}},\vec{k}\right)dt_{\text{rel}}$.

We refer the reader for more details on the nonequilibrium many-body physics, including the relations between different components of the GFs and their connection to observables, to Refs. [14, 19].

8.1.2 Path-Integral Approach

Since most of the nonequilibrium problems are solved numerically, due to complexity of the problem, we present main details on how it can be done in an often-used (numerical) path-integral approach (the discrete-time formulation for the GF equations in the second quantization case is very similar and can be easily done by using the formalism below). The main nontrivial thing in the path-integral formulation is to construct the time matrices that preserve the boundary conditions and other constraints (see below).

The path-integral calculations deal with the time-ordered GFs. Let us begin with the equilibrium formulation of the approach in the noninteracting case [20] with the Hamiltonian $H_0 = \sum_{\alpha} \varepsilon_\alpha a_\alpha^+(\tau) a_\alpha(\tau)$ (the interaction case will be discussed later). The equation for the equilibrium time-ordered GF in path-integral representation can be written as

$$G_{\alpha_1\alpha_2}(\tau_1,\tau_2) = \left\langle T a_{\alpha_1}(\tau_1) a_{\alpha_2}^+(\tau_2) \right\rangle = \frac{1}{Z} Tr\left[e^{-\beta\left(\widehat{H}-\mu\widehat{N}\right)} a_{\alpha_1}^H(\tau_1) a_{\alpha_2}^{H^+}(\tau_2) \right]$$
$$= \frac{1}{Z}\int D[\phi]D[\phi^*]\phi_{\alpha_1}(\tau_1)\phi_{\alpha_2}^*(\tau_2) e^{-\int_0^\beta d\tau\left[\sum_\alpha \phi_\alpha^*(\tau)\left(\frac{\partial}{\partial\tau}-\mu\right)\phi_\alpha(\tau)+H(\phi^*,\phi)\right]}, \tag{8.40}$$

where ϕ and ϕ^* are the Grassman fermionic fields, α are their state indices, and $H(\phi^*,\phi) = \sum_\alpha \varepsilon_\alpha \phi_\alpha^* \phi_\alpha$.

To get the matrix expression for the GF we divide the time interval $[0,\beta]$ on M pieces with step ε and write the expression for the evolution operator:

$$U\left(\phi_{af}^*, t_f; \phi_{ai}, t_i\right) = \left\langle \phi_{ai} | e^{-\frac{i}{\hbar}\widehat{H}\cdot(t_f-t_i)} | \phi_{ai} \right\rangle$$
$$= \int \prod_{k=1}^{M-1} \prod_\alpha \frac{D[\phi_{ak}]D\left[\phi_{ak}^*\right]}{N} e^{-A(\phi^*,\phi)}, \tag{8.41}$$

where

$$
\begin{aligned}
A(\phi^*,\phi) &= \sum_{k=1}^{M-1}\sum_{\alpha}\phi^*_{\alpha k}\phi_{\alpha k} - \sum_{k=1}^{M}\left(\sum_{\alpha}\phi^*_{\alpha k}\phi_{\alpha k-1} - \frac{i\varepsilon}{\hbar}H(\phi^*_{\alpha k},\phi_{\alpha k-1})\right) \\
&= -\sum_{\alpha}\phi^*_{\alpha M}\phi_{\alpha M-1} + \frac{i\varepsilon}{\hbar}H(\phi^*_{\alpha M},\phi_{\alpha M-1}) \\
&\quad - i\varepsilon\sum_{k=1}^{M-1}\left[i\sum_{\alpha}\phi^*_{\alpha k}\left(\frac{\phi_{\alpha k}-\phi_{\alpha k-1}}{\varepsilon}\right) - \frac{1}{\hbar}H(\phi^*_{\alpha k},\phi_{\alpha k-1})\right].
\end{aligned}
\tag{8.42}
$$

is the action.

The partition function in the path-integral formalism

$$
Z = Tre^{-\beta(\widehat{H}-\mu\widehat{N})} = \int \prod_{\alpha} D[\phi_\alpha]D[\phi^*_\alpha]e^{-\sum_{\alpha}\phi^*_\alpha\phi_\alpha}\left\langle \zeta\phi|e^{-\beta(\widehat{H}-\mu\widehat{N})}|\phi\right\rangle \tag{8.43}
$$

is calculated with the boundary conditions: $\phi_{\alpha 0} = \phi_\alpha$, $\phi^*_{\alpha M} = \zeta\phi_\alpha$ (where $\zeta = 1$ for bosons and $\zeta = -1$ for our case of fermions (with antiperiodic boundary conditions)). In detail,

$$
Z = \int \prod_{k=1}^{M}\prod_{\alpha}\frac{D[\phi_{\alpha k}]D[\phi^*_{\alpha k}]}{N}e^{-S(\phi^*,\phi)}, \tag{8.44}
$$

where

$$
\begin{aligned}
S(\phi^*,\phi) &= \varepsilon\sum_{k=2}^{M-1}\left[\sum_{\alpha}\phi^*_{\alpha k}\left\{\frac{(\phi_{\alpha k}-\phi_{\alpha k-1})}{\varepsilon} - \mu\phi_{\alpha k-1}\right\} + \frac{1}{\hbar}H(\phi^*_{\alpha k},\phi_{\alpha k-1})\right] \\
&\quad + \varepsilon\left[\sum_{\alpha}\phi^*_{\alpha 1}\left\{\frac{(\phi_{\alpha 1}-\zeta\phi_{\alpha M})}{\varepsilon} - \mu\zeta\phi_{\alpha M}\right\} + \frac{1}{\hbar}H(\phi^*_{\alpha 1},\zeta\phi_{\alpha M})\right].
\end{aligned}
\tag{8.45}
$$

Using the equations above one can write down the matrix expression for the GF and the partition function:

$$
G^{(0)}_{\alpha_1\alpha_2}(\tau_1,\tau_2) = \delta_{\alpha_1\alpha_2}S^{(\alpha)-1}_{\tau_1\tau_2}, \quad Z_0 = \prod_{\alpha}\left[detS^{(\alpha)}\right]^{-\xi}, \tag{8.46}
$$

where $S^{(\alpha)}$ is $M \times M$ matrix

$$S^{(\alpha)} = \begin{bmatrix} 1 & 0 & \dots & 0 & -\xi a_\alpha \\ -a_\alpha & 1 & \dots & 0 & 0 \\ 0 & -a_\alpha & \dots & \dots & \dots \\ \dots & \dots & \dots & 1 & 0 \\ 0 & 0 & \dots & -a_\alpha & 1 \end{bmatrix} \tag{8.47}$$

and

$$a_\alpha = 1 - \frac{\beta}{M}(\varepsilon_\alpha - \mu) = 1 - \Delta\tau(\varepsilon_\alpha - \mu). \tag{8.48}$$

One can easily check that the Eqs. (8.44)–(8.46) are correct by taking in these expressions the limit $M \to \infty$, which gives:

$$det S^{(\alpha)} = 1 - \xi a^M \to_{M\to\infty} 1 - \xi e^{-\beta(\varepsilon_\alpha - \mu)}, \tag{8.49}$$

$$Z_0 = \prod_\alpha \left[1 - \xi e^{-\beta(\varepsilon_\alpha - \mu)}\right]^{-\xi}, \tag{8.50}$$

$$\begin{aligned} G^{(0)}_{\alpha_1\alpha_2}(\tau_1,\tau_2) &= \delta_{\alpha_1\alpha_2} S^{(\alpha_1)-1}_{\tau_1\tau_2} \\ &= \delta_{\alpha_1\alpha_2} e^{-(\varepsilon_\alpha - \mu)(\tau_1 - \tau_2)} \left[\theta(\tau_1 - \tau_2)\left(1 + \frac{\xi}{e^{\beta(\varepsilon_\alpha - \mu)} - \xi}\right) + \theta(\tau_2 - \tau_1)\frac{\xi}{e^{\beta(\varepsilon_\alpha - \mu)} - \xi}\right]. \end{aligned} \tag{8.51}$$

These are familiar many-body theory results. In particular, for the fermions the time-ordered GF is

$$G^{(0)}_{\alpha_1\alpha_2}(\tau_1, \tau_2) = \delta_{\alpha_1\alpha_2} e^{-(\varepsilon_\alpha - \mu)(\tau_1 - \tau_2)}[\theta(\tau_1 - \tau_2)(1 - f(\varepsilon_\alpha - \mu)) - \theta(\tau_2 - \tau_1)f(\varepsilon_\alpha - \mu)], \tag{8.52}$$

where $f(\varepsilon_\alpha - \mu)$ is the Fermi distribution function.

Generalization of this equilibrium (Matsubara-time) formulation on the two- and three-branch cases is rather straightforward [21, 22]. The main difference in these cases from the previous one is the branch-dependent time step in front of a_α, Eq. (8.48). Namely, on the first branch, one has $\Delta\tau \to i\Delta t$, on the second: $\Delta\tau \to -i\Delta t$ (Δt is positive in both cases), and on the third branch $\Delta\tau$ remains the same (since the equilibrium formulation was done for the imaginary time).

In detail, for the Keldysh contour one has

$$
G^{(0)(\alpha)-1}_{t_1 t_2} = \begin{bmatrix}
1 & 0 & 0 & \vdots & 0 & 0 & 0 & 0 & \vdots & -\zeta e^{-\beta(\varepsilon_\alpha-\mu)} \\
-a_\alpha & 1 & 0 & \vdots & 0 & 0 & 0 & 0 & \vdots & 0 \\
0 & -a_\alpha & 1 & \vdots & 0 & 0 & 0 & 0 & \vdots & 0 \\
\vdots & \vdots & \vdots & \vdots & \vdots & \vdots & \vdots & \vdots & \vdots & \vdots \\
0 & 0 & 0 & \vdots & 1 & 0 & 0 & 0 & \vdots & 0 \\
0 & 0 & 0 & \vdots & 0 & 1 & 0 & 0 & \vdots & 0 \\
0 & 0 & 0 & \vdots & 0 & -\overline{a}_\alpha & 1 & 0 & \vdots & 0 \\
0 & 0 & 0 & \vdots & 0 & 0 & -\overline{a}_\alpha & 1 & \vdots & 0 \\
\vdots & \vdots & \vdots & \vdots & \vdots & \vdots & \vdots & \vdots & \vdots & \vdots \\
0 & 0 & 0 & \vdots & 0 & 0 & 0 & 0 & \vdots & 1
\end{bmatrix}, \tag{8.53}
$$

and for the Kadanoff–Baym contour:

$$
G^{(0)(\alpha)-1}_{t_1 t_2} = \begin{bmatrix}
1 & 0 & 0 & \vdots & 0 & 0 & 0 & 0 & \vdots & 0 & 0 & 0 & 0 & \vdots & -\xi a^M_\alpha \\
-a_\alpha & 1 & 0 & \vdots & 0 & 0 & 0 & 0 & \vdots & 0 & 0 & 0 & 0 & \vdots & 0 \\
0 & -a_\alpha & 1 & \vdots & 0 & 0 & 0 & 0 & \vdots & 0 & 0 & 0 & 0 & \vdots & 0 \\
\vdots & \vdots & \vdots & \vdots & \vdots & \vdots & \vdots & \vdots & \vdots & \vdots & \vdots & \vdots & \vdots & \vdots & \vdots \\
0 & 0 & 0 & \vdots & 1 & 0 & 0 & 0 & \vdots & 0 & 0 & 0 & 0 & \vdots & 0 \\
0 & 0 & 0 & \vdots & 0 & 1 & 0 & 0 & \vdots & 0 & 0 & 0 & 0 & \vdots & 0 \\
0 & 0 & 0 & \vdots & 0 & -\overline{a}_\alpha & 1 & 0 & \vdots & 0 & 0 & 0 & 0 & \vdots & 0 \\
0 & 0 & 0 & \vdots & 0 & 0 & -\overline{a}_\alpha & 1 & \vdots & 0 & 0 & 0 & 0 & \vdots & 0 \\
\vdots & \vdots & \vdots & \vdots & \vdots & \vdots & \vdots & \vdots & \vdots & \vdots & \vdots & \vdots & \vdots & \vdots & \vdots \\
0 & 0 & 0 & \vdots & 0 & 0 & 0 & 0 & \vdots & 1 & 0 & 0 & 0 & \vdots & 0 \\
0 & 0 & 0 & \vdots & 0 & 0 & 0 & 0 & \vdots & 0 & 1 & 0 & 0 & \vdots & 0 \\
0 & 0 & 0 & \vdots & 0 & 0 & 0 & 0 & \vdots & 0 & -a^M_\alpha & 1 & 0 & \vdots & 0 \\
0 & 0 & 0 & \vdots & 0 & 0 & 0 & 0 & \vdots & 0 & 0 & -a^M_\alpha & 1 & \vdots & 0 \\
\vdots & \vdots & \vdots & \vdots & \vdots & \vdots & \vdots & \vdots & \vdots & \vdots & \vdots & \vdots & \vdots & \vdots & \vdots \\
0 & 0 & 0 & \vdots & 0 & 0 & 0 & 0 & \vdots & 0 & 0 & 0 & 0 & \vdots & 1
\end{bmatrix}, \tag{8.54}
$$

where in both matrices

$$a_\alpha = 1 - i\Delta t(\varepsilon_\alpha - \mu), \tag{8.55}$$

$$\overline{a}_\alpha = 1 + i\Delta t(\varepsilon_\alpha - \mu), \tag{8.56}$$

and $a_\alpha^M = 1 - \frac{\beta}{M}(\varepsilon_\alpha - \mu)$ is the same as in Eq. (8.48). For better visualization of the structure of the matrices in Eqs. (8.53) and (8.54), we used bold shrift for the forward, backward, and Matsubara time blocks.

In the interacting case, in both Keldysh and Kadanoff–Baym representations one can write down the expression for the inverse interacting GF matrix by subtracting the self-energy matrix from the noninteracting GFs matrix defined by (8.53) or (8.54):

$$G_{t_1 t_2}^{(\alpha)-1} = G_{t_1 t_2}^{(0)(\alpha)-1} - \Sigma_{t_1 t_2}^{(\alpha)}. \tag{8.57}$$

The GF defined in Eq. (8.57) is used in further calculations.

8.2 Nonequilibrium DMFT

Now, after we became familiar with the main features of the nonequilibrium many-body formalism, we are ready to use it to formulate the nonequilibrium DMFT approach (we follow mostly the presentation line in Ref. [14]).

8.2.1 General Formulation

Let us assume that the Hubbard model includes an external time-dependent perturbation field, which we incorporate into the kinetic energy term (see below). In this case, the hopping parameters v_{ij} are time-dependent, and the kinetic part of the Hamiltonian becomes $-\sum_{i,j,\sigma} v_{ij}(t) c_{i\sigma}^+ c_{j\sigma}$. Since in the excited system, the time translation invariance is broken (except in some particular cases, like in long-lived, quasi-equilibrium, steady states), the nonequilibrium DMFT (we use abbreviation neDMFT below) GF, self-energy, and dynamical mean-field Δ are functions of two-time arguments t and t' on the Kadanoff–Baym or Keldysh contour, instead of one $t - t'$ variable on the real branch, as in the case of equilibrium.

The system of the neDMFT equations can be rather straightforwardly written by using the same arguments as in the equilibrium case. Namely, one can show that in infinite dimensions the electron self-energy is local in space,

$$\Sigma_{ij}(t,t') = \delta_{ij}\Sigma_i(t,t'). \tag{8.58}$$

In this case, the lattice GF satisfies the equation

$$(i\partial_t + \mu)G_{ij}(t,t') - \sum_l v_{il}(t)G_{lj}(t,t') - \int_C \Sigma_i(t,t'')G_{ij}(t'',t')dt'' = \delta_C(t,t')\delta_{ij}, \tag{8.59}$$

and one can immediately write down the formal solution for the inverse lattice GF:

$$G_{ij}^{-1}(t,t') = \left[\delta_{ij}(i\partial_t + \mu) - v_{ij}(t)\right]\delta_C(t,t') - \Sigma_i(t,t'). \tag{8.60}$$

The diagonal part of the inverse of this matrix is local GF $G_i(t,t')$.

Next, one can write the nonequilibrium version of the DMFT Dyson equation for the local functions:

$$G_i^{-1}(t,t') = \mathcal{G}_i^{-1}(t,t') - \Sigma_i(t,t'), \tag{8.61}$$

where the inverse dynamical mean-field function is

$$\mathcal{G}_i^{-1}(t,t') = (i\partial_t + \mu) - \Delta_i(t,t'). \tag{8.62}$$

Equation (8.61) can be formally solved as:

$$\mathcal{G}_i(t,t') = \int_C dt'[\delta + G_i\Sigma_i]^{-1}(t,t'')G_i(t'',t'). \tag{8.63}$$

Finally, the last (impurity) equation for the local GF $G_{ij}(t,t')$ has the same form as in the equilibrium case, but with the action

$$S_i = -\sum_\sigma \int_C dt \int_C dt' c_\sigma^+(t)\Delta_i(t,t')c_\sigma(t') - i\int_C dt H_{\text{loc}}(t), \tag{8.64}$$

where $H_{\text{loc}}(t)$ is the interacting part of the Hamiltonian. In the simplest case of one-orbital Hubbard model $H_{\text{loc}}(t) = Un_\uparrow(t)n_\downarrow(t)$.

Equations (8.60)–(8.62) and (8.64) are the system of neDMFT equations. In the next subsection, we discuss early developments and applications of this theory, and in Sect. 8.2.3 we proceed with recent advances in the field, including new results on the nonequilibrium properties of the Hubbard model.

8.2.2 *Initial Developments: Models and External Potentials*

To perform practical calculations, one needs to choose an external potential that will be incorporated into the neDMFT equations. This is a nontrivial task, since such a potential should be both gauge-invariant and the translational invariance should be preserved, a crucial assumption in DMFT (the self-energy is site-independent). The first neDMFT approach was proposed by Schmidt and Monien in 2002 [11]. The authors considered the case of the half-filled Hubbard model on Bethe lattice with a periodic scalar external potential term $e\sum_{i,\sigma}\Phi\left(\vec{R}_i, t\right)c^+_{i\sigma}c_{i\sigma}$, where $e\Phi\left(\vec{R}_i, t\right) = A\cos\left(\Omega t - \vec{q}\vec{R}_i\right)$, in the limit of small q's (the potential slowly changes in space). More explicitly, they focused on the properties of the nonequilibrium steady state of the system, since it is known that according to the Floquet theorem, a periodically driven system can reach such a state [23]. Since the field is periodic in time and each space coordinate, and the energy and momentum are conserved up to multiples of Ω and q [23], the authors derived the equations for the frequency and momentum-dependent GF and for the other DMFT functions. The momentum–frequency Fourier transformation for the steady-state GF has the following form:

$$G^{\alpha\beta}_{ij}(t, t') = \sum_{\vec{k},n} \int \frac{d\omega}{2\pi} e^{-i\omega t} e^{i(\omega - n\Omega)t'} e^{i\vec{k}\vec{R}_i} e^{-i\left(\vec{k}-n\vec{q}\right)\vec{R}_j} G^{\alpha\beta}_{on}\left(\omega, \vec{k}\right) \tag{8.65}$$

(α,β = $\pm$ are the time-branch indices on the Keldysh contour). Then, the Dyson equation that connects the GF, self-energy, and the dynamical mean-field $\mathcal{G}$ can be written in terms of the Fourier coefficients:

$$G^{\alpha\beta}_{on}\left(\omega, \vec{q}\right) = G^{(0)\alpha\beta}_{on}\left(\omega, \vec{q}\right) + \sum_{\gamma,\delta,l,m} \gamma\delta G^{\alpha\gamma}_{ol}\left(\omega, \vec{q}\right)\Sigma^{\gamma\delta}_{lm}\left(\omega, \vec{q}\right)G^{\delta\beta}_{mn}\left(\omega, \vec{q}\right). \tag{8.66}$$

The noninteracting GF $G^{(0)\alpha\beta}_{ij}(t, t')$ that define $G^{(0)\alpha\beta}_{on}\left(\omega, \vec{q}\right)$ can be found analytically and expressed in terms of the Bessel functions J_1. In the case of used in paper [11] Bethe lattice, the relation between the GF and the dynamical mean-field transform components can be found analytically:

$$\begin{aligned}\left(\mathcal{G}^{-1}\right)^{\alpha\beta}_{nm}\left(\omega, \vec{q}\right) &= \alpha\delta_{\alpha\beta} \\ &\times\left[(\omega - n\Omega)\delta_{nm} - \frac{A}{2}\left(\delta_{n,m+1} - \delta_{n,m-1}\right) - \alpha\beta t^2 \cos\left((n-m)q\right)G^{\alpha\beta}_{nm}\left(\omega, \vec{q}\right)\right].\end{aligned} \tag{8.67}$$

For the impurity solver, the IPT approximation was used. It gives:

$$\begin{aligned}\Sigma^{\alpha\beta}_{0n}\left(\omega, \vec{q}\right) &= \alpha\beta U^2 \int_{-\infty}^{\infty} \frac{d\omega_1}{2\pi} \\ &\times \int_{-\infty}^{\infty} \frac{d\omega_2}{2\pi} \sum_{m,l} \mathcal{G}^{\beta\alpha}_{l0}\left(\omega_1, \vec{q}\right) \mathcal{G}^{\alpha\beta}_{0,n+l+m}\left(\omega_1 + \omega_2, \vec{q}\right) \mathcal{G}^{\alpha\beta}_{0,-m}\left(\omega - \omega_2, \vec{q}\right).\end{aligned} \tag{8.68}$$

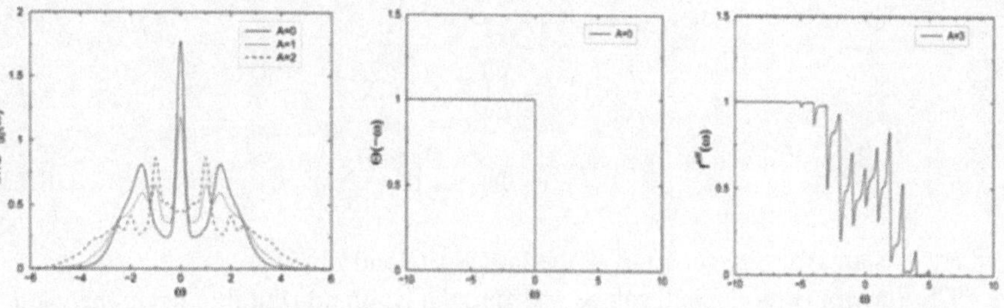

Fig. 8.2 The frequency dependence of the 00-component of ImG (left) and the effective charge distribution function (center and right) for different values of the field strength A at $U = 3$ (units $\Omega = 1$, see the text). From Ref. [11]

Using the equations above, the authors obtained several interesting results for the case of uniform fields. In particular, they found the field strength- (A-) dependence of the effective spectral density, defined as the average over harmonics:

$$A^{\mathrm{eff}}(\omega) = \frac{1}{\pi} ImG_{00}^{\mathrm{av}}(\omega) = \sum_l J_l^2(A) \frac{1}{\pi} Im\widetilde{G}(\omega - l\Omega) \tag{8.69}$$

($\widetilde{G}$ is the static GF). They also analyzed the dependence of the effective distribution function $f^{\mathrm{eff}}(\omega)$ on A. The distribution function is defined by an equation that connects it with the effective spectral density $A^{\mathrm{eff}}(\omega)$ and the 00 component of Keldysh GF:

$$G_{00}^{K}(\omega) = -2\pi i\left(1 - 2f^{\mathrm{eff}}(\omega)\right)A^{\mathrm{eff}}(\omega). \tag{8.70}$$

The results for $A^{\mathrm{eff}}(\omega)$ and $f^{\mathrm{eff}}(\omega)$ are shown in Fig. 8.2. As it follows from the left Figs. 8.2, in presence of perturbation the spectral function is more complex due to the sidebands weighted by the Bessel functions. With A increasing, the number of sidebands grows. Another remarkable result—a strong suppression of the zero-energy quasi-particle peak at large A's. The external potential also significantly modifies the effective distribution function (compare the central and right Figs. 8.2). Modifications of the distribution function at the energy side bands near the Fermi edge correspond to emission and absorption of energy quanta Ω [11].

Unfortunately, the model used in Ref. [11] does not capture correctly the dissipation mechanism, essential for the nonequilibrium steady states, and the Hamiltonian with the scalar external potential term $e\sum_{i,\sigma}\Phi\left(\vec{R}_i, t\right)c_{i\sigma}^{+}c_{i\sigma}$ is not gauge-invariant, i.e., the results may depend on the chosen gauge. In detail, the Hamiltonian should not change after the transformations

$$c_{j\sigma} \to c_{j\sigma} e^{\frac{ie}{\hbar c}\chi\left(\vec{R}_j,t\right)}, \tag{8.71}$$

$$\vec{A}\left(\vec{r},t\right) \to \vec{A}\left(\vec{r},t\right) + \vec{\nabla}\chi, \tag{8.72}$$

$$\Phi\left(\vec{r},t\right) \to \Phi\left(\vec{r},t\right) - \frac{\partial \chi}{\partial t}, \tag{8.73}$$

where χ is an arbitrary function of the lattice site and time.

The electromagnetic field can be added to the Hubbard Hamiltonian in the gauge-invariant way in the form of the Peierls substitution [24]

$$t_{ij} \to t_{ij\sigma} e^{-\frac{ie}{\hbar}\int_{\vec{R}_i}^{\vec{R}_j} d\vec{r}\vec{A}\left(\vec{r},t\right)} \equiv v_{ij}(t), \tag{8.74}$$

where $\vec{A}\left(\vec{r},t\right)$ is the vector potential. One can also include an additional scalar-potential term $e\sum_{i,\sigma}\Phi\left(\vec{R}_i,t\right)c^+_{i\sigma}c_{i\sigma}$ in the Hamiltonian. However, as discussed above, being used alone this term makes the Hamiltonian gauge non-invariant. Below we consider the case with no scalar potential.

Since DMFT corresponds to the spatially homogeneous case, the vector potential $\vec{A}\left(\vec{r},t\right)$ should be also spatially homogeneous. For such a potential, the Peierls substitution (8.74) corresponds to the following change of the free-electron dispersion in the kinetic part of the Hamiltonian:

$$\varepsilon\left(\vec{k}\right) \to \varepsilon\left(\vec{k} - \frac{e}{\hbar}\vec{A}(t)\right), \tag{8.75}$$

familiar from the classical electrodynamics. In the case of hypercubic lattice, it is natural to choose a symmetric vector potential pointing along the diagonal of the unit cell:

$$\vec{A}(t) = A(t)(1, 1, \ldots, 1). \tag{8.76}$$

This gives a simple formula for the dispersion (8.75):

$$\begin{aligned} \varepsilon\left(\vec{k} - \frac{e}{\hbar}\vec{A}(t)\right) &= -2t\sum_{\alpha=1}^{d} \cos\left(ak_\alpha - \frac{e}{\hbar}A(t)\right) \\ &= \varepsilon\left(\vec{k}\right)\cos\left(\frac{e}{\hbar}A(t)\right) + \bar{\varepsilon}\left(\vec{k}\right)\sin\left(\frac{e}{\hbar}A(t)\right), \end{aligned} \tag{8.77}$$

where

$$\overline{\varepsilon}\left(\overrightarrow{k}\right) = -\frac{2t^*}{\sqrt{2d}}\sum_{\alpha=1}^{d}\sin\left(ak_\alpha\right). \tag{8.78}$$

Now, since the "spectrum" depends on two functions ε and $\overline{\varepsilon}$, one can introduce two-energy DOS, $D(\varepsilon,\overline{\varepsilon}) = \sum_k \delta(\varepsilon - \varepsilon(k))\delta(\overline{\varepsilon} - \overline{\varepsilon}(k))$, which in the case of hypercubic lattice is a product of two Gaussian functions [25]:

$$D(\varepsilon,\overline{\varepsilon}) = \frac{1}{\sqrt{2\pi t^*}}e^{-\frac{\varepsilon^2}{2t^{*2}}} \times \frac{1}{\sqrt{2\pi t^*}}e^{-\frac{\overline{\varepsilon}^2}{2t^{*2}}}. \tag{8.79}$$

Analytical results for the DOS can be also obtained for some other high-dimensional lattices, like generalized fcc and bcc lattices, where $\varepsilon^{\mathrm{fcc}}\left(\overrightarrow{k}\right) = \frac{1}{2\sqrt{d(d-1)}} = \sum_{\alpha=2}^{d}\sum_{\beta=1}^{\alpha-1}\cos\left(k_\alpha\right)\cos\left(k_\beta\right)$ and $\varepsilon^{\mathrm{bcc}}\left(\overrightarrow{k}\right) = -\frac{8}{2\sqrt{d(d-1)(d-2)}} = \sum_{\alpha=3}^{d} \times \sum_{\beta=2}^{\alpha-1}\sum_{\gamma=1}^{\beta-1}\cos\left(k_\alpha\right)\cos\left(k_\beta\right)\cos\left(k_\gamma\right)$ [26].

For the first time, potential (8.76) and DOS (8.79) were used [25] to study the nonequilibrium properties of the infinite-dimensional Falicov–Kimball model, a simplified version of the Hubbard model with "frozen" spin-down electrons (zero hopping matrix elements for the spin-down electrons). The advantage of considering such an approximated case (that at the same time gives a lot of insights on the nonequilibrium properties of strongly correlated systems) is that the impurity problem can be solved exactly. In the discussion below, we use the notations $c_{i\uparrow} \equiv c_i$, $c_{i\downarrow} \equiv f_i$ for the electron spin operators. The exact solution of the Falicov–Kimball model is possible because the spin-down ("f-electron") density operator $f_i^+ f_i$ commutes with the Hamiltonian. Thus, $f_i^+ f_i$ corresponds to a conserved quantity and one can trace out the f-electronic states in the partition function by putting there $f_i^+ f_i = 0$ and 1 [27]. This allows one to calculate the impurity GF from the path-integral equation with the action (8.64) analytically:

$$\begin{aligned} G^{\mathrm{imp}}(t,t') &= (1 - \langle w_1\rangle)\left[(i\partial_t + \mu)\delta_C(t,t') - \Delta(t,t')\right]^{-1} \\ &\quad + \langle w_1\rangle\left[(i\partial_t + \mu - U)\delta_C(t,t') - \Delta(t,t')\right]^{-1}, \end{aligned} \tag{8.80}$$

where

$$\langle w_1\rangle = \frac{1}{N}\sum_i f_i^+ f_i \tag{8.81}$$

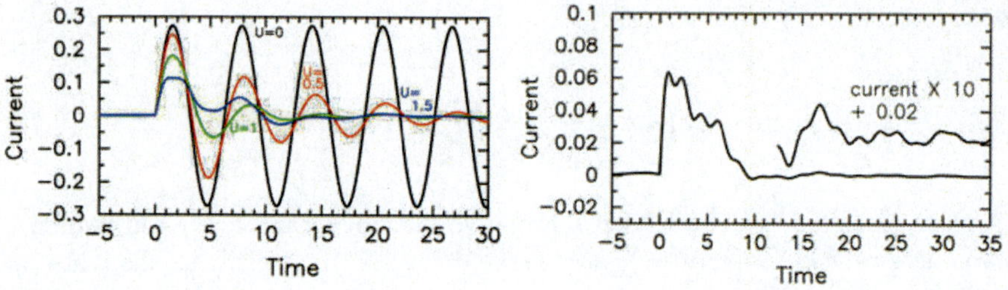

Fig. 8.3 The time-dependence of the current in the case of half-filled Falicov–Kimball model when constant electric field $E = t$ is turned on at time 0. Left: The results for different values of U/t (the critical value of MIT is $\sqrt{2}t$) and temperature $T = 0.1t$. Right: The same for the "deep" insulating phase ($\frac{U}{t} = 2$). (Reprinted Figs. 2 and 4 with permission from (J. K. Freericks, V. M. Turkowski, and V. Zlatić, Phys. Rev. Lett. **97**, 266408 (2006)). Copyright 2006 by the American Physical Society. https://doi.org/10.1103/PhysRevLett.97.266408)

is the average number of f-electrons per site. To get the neDMFT solution for the Falicov–Kimball model, one has to solve the system of Eqs. (8.80), (8.60), and (8.61).

Formulation of the gauge-invariant approach above was done in work [12], where it was also applied to study the transport properties of the Falicov–Kimball model on the hypercubic lattice perturbed by an external pulse. The current was calculated by using the formula

$$\langle j(t) \rangle = -ei \sum_{\vec{k}} v\left(\vec{k} - eEt\right) G^{<}\left(t, t, \vec{k}\right), \tag{8.82}$$

where the momentum summation can be substituted by the two-energy integration with the double-energy DOS (8.79). Indeed, since the velocity components are $v_\alpha\left(\vec{k}\right) = \frac{2t^*}{\sqrt{2d}} \sin\left(ak_\alpha - \frac{e}{\hbar}A(t)\right)$, one can perform the energy integration in Eq. (8.82) by using $\sum_{\alpha=1}^{d} \sin\left(ak_\alpha - \frac{e}{\hbar}A(t)\right) = \overline{\varepsilon}\left(\vec{k}\right) \cos\left(\frac{e}{\hbar}A(t)\right) - \varepsilon\left(\vec{k}\right) \sin\left(\frac{e}{\hbar}A(t)\right)$ (see Eqs. (8.78) and (8.79)). The results for the time-dependent current induced by a constant electric field turned on at zero time are shown in Fig. 8.3.

As it follows from this figure, the current demonstrates Bloch oscillations with frequency $\omega_B = eE$. The oscillations become more damped as U increases. However, the oscillations in the metallic phase (left Fig. 8.3) survive until surprisingly long times, ~$100/t$. As the system goes through the MIT, the magnitude of the current becomes suppressed due to correlation effects, though rather irregular small-amplitude "Bloch" oscillations still survive at rather long times (for a related results for the Hubbard model see Ref. [28] and the discussion below).

Thus, in Ref. [12] the authors performed the first consistent DMFT analysis of the role of correlation effects in the response of a strongly correlated system. This paper was followed by several other works where the conducting properties of the Falicov–Kimball model, including the Bloch oscillations, were analyzed in different cases

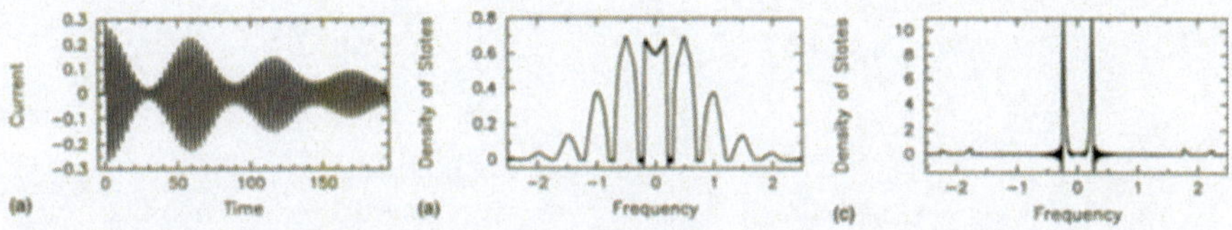

Fig. 8.4 Left: the same as in Fig. 8.3 for $U = 0.125t$, $E = 2t$. Center and right: local DOS in the Falicov–Kimball model in infinite dimensions in the long-time limit for $U = 0.5t$ and $E = 0.5t$ (center) and $E = 2t$ (right). (Reprinted Figs. 8a, 9a, and 9c with permission from (J.K. Freericks, Phys. Rev. B **77**, 075109 (2008)). Copyright 2008 by the American Physical Society. https://doi.org/10.1103/PhysRevB.77.075109)

[29, 30], including the steady-state limit [30]. In the last paper, it was shown that in case of a strong field the system demonstrates beatings in the Bloch oscillations with period $2\pi/U$ (Fig. 8.4).

It is easy to see from Fig. 8.4 that the beatings are the result of oscillations with two combined frequencies, $\omega_B \pm U/2 \equiv eE \pm U/2$. To confirm this conclusion, calculations of the long-time (steady state-limit) DOS were performed [30]. In the case of $U = 0$, the DOS is a set of delta-function peaks at frequencies eEn, where n is integer (Wannier–Stark ladder [25]). In the interacting case, the peaks at Bloch frequencies are broadened with the width ~U. As it follows from Fig. 8.4, at nonzero U the central band is split into two sub-bands centered at $\pm U/2$. The corresponding peaks at frequencies $\pm U/2$ become more pronounced and separated from the other peaks as E increases (central and right Fig. 8.4). Thus, it is obvious that the beats of the current take place due to the transitions between these sub-bands.

Similar to the work [30] structure of the steady-state DOS was obtained in Ref. [26], where a nonequilibrium Floquet DMFT formalism for systems in presence of an AC field was developed and applied to the Falicov–Kimball model. Among other results, the authors have found that as the field strength increases midgap states in the DOS emerge (Fig. 8.5). This phenomenon corresponds to a photoinduced insulator-to-metal transition. In the following paper of the group [31] it was shown that the optical conductivity in the steady state of the excited system exhibits dip and kink structures around the pulse frequency and that the system demonstrates a negative attenuation (gain) due to the inversion of population. Recently, it was shown with the real-space Floquet DMFT [32] that in the high-frequency limit the system can be effectively described by the Harper–Hofstadter Falicov–Kimball model.

Important results on the nonthermal charge equilibration in the Falicov-Kimball model were obtained by Eckstein and Kollar [33–35]. In particular, in Ref. [34] they have shown that after the interaction quench, the system reaches a nonthermal steady state. As results of the calculations at different Us show, after the interaction quench the double occupancy evolves to values different from the thermal-state ones (Fig. 8.6). Even for two initial states with the same energy (top and middle Fig. 8.6), the steady-state double occupancies are different. It suggests that memory effects play an important role in the equilibration.

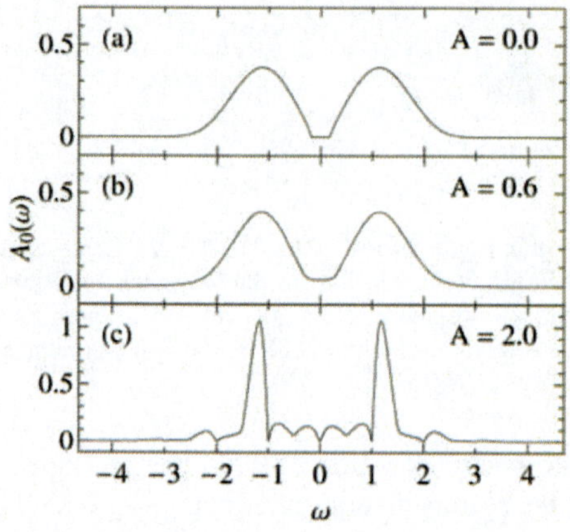

Fig. 8.5 Zero mode of the spectral function (see the r.h.s. of Eq. (8.65)) of the Falicov–Kimball model at half-filling and $U = 2.2t$ in presence of the external electric field with potential $A(t) = \frac{A}{ea}\sin(\Omega t)$ at different values of the field strength A. (Reprinted Fig. 7 with permission from (N. Tsuji, T. Oka, and H. Aoki, Phys. Rev. B 78, 235124 (2008)). Copyright 2008 by the American Physical Society. https://doi.org/10.1103/PhysRevB.78.235124)

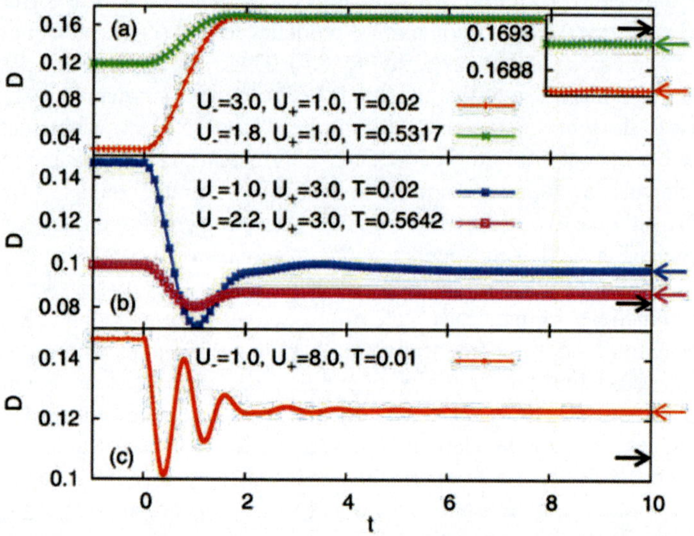

Fig. 8.6 Time-dependence of the double occupancy in the half-filled Falicov–Kimball model with semielliptic free-electron DOS after the interaction quench from U_- to U_+ for different values of U and temperatures (in units of the half-bandwidth). The arrows mark the double occupancy in the thermal state for the final U with the same density and internal energy, and the other arrows mark the long-time stationary double occupancies. In the inset—magnified dependencies of the double occupancy at long times. (Reprinted Fig. 1 with permission from (M. Eckstein, and M. Kollar, Phys. Rev. Lett. **100**, 120404 (2008)). Copyright 2008 by the American Physical Society. https://doi.org/10.1103/PhysRevLett.100.120404)

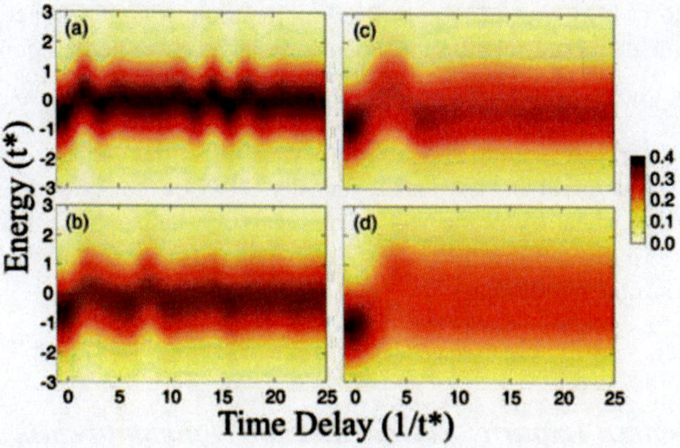

Fig. 8.7 Tr-PES intensity in the Falicov–Kimball model as a function of the probe pulse delay time in both metallic $U = 0.5t$ (**a**) and $U = 1.0t$ (**b**) and insulating $U = 1.5t > U_c = \sqrt{2}t$ (**c**), $U = 2t$ (**d**) phases. The calculations were performed at the field strength $E = 2t$ and the probe-pulse width $\delta = 1/t$. (Reprinted Fig. 2 with permission from (B. Moritz, T.P. Devereaux, and J.K. Freericks, Phys. Rev. B **81**, 165112 (2010)). Copyright 2010 by the American Physical Society. https://doi.org/10.1103/PhysRevB.81.165112)

In another related study [36], the authors analyzed dependence of the excitation spectrum of the Falicov–Kimball under slow change of the hopping and showed that there is a qualitative difference in behaviors of gapless and gapped systems: contrary to the gapped case, in gapless systems at sufficiently smooth ramps the spectrum asymptotic are ramp-independent and depends only on the initial spectrum of the system.

Before moving to the nonequilibrium properties of the Hubbard model, the main model of this book, we would like to mention the results of some other, last decade, studies of the Falicov–Kimball model. The time-resolved photoemission spectrum (tr-PES) and relaxation of the Falicov-Kimball model were considered in Refs. [37, 38] (for over-review, see Ref. [39]). In Ref. [37], the tr-PES response function was calculated in the case of applied constant pump-pulse and Gaussian probe-pulse fields. It was demonstrated that the tr-PES spectra of metal and insulating systems are quite different (Fig. 8.7). In Ref. [40], dynamics in the 2D Falicov–Kimball model after the interaction quench was studied by using a nonequilibrium DCA, a "cousin" of neDMFT. Though DCA did not give thermalization in the system, it was found that a quasi-thermal distribution is reached within the energy intervals (intervals with different chemical potentials) corresponding to the momentum patch-dependent sub-bands of the spectral function. In Ref. [41] (see also Ref. [39]) nonequilibrium properties of the driven by electric field Falicov–Kimball model in the charge-density wave-ordered state were analyzed. Among several interesting results, there was found emergence of subgap states in the DOS as temperature increased. It was argued that such states may significantly affect the transport properties of correlated systems, especially in the case of applied strong DC electric fields. In Ref. [42], out-of-time-ordered correlation (OTOC) functions in the Falicov–Kimball model in infinite dimensions in cases of local equilibrium and out-of-nonequilibrium were calculated. The OTOC functions, $-\left\langle \left[\widehat{A}(t), \widehat{B}(0)\right]^2 \right\rangle$, help to quantify how strong the time-evolving quantities depend on the initial conditions. It was found that the density–density OTOC is the largest at intermediate Us, close to the critical U_c (of the MIT). This result may help to better understand the dependence of dynamics in strongly correlated materials near or in the strange-metal phase on the initial conditions.

8.2.3 *Modern Impurity Solvers and the Nonequilibrium Properties of the Hubbard Model*

The multi-orbital impurity problem in the case of the Hubbard model is defined by the following action:

$$S = -\sum_{\alpha\beta} \int_C dt \int_C dt' c_\alpha^+(t) \Delta_{\alpha\beta}(t,t') c_\beta(t') - i \int_C dt H_{\text{loc}}(t), \tag{8.83}$$

where the function under the last integral is the interacting part of the Hamiltonian $H_{\text{loc}}(t) = \sum\limits_{\alpha,\beta} U^{\alpha\beta} n_\alpha(t) n_\beta(t)$ and $\Delta_{\alpha\beta}(t,t')$ is the dynamical mean-field matrix connected with the dynamical mean-field function as:

$$\mathcal{G}_{\alpha\beta}^{-1}(t,t') = (i\partial_t + \mu)\delta_{\alpha\beta} - \Delta_{\alpha\beta}(t,t') \tag{8.84}$$

(we use combined spin-orbital indices).

Action (8.83) can be derived from the impurity Hamiltonian

$$H_{\text{imp}}(t) = \sum_{\alpha,\beta} U^{\alpha\beta} n_\alpha(t) n_\beta(t) + \sum_\nu \varepsilon^\nu(t) c_\nu^+ c_\nu + \sum_\nu \left[V^{p\nu}(t) d_p^+ c_\nu + h.c. \right] \tag{8.85}$$

by integrating out the bath states c_ν. This gives the hybridization function

$$\Delta_{\alpha\beta}(t,t') = \sum_\nu V^{\alpha\nu}(t) G_\nu^{(0)}(t,t') V^{\beta\nu *}(t'), \tag{8.86}$$

where

$$G_{\nu}^{(0)}(t,t') = \text{i}f(\varepsilon^{\nu}(0) - \theta_C(t,t'))e^{-i\int_{t'}^{t} d\bar{t}\varepsilon^{\nu}(\bar{t})} \tag{8.87}$$

is the GF of noninteracting bath electrons. Thus, the Hamiltonian (8.85) transforms into the Hamiltonian that has two terms: the term that describes interaction of d-electrons with bath field $\Delta_{\alpha\beta}(t,t')$ and the Coulomb interaction term $H_{\text{loc}}(t)$. As a result, the corresponding action has the form (8.83).

Unfortunately, contrary to the case of the Falicov–Kimball model, the impurity problem for the Hubbard model cannot be solved analytically. Thus, reliable nonequilibrium numerical tools are needed. Several nonequilibrium DMFT solvers were developed, among them the state-of-art nonequilibrium CT-QMC solver [42, 43]. With the last approach, the neDMFT results for the Hubbard model acquired the needed level of accuracy. Before proceeding with discussion of the nonequilibrium properties of the Hubbard model, we rather briefly describe the main details of nonequilibrium CT-QMC approach and mention some other alternative tools. Namely, we discuss the weak- and strong coupling–expansion algorithms for CT-QMC. In the weak-coupling expansion CT-QMC [43–47], one uses the following local part of the Hamiltonian:

$$H_{\text{loc}}(t) = U(t)\left(n_{\uparrow} - \alpha_{\uparrow}\right)\left(n_{\downarrow} - \alpha_{\downarrow}\right) - \sum_{\sigma}(\mu - U(t)\alpha_{\bar{\sigma}})n_{\sigma}, \tag{8.88}$$

where in general U is time-dependent (e.g., in the case of a quenched interaction) and the shifts of the spin densities and chemical potential are introduced to avoid the sign problem. Then, in the impurity action (8.83) for the interacting part one puts

$$S_{\text{int}} = -i\int_C H_{\text{int}}(t)dt = -i\int_C U(t)\left(n_{\uparrow} - \alpha_{\uparrow}\right)\left(n_{\downarrow} - \alpha_{\downarrow}\right)dt, \tag{8.89}$$

and expanding the partition function in terms of S_{int}

$$Z = \sum_{n=0}^{\infty}\frac{(-1)^n}{n!}\int_C dt_1\ldots\int_C dt_n Tr\left[T_C e^{S_0} H_{\text{int}}(t_1)\ldots H_{\text{int}}(t_n)\right] \tag{8.90}$$

one obtains

$$\frac{Z}{Z_0} = \sum_{n=0}^{\infty} \times\frac{(-1)^n}{n!}\int_C dt_1\ldots\int_C dt_n U(t_1)\ldots U(t_n)\prod_{\sigma}\langle(n_\sigma(t_1) - \alpha_\sigma)\ldots(n_\sigma(t_n) - \alpha_\sigma)\rangle_{S_0}, \tag{8.91}$$

where $Z_0 = Tr[T_C e^{S_0}]$. Integration in Eq. (8.91) can be rewritten as a sum of weights w of the configurations $c = \{t_1 < \ldots < t_n\}$:

$$\frac{Z}{Z_0} = \sum_c w(c) \equiv \sum_c (-i)^n \int_C dt_1 \ldots \int_C dt_n U(t_1) \ldots U(t_n) \prod_\sigma detM_\sigma^{-1}, \quad (8.92)$$

where

$$M_{\sigma ij}^{-1} = -i\mathcal{G}_\sigma(t_i, t_j) - \alpha_\sigma \delta_{ij}. \quad (8.93)$$

Then, one can update the configuration in Eq. (8.92) by using the following equations for the proposed probabilities: $p^{\text{prob}}(n - 1 \to n) = dt_n/(2t_{\max} + \beta)$ (in the denominator—the length of the Kadanoff–Baym time contour) to add an nth interaction vertex at time t_n and $p^{\text{prob}}(n \to n - 1) = 1/n$ to remove nth vertex at time t_n. Similar to the equilibrium case, one can show (by multiplying the probabilities by the corresponding weights w) that the move acceptance ratio is

$$R = |U(t_n)| \frac{2t_{\max} + \beta}{n} \prod_\sigma \frac{detM_\sigma^{(n)-1}}{detM_\sigma^{(n-1)-1}}. \quad (8.94)$$

Next, to calculate GF (or other operator $\widehat{O}$ averages), one performs an expansion in the corresponding formulas similar to (8.90) with inserted product $d_\sigma d_\sigma^+$ (or other operators), and to measure an observable by Monte-Carlo sampling, one uses the formula $\left\langle \widehat{O} \right\rangle_{\text{MC}} = \sum_c \frac{O_c}{w_c}$, where w_c are the weights in the sum in the partition function (8.92) and O_c are the corresponding weights in the average expansion for $\widehat{O}$. The formula to obtain GF after the MC measurements is

$$G_\sigma(t, t') = \mathcal{G}_\sigma(t, t') + \int_C dt_1 \int_C dt_2 \mathcal{G}_\sigma(t, t_1)\mathcal{G}_\sigma(t_2, t') \left\langle i \sum_{i,j=1}^{n} \delta_C(t_1, t_i) M_{\sigma ij}(t, t_j) \delta_C(t_2, t_j) \right\rangle_{\text{MC}} \quad (8.95)$$

(we refer the reader for more details to papers [14, 43–47]).

In a similar way, one can perform calculations by using a strong-coupling expansion [44, 46, 48] in powers of the hybridization part of the impurity action. The partition function in this case is:

$$Z = \sum_{n=0}^{\infty} \frac{(-1)^n}{n!} \int_C dt_1 \ldots \int_C dt_n Tr\left[T_C e^{S_{\text{loc}} + S_{\text{bath}}} H_{\text{hyb}}(t_1) \ldots H_{\text{hyb}}(t_n)\right], \quad (8.96)$$

where $H_{\text{hyb}}(t) = \sum_{\nu}\left[V^{p\nu}(t)d_p^{+}c_{\nu} + h.c.\right]$. The trace over the noninteracting bath electrons in Eq. (8.96) can be easily evaluated. It gives $det\Delta^{-1}$, where Δ^{-1} is the inverse of the hybridization function matrix $\Delta_{\alpha_i'\beta_j}(t_i', t_j)$ that depends on time and flavor (spin, orbital,. . .) indices. The elements of this matrix can be interpreted as the annihilation of flavor β state at time t_j and creation of α' state at time t_i'. Thus, one can write the partition function in the form of the first sum in (8.92), with the weights of configurations with n creation and annihilation operators:

$$w(\{(t_1',p_1'),\ldots,(t_n',p_n');(t_1,p_1),\ldots(t_1,p_1)\})$$
$$= Tr_{\text{loc}}\left[T_C e^{-i\int_C d\bar{t}H_{\text{loc}}(\bar{t})} d_{p_1'}^{+}(t_1')d_{p_1}(t_1)\ldots d_{p_n'}^{+}(t_n')d_{p_1}(t_n)\right] dt_1\ldots dt_n' \frac{(-1)^n}{(n!)^2} det\Delta^{-1}, \tag{8.97}$$

where the trace over the impurity states can be calculated explicitly by using the matrix representation with the eigenbasis of $H_{\text{loc}}(0)$.

Recently, a modified hybridization–expansion CT-QMC solver CT-1/2-HYB-QMC was proposed [49], where the diagrammatic expansion is performed only in one of the spin channels, and the resulting opposite-spin single-particle problem is solved semi-analytically for each expansion diagram. Such an approximated scheme makes the dynamical sign problem less severe due to a smaller order of sampled diagrams and allows one to double the length of the real branches of the contour, i.e., to perform the analysis for much longer times comparing to standard CT-QMC. Two other popular solvers used in the neDMFT are weak-coupling [11, 50, 51] and strong-coupling PTs [52]. In the weak-coupling case, the most-often used approximation is the second-order PT, and in the strong-coupling case, it is the lowest order NCA, where the diagrams without crossing hybridization lines are summed up, and its simplest extension—OCA with one crossing hybridization line (for more details, see Ref. [14]). NCA and OCA propagators can be also used in the numerically exact Monte Carlo calculations [49, 53]. Rather recently, two new neDMFT solvers were developed: DMRG matrix product state (MPS) [54] and multiconfiguration time-dependent Hartree [55] impurity solvers.

Accuracy of different approximations is an important issue, even more clamant in the nonequilibrium case. One of the ways to test the accuracy of the solver is to compare the numerical results for the nonequilibrium GF and self-energy spectral moments with the exact analytical results [56, 57].

Below, we summarize the main results on the nonequilibrium properties for the Hubbard model found with neDMFT by following the order and presentation of review [14]. In each sub-subsection, we begin with results presented in work [14] and proceed with new results obtained in the last 6 years.

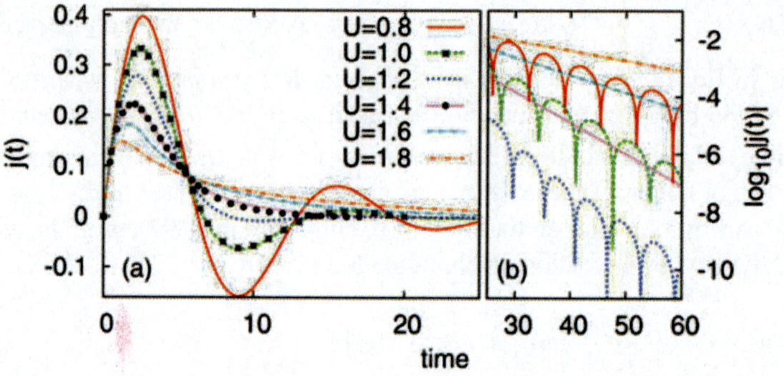

Fig. 8.8 Left: neDMFT results for the current as function of time in the infinite-dimensional Hubbard model on the hypercubic lattice after the constant field $E = 0.5t$ is turned in the case of different values of U ($T = 0.1t$, IPT solver). Right: the same in the logarithmic scale at longer times. (Reprinted Figs. 1a and 1b with permission from (M. Eckstein, and P. Werner, Phys. Rev. Lett. **107**, 186,406 (2011)). Copyright 2011 by the American Physical Society. https://doi.org/10.1103/PhysRevLett.107.186406)

8.2.3.1 Steady State in Isolated System

The Falicov–Kimball model results on the steady state of non-dissipative system were extended on the case of the Hubbard model in a static field in Ref. [58]. In this work, several important results were found: splitting of the Wannier–Stark peaks at strong field E, broadening of the Wannier–Stark ladder and evolution of the DOS toward the equilibrium DOS with U increasing. Calculations of the time-dependent current demonstrated [28] that in the case of weak fields the linear response behavior sets up only when the interaction exceeds the critical value (see Fig. 8.8). Another early study [59] showed that a steep gradient of the field potential enhances the carrier localization and, hence, increases the strength of the local carrier–carrier interaction. In the case of strong external field along one of the crystallographic directions the localization results in dimensional reduction. The reason for this is that when a strong field is applied, the change of the potential between the neighboring sites in the field direction is larger than any other energy scale in the system.

8.2.3.2 Steady-State Current in Dissipative Systems

It is important to remember that Bloch oscillations are a transient state, and the stationary state can be reached only when the system is coupled to a bath, otherwise, Joules heat leads to a time-dependent change of energy [14]. It was shown in Ref. [51] that in order to have a nonzero stationary current at long times the system needs to be coupled to an external bath, otherwise the current will decay to zero. Moreover, in the case of dissipation the electric-field energy is balanced by the dissipation of

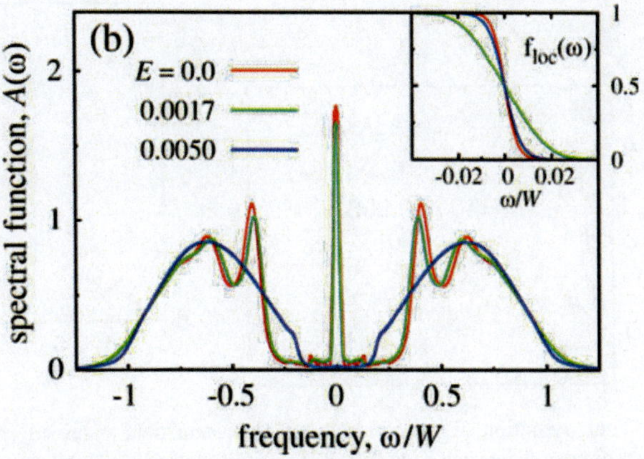

Fig. 8.9 neDMFT results for the spectral (main figure) and the distribution (inset) functions in the 3D Hubbard model in the case of different external fields at $U = 1.225W$, self-energy damping (dissipation) parameter $\Gamma = 0.00167W$ and bath temperature $T_b = 0.00167W$. (Reprinted Fig. 3b with permission from J. Li, C. Aron, G. Kotliar, and J.E. Han, Phys. Rev. Lett. **114**, 226, 403(2015). Copyright 2015 by the American Physical Society. https://doi.org/10.1103/PhysRevLett.114.226403)

energy due to the electron–electron and the electron-bath interaction, which results in a finite-temperature electron distribution and the finite current at long times. Quite counterintuitively, it was found that in the dissipative 3D Hubbard model at Us close to the critical value of MIT the electric field can transform a metallic state into an insulating steady state [60] (Fig. 8.9).

8.2.3.3 Dielectric Breakdown

Another important effect of the DC field is breakdown of the insulating phase due to creation of the doublon–holon pairs when the strength of the field F is above a threshold value F_{th}. Tunneling between sites at distance l, the electron gains energy lF which may be transformed in production of a doublon–holon pair. This results in current $\Gamma_{dh} \sim l\dot{d}$, where $\dot{d}$ is the rate of the production of the pair. For weak fields, the effect of tunneling takes place over many sites, with exponentially small rate, $j{\sim}F \cdot \exp\left[-F_{\text{th}}/F\right]$ [14]. The corresponding analysis for the Hubbard model with no dissipation was performed in Ref. [61] and later it was extended on the dissipative case [62]. A typical dependence of the current on the strength of the field in the insulating phase and the dependence $F_{\text{th}}(U)$ are shown in Fig. 8.10. As it was mentioned above, in the case of no coupling to thermostat the current can be only quasi-steady, since the energy increases with time proportional to F and the number of doublon–holon pairs grows.

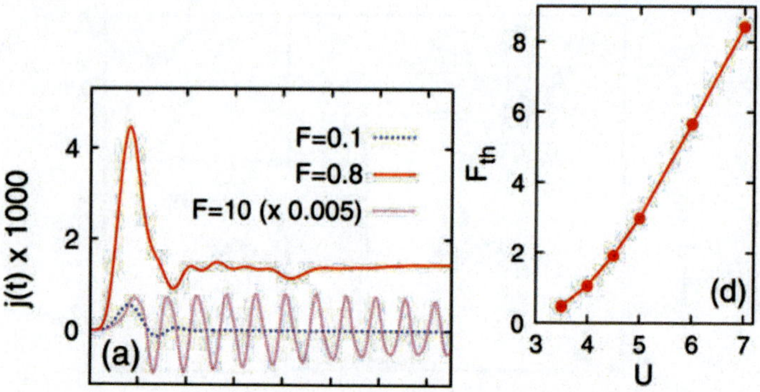

Fig. 8.10 Left: Time evolution of the current after a constant field is turned on in the case of different field magnitudes. Time changes from 0 to 14.5/t (half-filled Hubbard model, Gaussian DOS, $U = 5t$, $T = 0.1t$, NCA solver). Right: Dependence of the threshold electric field on U obtained with the OCA solver. (Reprinted Figs. 1a and 3d with permission from (M. Eckstein, T. Oka, and P. Werner, Phys. Rev. Lett. **105**, 146404 (2010)). Copyright 2010 by the American Physical Society. https://doi.org/10.1103/PhysRevLett.105.146404)

On the first sight, the result found in Ref. [62] is surprising—the doublon–holon part of the current almost does not depend on time, while the number of carriers increases. As it was argued in this work, the measured current consists mostly of the tunneling part $F{\cdot}\exp\,[-F_{\rm th}/F]$, therefore the thermally-excited and field-generated carriers must quickly reach the infinite-T distribution with zero average kinetic energy, and hence—with zero mobility. Indeed, as the authors showed by calculating the spectral function and comparing the electric current with the doublon–holon creation rate, the doublon–holon current becomes almost time-independent at long times. On the other hand, it was also demonstrated that in presence of thermostat carriers acquire a finite temperature, and hence—a finite-temperature distribution function and a finite kinetic energy. In this case, the current increases with time proportional to the number of excited carriers generated by the field.

8.2.3.4 Photoexcitations and Photodoping

The lifetime of doublons and holons is long, their creation (destruction), and hence change of the interaction energy to the quasi-equilibrium value can happen due to emission of spin excitations or changing kinetic energy of other excitations. As a result, thermalization of a photoexcited Mott insulator is rather slow [14]. It was found in Ref. [63] that the time evolution of the double occupancy after the system is excited with pulse $E_0 \sin{(\Omega t)} e^{-(t-2)^2}$ changes from a strongly damped transient regime at short times to an exponential thermalization at long times (Fig. 8.11). Time dependence of the double occupancy in Fig. 8.11 can be fit by an analytical expression

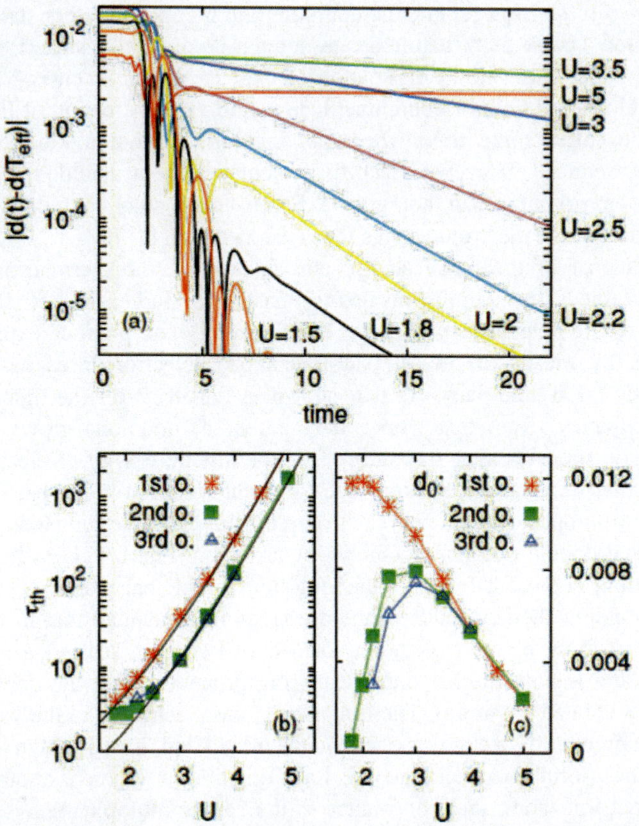

Fig. 8.11 (**a**) Time dependence of the double occupancy in the Hubbard at different values of U after a pulse with frequency $\Omega = \pi U/2$ is applied (Gaussian free-electron DOS, $T = 0.2t$, OCA solver). The curves correspond to the following values of E_0/t (chosen to give the effective temperature $T_{\text{eff}} = 0.5t$; for U changing from $5t$ to $1.5t$): 3.6701, 2.4397, 2.1325, 1.9123, 1.8198, 1.7761, 1.6632, and 1.2765. (**b**) Thermalization time and (**c**) d_0 as functions of U in the case of different solvers – OCA (squares), NCA (stars), and the third-order hybridization-expansion PT (triangles). (Reprinted Fig. 2 with permission from (M. Eckstein and P. Werner, Phys. Rev. B **84**, 035122 (2011)). Copyright 2011 by the American Physical Society. https://doi.org/10.1103/PhysRevB.84.035122)

$$d(t) = d(T_{\text{eff}}) + d_0 e^{-\frac{t}{\tau_{\text{th}}}}, \tag{8.98}$$

where $d(T_{\text{eff}})$ is the final value of double occupancy that can be obtained as the thermal expectation value for given pulse (T_{eff} is the corresponding effective temperature), and τ_{th} is the thermalization time. Thus, after an initial transient evolution, the double occupancy exponentially decreases. On the other hand, parameter τ_{th} exponentially increases with U (Fig. 8.11b), while d_0 that defines the dynamics of the initial (rapid) relaxation process have a more complex dependence on U (Fig. 8.11c).

Therefore, in the Mott regime the doublons and holons are rather stable (with ps lifetimes), and a pulse perturbation creates a long-lived metallic state. Long lifetime of these quasiparticles allows dissipation of their large kinetic energy to phonons before doublons and holons recombine. One would expect creation of Fermi-liquid quasiparticles after some time when the doublon and holon kinetic energy is significantly reduced. However, neDMFT calculations that included coupling to a dissipative environment did not give a Fermi-liquid state, but demonstrated a reconstruction of the electronic states [64] (see also Ref. [14]).

The influence of the pulse energy and fluence on the thermalization in the Hubbard model in the initially insulating phase was studied in Ref. [65]. It was found that excited electrons with large kinetic energy can produce extra doublon-holon pairs by means of the "impact-ionization" (electron loses its energy by creating a doublon-hole pair). As it is shown in Fig. 8.12 for the time-dependent double occupancy (from this work), the number of doublons grows during the thermalization process (after the pulse), i.e., the kinetic energy of electrons transforms into the potential (interaction) energy of the doublon-holon pairs. From the value of the pumped energy one can compute the effective temperature and the expected final double occupancy (shown by arrows in Fig. 8.12). As the results of the calculations show, for $U = 3.5t$ the long-time double occupancy is smaller than the corresponding thermal value. Thus, the authors concluded that in the case of strong perturbation the thermalization consists of two time scales—a fast (impact ionization) and a slow (higher order scattering processes after the doublon-holon pairs are not created anymore). The last stage is more sensitive to the values of the gap and the number of excited carriers. On the other hand, it was shown in Ref. [66] that in the multi-orbital case of materials $LaVO_3$ and $YTiO_3$ there is another channel of the impact ionization that contributes to the carrier multiplication—local spin-state transitions.

It was also found in Ref. [67] that presence of doublons and holons in the excited system creates low-energy screening modes and reduces the (Mott) gap. The coupling of the excited electrons to the low-energy modes accelerates the thermalization process. An important exact result related to the system relaxation was obtained recently by Kemper, Krishnamurthy, and Freericks [68] who proved that populations of the pulse-excited system do not change if the GF does not depend on the average time. This result is not only very helpful for the analysis of a (long-time) response, but also can be used to test the accuracy of the numerical propagation in time and the accuracy of other nonequilibrium tools for strongly correlated systems.

8.2.3.5 High-Harmonic Generation

Nonlinear response, including non-linear emission, is an important part of the dynamics of the system. In Refs. [69, 70], high-harmonic generation (HHG) (generation of harmonics with frequencies integer of the pulse frequency due to nonlinear effects) was studied in the case of Hubbard model in presence of field $E(t) = -E_0 \sin(\Omega t)$ by using Floquet DMFT. The HHG spectrum was calculated for the time-dependent current $\vec{j}(t)$. Namely, using the result for the current, the

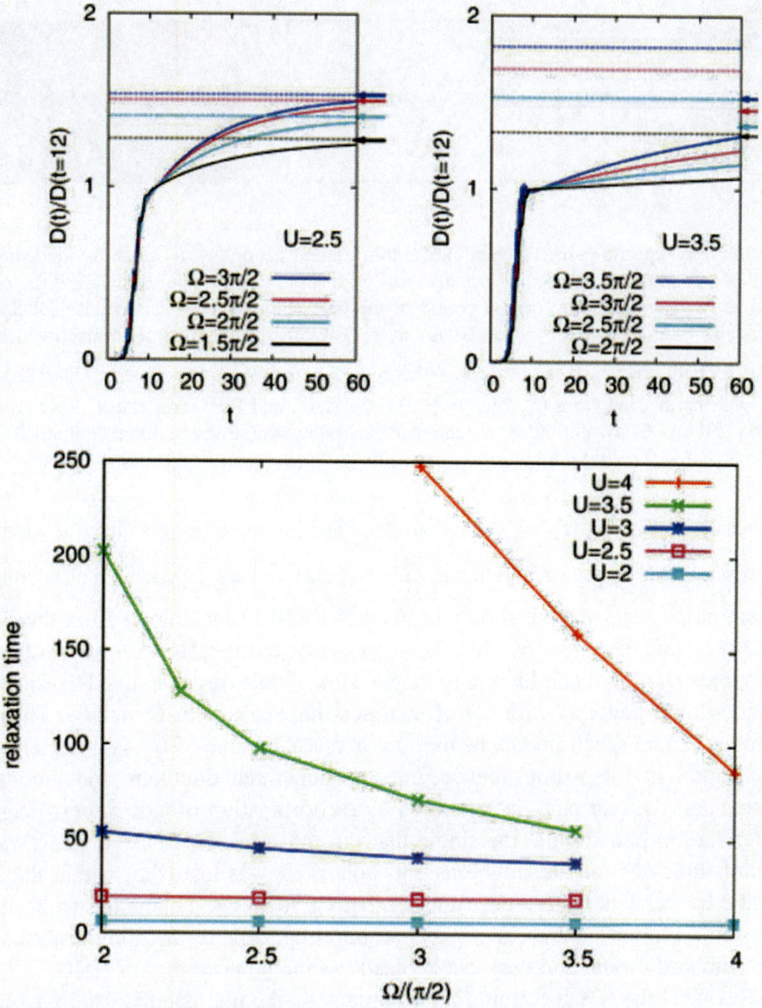

Fig. 8.12 Top: time evolution of the normalized doublon density in the Hubbard model at different *U*s (below and above the critical value) and at different values of the pulse frequency (horizontal lines are the expected thermal values; arrows indicate extrapolated long-time values that come from the exponential fit (bottom figure)). Bottom: Relaxation times τ obtained by fitting the normalized double occupancy to $a + be^{-t/\tau}$ (Gaussian DOS, initial temperature $= 0.2t$, Gaussian pulse with duration $\sqrt{6}/W$, NCA solver). (Reprinted Fig. 2 with permission from (P. Werner, K. Held, and M. Eckstein, Phys. Rev. B **90**, 235102 (2014)). Copyright 2014 by the American Physical Society. https://doi.org/10.1103/PhysRevB.90.235102)

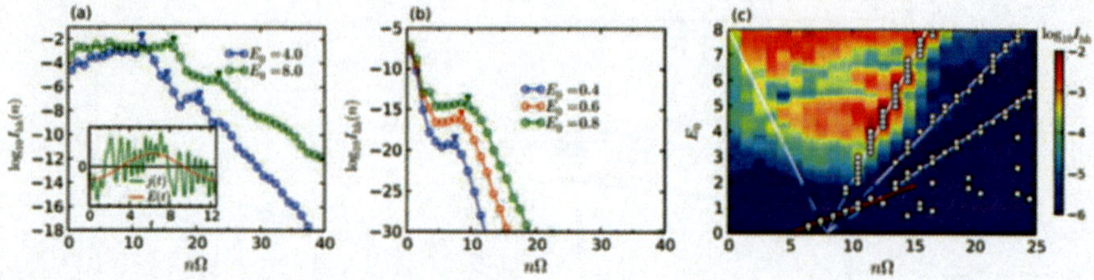

Fig. 8.13 HHG spectra in the Hubbard model for different strong (**a**) and weak (**b**) field strengths E_0 and as a continuous function of the strength (**c**). Inset in (**a**)—the pulse (red) and the current (green) as functions of time for one period of the field (Floquet DMFT, Gaussian DOS, pulse frequency $\Omega = 0.5t$, $U = 8t$, $T = 0.5t$, NCA solver). The calculations were performed by including the Buttiker bath imaginary part of the self-energy $Im\Sigma_{bath}(\omega) = -0.06t\sqrt{1-(\omega/5)^2}$. (Reprinted Fig. 1 with permission from (Y. Murakami, M. Eckstein, and P. Werner, Phys. Rev. Lett **121**, 057405 (2018)). Copyright 2018 by the American Physical Society. https://doi.org/10.1103/PhysRevLett.121.057405)

dipole acceleration $d\vec{j}(t)/dt$ was calculated. The last quantity was used to calculate the emission intensity at frequencies $n\Omega$: $I_{hh}(n\Omega) = \left|n\Omega\,\vec{j}(n\Omega)\right|^2$. The results for different pulse strengths are shown in Fig. 8.13 [69]. As it follows from this figure and other results of works [69, 70], the main source of the HHG in Mott insulators is the doublon–holon recombination. In the strong-field regime, the HH spectrum shows multiple plateaus with cutoff energies that scale with U and E_0. Thus, the experimental HH spectrum can be used to extract the value of U. As it is mentioned in the paper, in the strong-field regime the generated doublons and holons are localized and the mth plato is generated by recombination of mth nearest-neighbor doublon–holon pairs, while the single plato in the weak-field case is generated by recombination of itinerant doublons and holons. It was also shown that the Mott insulator has a wider HH spectrum comparing to a semiconductor with similar dispersion, i.e., correlation effect plays an important role in the nonlinear response of the Hubbard model and they can be used to tune new sources of HHG.

In Ref. [71] the HH spectrum for two systems—the one-band Holstein–Hubbard model for coupled electrons and plasmons and a two-orbital Hubbard model with the Hund's coupling—was calculated. In the first case, the authors wanted to analyze the role of the dynamical screening of the Coulomb interaction, and in the second—the role of local spin excitations in the HHG. Several interesting results were found. In particular, for the Hubbard–Holstein model it was shown that the spectrum reveals information about the bare and screened U and about the plasmon frequency, and in the two-orbital case there are string states that consist of local spin excitations and lead to an increase of the radiation intensity.

8.2.3.6 AC Fields and Isolated System

An AC field applied to an isolated system can also result in a nontrivial response. In particular, as it was shown in Ref. [72], the "nonadiabatic" switch-on of the AC field can produce a dynamical band-flipping (inverting the order of bands), corresponding to a negative-temperature distribution, and to change the effective interaction from repulsive to attractive, that may generate superconductivity [72]! In Ref. [73] it was shown analytically that in a periodically-driven system a quasiperiodic time evolution synchronized with driving force can emerge not only in the noninteracting, but also in a weakly-interacting system, with separated synchronization and evolution (to the state $T = \infty$) timescales. This situation, in some way, resembles the pre-thermalization dynamics in quenched systems. The statements above were supported by numerical results on the example of the Hubbard model with periodic-in-time Coulomb repulsion.

In Ref. [74], it was shown that there are long-lived prethermalized states in the insulating phase of the Hubbard model at presence of a periodic perturbation, including the resonance driving case, with large number of doublons and holons and a nonzero current. It was also shown that in the high driving-frequency limit, which corresponds to the noninteracting electrons, the problem of the dynamics of double occupancy can be mapped on a problem described by a simple effective Hamiltonian. Using the next-order correction to the high-frequency limit in the Hamiltonian allows one to describe the finite-frequency prethermalization plateau (in the case of near-resonant driving, see Ref. [75], where an asymmetric excitation spectrum with respect to the detuning was found).

The steady states of Mott insulator in the case of DC and AC fields were analyzed in Ref. [76]. In the case of AC field, the analysis was performed for a wide range of the field strengths and frequencies, and it was found that the response of the system changes from a field-induced tunneling (low frequencies) to a nonequilibrium multi-photon absorption (high frequencies). In the intermediate frequencies regime, it was found that the single-electron spectrum includes both Floquet subbands and Wannier–Stark peaks that result in a nontrivial dependencies of the current and of the double occupancy on time as the field strength changes.

Quite remarkable results were found in Ref. [77], where it was shown that in the periodically-modulated case with moderate interaction the system thermalizes to an infinite-temperature state with no oscillations, while in the case of strong interactions the external field results in a quasi-stationary long-lived (Floquet prethermalized) state synchronized with drive oscillations. It was shown that the properties of this state strongly depend on the drive frequency. For instance, there exists a critical frequency that gives a rapid thermalization despite strong electron–electron interactions (Fig. 8.14). At critical frequency, the Floquet prethermal state is melted allowing the system to thermalize to the infinite-temperature state. The authors called this phenomenon resonant thermalization, since at this frequency the periodic modulation of interaction is resonant with the energy of doublon excitations, which favors doublon creation (increase of double occupancy). Otherwise, doublon

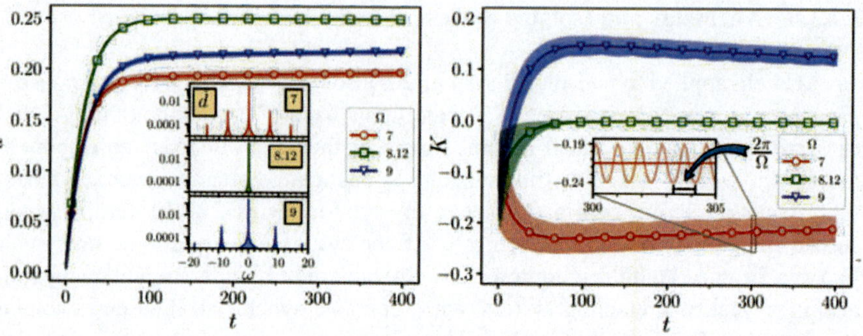

Fig. 8.14 Left: Double occupancy as function of time in the Hubbard model at different drive frequencies of the interaction potential (the critical frequency $\Omega = 8.12V$). Shaded areas correspond to the oscillating part of the double occupancy. Inset: the frequency dependencies of the double occupancy. Right: Kinetic energy as function of time for different drive frequencies. The calculations were performed by using NCA solver, Bethe lattice DOS $\rho_0(\varepsilon) = \sqrt{4V^2 - \varepsilon^2}/(2\pi V)^2$, $T = 0.05V$, and time-dependent interaction $U(t) = 8V + 2V\sin(\Omega t)$ (the static critical interaction is $U_c = 4.2V$). (Reprinted left and center Figs. 2 with permission from (F. Peronaci, M. Schiró, and O. Parcollet, Phys. Rev. Lett. **120**, 197601 (2018)). Copyright 2018 by the American Physical Society. https://doi.org/10.1103/PhysRevLett.120.197601)

creation would be suppressed by the large average interaction. As the authors demonstrate using a proposed periodic Schrieffer–Wolf transformation, the resonant thermalization is a non-perturbative effect.

Melting of the long-range AFM order in the periodically driven Hubbard model in the case of different ratios between the driving frequency, Coulomb repulsion and hopping were analyzed in Ref. [78]. In, probably, the most interesting—resonance—case with frequency $n\omega = U > t$ (n is integer), one can reach an effective noninteracting regime $U = 0$, moving from the large-U dynamics scenario (transfer of energy from the charge excitations to spins) to the small U one (dynamics of residual quasiparticles). Another (saddle point + fluctuations) approach was used to analyze the response of an AFM state in the case of 2D half-filled Hubbard model to a periodic perturbation [79]. It was found that when the driving frequency is below the gap, a discontinuous transition to a time-dependent solution takes place. It was also found that generated magnon states are nonthermally distributed even at driving frequencies much larger than the gap. In addition, when the drive amplitude is above the critical value and the frequency approached to zero (that leads to the destruction of the AFM phase caused by collective excitations), it was found that the low-energy magnon distribution diverges.

8.2.3.7 AC Fields and Open Systems

As it was shown in Ref. [80], a stationary and periodic-in-time field can modify the values of effective hopping due to coupling of the field to the current. In the last paper, the Floquet formalism was generalized and applied for the case of periodic-in-

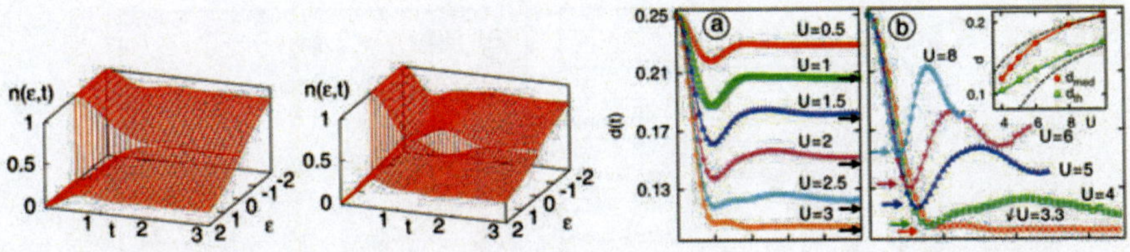

Fig. 8.15 Left two figures: evolution of the momentum distribution functions for quenches $U = 0 \to 3$ eV (left) and $U = 0 \to 5$ eV (right) (half-filled Hubbard model, semi-circular free-electron DOS, energy units—half-bandwidth, weak-coupling CT-QMC solver). Right: evolution of the double occupancy after the interaction is quenched from 0 to different finite values of U (times from 0 to 5 (left) and from 0 to 3.5 (right)). Inset: different estimations of the stationary value of the double occupancy (for details, see Refs. [13, 14]). (Reprinted Figs. 1, 2a, and 2b with permission from (M. Eckstein, M. Kollar, and P. Werner, Phys. Rev. Lett. **103**, 056403 (2009)). Copyright 2009 by the American Physical Society. https://doi.org/10.1103/PhysRevLett.103.056403)

time fields and it was shown that the system undergoes an insulator-to-metal transition (generated by excitations of charges into the upper Hubbard band) when the frequency of the external field is larger than the value of the gap.

8.2.3.8 Quenches, Relaxation and Thermalization

Analysis of the dynamics in the Hubbard model after an interaction quench [13, 50, 81–84] was another important application of neDMFT, in particular, because the theoretical results can be validated in experiments, e.g., on cold atom systems. In the case of PM state, it was found [13, 50] that the relaxation dynamics depends on the value of the final U showing different regimes at small and large values of U, and an intermediate regime with U around 3.2 (in the units of the half-bandwidth for the used semi-circular free-electron DOS). Except the last case, the system does not relax directly to the thermal state, but enters into a metastable regime. The time dependencies of the momentum distribution function and double occupancy after the quench at different values of U are shown in Fig. 8.15. As it follows from this figure, the dynamical phase transition takes place at the same time scale (~2) for all Us. For large Us (right Fig. 8.15), the double occupancy shows a damped collapse and subsequent oscillations with the "beat" period ~$2\pi/U$. The thermal state is reached at longer times [63], not accessible to the weak-coupling CT-QMC approach used in works [13, 50]. Another important, though a rather expected, finding of Ref. [50] that followed from comparison of the IPT and weak-interaction CT-QMC results is that IPT is a rather good approximation at short times (since it works up to times t/U^2 [29]).

In Ref. [84], a relaxation of the hot-electron system after a hopping quench (increase) was analyzed. The Coulomb interaction was chosen to be near the critical value, thus the main question was details of the relaxation to the Fermi Liquid state. There was found an electronic bottleneck time before which the charge and the electronic degrees of freedom are in a nonthermal state (with no effective

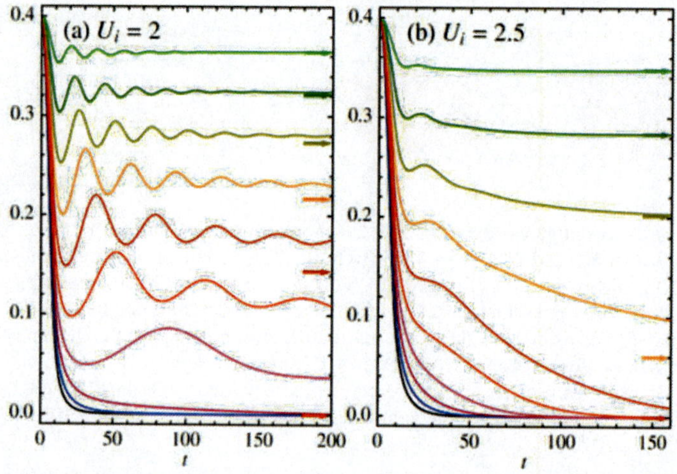

Fig. 8.16 Time evolution of magnetization after the interaction is quenched from $U_i = 2t^*$ to $U_f = t^*, \ldots, 1.9t^*$ (**a**) and from $U_i = 2.5t^*$ to $U_f = 1.5t^*, \ldots, 2.5t^*$ (**b**) (bottom to top). Arrows indicate expected thermalized values of magnetization (semielliptic DOS, time in units $1/t^*$, the third-order weak-coupling PT solver). (Reprinted Fig. 2 with permission from (N. Tsuji, M. Eckstein, and P. Werner, Phys. Rev. Lett. **110**, 136404 (2013)). Copyright 2013 by the American Physical Society. https://doi.org/10.1103/PhysRevLett.110.136404)

temperature). Another interesting finding of work [84]—the electronic relaxation is related to the spinon lifetime that is much longer than the fs hopping time, i.e., there is no rapid thermalization in metal. This corresponds to a long-lived bad-metal state that exists before the quasiparticle peak in the spectral function (and, hence, quasiparticles) is formed.

8.2.3.9 Interaction Quench in Presence of a Long-Range Order

As the next step, the effect of the interaction quench was studied in the case of an (AFM) ordered system [81, 82]. It was found that in presence of order, the relaxation is qualitatively different from the non-ordered case. In particular, after a quenched decrease of U, the magnetization quickly decreases and starts to coherently oscillate with a slow decay [82] (Fig. 8.16a). At small values of final U_f ($U_f \leq 1.2t^*$ at initial $U_i = 2t^*$ and $U_f \leq 1.9t^*$ at initial $U_i \leq 2.5t^*$) there are no oscillations and the magnetization exponentially decreases to zero. As final U decreases, it was found that the center of oscillations of the magnetization is different from the estimated thermalized value, suggesting that the system becomes trapped in a nonequilibrium quasi-steady state. At larger initial and final Us (Fig. 8.16b, see also Ref. [81]), the magnetization also does not immediately decay to zero after the quench, but remains trapped at a rather large value, with an exponentially long lifetime. It was argued in Ref. [81] that the (increased-U) quench-generated trapped nonthermal state is

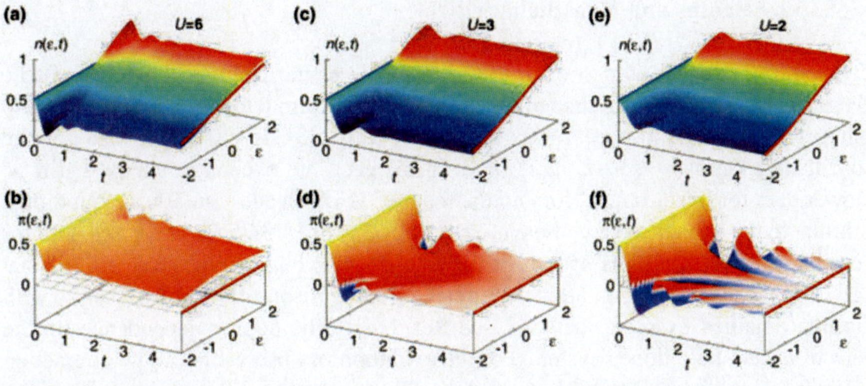

Fig. 8.17 Diagonal (top row) versus non-diagonal (bottom row) momentum distributions in the Hubbard model as functions of time and energy at different values of U (Bethe lattice, half-filling, DMGR solver). (From Ref. [86])

analogous to a quasistationary photodoped state with long-lived doublons. In the case of low effective doping this excited state demonstrates an AFM order even if the expected thermal state is PM. When the density of carriers is larger than a critical value, there is no trapping and the magnetization quickly relaxes to zero as $t^{-1/2}$ [82]. The authors of work [81] also argued that the obtained results can be used to describe some nonequilibrium properties of superconductors. Namely, it is known that at half-filling the AFM phase of the repulsive Hubbard model can be mapped on the s-wave superconducting phase of the attractive Hubbard model by the particle–hole transformation for the spin-up electrons in the bi-partite lattice [85]. Thus, long-lived nonthermal AFM state suggests a possibility of a nonthermal long-lived superconducting state created by a quench of the attractive interaction, similar to the one observed in cuprates (though we are discussing the s-wave case; for references, see work [81]). Due to analogy with the photoinduced nonthermal AMF state, the corresponding superconducting state is created by breakup of photoinduced pairs after decrease of the strong attractive interaction. The stability of this state might be explained by a slow recombination of unpaired electrons.

Difference between melting of the long-range AFM order in weakly and strongly interacting systems after the quench of hopping from zero to a finite value was explored in work [86]. It was shown that in the strongly correlated (Mott insulator) case, the melting of the long-range order occurs due to the transfer of excited charge energy to the spin background with preserved local spins and their exchange coupling, while in the weakly interacting (metallic) case the local moments decay destroying the long-range order. In the last case, the dynamics is dominated by residual quasiparticles, as one can see from the oscillating off-diagonal (with nesting pair of momenta) components of the momentum distribution $Re\left\langle c^{+}_{\vec{k}} c_{\vec{k}+(\pi,\ldots,\pi)}\right\rangle$ (Fig. 8.17).

8.2.3.10 Ramps and Nonadiabaticity

To minimize the excitation energy, the response of the Hubbard model on Bethe lattice at the ramps (slow change) of the interaction from 0 to U with different ramp shapes at fixed ramp time τ was studied in Ref. [83]. It was found that for the oscillating ramps, ~ sin $(2\pi n\omega\tau)$ (n is an integer), the excitation energies can be lower than for linear ramps (for details, see Ref. [14]). In addition, it was found that, similar to the quenched case, there is a critical value of U (3.75 for linear and 4.25 for oscillating ramp (in units of half of the free-electron bandwidth) at $\tau = 1.25$) that gives rapid thermalization after the ramp (for comparison of the above results with results obtained by other methods, see Ref. [14]). The doping dependence (in the electron- and hole-doped systems) of the evolution of observables at the interaction ramp was studied in Ref. [87]. It was found that in the case of weak interaction, the local and nonlocal (momentum-dependent) observables evolve to a thermal state that depends on the doping and the duration of ramp.

8.2.3.11 Optically-Tuned Physical Properties

A generalized neDMFT Keldysh formalism was proposed [88] to study the Loschmidt echo, whose nonanalytic behavior is a sign of the dynamical phase transition. The mutual influence of the AFM order and excited charge population on each other was studied in Refs. [89–91] (for a review, see Ref. [92]). In Ref. [90], it was shown that AFM correlations strongly affect the charge dynamics, since the spin background absorbs the energy of excited carriers at the ultrafast time scale, resulting in ultrafast transport and the electron–hole separation, while in the PM state at presence of strong fields, the carriers become localized. Thus, the photovoltaic properties of the AFM materials can be tuned by using this interplay between the charge and spin degrees of freedom.

The authors of Ref. [91] found that periodic electric fields can be used to reversibly control the exchange interaction J_{ex} in AFM Mott insulators in a transverse magnetic field (Fig. 8.18). In the case of small E_0s, the exchange interaction can be enhanced (reduced) at frequencies below (above) the Mott gap. For large E_0s, the sign of J_{ex} can be changed resulting in a time reversal of spin dynamics. It must be noted that obtained results on the tuning of the exchange energy are still valid for a transient state, and to study the quasistationary state longer simulation times are needed. Thus, as it follows from the results of the paper [91], pulses might be used to tune the magnetic and the dynamical properties of materials.

In Ref. [89], it was shown that photoexcited carriers in an AFM state modify the exchange interaction within a few hopping times of the pulse, which might be a new way to control magnetic properties without an external magnetic field, by using a pulse perturbation. Another example of possible tuning of the properties of correlated systems by light is the reduction of their temperature by photodoping [93, 94]. As it was proposed in Ref. [93], absorbed pulses can induce an evaporative

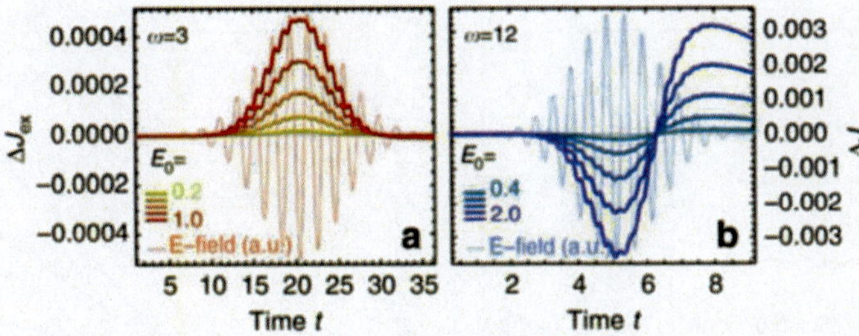

Fig. 8.18 Time dependence of the change of the exchange interaction J_{ex} in the Hubbard model, extracted from the time evolution of spin degrees of freedom during the pulse, at different values of the field strength E_0 and for frequencies ω below and above the bandgap (Gaussian DOS, $T = 0.025$, $U = 10$, pulse duration $15\pi/(2.1\omega)$ (pulse is shown), units $t^* = 1$, NCA solver). (From Ref. [91])

hole cooling. Namely, since as the authors suggested the entropy of a filled narrow band grows rapidly due to the isentropic (constant-entropy) transfer of holes from the doped Mott insulator. The suggestion was confirmed with the neDMFT calculations in the case of appropriate chirped pulses. It was shown that such a cooling may result in the emergence of the AFM order when the system goes below the Néel temperature (see Fig. 8.19).

In the following paper [94], the authors showed that by "photo-cooling" in the case of particle–hole symmetric setup one can get a sharp Drude peak in the optical conductivity and even superconducting phases (see also Refs. [95, 96] and Fig. 8.20)—the standard one at negative effective temperature and an η-type superconductivity (with a staggered order parameter $(-1)^i\left\langle c_\uparrow^+ c_\downarrow^+ \right\rangle$ with different signs on different sublattices). The last type of superconductivity can be created in the case of a large doublon–holon concentration at a positive effective temperature (for a more detailed discussion of the superconducting case, see below).

In another recent work [97], it was shown that changing the coupling of the system to a fermion bath in time, one can switch the electronic properties of the fully frustrated single-band Hubbard model between the metallic and insulating states that can be applied in memristor technologies, and in paper [98] it was demonstrated that doublon interaction can be changed from an attractive to zero and a repulsive by changing the power of the AC field.

8.2.3.12 Tuning Superconducting State

Possibility of a photodoping-induced transition from normal to the η − pairing superconducting state was also explored in works [95, 96, 99]. The nonequilibrium temperature-doping phase diagram of the photodoped Hubbard model is shown in Fig. 8.20 (from Ref. [96]). As it follows from this diagram, a variety of phases can be realized with photodoping, including the exotic η − pairing superconducting phase.

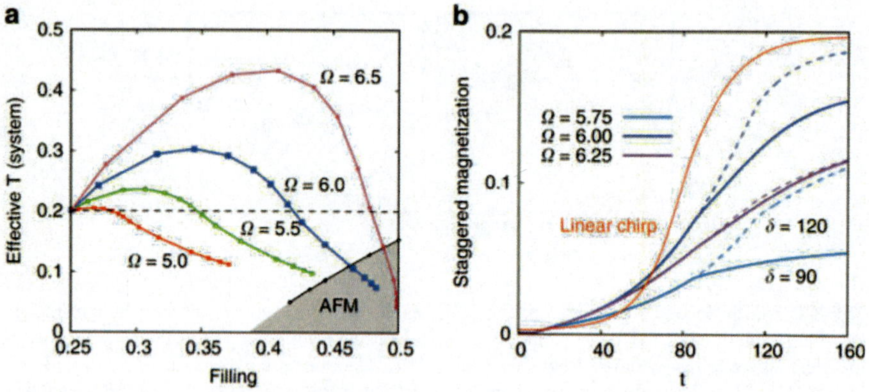

Fig. 8.19 Left: Effective temperature of photodoped Hubbard on Bethe lattice as a function of doping at different chirped-pulse frequencies. The dashed line corresponds to temperature of the initial quarter-filled state. Different points on the curves correspond to different pulse durations δ (pulse magnitude 0.8 (in units of hoping energy)). The dashed area with black line defines the AFM phase at equilibrium. Right: the corresponding staggered magnetizations for different pulse frequencies and pulse durations in presence of the staggered magnetic field $h = 0.001$ (for the pulse parameters and other details, see Ref. [93]). (From Ref. [93])

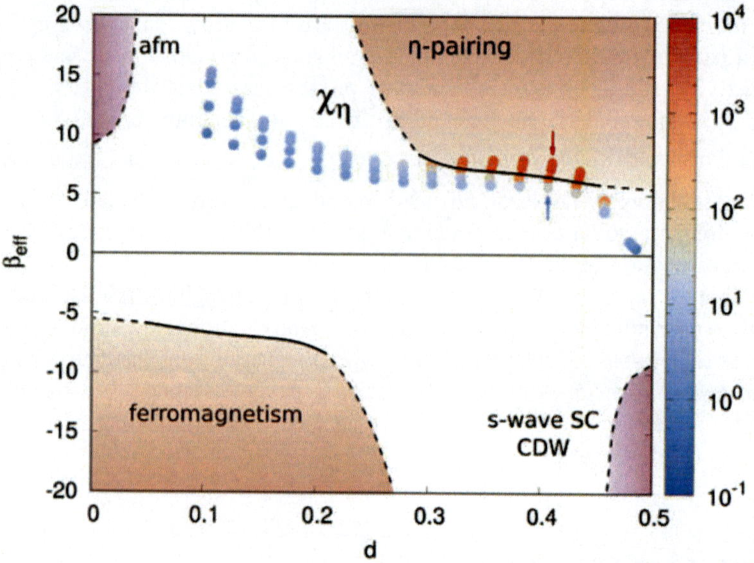

Fig. 8.20 A semiquantitative doping-inverse effective temperature phase diagram of the Hubbard model out-of-equilibrium. The photodoped states where prepared by shifting the chemical potential (for details, see Ref. [96]). The negative temperature region is obtained from reflection of the positive region. The results for susceptibility $\chi_\eta = -\langle \eta_x \rangle / h_x$, where $\eta_x = \frac{1}{2}\left(c_\uparrow c_\downarrow + c_\downarrow c_\uparrow\right)$ is the pairing field operator and $h_x = 0.0001$ is a small staggering field, are also shown. Calculations were performed in the case of Bethe lattice with bandwidth 4, Coulomb interaction $U = 8$, semielliptic bath DOS with peaks at $\pm U/2$, bath coupling $\Gamma = 0.05$ (units $t = 1$), by using the NCA solver. (From Ref. [96])

As the authors of the work [96] show, the optical conductivity in the superconducting phase has metallic behavior, with a peak at zero frequency and negative values at large frequencies. As it was suggested by the authors, these features can be used to detect the η – pairing phase in pump-probe experiments. In a related study [100] it was shown that combination of drive and dissipation leads to nonequilibrium steady states with a number of doublons much larger than the maximum thermal-equilibrium value. From analytical arguments, using the Floquet Hamiltonian, the authors explained this effect to happen due to an enhancement of local pairing correlations that possibly can also give an η-pairing phase.

8.2.3.13 Extensions of neDMFT: Clusters, Multi-layers, Multi-bands

Several extensions of the neDMF approach have been made. A cluster neDMFT approach was formulated in Ref. [101], where it was also used to study the time-dependent momentum distribution functions in the 1D and 2D Hubbard models after a quench of the interaction. Also, a real-space version of neDMFT was formulated [102] and used (see, e.g., Ref. [103]) to study electronic and transport properties of heterostructures between metallic leads.

In recent years, a significant progress has been made in understanding the nonequilibrium properties of multi-orbital systems by using multi-orbital neDMFT, though the number of bands in the analysis is still small due to high computational cost: only two- [104–108] and three- [66, 109, 110] band systems were studied so far. Multi-orbital models allow one to study the role of Hund coupling, a very important part of the interaction, in the nonequilibrium properties of correlated materials. The role of this coupling was analyzed, in particular, in Ref. [104], where the carrier relaxation in photodoped two-band half-filled Hubbard model was studied with (real-time) neDMFT. It was shown that the ultrafast carrier relaxation in the system is driven by emission of local spin excitations and the relaxation time is inversely proportional to the Hund coupling J_H. It was also found that in the case of strong Hound coupling, in addition to photogenerated side bands there is another effect—the photodoped populations split into several resonances. Moreover, strong Hund coupling produces two other effects in the dynamics of local states—thermal blocking (at $T > J_H$) and kinetic freezing (J_H larger than the bandwidth W). In detail, since at higher temperatures most of the high-energy states (energy scale J_H) are occupied, the possibilities to dynamically change the kinetic energy are very limited. Next, since the low-energy excited states are mostly doublons, it means that at high temperatures, the dynamics of low-energy doublons is thermally blocked. In other words, thermal blocking takes place when the initial T competes with Hund's splitting of the local states, limiting the dynamics to thermally distributed local spin excitations. The kinetic freezing effect is a result of competition between the Hund's coupling and the kinetic (defined by the electron bandwidth W) effects. When J_H is of order or larger than W, the available kinetic energy is smaller than J_H, the energy required to generate local spin excitations. Thus, the dynamics of local spin excitations is frozen (kinetic freezing). In addition,

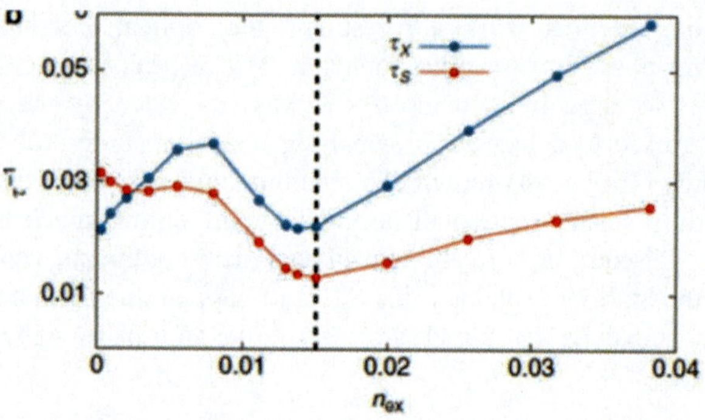

Fig. 8.21 Relaxation rates for spin and orbital order parameters (the last defined as difference in orbital occupancies) in three-quarter-filled two-orbital Hubbard model (a model of $KCuO_3$) as functions of the excitation density after a laser pulse excitation ($U = 7t, J_H = 0.7t$, initial temperature $T = 0.01t$, NCA solver). (From Ref. [107])

in Ref. [106] it was shown that in the two-band Hubbard model with Hund's coupling strong electric fields can lead to a transient localization of electrons with formed atomic-like spin multiplets that depend on J_H. Experimental analysis of these states can be used to "measure" the Hund's coupling in materials.

Another study of the two-band Hubbard model [107] showed that photoexcitation can generate spin-orbital interwined hidden order that can be totally different from the initial equilibrium state, and that the system can evolve to a hidden interwined-order state through a nonthermal partial melting of different orders. The melting is mediated by photoinduced charge excitations in presence of strong spin-orbital exchange interactions. Analysis of the spin- and orbital-order (described by the corresponding state occupancies) relaxation rates as functions of the (charge) excitation density n_{ex} (proportional to the change of the pulse strength) showed, in particular, that the spin relaxation slows down as n_{ex} increases (Fig. 8.21). This corresponds to a slower melting of spin order that survives at the end, while the antiferro-orbital order melts. Thus, as the results of the work [107] show, orbital degrees of freedom can play a crucial role in the ultrafast dynamics of materials.

Another exciting topic is a photodoping-induced superconductivity in the two-orbital Hubbard model. It was shown in Ref. [105] that in this model at large U the pairing susceptibility can be significantly enhanced by photodoping that may result in a superconducting state. In Ref. [108], it was demonstrated that the crystal-field quenches in the vicinity of the high-to-low spin transition in the two-orbital Hubbard model at high initial temperatures (and with included Hund coupling) can generate another type of condensed state—an excitonic condensate.

Finally, we mention that in Ref. [109] multi-band GW + EDMFT formalism (see Chap. 10) was extended on the multi-orbital case and was applied to study the dynamics of the Emery model of a three-band charge-transfer insulator, relevant to cuprate superconductors.

8.2.3.14 Electron–Phonon Interaction

Recently, the neDMFT was also generalized and applied to analyze the nonequilibrium response of coupled "non-correlated electron"-phonon (Holstein model) systems [111–113] and the Hubbard model with an additional electron–phonon interaction (Holstein–Hubbard model) [114–116]. Since phonons play an important role in relaxation of strongly correlated and other materials, we begin with the case of Holstein model, though it does not include the Coulomb interaction between electrons. In Ref. [111], the authors studied relaxation of the Holstein model after the quench of interaction and it was found that in the weak-coupling regime there is a qualitative change in the thermalization dynamics of the system as the interaction increases: at lower values of the electron–phonon interaction the phonon dynamics is damped before the electrons are thermalized. In the case of strong interaction, the electrons relax faster than the phonon damping takes place. In work [112], details of formation of polaronic states, important for transient dynamics in metals, were studied in the case of one-electron Holstein model that was solved exactly. It was found that in the adiabatic regime at small phonon frequencies the initial coherent phonon oscillations are strongly damped, and as a result—the excited state of the system is a mixture of excited polaron and metastable delocalized states. In Ref. [113], it was shown that the excited phonon populations in the Holstein model suppress the time-dependent decay rates of the excited electrons due to strong phonon absorption. The increased phonon occupation also results in the value of the nonequilibrium decay rate that is different from the equilibrium scattering rate.

In the case of more relevant to this book strongly correlated Holstein–Hubbard model, several important findings were also reported. In particular, in Ref. [115] a critical analysis of the standard assumptions about the relaxation of excited electrons was performed and it was demonstrated that some assumptions, like "the self-energy governs the relaxation rate" and "the electron subsystem rapidly thermalizes before being driven by electron-phonon scattering to the equilibrium state," are not necessarily valid. In Ref. [116], possibility of the insulator-to-metal transition generated by a sudden switch-on of the electron-phonon interaction was studied by using two different solvers. In the case of the Slave-Rotor solver, it was found that above a critical value of the (quenched) electron–phonon coupling the system evolves into a nonthermal (transient) metallic state with coherent quasiparticles. On the other hand, by using the OCA solver it was shown that the system evolves in a bad-metal state (the evolutions of the double occupancy after the quench are shown in Fig. 8.22). However, the main results obtained by both approaches are qualitatively similar—the adiabatic switching of the electron–phonon coupling g results in a reduced effective interaction between electrons $\left(U_{\text{eff}}(\omega) = U - 2g\omega_0^2/\left(\omega_0^2 - \omega^2\right)\right.$, where ω_0 is phonon frequency), and drives the system to a "more metallic" state.

To summarize, neDMFT approach helped to find different important nonequilibrium properties of the Hubbard and post-Hubbard models that can be classified by using the field strength–frequency diagram in the review by Aoki et al. [14] (Fig. 8.23). Important results include established stages of the dynamics,

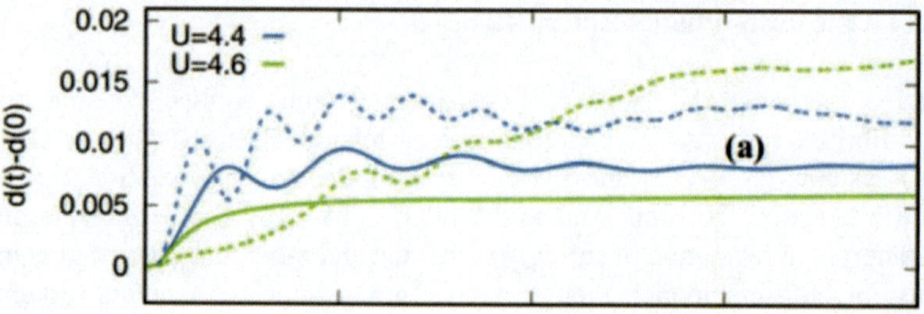

Fig. 8.22 Time evolution of the double occupancy in the Holstein–Hubbard model after the quench of the electron–phonon interaction in the case of different values of U. The results were obtained with the OCA (initial temperature $T = 1/(30t)$, solid lines) and the Slave-Rotor ($T = 0.05t$, dashed lines) solvers. Higher post-transient double occupancies correspond to a nonthermal-metal and lower—to a bad-metal state (hypercubic lattice, phonon frequency $\omega_0 = 0.2t$, final electron-phonon coupling $g = 0.44t$, time changes from 0 to $40/t$). (Reprinted Fig. 7a with permission from (S. Sayyad, R. Žitko, H.U.R. Strand, P. Werner, and D. Golež, Phys. Rev. B **99**, 045118 (2019)). Copyright 2019 by the American Physical Society. https://doi.org/10.1103/PhysRevB.99.045118)

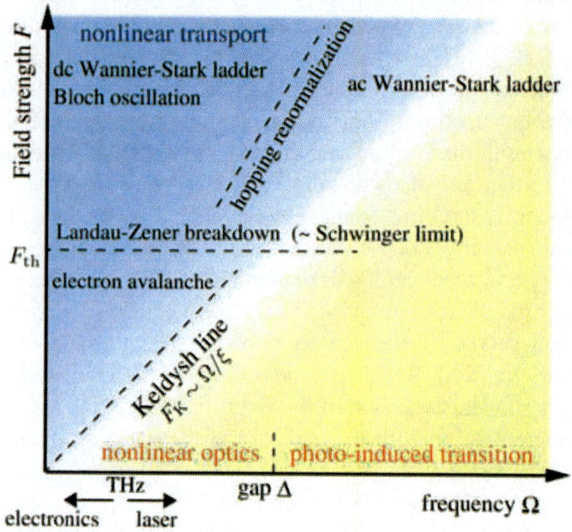

Fig. 8.23 Different regimes of the excited systems as functions of the applied field strength F and frequency Ω. (Reprinted Fig. 17 with permission from (H. Aoki, N. Tsuji, M. Eckstein, M. Kollar, T. Oka, P. Werner, Rev. Mod. Phys. **86**, 779 (2014)). Copyright 2014 by the American Physical Society. https://doi.org/10.1103/RevModPhys.86.779)

including relaxation to steady-state regimes and to different phases that depend on perturbations, and the quantified role of the correlation effects in these processes. Another important result—a possibility to tune properties of the materials by a combined effect of changing the correlation strength (e.g., by quench of U or hopping) and the strength and shape of the field.

Thus, as we demonstrated neDMFT is a powerful tool to analyze nonequilibrium properties of strongly correlated systems, even though at the moment it suffers from some computational limitations that do not allow one to fully analyze the nonequilibrium properties of real materials. The main limitations are small size of the systems (number of orbitals and atoms in the unit cell) and non-accessible long (equilibration) times. These difficulties will be run over in the future with further enhancement of the computer power.

To conclude this chapter, we mention some other, alternative to neDMFT, many-body approaches to nonequilibrium strongly correlated systems developed in recent years—time-dependent many-variable variational Monte Carlo method [117], nonequilibrium self-energy functional approach [118], quantum Boltzmann equation approach [119], nonequilibrium dynamical cluster theory [101, 120–122], HF + EDMFT (extended DMFT) [110], GW + EDMFT for charge-transfer insulators [110] and neDMFT extended for bosonic lattice models [123]. Finally, as we will discuss in Chap. 12, there is an alternative to neDMFT *ab initio* approach, a technically much simpler and cheaper time-dependent density-functional theory (TDDFT).

References

1. Servin, R., Jin-Hyung, P., In-yeal, L., Jeong Min, B., Kyung Soo, Y., Gil-Ho, K.: J. Phys. D. Appl. Phys. **47**, 295101 (2014)
2. Markov, P., Marvel, R.E., Conley, H.J., Miller, K.J., Haglund Jr., R.F., Weiss, S.M.: ACS Photonics. **2**, 1175 (2015)
3. Merced, E., Torres, D., Tan, X., Sepúlveda, N.: J. Microelectromechan. Syst. **24**, 100 (2015)
4. Liu, C., Balin, I., Magdassi, S., Abdulhalim, I., Long, Y.: Opt. Express. **3**, A124 (2015)
5. Nie, G., Zhang, L., Lei, J., Yang, L., Zhang, Z., Lu, X., Wang, C.: J. Mater. Chem. A. **2**, 2910 (2014)
6. Yang, S., Gong, Y., Liu, Z., Zhan, L., Hashim, D.P., Ma, L., Vajtai, R., Ajayan, P.M.: Nano Lett. **13**, 1596 (2013)
7. Hilton, D.J., Prasankumar, R.P., Fourmaux, S., Cavalleri, A., Brassard, D., El Khakani, M.A., Kieffer, J.C., Taylor, A.J., Averitt, R.D.: Phys. Rev. Lett. **99**, 226401 (2007)
8. Bigot, J.Y., Vomir, M.: Ann. Phys. **525**, 2 (2013)
9. Wegkamp, D., Herzog, M., Xian, L., Gatti, M., Cudazzo, P., McGahan, C.L., Marvel, R.E., Haglund Jr., R.F., Rubio, A., Wolf, M., Stähler, J.: Phys. Rev. Lett. **113**, 216401 (2014)
10. Otto, M.R., Rene de Cotret, L.P., Valverde-Chavez, D.A., Tiwari, K.L., Emond, N., Chaker, M., Cooke, D.G., Siwick, B.J.: PNAS. **116**, 450 (2018)
11. Schmidt, P., Monien, H.: preprint arXiv:cond-mat/0202046 (2002)
12. Freericks, J.K., Turkowski, V.M., Zlatić, V.: Phys. Rev. Lett. **97**, 266408 (2006)
13. Eckstein, M., Kollar, M., Werner, P.: Phys. Rev. Lett. **103**, 056403 (2009)
14. Aoki, H., Tsuji, N., Eckstein, M., Kollar, M., Oka, T., Werner, P.: Rev. Mod. Phys. **86**, 779 (2014)
15. Keldysh, K.V., Eksp, Z.: Teor. Fiz. **47**, 1515 (1964).; Sov. Phys. JETP 20, 1018 (1965)
16. Schwinger, J.: J. Math. Phys. **2**, 407 (1961)
17. Kleinert, H.: Path Integrals in Quantum Mechanics, Statistics, Polymer Physics, and Financial Markets. World Scientific Publishers, Singapore (2009)
18. Kadanoff, L.P., Baym, G.: Quantum Statistical Mechanics. Advanced Books Classics, CRC Press, Boca Raton (1994)
19. Stefanucci, G., van Leeuwen, R.: Nonequilibrium Many-Body Theory of Quantum Systems: A Modern Introduction. Cambridge University Press, Cambridge (2013)
20. Negele, J.W., Orland, H.: Quantum Many-Particle Systems (Frontiers in Physics, 1988)

21. Kamenev, A.: Field Theory of Non-Equilibrium Systems. Cambridge University Press, Cambridge (2011)
22. Kamenev, A.: Keldysh Technique", in The Capri Spring School on Transport in Nanostructures (Capri, Italy, 2009)
23. P. Hanggi, Quantum Transport and Dissipation, Driven Quantum Systems (1998)
24. Peierls, R.E.: Z. Phys. **80**, 763 (1933)
25. Turkowski, V., Freericks, J.K.: Phys. Rev. B. **71**, 085104 (2005)
26. Tsuji, N., Oka, T., Aoki, H.: Phys. Rev. B. **78**, 235124 (2008)
27. Freericks, J.K., Zlatic, V.: Rev. Mod. Phys. **75**, 1333 (2003)
28. Eckstein, M., Werner, P.: Phys. Rev. Lett. **107**, 186406 (2011)
29. Turkowski, V., Freericks, J.K.: Phys. Rev. B. **75**, 125110 (2007)
30. Freericks, J.K.: Phys. Rev. B. **77**, 075109 (2008)
31. Tsuji, N., Oka, T., Aoki, H.: Phys. Rev. Lett. **103**, 047403 (2009)
32. Qin, T., Hofstetter, W.: Phys. Rev. B. **96**, 075134 (2017)
33. Eckstein, M., Kollar, M.: Phys. Rev. B. **78**, 245113 (2008)
34. Eckstein, M., Kollar, M.: Phys. Rev. Lett. **100**, 120404 (2008)
35. Eckstein, M., Kollar, M.: Phys. Rev. B. **78**, 205119 (2008)
36. Eckstein, M., Kollar, M.: New J. Phys. **12**, 055012 (2010)
37. Moritz, B., Devereaux, T.P., Freericks, J.K.: Phys. Rev. B. **81**, 165112 (2010)
38. Moritz, B., Kemper, A.F., Sentef, M., Devereaux, T.P., Freericks, J.K.: Phys. Rev. Lett. **111**, 077401 (2013)
39. Freericks, J.K., Matveev, O.P., Shen, W., Shvaika, A.M., Devereaux, T.P.: Phys. Scr. **92**, 034007 (2017)
40. Herrmann, A.J., Tsuji, N., Eckstein, M., Werner, P.: Phys. Rev. B. **94**, 245114 (2016)
41. Matveev, O.P., Shvaika, A.M., Devereaux, T.P., Freericks, J.K.: Phys. Rev. B. **93**, 045110 (2016)
42. Tsuji, N., Werner, P., Ueda, M.: Phys. Rev. A. **95**, 011601 (2017)
43. Rubtsov, A.N., Savkin, V.V., Lichtenstein, A.I.: Phys. Rev. B. **72**, 035122 (2005)
44. Mühlbacher, L., Rabani, E.: Phys. Rev. Lett. **100**, 176403 (2008)
45. Gull, E., Werner, P., Parcollet, O., Troyer, M.: Europhys. Lett. **82**, 57003 (2008)
46. Werner, P., Oka, T., Millis, A.J.: Phys. Rev. B. **79**, 035320 (2009)
47. Werner, P., Oka, T., Eckstein, M., Millis, A.J.: Phys. Rev. B. **81**, 035108 (2010)
48. Schiró, M., Fabrizio, M.: Phys. Rev. B. **79**, 153302 (2009)
49. Dong, Q.: Non-Equilibrium Dynamics with Quantum Monte Carlo, PhD Thesis (University of Michigan, 2017)
50. Eckstein, M., Kollar, M., Werner, P.: Phys. Rev. B. **81**, 115131 (2010)
51. Amaricci, A., Weber, C., Capone, M., Kotliar, G.: Phys. Rev. B. **86**, 085110 (2012)
52. Eckstein, M., Werner, P.: Phys. Rev. B. **82**, 115115 (2010)
53. Gull, E., Reichman, D.R., Millis, A.J.: Phys. Rev. B. **84**, 085134 (2011)
54. Wolf, F.A., McCulloch, I.P., Schollwock, U.: Phys. Rev. B. **90**, 235131 (2014)
55. Balzer, K., Li, Z., Vendrell, O., Eckstein, M.: Phys. Rev. B. **91**, 045136 (2015)
56. Turkowski, V.M., Freericks, J.K.: Phys. Rev. B. **73**, 075108. 73, 209902(E) (2006)
57. Turkowski, V., Freericks, J.K.: Phys. Rev. B. **77**, 205102 (2008).; 82, 119904(E) (2010)
58. Joura, A.V., Freericks, J.K., Pruschke, T.: Phys. Rev. Lett. **101**, 196401 (2008)
59. Aron, C., Kotliar, G., Weber, C.: Phys. Rev. Lett. **108**, 086401 (2012)
60. Li, J., Aron, C., Kotliar, G., Han, J.E.: Phys. Rev. Lett. **114**, 226403 (2015)
61. Eckstein, M., Oka, T., Werner, P.: Phys. Rev. Lett. **105**, 146404 (2010)
62. Eckstein, M., Werner, P.: J. Phys. Conf. Ser. **427**, 012005 (2013)
63. Eckstein, M., Werner, P.: Phys. Rev. B. **84**, 035122 (2011)
64. Eckstein, M., Werner, P.: Phys. Rev. Lett. **110**, 126401 (2013)
65. Werner, P., Held, K., Eckstein, M.: Phys. Rev. B. **90**, 235102 (2014)
66. Petocchi, F., Beck, S., Ederer, C., Werner, P.: Phys. Rev. B. **100**, 075147 (2019)
67. Golez, D., Eckstein, M., Werner, P.: Phys. Rev. B. **92**, 195123 (2015)

68. Kemper, A.F., Krishnamurthy, H.R., Freericks, J.K.: Fortschr. Phys. **65**, 1600042 (2017)
69. Murakami, Y., Eckstein, M., Werner, P.: Phys. Rev. Lett. **121**, 057405 (2018)
70. Murakami, Y., Eckstein, M., Werner, P.: JPS Conf. Proc. **30**, 011183 (2020)
71. Lysne, M., Murakami, Y., Werner, P.: Phys. Rev. B. **101**, 195139 (2020)
72. Tsuji, N., Oka, T., Werner, P., Aoki, H.: Phys. Rev. Lett. **106**, 236401 (2011)
73. Canovi, E., Kollar, M., Eckstein, M.: Phys. Rev. E. **93**, 012130 (2016)
74. Herrmann, A., Murakami, Y., Eckstein, M., Werner, P.: Eur. Phys. Lett. **120**, 57001 (2017)
75. Sandholzer, K., Murakami, Y., Görg, F., Minguzzi, J., Messer, M., Desbuquois, R., Eckstein, M., Werner, P., Esslinger, T.: Phys. Rev. Lett. **123**, 193602 (2019)
76. Murakami, Y., Werner, P.: Phys. Rev. B. **98**, 075102 (2018)
77. Peronaci, F., Schiró, M., Parcollet, O.: Phys. Rev. Lett. **120**, 197601 (2018)
78. Mendoza-Arenas, J.J., Gomez-Ruiz, F.J., Eckstein, M., Jaksch, D., Clark, S.R.: Ann. Phys. (Berlin). **529**, 1700024 (2017)
79. Walldorf, N., Kennes, D.M., Paaske, J., Millis, A.J.: Phys. Rev. B. **100**, 121110 (2019)
80. Lubatsch, A., Kroha, J.: Ann. Phys. (Berlin). **18**, 863 (2009)
81. Werner, P., Tsuji, N., Eckstein, M.: Phys. Rev. B. **86**, 205101 (2012)
82. Tsuji, N., Eckstein, M., Werner, P.: Phys. Rev. Lett. **110**, 136404 (2013)
83. Eurich, N., Eckstein, M., Werner, P.: Phys. Rev. B. **83**, 155122 (2011)
84. Sayyad, S., Eckstein, M.: Phys. Rev. Lett. **117**, 096403 (2016)
85. Shiba, H.: Prog. Theor. Phys. **48**, 2171 (1972)
86. Balzer, K., Wolf, F.A., McCulloch, I.P., Werner, P., Eckstein, M.: Phys. Rev. X. **5**, 031039 (2015)
87. Sayyad, S., Tsuji, N., Vaezi, A., Capone, M., Eckstein, M., Aoki, H.: Phys. Rev. B. **99**, 165132 (2019)
88. Canovi, E., Werner, P., Eckstein, M.: Phys. Rev. Lett. **113**, 265702 (2014)
89. Mentink, J.H., Eckstein, M.: Phys. Rev. Lett. **113**, 057201 (2014)
90. Eckstein, M., Werner, P.: Phys. Rev. Lett. **113**, 076405 (2014)
91. Mentink, J.H., Balzer, K., Eckstein, M.: Nat. Comm. **6**, 6708 (2015)
92. Mentink, J.H.: J. Phys. Condens. Matter. **29**, 453001 (2017)
93. Werner, P., Eckstein, M., Müller, M., Refael, G.: Nat. Comm. **10**, 5556 (2019)
94. Werner, P., Li, J., Golež, D., Eckstein, M.: Phys. Rev. B. **100**, 155130 (2019)
95. Dasari, N., Eckstein, M.: Phys. Rev. B. **98**, 235149 (2018)
96. Li, J., Golez, D., Werner, P., Eckstein, M.: preprint arXiv:1908.08693v1 (2019)
97. Kleinhenz, J., Krivenko, I., Cohen, G., Gull, E.: preprint arXiv:2004.04669v1 (2020)
98. Valmispild, V.N., Dutreix, C., Eckstein, M., Katsnelson, M.I., Lichtenstein, A.I., Stepanov, E. A.: preprint arXiv:2005.12812v1 (2020)
99. Li, J., Golez, D., Werner, P., Eckstein, M.: Mod. Phys. Let. B. **34**, 2040054 (2020)
100. Peronaci, F., Parcollet, O., Schiró, M.: Phys. Rev. B. **101**, 161101 (2020)
101. Tsuji, N., Barmettler, P., Aoki, H., Werner, P.: Phys. Rev. B. **90** (075117, 2014)
102. Okamoto, S.: Phys. Rev. B. **76**, 035105 (2007)
103. Titvinidze, I., Sorantin, M.E., Dorda, A., von der Linden, W., Arrigoni, E.: Phys. Rev. B. **98**, 035146 (2018)
104. Strand, H.U.R., Golež, D., Eckstein, M., Werner, P.: Phys. Rev. B. **96**, 165104 (2017)
105. Werner, P., Strand, H.U.R., Hoshino, S., Murakami, Y., Eckstein, M.: Phys. Rev. B. **97**, 165119 (2018)
106. Dasari, N., Li, J., Werner, P., Eckstein, M.: Phys. Rev. B. **101**, 161107 (2020)
107. Li, J., Strand, H.U.R., Werner, P., Eckstein, M.: Nat. Comm. **9**, 4581 (2018)
108. Werner, P., Murakami, Y.: preprint arXiv:2006.02096v2 (2020)
109. Golež, D., Eckstein, M., Werner, P.: Phys. Rev. B. **100**, 235117 (2019)
110. Golež, D., Boehnke, L., Eckstein, M., Werner, P.: Phys. Rev. B. **100**, 041111 (2019)
111. Murakami, Y., Werner, P., Tsuji, N., Aoki, H.: Phys. Rev. B. **91**, 045128 (2015)
112. Sayyad, S., Eckstein, M.: Phys. Rev. B. **91**, 104301 (2015)

113. Abdurazakov, O., Nevola, D., Rustagi, A., Freericks, J.K., Dougherty, D.B., Kemper, A.F.: Phys. Rev. B. **98**, 245110 (2018)
114. Kemper, A.F., Freericks, J.K.: Entropy. **18**, 180 (2016)
115. Kemper, A.F., Abdurazakov, O., Freericks, J.K.: Phys. Rev. X. **8**, 041009 (2018)
116. Sayyad, S., Žitko, R., Strand, H.U.R., Werner, P., Golež, D.: Phys. Rev. B. **99**, 045118 (2019)
117. Ido, K., Ohgoe, T., Imada, M.: Phys. Rev. B. **92**, 245106 (2015)
118. Hofmann, F., Eckstein, M., Potthoff, M.: Phys. Rev. B. **93**, 235104 (2016)
119. Wais, M., Eckstein, M., Fischer, R., Werner, P., Battiato, M., Held, K.: Phys. Rev. B. **98**, 134312 (2018)
120. Eckstein, M., Werner, P.: Sci. Rep. **6**, 21235 (2016)
121. Bittner, N., Golež, D., Eckstein, M., Werner, P.: Phys. Rev. B. **101**, 085127 (2020)
122. Bittner, N., Golez, D., Eckstein, M., Werner, P.: preprint arXiv:2005.11722v1 (2020)
123. Strand, H.U.R., Eckstein, M., Werner, P.: Phys. Rev. X. **5**, 011038 (2015)

Chapter 9
Extensions of DMFT to the Nonlocal Case

9.1 Different Nonlocal Extensions of DMFT

Nonlocal correlations play an important role in condensed matter physics. Examples include cuprate superconductivity, systems close to a phase transition, low-dimensional systems, etc. (for an over-review, see Ref. [1]). Thus, generalization of DMFT approach on the case with nonlocal correlations is an important task.

As it was established in chapter 3, the noninteracting nonlocal (nearest-site) GFs decay rapidly as the distance between the sites increases, $G_{ij}^{(0)} \sim d^{-|i-j|/2}$, and the nonlocal correlations become less and less important as dimensionality grows and finally become irrelevant in the limit of infinite dimension where GFs are local in space, i.e., $G_{ij}^{(0)}$ and other GFs are nonzero only when $i = j$. This is the DMFT approximation, where the Luttinger–Ward functional is built by using only diagrams with local vertices. Thus, the lowest-order nonlocal correction to DMFT can be obtained if one takes into account in the Luttinger–Ward functional the diagrams that contain vertices with nearest-neighbor sites. Then, one can obtain from this functional a nonzero nonlocal self-energy:

$$\Sigma_{ij} = \frac{\delta \Phi}{\delta G_{ij}} \; (i \neq j), \tag{9.1}$$

that is $1/d$-corrected DMFT self-energy. Next, if one includes nonlocal vertices with sites separated by more than one interatomic distance, the correction will be of order $1/d^n$, where n is the number of interatomic distances. This is a systematic, though complicated, way to calculate nonlocal corrections to the self-energy, which gives the exact result as $n \to \infty$. Taking into account the fact that we consider the case of large interatomic coordination, i.e., large d, the lowest order correction, $1/d$, can be regarded as a reasonable approximation. The $1/d$-expansion formalism was first derived by Schiller and Ingersent [2]. In their approximation, one considers all local diagrams and diagrams with vertex sites separated no further than one

V. Turkowski, *Dynamical Mean-Field Theory for Strongly Correlated Materials*,
https://doi.org/10.1007/978-3-030-64904-3_9

interatomic distance. Then, the Luttinger–Ward functional consists of two parts: the nearest-neighbor site nonlocal diagrams and the local diagrams:

$$\Phi(G) = \sum_{\langle i,j\rangle,\sigma} \Phi_{2-\text{imp}}\left[G_{ii\sigma}, G_{ij\sigma}, G_{ji\sigma}, G_{jj\sigma}\right] - (2d-1)\sum_{i,\sigma}\Phi_{1-\text{imp}}\left[G_{ii\sigma}\right]. \tag{9.2}$$

Using the methodology of DMFT and mapping the problem on one- and two-site impurity problems, these functionals can be generated from the one- and two-site impurity actions:

$$\begin{aligned} S_1 = &-\int_0^\beta d\tau \int_0^\beta d\tau' \sum_\sigma c_{1\sigma}^+(\tau)\mathcal{G}_{1,\sigma}^{-1}(\tau-\tau')c_{1\sigma}(\tau') \\ &+ U\int_0^\beta d\tau c_{1\uparrow}^+(\tau)c_{1\uparrow}(\tau)c_{1\downarrow}^+(\tau)c_{1\downarrow}(\tau), \end{aligned} \tag{9.3}$$

$$\begin{aligned} S_2 = &-\int_0^\beta d\tau \int_0^\beta d\tau' \sum_{i,j=1}^{2}\sum_\sigma c_{i\sigma}^+(\tau)\left[\mathcal{G}_{2,\sigma}^{-1}(\tau-\tau')\right]_{ij}c_{j\sigma}(\tau') \\ &+ U\sum_{i=1}^{2}\int_0^\beta d\tau c_{1\uparrow}^+(\tau)c_{1\uparrow}(\tau)c_{1\downarrow}^+(\tau)c_{1\downarrow}(\tau), \end{aligned} \tag{9.4}$$

where $\mathcal{G}_{1,\sigma}$ is a one-site and (two-by-two matrix) $\mathcal{G}_{2,\sigma}$ is a two-site dynamical mean-field function. These impurity equations are complemented by the Dyson equations for the local ($G_{11,\ \sigma}$) and the two-site (matrix) ($\widehat{G}_\sigma$) impurity GFs

$$G_{11,\sigma}^{-1}(i\omega_n) = \mathcal{G}_{1,\sigma}^{-1}(i\omega_n) - \Sigma_{1,\sigma}(i\omega_n), \tag{9.5}$$

$$\left[G_\sigma^{-1}(i\omega_n)\right]_{ij} = \left[\mathcal{G}_{2,\sigma}^{-1}(i\omega_n)\right]_{ij} - \Sigma_{ij,\sigma}(i\omega_n), \tag{9.6}$$

and equations that connect the lattice local GFs and self-energies (see Chap. 3). Solution of this system of equations gives the nonlocal self-energy $\Sigma_{ij,\ \sigma}(i\omega_n)$ and nonlocal GF (along with the local ones). The authors applied the approach to the Falicov-Kimball model and found that the local self-energy approximation works rather well even in 2D case (which is not true in many other cases, including the Hubbard model).

Systematic nonlocal expansions of DMFT were performed in the strong-coupling case by using the auxiliary field [3] and the Hubbard operator representation [4] approaches. Other widely used nonlocal self-energy approaches are the DCA and the cellular (CDMFT) extensions of DMFT (Refs. [5] and [6, 7], correspondingly), which were successfully applied to several problems, including the pseudogap phase and unconventional superconductivity in the 2D Hubbard model. An alternative way to include the nonlocal contribution to the DMFT self-energy is to add to the local self-energy another, momentum-dependent, part obtained with a “nonlocal” approach, like GW (for GW+ DMFT approach, see Refs. [8, 9]), spin-fermion model (spin-fluctuation contribution to the self-energy) [10], and fluctuation-exchange (FLEX) approach [11] (see Ref. [1] for a review). All these approaches suffer from some limitations, like a low-order expansion, small number of k-points, or uncontrolled *ad hoc* approximations.

Another way to include nonlocal correlations in DMFT, that can be realized in practice, is to use two-particle (2P) formalism and to keep only the local in space vertices in the Luttinger–Ward functional, but to take into account their frequency dependence (i.e., to assume that the vertices are dynamical). This approach is based on using 2P GFs, which allow one to take into account momentum correction to the DMFT self-energy (see next section). Before going into details of this approach, we give a semiquantitative idea of why the nonlocal theory has to include 2P GFs by deriving and analyzing an equation for the momentum-dependent self-energy in the Hubbard model in the PM phase. For this, let us rewrite the Hubbard Hamiltonian in momentum representation:

$$H=\sum_{\vec{k},\sigma}\varepsilon\left(\vec{k}\right)c^{+}_{\vec{k}\sigma}c_{\vec{k}\sigma}+\frac{U}{N}\sum_{\vec{k},\vec{p},\vec{q}}c^{+}_{\vec{p}-\vec{q}\uparrow}c^{+}_{\vec{k}+\vec{q}\downarrow}c_{\vec{k}\downarrow}c_{\vec{p}\uparrow} \tag{9.7}$$

(N is number of sites) and write down the equation of motion for the momentum-dependent GF $G_{\vec{k}\uparrow}(t-t')=-i\left\langle Tc_{\vec{k}\uparrow}(t)\,c^{+}_{\vec{k}\uparrow}(t')\right\rangle$ (we consider the spin-up case; the analysis in the sin-down case is practically the same). This can be done by using the Heisenberg equation of motion for the electron operators in the Hamiltonian (9.7) and by substituting the result into the expression for the derivative of GF with respect to time (see subsection 2.2.1). The result is

$$i\frac{\partial G_{\vec{k}\uparrow}(t-t')}{\partial t}-\varepsilon\left(\vec{k}\right)G_{\vec{k}\uparrow}(t-t')+i\frac{U}{N}\sum_{\vec{p},q}\left\langle Tc^{+}_{\vec{p}+\vec{q}\downarrow}c_{\vec{p}\downarrow}c_{\vec{k}+\vec{q}\uparrow}(t)c^{+}_{\vec{k}\uparrow}(t')\right\rangle=\delta(t-t'). \tag{9.8}$$

Comparing the last equation with the Dyson equation that connects the GF and the self-energy

$$\left(i\frac{\partial}{\partial t}-\varepsilon\left(\vec{k}\right)\right)G_{\vec{k}\uparrow}(t-t')-\int\Sigma_{\vec{k}\uparrow}(t-t'')G_{\vec{k}\uparrow}(t''-t')dt''=\delta(t-t'), \tag{9.9}$$

one gets

$$\int\Sigma_{\vec{k}\uparrow}(t-t'')G_{\vec{k}\uparrow}(t''-t')dt''=\frac{U}{N}\sum_{\vec{p},\vec{q}}(-i)\left\langle Tc^{+}_{\vec{p}+\vec{q}\downarrow}c_{\vec{p}\downarrow}c_{\vec{k}+\vec{q}\uparrow}(t)c^{+}_{\vec{k}\uparrow}(t')\right\rangle, \tag{9.10}$$

or after Fourier transformation,

$$\Sigma_{\vec{k}\uparrow}(\omega)G_{\vec{k}\uparrow}(\omega) \equiv \frac{\Sigma_{\vec{k}\uparrow}(\omega)}{\omega - \varepsilon\left(\vec{k}\right) - \Sigma_{\vec{k}\uparrow}(\omega)}$$

$$= \int e^{-i\omega(t-t')}dt \frac{U}{N}\sum_{\vec{p},\vec{q}}(-i)\left\langle Tc_{\vec{k}+\vec{q}\uparrow}c^{+}_{\vec{p}+\vec{q}\downarrow}c_{\vec{p}\downarrow}(t)c^{+}_{\vec{k}\uparrow}(t')\right\rangle. \tag{9.11}$$

Thus, as it follows from the last equation, the momentum-dependent self-energy is defined by 2P GF $(-i)\left\langle Tc^{+}_{\vec{p}+\vec{q}\downarrow}c_{\vec{p}\downarrow}c_{\vec{k}+\vec{q}\uparrow}(t)c^{+}_{\vec{k}\uparrow}(t')\right\rangle$. Though this 2P GF is momentum-dependent, i.e., are nonlocal in space, one can still approximate the vertices in the diagrams by local functions. To summarize the above-said the approach should include at least 2P GFs and possibly different-order (n-particle) dynamical vertices. The development of such a methodology, called Dynamical Vertex Approach (from this place we use abbreviation DΓA for this approach) started with formulation of the mentioned in the previous chapter Dynamical Vertex Approximation algorithm with 2P approximation for the dynamical vertices [12] (see also Ref. [13] for a parallel, same-spirit, development). In DΓA, one gets the 2P dynamical vertex correction to the nonlocal susceptibility that is used to calculate the nonlocal correction to the self-energy. The approach can be characterized as an expansion of Feynman diagrams in terms of the locality of diagrams (not in traditional orders of interaction), which gives the exact solution for $n \to \infty$.

Similar in spirit to DΓA, the dual fermion (DF) approach was also proposed [14] to include nonlocal correlations into DMFT. In this method, the lattice problem is mapped on a solvable local-reference system coupled to nonlocal degrees of freedom. Then, in the new model after introducing dual fields to decouple the impurity, the Hubbard–Stratonovich transformation is made and to solve the resulting problem a PT theory is applied. Though in general dual fermions interact through the $n \geq 2$-particle vertex functions of the local reference system, in practical calculations, similar to DΓA, $n > 2$-particle vertices, are neglected (for recent developments and applications of the static and the nonequilibrium DF approaches, see Refs. [15] and [16, 17], correspondingly). Several other developments of the DΓA and DF approaches followed, like a combined DΓA-DF one-particle irreducible (1PI) vertex functional approach [18], DMFT to fRG (DMF2RG) [19–25] that uses fRG to generate the nonlocal diagrams, and the triply-irreducible local expansion (TRILEX) [26] based on the three-point fermion–boson vertex.

Other new tools include extended DMFT (EDMFT) [27, 28] that deals with the local correlations induced by nonlocal interactions and the last are mapped onto local bosonic degrees of freedom in the framework of the dual boson (DB) approach [29, 30] (recently generalized on the nonequilibrium case [31]) that includes long-range bosonic modes (for over-review, see Ref. [1]).

Since DΓA is probably the most popular and one of the most powerful nonlocal extensions of DMFT, in this chapter we mostly discuss this method paying a special attention to its relation to the many-body state-of-art Bethe–Salpeter equation (BSE) approach. In a large part of the chapter, we follow the presentation in Ref. [1].

9.2 Dynamical Vertex Approximation (DΓA)

Before discussing DΓA approach, we summarize the main elements of the 2P GF formalism, used in DΓA, beginning with definition of the n-particle GF:

$$G^{(n)}_{i_1\ldots i_{2n}}(\tau_1,\ldots,\tau_{2n}) = (-1)^n\left\langle T_\tau c_{i_1}(\tau_1)c^+_{i_2}(\tau_2)c_{i_3}(\tau_3)c^+_{i_4}(\tau_4)\ldots c_{i_{2n-1}}(\tau_{2n-1})c^+_{i_{2n}}(\tau_{2n})\right\rangle, \tag{9.12}$$

where indices i_j correspond to the coordinate/momentum, orbital, spin, and other degrees of freedom. Since we consider for simplicity the PM case, where the single-particle GF is spin-diagonal and spin-invariant, the spin index will be dropped from the notations in single-particle GF. In the case of 2P GF, the momentum and spin conservation and the spin-flip invariance allow one to significantly reduce the number of spin indices. Moreover, due to the lattice- and time-translational invariance, one of the momentum and time indices can be put equal to zero. Then, the definition of the 2P GF in the momentum-time representation can be written as:

$$\begin{aligned} G^{(2)}_{\vec{k}\sigma,\vec{k}+\vec{q}\sigma,\vec{k}'+\vec{q}\sigma',\vec{k}'\sigma'}(\tau_1,\tau_2,\tau_3,0) &= \left\langle T_\tau c_{\vec{k}\sigma}(\tau_1)c^+_{\vec{k}+\vec{q}\sigma}(\tau_2)c_{\vec{k}'+\vec{q}\sigma'}(\tau_3)c^+_{\vec{k}'\sigma'}(0)\right\rangle \\ &\equiv G^{(2)}_{\sigma\sigma',\vec{k}\vec{k}'\vec{q}}(\tau_1,\tau_2,\tau_3). \end{aligned} \tag{9.13}$$

To obtain the frequency-dependent 2P GF, the frequency transformation of GF (9.13) can be performed in the following way:

$$G^{(2)}_{\sigma\sigma',\vec{k}\vec{k}'\vec{q}}(\nu,\nu',\omega) = \int_0^\beta d\tau_1 e^{i\nu\tau_1}\int_0^\beta d\tau_2 e^{-i(\nu+\omega)\tau_2}\int_0^\beta d\tau_3 e^{i(\nu'+\omega)\tau_3}G^{(2)}_{\sigma\sigma',\vec{k}\vec{k}'\vec{q}}(\tau_1,\tau_2,\tau_3), \tag{9.14}$$

where the frequencies are assigned to Matsubara times and are chosen in the form of particle-hole representation, which corresponds to 2P GF as a scattering amplitude of particle and hole with total momentum and energy $\left(\omega,\vec{q}\right)$ (Fig. 9.1a). Alternatively, one can regard the 2P GF (9.13) to be in the particle–particle representation (with the momenta and frequencies defined in right Fig. 9.1. The first picture is more convenient for the case of strong particle–hole fluctuations, and the second—when the particle–particle fluctuations are strong.

Due to spin conservation, out of $2^4 = 16$ spin–index combinations of the 2P GF only six are possible. Out of the remaining six, two combinations $\sigma\bar{\sigma}\sigma\bar{\sigma}$ can be expressed in terms of $\sigma\sigma\bar{\sigma}\bar{\sigma}$. Thus, only four independent combinations of the spin indices in the 2P GF remain—↑↑↓↓, ↓↓↑↑, ↑↓↑↓, ↓↑↓↑. There are additional relations between the spin components due to SU(2) symmetry, but it is convenient

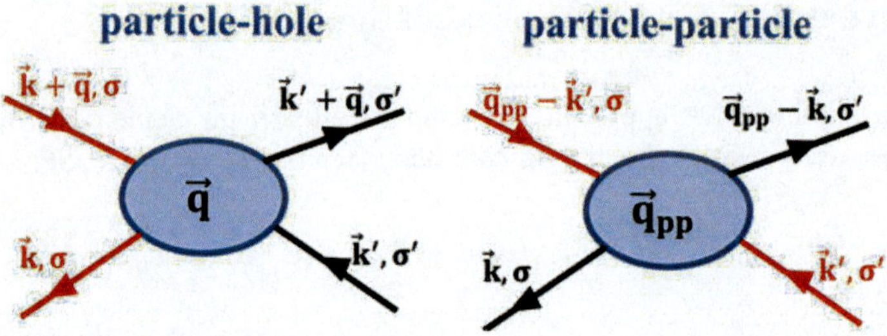

Fig. 9.1 2P GF in the particle-hole (left) and particle–particle (right) representations. The red arrows denote the "incoming" particle and hole in the left figure and the "incoming" two particles in the right figure. The right diagram can be obtained from the left one by the following transformation of the frequency–momentum variables: $\vec{q} \to \vec{q}_{pp} = \vec{q} + \vec{k} + \vec{k}'$, $\vec{k}_{pp} = \vec{k}$, $\vec{k}'_{pp} = \vec{k}'$. Here and in the next three figures, we use notations of Ref. [1]

to use two components $\uparrow\uparrow$ and $\uparrow\downarrow$ (or equivalent flipped combinations) that describe these four combinations, as in Eq. (9.13). From the results for the single-particle and 2P GFs, one can calculate the generalized susceptibility

$$\chi_{\sigma\sigma',\vec{k}\vec{k}'\vec{q}}(\nu,\nu',\omega) = G^{(2)}_{\sigma\sigma',\vec{k}\vec{k}'\vec{q}}(\nu,\nu',\omega) - \beta G_{\vec{k}}(\nu)G_{\vec{k}'}(\nu')\delta_{\omega 0}\delta_{q0}, \tag{9.15}$$

and its charge (spin) components

$$\chi^{c(s)}_{\vec{k}\vec{k}'\vec{q}}(\nu,\nu',\omega) = \chi_{\uparrow\uparrow\vec{k}\vec{k}'\vec{q}}(\nu,\nu',\omega) \pm \chi_{\uparrow\downarrow\vec{k}\vec{k}'\vec{q}}(\nu,\nu',\omega). \tag{9.16}$$

From these functions, one can obtain the physical response functions of interest by summing over $\vec{k}, \vec{k}'$ and ν, ν':

$$\chi^{c(s)}_{\vec{q}}(\omega) = \sum_{\vec{k}\vec{k}',\nu,\nu'} \chi^{c(s)}_{\vec{k}\vec{k}'\vec{q}}(\nu,\nu',\omega). \tag{9.17}$$

Thus, since the calculations of single-particle GF is a relatively simple task, the problem reduces to finding the generalized susceptibility $\chi_{\sigma\sigma',\vec{k}\vec{k}'\vec{q}}(\nu,\nu',\omega)$, defined by the 2P GF $G^{(2)}_{\sigma\sigma',\vec{k}\vec{k}'\vec{q}}(\nu,\nu',\omega)$ (in Eq. (9.15)), or the charge and the spin components of the generalized susceptibility Eq. (9.17). To accomplish the last task, it is convenient to rewrite the formulas for the charge and spin components of generalized susceptibility by separating the term with the product of the single-particle GFs (that describes separate propagation of electrons and holes) and the two-particle term (propagation with the electron–hole scatterings) written as a product of in- and outgoing electron and hole single-particle GFs and the two-particle vertex function $F^{c(s)}_{\vec{k}\vec{k}'\vec{q}}$ (see also Fig. 9.2):

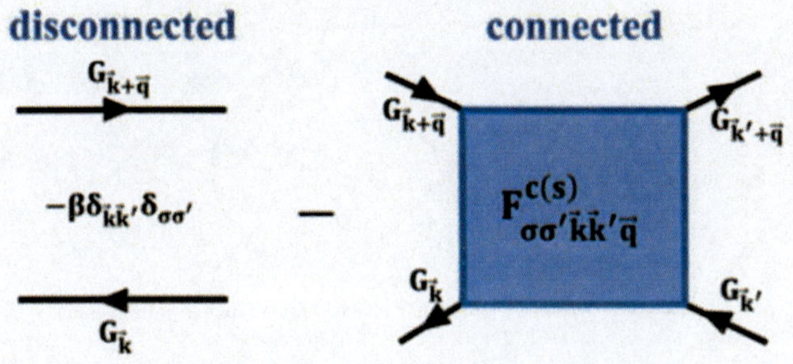

Fig. 9.2 The generalized (spin-dependent) susceptibility $\chi^{c(s)}_{\vec{k}\vec{k}'\vec{q}}$ can be represented as a sum of the disconnected (independent propagation of the electron-hole pair) and connected (the electron-hole scattering) parts (see Eq. 9.18)

$$\chi^{c(s)}_{\vec{k}\vec{k}'\vec{q}}(\nu,\nu',\omega) = -\beta G_{\vec{k}}(\nu)G_{\vec{k}+\vec{q}}(\nu+\omega)\delta_{\nu\nu'}\delta_{\vec{k}\vec{k}'}$$
$$- G_{\vec{k}}(\nu)G_{\vec{k}+\vec{q}}(\nu+\omega)F^{c(s)}_{\vec{k}\vec{k}'\vec{q}}(\nu,\nu',\omega)G_{\vec{k}'}(\nu')G_{\vec{k}'+\vec{q}}(\nu'+\omega), \quad (9.18)$$

where the 2P vertex function $F^{c(s)}_{\vec{k}\vec{k}'\vec{q}}(\nu,\nu',\omega)$ approahes $+U$ as $U \to 0$. In the Fermi liquid regime, $F^{c(s)}_{\vec{k}\vec{k}'\vec{q}}(\nu,\nu',\omega)$ is proportional to the quasiparticle scattering amplitude.

Topologically, all diagrams that contribute to $F^{c(s)}$ can be divided into four different classes: irreducible $\Lambda^{c(s)}_{\vec{k}\vec{k}'\vec{q}}(\nu,\nu',\omega)$ and reducible in one of the channels—particle–hole (*ph*) $\Phi^{c(s);\mathrm{ph}}_{\vec{k}\vec{k}'\vec{q}}(\nu,\nu',\omega)$, vertical particle–hole $(\overline{\mathrm{ph}})$ $\Phi^{c(s);\overline{\mathrm{ph}}}_{\vec{k}\vec{k}'\vec{q}}(\nu,\nu',\omega)$ and particle–particle (*pp*) $\Phi^{c(s);\mathrm{pp}}_{\vec{k}\vec{k}'\vec{q}}(\nu,\nu',\omega)$:

$$F^{c(s)}_{\vec{k}\vec{k}'\vec{q}}(\nu,\nu',\omega) = \Lambda^{c(s)}_{\vec{k}\vec{k}'\vec{q}}(\nu,\nu',\omega) + \Phi^{c(s);ph}_{\vec{k}\vec{k}'\vec{q}}(\nu,\nu',\omega) + \Phi^{c(s);\overline{ph}}_{\vec{k}\vec{k}'\vec{q}}(\nu,\nu',\omega)$$
$$+ \Phi^{c(s);pp}_{\vec{k}\vec{k}'\vec{q}}(\nu,\nu',\omega). \quad (9.19)$$

Such a decomposition of the diagrams is so-called parquet decomposition (see also Fig. 9.3). The diagram is called reducible in ph ($\overline{ph}$, *pp*) channel if it can be split into two parts by cutting two lines corresponding to a *ph* (vertical *ph, pp*) pair. Decomposition (9.19) is a fundamental part of the parquet equations formalism. To shorten notations, we will use indices $r = c$, s, and $l = ph, \overline{ph}, pp$.

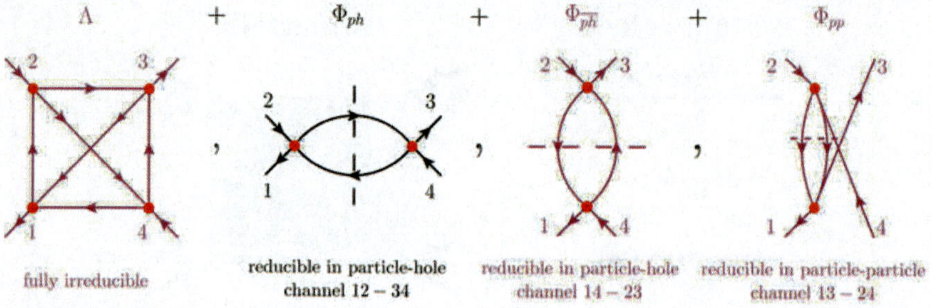

Fig. 9.3 Parquet decomposition of the one-particle irreducible vertex F into its 2P fully irreducible part Λ and three parts reducible in different channels. (Reprinted Fig. 5 with permission from (G. Rohringer, H. Hafermann, A. Toschi, A. A. Katanin, A. E. Antipov, M. I. Katsnelson, A. I. Lichtenstein, A. N. Rubtsov, and K. Held Rev. Mod. Phys. **90**, 025003 (2018). Copyright 2018 by the American Physical Society. https://doi.org/10.1103/RevModPhys.90.025003)

Let us find the relation between the parquet equation approach and the often-used BSE formalism that corresponds to an infinite summation of ladder diagrams. The BSE susceptibility is defined by equation

$$\chi^{r}_{\vec{k}\vec{k}'\vec{q}}(\nu,\nu',\omega) = \chi^{(0)}_{\vec{k}\vec{k}'\vec{q}}(\nu,\nu',\omega) - \sum_{\vec{k}_1\vec{k}_2,\nu_1,\nu_2} \chi^{(0)}_{\vec{k}\vec{k}_1\vec{q}}(\nu,\nu_1,\omega)\Gamma^{lr}_{\vec{k}_1\vec{k}_2\vec{q}}(\nu_1,\nu_2,\omega)\chi^{r}_{\vec{k}_2\vec{k}'\vec{q}}(\nu_2,\nu',\omega) \tag{9.20}$$

with the bare "bubble" susceptibility

$$\chi^{(0)}_{\vec{k}\vec{k}_1\vec{q}}(\nu,\nu_1,\omega) = -\beta G_{\vec{k}}(\nu)\mathrm{G}_{\vec{\mathrm{k}}+\vec{\mathrm{q}}}(\nu+\omega)\delta_{\nu\nu_1}\delta_{\vec{k}\vec{k}_1} \tag{9.21}$$

(see also Eq. (9.18)) and 2P irreducible in channel l vertices $\Gamma^{lr}_{\vec{k}_1\vec{k}_2\vec{q}}(\nu_1,\nu_2,\omega)$. The last functions are related to the F vertex as

$$F^{r}_{\vec{k}_1\vec{k}_2\vec{q}}(\nu_1,\nu_2,\omega) = \Gamma^{lr}_{\vec{k}_1\vec{k}_2\vec{q}}(\nu_1,\nu_2,\omega) + \Phi^{lr}_{\vec{k}_1\vec{k}_2\vec{q}}(\nu_1,\nu_2,\omega), \tag{9.22}$$

i.e., F is divided in irreducible-in-channel-l $\Gamma^{lr}_{\vec{k}_1\vec{k}_2\vec{q}}$ and reducible $\Phi^{lr}_{\vec{k}_1\vec{k}_2\vec{q}}$ parts (shown in Fig. 9.3). Using the last equation, one can establish the relationship between the parquet equation and the BSE approaches in terms of the relation between their vertices. For example, in the *ph* channel one has:

$$F^{r}_{\vec{k}_1\vec{k}_2\vec{q}}(\nu_1,\nu_2,\omega)=\Gamma^{phr}_{\vec{k}_1\vec{k}_2\vec{q}}(\nu_1,\nu_2,\omega)$$
$$+\sum_{\vec{k}_3,\nu_3}\Gamma^{phr}_{\vec{k}\vec{k}_3\vec{q}}(\nu,\nu_3,\omega)G_{\vec{k}_3}(\nu_3)G_{\vec{k}_3+\vec{q}}(\nu_3+\omega)F^{r}_{\vec{k}_3\vec{k}_2\vec{q}}(\nu_3,\nu_2,\omega) \tag{9.23}$$

It is a great simplification that due to the SU(2) symmetry the charge and spin sectors in this equation do not couple.

Equations (9.19), (9.22), and (9.23) are a system of three equations for three vertices F, Γ^l, Λ, and the single-particle GF. Thus, to be able to find the solution, three equations for vertices have to be complemented by one more equation. This equation is the Dyson equation for the single-particle GF:

$$G^{-1}_{\vec{k}}(\nu)=G^{(0)-1}_{\vec{k}}(\nu)-\Sigma_{\vec{k}}(\nu). \tag{9.24}$$

We consider the case of momentum-dependent self-energy, i.e., we assume that the self-energy is a function of one of the vertex functions that gives the momentum dependence of Σ (it is enough to express the self-energy in terms of one of the vertices, since each of them can be expressed in terms of the others). The equation that expresses the self-energy in terms of a vertex can be found from the Heisenberg equation of motion for the single-particle GF. For the F vertex it reads:

$$\Sigma_{\vec{k}}(\nu)=U\frac{n}{2}$$
$$-U\sum_{\nu',\omega,\vec{k}',\vec{q}}F^{\uparrow\downarrow}_{\vec{k}\vec{k}'\vec{q}}(\nu,\nu',\omega)G_{\vec{k}'}(\nu')G_{\vec{k}'+\vec{q}}(\nu'+\omega)G_{\vec{k}+\vec{q}}(\nu+\omega). \tag{9.25}$$

Once one of the three functions F, Γ^l, and Λ is known, the system of Eqs. (9.19), (9.22)–(9.25) can be solved. In the parquet approximation, one fixes Λ, by putting $\Lambda^{c(s)}_{\vec{k}\vec{k}'\vec{q}}(\nu,\nu',\omega)=U$ [32], and in parquet DΓA $\Lambda^{c(s)}_{\vec{k}\vec{k}'\vec{q}}(\nu,\nu',\omega)=\Lambda^{c(s)}(\nu,\nu',\omega)$ (a momentum-independent vertex, see below). Equation for the self-energy (9.25) is one of the key equations in this chapter. It shows how the momentum "enters" into the DMFT electron self-energy through a vertex function. In the case $F^{\uparrow\downarrow}_{\vec{k}\vec{k}'\vec{q}}(\nu,\nu',\omega)=U$, self-energy Eq.(9.25) is equal to the second-order IPT self-energy (Chap. 6).

To solve the system (9.19)–(9.25) is a complicated task because the equations contain many variables. Using all possible symmetries significantly simplifies the solution (for details on symmetries, see Ref. [1]). Another simplification that makes the solution possible is to neglect the momentum dependence of the vertex functions (the key assumption in the DΓA, see below). Let us consider the structure of the diagrams in this case on the example of F vertex.

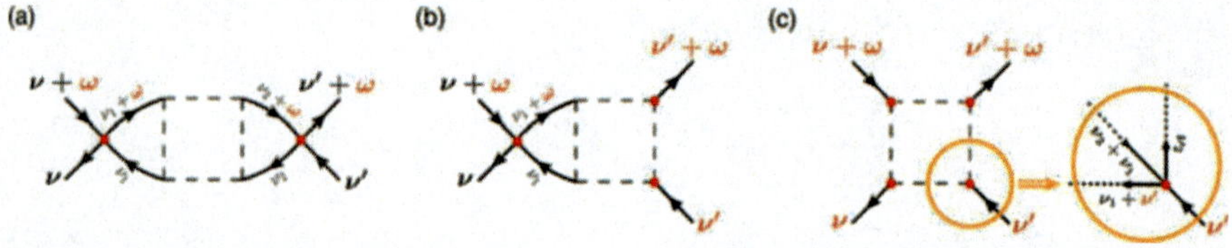

Fig. 9.4 Different types of diagrams and their frequency dependence: (**a**) the left and the right pairs of the external lines are attached to the same two bare interaction vertices, (**b**) only the left external lines are connected to the same bare vertex, (**c**) all external lines enter at different bare vertices. Diagrams depend on the external frequencies marked in red. (Reprinted Fig. 7 with permission from (G. Rohringer, H. Hafermann, A. Toschi, A. A. Katanin, A. E. Antipov, M. I. Katsnelson, A. I. Lichtenstein, A. N. Rubtsov, and K. Held Rev. Mod. Phys. **90**, 025003 (2018)). Copyright 2018 by the American Physical Society. https://doi.org/10.1103/RevModPhys.90.025003)

In general case, the vertex depends on three frequencies. There are three different types of the vertex F diagrams with respect to the external lines connected to the bare interaction vertices: (a) the left and the right pairs of external lines are attached to the same two bare vertices, (b) only the left external lines are connected to the same vertex, and (c) all four lines are connected to different vertices. The frequency dependence of these diagrams is different. The types of vertex diagrams are summarized in Fig. 9.4, where the (external) frequency arguments of the diagrams are shown in red. The first two types of diagrams are particle-reducible, while the third one, which defines Λ, is irreducible. Contribution of the last diagram to the vertex function depends on three frequencies (Fig. 9.4c, the encircled part), thus it should decay as frequency increases in all three directions in the frequency space. This is a helpful property since in DMFT one needs to consider only the low-frequency contribution of the local vertices [33]. Thus, the asymptotic dependence of F comes from reducible terms only, which is a great simplification in the calculations. On the other hand, the low-frequency dependence of the reducible parts can be rather complicated for some values of the parameters of the Hubbard model. For example, they can be divergent even for not very large Us (see Fig. 9.5 for case of the Anderson impurity model). As it was proved for the Falicov–Kimball model, these divergences are not a result of computational accuracy, but a reflection of breakdown of the perturbation theory. In general, the physical origins for the vertex divergencies in the Hubbard model are still a topic of debate (see Ref. [1] and references for both models therein).

In Ref. [34] it was shown that the breakdown of the perturbation expansion for the many-electron problem results in divergences of irreducible vertex functions. Physically, the divergence(s) corresponds to change of the effective electronic interaction from repulsive to attractive (flipping of the sign in front of the interaction potential) in some interacting channels. From the analysis of susceptibilities, it was also found that entering the non-perturbative regime causes an enhancement of the charge response that leads to phase-separation instabilities near MIT. Though, such a flip of the interaction cannot produce s-wave superconductivity or charge-density wave instabilities. In DMFT, the problem of divergencies of vertices can be eliminated by

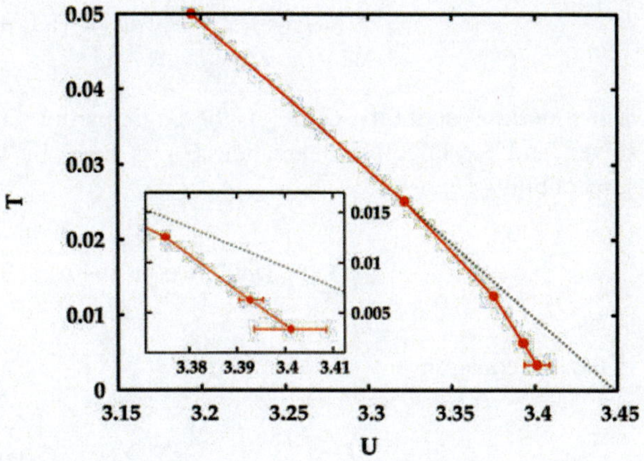

Fig. 9.5 A vertex divergence line in the Anderson impurity model. Inset: a region near zero-temperature. (Reprinted Fig. 8 with permission from (P. Chalupa, P. Gunacker, T. Schäfer, K. Held, and A. Toschi, Phys. Rev. B **97**, 245136 (2018)). Copyright 2018 by the American Physical Society. https://doi.org/10.1103/PhysRevB.97.245136)

the Bethe–Salpeter resummation of the ladder diagrams (for a more extended discussion on the vertex diagram divergencies, see Ref. [1] and references in this review).

Several approaches have been used to calculate the local dynamical vertices, including IPT [13], ED [35], NRG [36, 37], functional RG [38–40], and CT-QMC [41, 42]. In all these approaches, one basically calculates the 2P GF, similar to the calculations of the single-particle GF, from the impurity equation with different solvers. Then, from the results for the 2P GF one obtains the generalized susceptibility Eq. (9.15) and then, by using, e.g., Eq. (9.18),—the vertex function. Since the solver plays an important role in obtaining the vertex correction to DMFT, in the next section we give details on some of the most important solvers used in the 2P case.

9.3 Solvers for 2P GFs

Below, we summarize details of the calculation of the 2P GF or generalized susceptibility by using the IPT [13] and the strong- and weak-coupling expansion CT-QMC (HYB-CT-QMC and INT-CT-QMC) formalisms [42, 43]. In the IPT approach [13], Kusunose proposed to calculate the charge component of the local susceptibility in the ladder approximation in the charge channel as:

$$\chi_{ch}(i\omega_n) = \frac{\chi^{(0)}(i\omega_n)}{1 + U_{ch}\chi^{(0)}(i\omega_n)}, \tag{9.26}$$

where

$$\chi^{(0)}(i\omega_n) = -T\sum_m G(i\omega_m)G(i\omega_n + i\omega_m) \tag{9.27}$$

is one-loop susceptibility, $G(i\omega_m)$ is the single-particle GF obtained by using the IPT solver, and the effective interaction U_{ch} is fixed by the exact sum rule for the susceptibility

$$T\sum_n \chi_{ch}(i\omega_n) = \frac{1}{2}[n(1-n) + 2D]. \tag{9.28}$$

In the last equation,

$$D = \left(\frac{n}{2}\right)^2 + \frac{T}{U}\sum_n \Sigma(i\omega_n)G(i\omega_n) \tag{9.29}$$

is the double occupancy obtained with IPT. Once the susceptibility is calculated, one obtains the one-frequency local vertex function $\Gamma^{ch}(i\omega_n) = \chi^{(0)-1}(i\omega_n) - \chi_{ch}^{-1}(i\omega_n)$, substitutes it in the BSE (9.20) instead of $\Gamma^{lr}_{\vec{k}_1\vec{k}_2\vec{q}}(\nu_1, \nu_2, \omega)$, obtains the three-frequency susceptibility and then—the vertex function $F^{c(s)}_{\vec{k}\vec{k}'\vec{q}}(\nu, \nu', \omega)$ from Eq. (9.18). For the spin channel, the procedure is similar, with the substitution $U_{ch} \to -U_{sp}$ in Eq. (9.26) (see Ref. [13]).

The scheme above is an approximate solution. Now, let us give some details of the calculation of the local susceptibility, which defines the vertex function, by using a numerically exact scheme—the 2P generalized HYB-CT-QMC approach. In this QMC approach, the single particle and the 2P GFs are measured in the time interval $[0, \beta]$ as:

$$\begin{aligned} G_{ab}(\tau - \tau') &\equiv -\langle T_\tau a_a(\tau) c_b^+(\tau')\rangle \\ &= -T\left\langle \sum_{\alpha,\beta=1}^{k_C} M^C_{\beta\alpha}\delta^-\left(\tau - \tau', \tau^e_\alpha - \tau^s_\beta\right)\delta_{a\alpha}\delta_{b\beta}\right\rangle_{MC}, \end{aligned} \tag{9.30}$$

$$\begin{aligned} G_{abcd}(\tau_a, \tau_b, \tau_c, \tau_d) &\equiv \langle T_\tau a_a(\tau_a) c_b^+(\tau_b) a_c(\tau_c) c_d^+(\tau_d)\rangle \\ &= T\left\langle \sum_{\alpha,\beta,\gamma,\delta=1}^{k_C} \left(M^C_{\beta\alpha}M^C_{\delta\gamma} - M^C_{\delta\alpha}M^C_{\beta\gamma}\right)\delta^-\left(\tau_{ab}, \tau^e_\alpha - \tau^s_\beta\right)\right. \\ &\left.\delta^-\left(\tau_{cd}, \tau^e_\gamma - \tau^s_\delta\right)\delta^+\left(\tau_{ad}, \tau^e_\alpha - \tau^s_\delta\right)\delta_{a\alpha}\delta_{b\beta}\delta_{c\gamma}\delta_{d\delta}\right\rangle_{MC} \end{aligned} \tag{9.31}$$

In Eqs. (9.30) and (9.31), $M^C_{\alpha\beta}$ are the matrix elements of the inverse of the hybridization function $\Delta_{\alpha\beta}$ (with removed αth raw and βth column) for the configuration corresponding to a given segment of the occupied fermionic states with starting points τ_i^s and ending points τ_i^e $(i = 1, \ldots, k_C)$ (see Refs. [44, 45] and Chap. 6). The delta functions in Eqs. (9.30) and (9.31) are defined as

$$\delta^-(\tau, \tau') = sgn\,(\tau')\delta(\tau - \tau' - \theta(-\tau')\beta), \tag{9.32}$$

$$\delta^+(\tau, \tau') = \delta(\tau - \tau' - \theta(-\tau')\beta). \tag{9.33}$$

Recently, equations of motion for the GFs for the multi-orbital Anderson impurity model that relate the one-particle and 2P GFs to up to six-particle correlators were derived [46]. They were implemented in HYB-CTQMC and the results of the calculations showed that using this tool one can get rid of high noise levels at large Matsubara frequencies, the problem known also in the one-particle GF case.

In the weak-interaction expansion case (INT-CT-QMC), the procedure is similar to the one for the one-particle calculations (see Chap. 6). Namely, one defines the GFs that contribute to the U-expansion terms for the 2P GF (top line in Eq. (9.31), for simplicity we consider the PM case):

$$G_{\sigma\tau_1,\ldots,\tau_k}(\tau, \tau', \bar{\tau}, \bar{\tau}') = -\frac{\langle T_\tau d_\sigma(\tau) d_\sigma^+(\tau') d_\sigma(\bar{\tau}) d_\sigma^+(\bar{\tau}') n_\uparrow(\tau_1) n_\downarrow(\tau_1) \ldots n_\uparrow(\tau_k) n_\downarrow(\tau_k)\rangle_0}{\langle n_\uparrow(\tau_1) n_\downarrow(\tau_1) \ldots n_\uparrow(\tau_k) n_\downarrow(\tau_k)\rangle_0}. \tag{9.34}$$

Then, one can get the 2P GF by averaging GFs (9.34) over configurations:

$$G_\sigma(\tau, \tau', \bar{\tau}, \bar{\tau}') = \langle G_{\sigma\tau_1,\ldots,\tau_k}(\tau, \tau', \bar{\tau}, \bar{\tau}')\rangle_{MC}. \tag{9.35}$$

Explicitly, the functions (9.34) in the average brackets in Eq. (9.35) are

$$-\det \begin{bmatrix} \mathcal{G}_\sigma(\tau - \tau') & \mathcal{G}_\sigma(\tau - \bar{\tau}') & \mathcal{G}_\sigma(\tau - \tau_1) & \mathcal{G}_\sigma(\tau - \tau_2) & \vdots & \mathcal{G}_\sigma(\tau - \tau_k) \\ \mathcal{G}_\sigma(\bar{\tau} - \tau') & \mathcal{G}_\sigma(\bar{\tau} - \bar{\tau}') & \mathcal{G}_\sigma(\bar{\tau} - \tau_1) & \mathcal{G}_\sigma(\bar{\tau} - \tau_2) & \vdots & \mathcal{G}_\sigma(\bar{\tau} - \tau_k) \\ \mathcal{G}_\sigma(\tau_1 - \tau') & \mathcal{G}_\sigma(\tau_1 - \bar{\tau}') & \mathcal{G}_\sigma(0) - \alpha_\sigma & \mathcal{G}_\sigma(\tau_1 - \tau_2) & \vdots & \mathcal{G}_\sigma(\tau_1 - \tau_k) \\ \mathcal{G}_\sigma(\tau_2 - \tau') & \mathcal{G}_\sigma(\tau_2 - \bar{\tau}') & \mathcal{G}_\sigma(\tau_2 - \tau_1) & \mathcal{G}_\sigma(0) - \alpha_\sigma & \vdots & \mathcal{G}_\sigma(\tau_2 - \tau_k) \\ \vdots & \vdots & \vdots & \vdots & \vdots & \vdots \\ \mathcal{G}_\sigma(\tau_k - \tau') & \mathcal{G}_\sigma(\tau_k - \bar{\tau}') & \mathcal{G}_\sigma(\tau_k - \tau_1) & \mathcal{G}_\sigma(\tau_k - \tau_2) & \vdots & \mathcal{G}_\sigma(0) - \alpha_\sigma \end{bmatrix} \tag{9.36}$$

(in Eq.(9.34), according to the Wick theorem, the denominator cancels with the disconnected diagrams in the numerator).

Now, when we derived the parquet equations and defined how to calculate the dynamical local-in-space vertices (by using the local three-frequency susceptibility) with different approaches, we move to a discussion of main details of the DΓA approach.

9.4 DΓA: Parquet and Ladder Approximations

In DMFT, the skeleton diagrams for self-energy are local (Fig. 9.6). Since the self-energy is an one-particle (irreducible) vertex, Toschi, Katanin, and Held suggested [12] that it is natural to generalize DMFT on the case where the n-particle vertices are local and the vertices (including the self-energy) with the lower number of particles (lower-order vertices) can be nonlocal. Thus, taking the limit $n \to \infty$ one can obtain the exact solution with non-local vertices in all diagrams. However, usually the calculations are limited by case $n = 2$, and $n = 3$ case is sometimes used to estimate the accuracy of the $n = 2$ (2P) approximation [47].

Local-in-space Λ showed itself to be a good approximation even in the 2D case in the parameter region with strong nonlocal fluctuations, as DCA calculations show [40]. On the other hand, purely local Λ corresponds to absence of any ladder diagrams, i.e., to absence of collective (spin and charge density wave and other) excitations responsible for strong nonlocal correlations. At the same time, in most calculations it is assumed that vertex F is momentum-dependent. At present status of the theory, magnons and other nonlocal correlation effects are usually described in ladder approximation in terms of the bare U, which is, strictly speaking, appropriate only in the case of weak coupling. Since local Λ is also often small, DΓA with such a

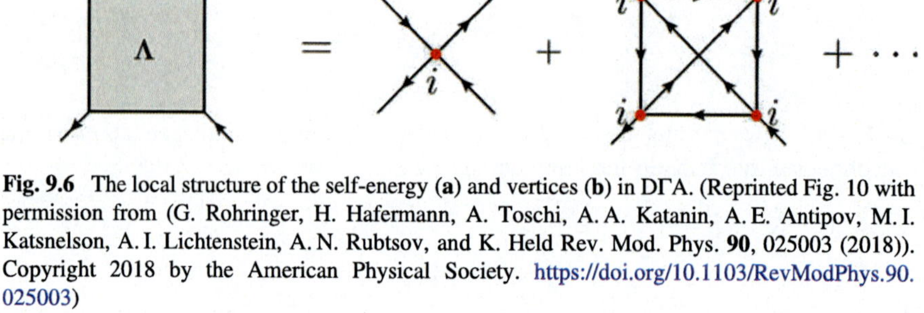

Fig. 9.6 The local structure of the self-energy (**a**) and vertices (**b**) in DΓA. (Reprinted Fig. 10 with permission from (G. Rohringer, H. Hafermann, A. Toschi, A. A. Katanin, A. E. Antipov, M. I. Katsnelson, A. I. Lichtenstein, A. N. Rubtsov, and K. Held Rev. Mod. Phys. **90**, 025003 (2018)). Copyright 2018 by the American Physical Society. https://doi.org/10.1103/RevModPhys.90.025003)

local Λ instead of U allows one to describe at the same time effects generated by strong local correlations, like Mott transitions, and by nonlocal correlations, like collective excitations, at all length scales.

9.4.1 Parquet DΓA

In parquet DΓA, one obtains F and the electron self-energy for a finite-dimensional system (using the parquet equations and BSE) from the local Λ obtained by solving the Anderson impurity problem. Even though there are several efficient implementations of the parquet DΓA equations applied to different systems [48, 49], the approach is rather costly, especially in the case near phase transitions (where many k-points are needed) and in the multiorbital systems (real materials).

9.4.2 Ladder DΓA

Thus, for these cases, further approximation(s) is (are) needed. The most straightforward approximation is the ladder approximation [12, 13, 50], where only one of the reducible terms Φ is assumed to be momentum-dependent (this term can be with $r = s$ in particle–hole sectors for magnetic states, or $r = \uparrow\downarrow$ and $l = pp$ for superconductivity, etc.). In this case, parquet equations reduce to the BSEs.

As an example, let us derive the ladder DΓA equations in the magnetic case with $l = ph$ or $\overline{ph}$, where Λ and Φ^{pp} are local, which gives (see Eq. (9.19))

$$F^{\uparrow\downarrow}_{\vec{k}\vec{k}'\vec{q}}(\nu,\nu',\omega) = \Lambda^{\uparrow\downarrow}(\nu,\nu',\omega) + \Phi^{\uparrow\downarrow \mathrm{ph}}_{\vec{k}\vec{k}'\vec{q}}(\nu,\nu',\omega) + \Phi^{\uparrow\downarrow;\overline{\mathrm{ph}}}_{\vec{k}\vec{k}'\vec{q}}(\nu,\nu',\omega) + \Phi^{\uparrow\downarrow \mathrm{pp}}(\nu,\nu',\omega). \tag{9.37}$$

Though the corresponding 2P vertices still could be nonlocal, it is assumed that Γ^{ph} and $\Gamma^{\overline{ph}}$ are momentum-independent. This gives for $l = ph$ (Eq. (9.23)):

$$F^{(lad)r}_{\vec{q}}(\nu,\nu',\omega) = \Gamma^{r}(\nu,\nu',\omega) + \Phi^{r}_{\vec{q}}(\nu,\nu',\omega) = \Gamma^{phr}(\nu,\nu',\omega) + \sum_{\vec{k}_1,\nu_1} \Gamma^{phr}(\nu,\nu_1,\omega) G^{(loc)}_{\vec{k}_1}(\nu_1) G^{(loc)}_{\vec{k}_1+\vec{q}}(\nu_1+\omega) F^{(lad)r}_{\vec{q}}(\nu_1,\nu',\omega), \tag{9.38}$$

where $r = c, s$, $G^{(loc)}_{\vec{k}_1}(\nu_1) = 1/\left[G^{(0)-1}_{\vec{k}_1}(\nu_1) - \Sigma(\nu_1)\right]$ (and the same for $l = \overline{ph}$ with $r = s$). Combining Eqs. (9.37) and (9.38), one gets

$$
\begin{aligned}
F^{\uparrow\downarrow}_{\vec{k}\vec{k}'\vec{q}}(\nu,\nu',\omega) = & \frac{1}{2}\left(F^{(lad)c}_{\vec{q}}(\nu,\nu',\omega) - F^{(lad)s}_{\vec{q}}(\nu,\nu',\omega)\right) \\
& - F^{(lad)s}_{\vec{k}'-\vec{k}}(\nu,\nu+\omega,\nu'-\nu) \\
& - \frac{1}{2}\left(F^{(lad)c}(\nu,\nu',\omega) - F^{(lad)s}(\nu,\nu',\omega)\right),
\end{aligned} \tag{9.39}
$$

where the last-line local terms give the double-counting correction to the local contributions to the vertex. Substitution of this result into Eq. (9.25) gives the momentum-dependent self-energy.

It should be noted that though the ladder DΓA greatly simplifies the calculations comparing to the parquet DΓA, the susceptibility obtained by using Eq. (9.38) violates some sum rules, e.g., $\sum\limits_{\vec{k},\omega}\chi^{\uparrow\uparrow}_{\vec{k}}(\omega) = \frac{n}{2}\left(1 - \frac{n}{2}\right)$, which results in a wrong high-frequency asymptotic of the self-energy [1, 12, 51]. To remove this problem, one can introduce into the formula for susceptibility a Moriya-type [52] screening length λ_r term: $\chi^{r-1}_{\vec{k}}(\omega) \rightarrow \chi^{r-1}_{\vec{k}}(\omega) + \lambda_r$, which also corrects the self-energy [51, 53]. Such a renormalization of the inverse susceptibility can be justified by arguing that DMFT usually overestimates the correlation lengths [1]. Another potential problem—divergencies of Γ^r—does not exist in DΓA, since $\Gamma^r = F^r/(1 + F^r G^{(loc)} G^{(loc)})$ and the BSE (9.20) can be reformulated in terms of the vertex F^r [1, 54].

Let us mention some generalizations of the DΓA approach on different cases: multiorbitals [55] (an *ab initio* DΓA code to study multiorbital materials was developed in Ref. [56]), one-atom case [57], the case of molecules [58], and the attractive Hubbard model [59, 60]. Finally, let us mention some other new developments in the field. A truncated unity parquet solver (TUPS) that solves the parquet equations using a truncated form-factor basis for the fermionic momenta (that allows to treat fluctuations from different scattering channels on equal footing) was recently implemented [61]. Using this approach, it was shown that the parquet approximation might satisfy the Mermin–Wagner theorem [62]. A recently developed approach based on classification of diagrams according to the principle of single-boson exchange [63, 64] seems to be a powerful alternative to the parquet decomposition. As it was shown in Ref. [63], the single-boson exchange decomposition gives a closed set of equations for the Hedin three-leg vertex, the polarization and the electronic self-energy. A fermion-boson vertex approach to calculate the 2P correlation functions in presence of an external field was proposed in Ref. [65]. Analytical expressions for the momentum-dependent susceptibilities for the solution of the BSE in the case of strong coupling were obtained in Ref. [66].

9.5 DΓA *versus* DMFT for 3D and 2D Systems

We limit the discussion by the 3D and 2D cases, since in the 1D case with a low coordination number there are more appropriate approaches to solve the problem than DMFT (especially, the Bethe ansatz), thus 1D systems are basically beyond the scope of DMFT where it is expected to give rather meaningless results. Therefore, there is no reason to compare 1D DΓA and DMFT solutions.

9.5.1 3D Case

The magnetic phase diagram of the 3D Hubbard model obtained with DMFT, DΓA (ladder approximation), and several other nonlocal approaches are shown in Fig. 9.7. As it follows from this figure, due to spin fluctuations DΓA (and also DF) significantly reduces the DMFT Néel temperature, especially in the strong-coupling regime, where it agrees with the results of the Heisenberg model, contrary to DMFT and several other methods. For intermediate *U*s (~8t–10t), the DΓA and DF results agree well with the DCA and diagrammatic determinant Monte Carlo results. Overall, DΓA and DF results for the Néel temperature agree well. The DΓA critical exponents for the susceptibility and correlation length are in agreement with

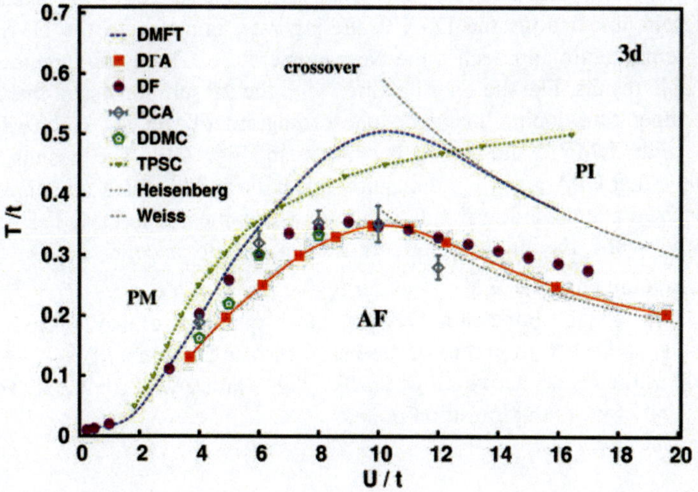

Fig. 9.7 The temperature-U magnetic phase diagram of the 3D Hubbard model obtained with DMFT, DΓA, and other nonlocal approaches (see also work [1] and references therein). (Reprinted Fig. 22 with permission from (G. Rohringer, H. Hafermann, A. Toschi, A. A. Katanin, A. E. Antipov, M. I. Katsnelson, A. I. Lichtenstein, A. N. Rubtsov, and K. Held Rev. Mod. Phys. **90**, 025003 (2018)). Copyright 2018 by the American Physical Society. https://doi.org/10.1103/RevModPhys.90.025003)

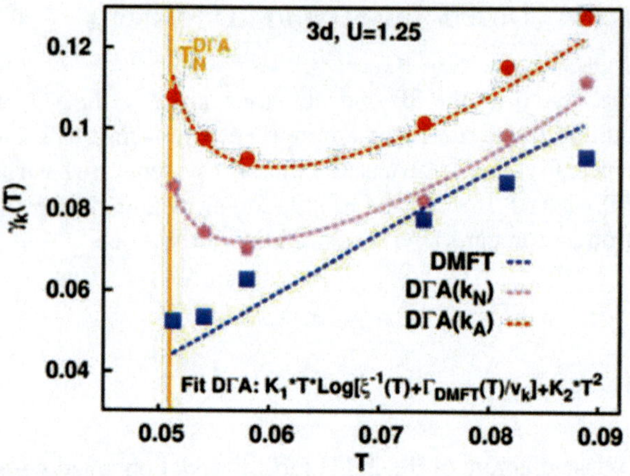

Fig. 9.8 The DMFT and DΓA results for the scattering rate as function of *T/t* in the 3D Hubbard model for the nodal, $\vec{k}_N = (\pi/2, \pi/2, \pi/2)$, and antinodal, $\vec{k}_A = (0, \pi/2, \pi)$, *k*-points (for details, see Ref. [53]). (Reprinted top Fig. 4 with permission from (G. Rohringer, and A. Toschi, Phys. Rev. B **94**, 125144 (2016)). Copyright 2016 by the American Physical Society. https://doi.org/10.1103/PhysRevB.94.125144)

universality class of the 3D Heisenberg model, and are rather different from the DMFT results [1].

Temperature dependencies of the DMFT and DΓA scattering rates are also rather different (Fig. 9.8). In particular, due to an enhanced scattering of electrons on nonlocal spin fluctuations the DΓA scattering rate, contrary to the DMFT one, grows as temperature approaches the Néel temperature. This is in agreement with other reliable results, like the ones obtained with the 2P self-consistent theory [67].

The temperature-doping magnetic phase diagram of the 3D Hubbard model obtained with DMFT and DΓA is shown in Fig. 9.9. The results of the Figure show that DΓA gives a significantly lower T_N and the DΓA quantum critical point is located at a much lower value of doping as compared to DMFT. Even more importantly, when the nonlocal effects are taken into account the AF pattern becomes incommensurate with an ordering vector $\vec{Q} = (\pi, \pi, \pi - Q_z)$ (see also Refs. [1, 68], where connection between these quantum critical properties, the Kohn points, and other properties of the Fermi surface is discussed).

Overall, as the results above show, DMFT works pretty well for the 3D Hubbard model, except close to the transition points.

9.5.2 2D Case

The nonlocal corrections play a more dramatic role in the 2D case. Though DMFT gives a finite-temperature mean-field transition to the AFM state in the 2D Hubbard

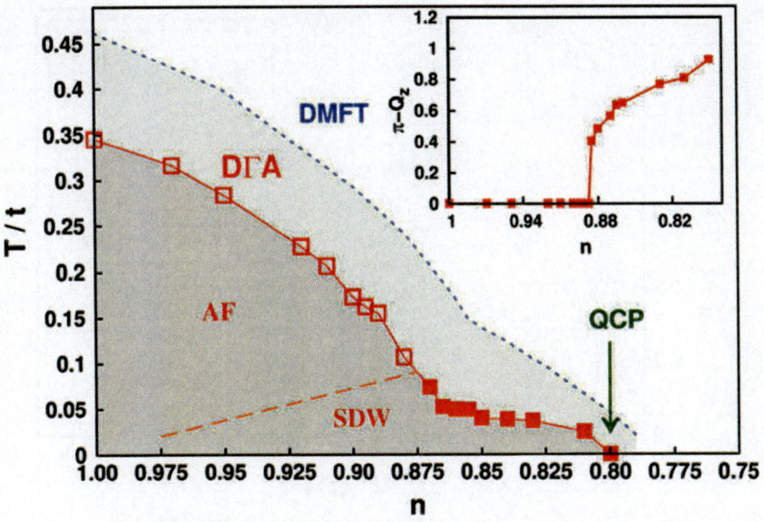

Fig. 9.9 Temperature-doping magnetic phase diagram of the 3D Hubbard model at $U = 4\sqrt{6}t$ obtained with the DMFT and DΓA. Open squares correspond to commensurate AFM and filled—incommensurate spin-density wave (SDW) ordering obtained with DGA. Inset: Doping dependence of the degree of incommensurability, i.e., "deviation" from the AFM order (see Ref. [68] for more details). (Reprinted Fig. 1 with permission from (T. Schafer, A.A. Katanin, K. Held, and A. Toschi, Phys. Rev. Lett. 119, 046402 (2017)). Copyright 2017 by the American Physical Society. https://doi.org/10.1103/PhysRevLett.119.046402)

model, the Mermin–Wagner theorem [62] states that at finite temperatures the long-range order in 2D systems with a short-range interaction is impossible, i.e., that continuous symmetries cannot be spontaneously broken in this case. Thus, fluctuations that destroy the mean-field order should play an important role at long length scales. Below, we consider the case of the 2D Hubbard model on the square lattice (for triangular lattice results, see Ref. [1]).

Let us begin with cases of half-filling. Possibility of MIT in the 2D Hubbard model is a topic of active debate for more than 40 years, beginning from the early work by Castellani et al. [69]. There is general agreement that the ground state of the 2D Hubbard model at half-filling is an AFM insulating state. When the PM order is enforced, DMFT gives a first-order MIT transition, with the boundary curve ending with a second-order critical point at $U_c \simeq 10t$ (Fig. 9.10) [71, 72]. Inclusion of short-range AF correlations with different methods [72–74] significantly reduces the critical U (to ~$6t$) and the width of the coexistence region. Moreover, in the PI phase the slope of the transition line also changes due to strong reduction of the entropy. Taken into account long-range AF fluctuations with ladder DΓA, ladder DF [75], and several other approaches (see Ref. [1]) transform the MIT line into a crossover area at small Us (shown in color in Fig. 9.10). The origin of this low-T insulating phase is different from the one obtained with DMFT (see below).

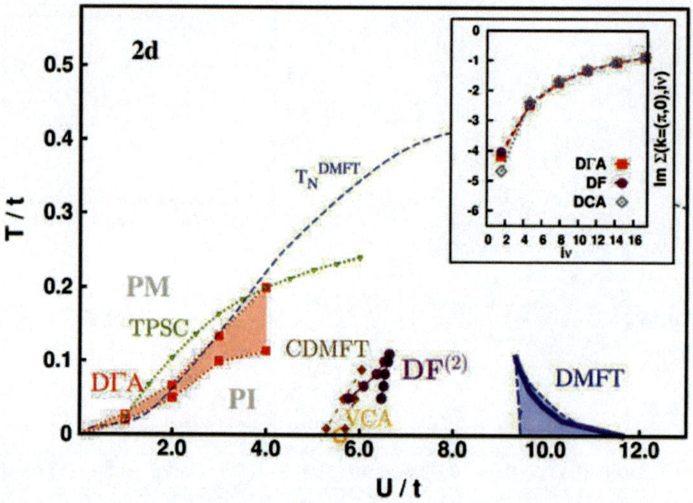

Fig. 9.10 Main figure: Temperature-U phase diagram of the 2D square-lattice Hubbard model at half-filling as obtained with different approaches (see the text). The dashed blue line is the DMFT Néel temperature. Inset: the electron self-energy as a function of the Matsubara frequency at the antinodal point $(\pi, 0)$ at $U = 8t$ and $T = 0.5t$ obtained with different approaches, showing a good agreement of the solutions (from Ref. [70]). (Reprinted Fig. 26 with permission from (G. Rohringer, H. Hafermann, A. Toschi, A. A. Katanin, A. E. Antipov, M. I. Katsnelson, A. I. Lichtenstein, A. N. Rubtsov, and K. Held Rev. Mod. Phys. 90, 025003 (2018)). Copyright 2018 by the American Physical Society. https://doi.org/10.1103/RevModPhys.90.025003)

The DΓA results for the spin susceptibility [53] (Figs. 9.11c and d) show that the susceptibility as function of distance demonstrates AF oscillations. At high temperatures (PM phase), the oscillations decay over a few unit cells (decay length $\xi \simeq 4$), while in the PI regime, the oscillations survive over a thousand of the lattice spacings ($\xi \simeq 1000$). Thus, at a weak coupling one needs long-scale AF fluctuations to open the spectral gap. Comparison of the DMFT and DΓA kinetic energies at different Us (Figs. 9.11e and f) shows that at large Us AFM fluctuations lead to a kinetic (Heisenberg superexchange) energy gain as compared to the (PM) DMFT solution. This is opposite to the results for weak interaction, where the DΓA kinetic energy is smaller (but potential energy larger due to AFM fluctuations, like in a Slater AFM) as compared to the DMFT solution [53, 75].

Another important area of application of the DΓA is response of the 2D Hubbard model to a perturbation by sources of light. In standard semiconductors, the excitations that arise due to the solid-light interaction are exciton polaritons. In Ref. [76] it was shown that in the half-filled 2D Hubbard model the corresponding excitations are π-tons—glued by AFM and CDW fluctuations (the strongest at the k-point $(\pi, \ldots, \pi)$) two particle-two hole pairs coupled to the light. As it was demonstrated in paper [76], π-tons yield the leading vertex correction to the optical conductivity in both insulating and metallic phases (the results for the optical conductivity and the

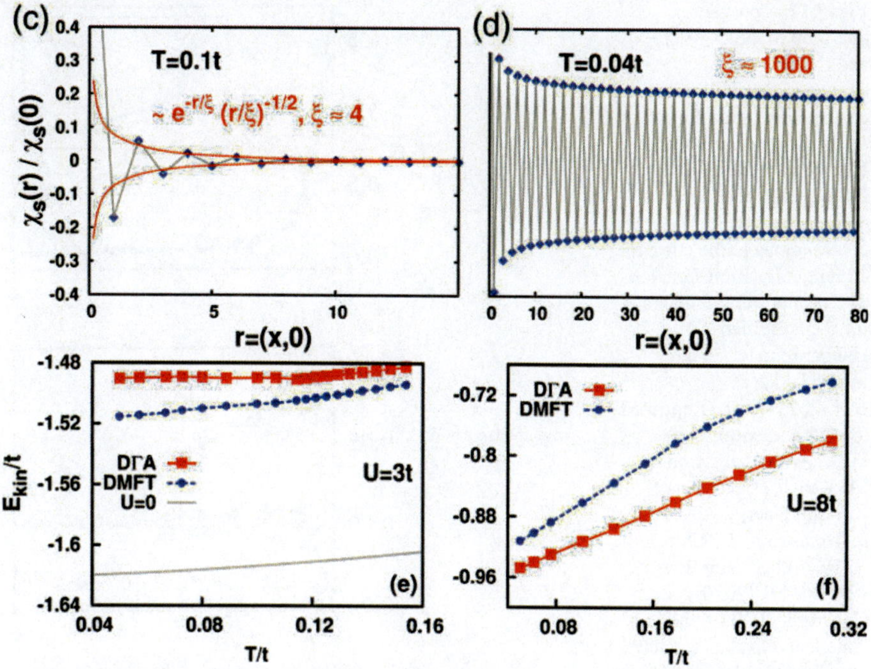

Fig. 9.11 Top: Ladder DΓA real-space spin susceptibility at $U = 2t$ in the PM (**c**) and PI (**d**) phase in the half-filled 2D Hubbard model. Bottom: DMFT and DΓA kinetic energies of the half-filled 2D Hubbard model as function of temperature at (**e**) $U = 3t$ (**e**) and (**f**) $U = 8t$. From Ref. [1] that reproduces original figures (**e**), (**f**) from Ref. [53]. (Reprinted Figs. 28c–f with permission from (G. Rohringer, H. Hafermann, A. Toschi, A. A. Katanin, A. E. Antipov, M. I. Katsnelson, A. I. Lichtenstein, A. N. Rubtsov, and K. Held Rev. Mod. Phys. 90, 025003 (2018)). Copyright 2018 by the American Physical Society. https://doi.org/10.1103/RevModPhys.90.025003. Reprinted Fig. 28e, 28f with permission from (G. Rohringer, and A. Toschi, Phys. Rev. B **94**, 125144 (2016)). Copyright 2016 by the American Physical Society. https://doi.org/10.1103/PhysRevB.94.125144)

current–current correlation function with contributions from different vertex–correction channels are shown in Fig. 9.12).

Away from half-filling, a large part of studies on the 2D Hubbard model was on possible superconductivity, the problem related to cuprates. It was established that in addition to local dynamical fluctuations, spatial fluctuations are needed to obtain superconductivity [1]. We discussed the cluster DMFT superconducting solutions of the 2D Hubbard model in Chap. 5. In the nonlocal approaches, superconducting solution was also found. In particular, DΓA [77] and several other approaches [11, 78] give a superconducting dome region in the temperature-doping phase diagram (contrary to DF [79–82]; see, however, recent results in Ref. [83], where a superconducting dome region in the phase diagram was also obtained in the case of dual parquet scheme with DF representation). It was shown in Ref. [77] that the frequency dependence of the local vertex is needed to obtain the experimentally observed T_c.

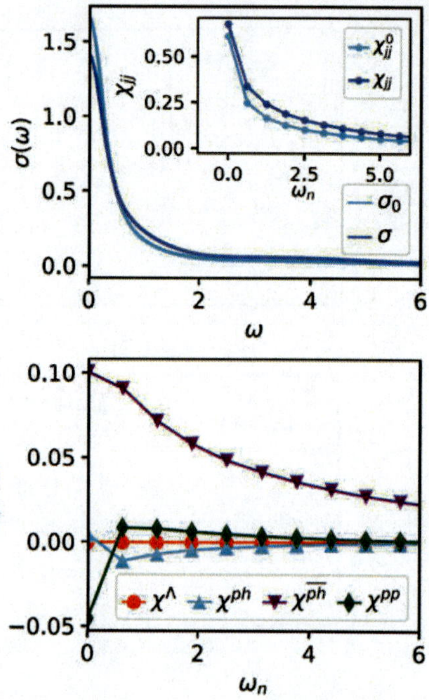

Fig. 9.12 Top: the one-loop (bubble-, "zero"-) and full optical conductivities and current-current correlation functions (inset) as functions of the Matsubara frequency. Bottom: different vertex correction–channel contributions to the current–current correlation function. The results were obtained for the half-filled 2D Hubbard model on the square lattice at $U = 4t$, $T = 0.1t$. (Reprinted top left and bottom left Figs. 2 with permission from (A. Kauch, P. Pudleiner, K. Astleithner, P. Thunström, T. Ribic, and K. Held Phys. Rev. Lett. **124**, 047401 (2020)). Copyright 2020 by the American Physical Society. https://doi.org/10.1103/PhysRevLett.124.047401)

For studies of the pseudogap phase (that shapes the dome curve) with a nonlocal approach, we refer the reader to work [84], where charge-ordering and some other effects of doping were analyzed by using the local-vertex correction. Role of the vertex correction was also studied in the case of "normal" conductivity in the 2D Hubbard model away from half-filling [85] and it was found that the vertex correction gives a large contribution to conductivity, except the high-temperature case where the local self-energy might be a good approximation. Other calculations away from half-filling showed that taken into account local dynamical vertex strongly affects the nonlocal spin correlations, i.e., the momentum-dependent susceptibility [86]. Contrary to RPA and several other approximations that give commensurate Néel AFM state in the 2D Hubbard model, the vertex corrections favor an incommensurate ordering wave vectors away from (π,π).

References

1. Rohringer, G., Hafermann, H., Toschi, A., Katanin, A.A., Antipov, A.E., Katsnelson, M.I., Lichtenstein, A.I., Rubtsov, A.N., Held, K.: Rev. Mod. Phys. **90**, 025003 (2018)
2. Schiller, A., Ingersent, K.: Phys. Rev. Lett. **75**, 113 (1995)
3. Pairault, S., Senechal, D., Tremblay, A.-M.S.: Phys. Rev. Lett. **80**, 5389 (1998)
4. Stanescu, T.D., Kotliar, G.: Phys. Rev. B. **70**, 205112 (2004)

5. Hettler, M.H., Tahvildar-Zadeh, A.N., Jarrell, M., Pruschke, T., Krishnamurthy, H.R.: Phys. Rev. B. **58**, R7475 (1998)
6. Lichtenstein, A.I., Katsnelson, M.I.: Phys. Rev. B. **62**, R9283 (2000)
7. Kotliar, G., Savrasov, S.Y., Pálsson, G., Biroli, G.: Phys. Rev. Lett. **87**, 186401 (2001)
8. Sun, P., Kotliar, G.: Phys. Rev. B. **66**, 085120 (2002)
9. Biermann, S., Aryasetiawan, F., Georges, A.: Phys. Rev. Lett. **90**, 086402 (2003)
10. Sadovskii, M.V., Nekrasov, I.A., Kuchinskii, E.Z., Pruschke, T., Anisimov, V.I.: Phys. Rev. B. **72**, 155105 (2005)
11. Kitatani, M., Tsuji, N., Aoki, H.: Phys. Rev. B. **92**, 085104 (2015)
12. Toschi, A., Katanin, A.A., Held, K.: Phys. Rev. B. **75**, 045118 (2007)
13. Kusunose, H.: J. Phys. Soc. Jpn. **75**, 054713 (2006)
14. Rubtsov, A.N., Katsnelson, M.I., Lichtenstein, A.I.: Phys. Rev. B. **77**, 033101 (2008)
15. van Loon, E.G.C.P., Katsnelson, M.I., Hafermann, H.: Phys. Rev. B. **98**, 155117 (2018)
16. Chen, F., Cohen, G., Galperin, M.: Phys. Rev. Lett. **122**, 186803 (2019)
17. Zhou, C., Guo, H.: Phys. Rev. B. **99**, 075414 (2019)
18. Rohringer, G.A., Toschi, H., Hafermann, K., Held, V., Anisimov, I., Katanin, A.A.: Phys. Rev. B. **88**, 115112 (2013)
19. Taranto, C., Andergassen, S., Bauer, J., Held, K., Katanin, A., Metzner, W., Rohringer, G., Toschi, A.: Phys. Rev. Lett. **112**, 196402 (2014)
20. Eckhardt, C.J., Schober, G.A.H., Ehrlich, J., Honerkamp, C.: Phys. Rev. B. **98**, 075143 (2018)
21. Vilardi, D., Taranto, C., Metzner, W.: Phys. Rev. B. **96**, 235110 (2017)
22. Tagliavini, A., Hille, C., Kugler, F.B., Andergassen, S., Toschi, A., Honerkamp, C.: SciPost Phys. **6**, 009 (2019)
23. Vilardi, D., Taranto, C., Metzner, W.: Phys. Rev. B. **99**, 104501 (2019)
24. Katanin, A.A.: Phys. Rev. B. **99**, 115112 (2019)
25. Kugler, F.B., von Delft, J.: New J. Phys. **20**, 123029 (2018).; Erratum: 20, 123029 (2018)
26. Ayral, T., Parcollet, O.: Phys. Rev. B. **92**, 115109 (2015)
27. Si, Q., Smith, J.L.: Phys. Rev. Lett. **77**, 3391 (1996)
28. Chitra, R., Kotliar, G.: Phys. Rev. Lett. **84**, 3678 (2000)
29. Rubtsov, A.N., Katsnelson, M.I., Lichtenstein, A.I.: Ann. Phys. (Amsterdam). **327**, 1320 (2012)
30. Peters, L., van Loon, E.G.C.P., Rubtsov, A.N., Lichtenstein, A.I., Katsnelson, M.I., Stepanov, E.A.: Phys. Rev. B. **100**, 165128 (2019)
31. Chen, F., Katsnelson, M.I., Galperin, M.: Phys. Rev. B. **101**, 235439 (2020)
32. Bickers, N.E.: Theoretical Methods for Strongly Correlated Electrons. Springer-Verlag, New York (2004)
33. Wentzell, N., Li, G., Tagliavini, A., Taranto, C., Rohringer, G., Held, K., Toschi, A., Andergassen, S.: arXiv:1610.06520 (2006)
34. Reitner, M., Chalupa, P., Del Re, L., Springer, D., Ciuchi, S., Sangiovanni, G., Toschi, A.: preprint arXiv:2002.12869v1 (2020)
35. Georges, A., Kotliar, G., Krauth, W., Rozenberg, M.J.: Rev. Mod. Phys. **68**, 13 (1996)
36. Bulla, R., Hewson, A.C., Pruschke, T.: J. Phys. Condens. Matter. **10**, 8365 (1998)
37. Bulla, R., Costi, T., Pruschke, T.: Rev. Mod. Phys. **80**, 395 (2008)
38. Kinza, M., Honerkamp, C.: Phys. Rev. B. **88**, 195136 (2013)
39. Tahvildar-Zadeh, A.N., Freericks, J.K., Jarrell, M.: Phys. Rev. B. **55**, 942 (1997)
40. Maier, T.A., Jarrell, M.S., Scalapino, D.J.: Phys. Rev. Lett. **96**, 047005 (2006)
41. Gull, E., Werner, P., Millis, A., Troyer, M.: Phys. Rev. B. **76**, 235123 (2007)
42. Hafermann, H., Patton, K.R., Werner, P.: Phys. Rev. B. **85**, 205106 (2012)
43. Gull, E., Millis, A.J., Lichtenstein, A.I., Rubtsov, A.N., Troyer, M., Werner, P.: Rev. Mod. Phys. **83**, 349 (2011)
44. Werner, P., Comanac, A., de Medici, L., Troyer, M., Millis, A.J.: Phys. Rev. Lett. **97**, 076405 (2006)
45. Werner, P., Millis, A.J.: Phys. Rev. B. **74**, 155107 (2006)

46. Kaufmann, J., Gunacker, P., Kowalski, A., Sangiovanni, G., Held, K.: Phys. Rev. B. **100**, 075119 (2019)
47. Ribic, T., Gunacker, P., Iskakov, S., Wallerberger, M., Rohringer, G., Rubtsov, A.N., Gull, E., Held, K.: Phys. Rev. B. **96**, 235127 (2017)
48. Valli, A., Schaefer, T., Thunström, P., Rohringer, G., Andergassen, S., Sangiovanni, G., Held, K., Toschi, A.: Phys. Rev. B. **91**, 115115 (2015)
49. Li, G., Wentzell, N., Pudleiner, P., Thunström, P., Held, K.: Phys. Rev. B. **93**, 165103 (2016)
50. Slezak, C., Jarrell, M., Maier, T., Deisz, J.: J. Phys. Condens. Matter. **21**, 435604 (2009)
51. Katanin, A.A., Toschi, A., Held, K.: Phys. Rev. B. **80**, 075104 (2009)
52. Moriya, T.: In: Cardona, M., Fulde, P., Queisser, H.-J. (eds.) Spin Fluctuations in Itinerant Electron Magnetism. Springer Verlag, Berlin (1985)
53. Rohringer, G., Toschi, A.: Phys. Rev. B. **94**, 125144 (2016)
54. Rohringer, G.: "New Routes Towards a Theoretical Treatment of Nonlocal Electronic Correlations," Ph.D. thesis (Vienna University of Technology, 2013)
55. Galler, A., Thunström, P., Gunacker, P., Tomczak, J.M., Held, K.: Phys. Rev. B. **95**, 115107 (2017)
56. Gallera, A., Thunström, P., Kaufmann, J., Pickem, M., Tomczak, J.M., Held, K.: Comp. Phys. Commun. **245**, 106847 (2019)
57. Thunström, P., Gunnarsson, O., Ciuchi, S., Rohringer, G.: Phys. Rev. B. **98**, 235107 (2018)
58. Pudleiner, P., Thunström, P., Valli, A., Kauch, A., Li, G., Held, K.: Phys. Rev. B. **99**, 125111 (2019)
59. Del Re, L., Capone, M., Toschi, A.: Phys. Rev. B. **99**, 045137 (2019)
60. Springer, D., Chalupa, P., Ciuchi, S., Sangiovanni, G., Toschi, A.: Phys. Rev. B. **101**, 155148 (2020)
61. Eckhardt, C.J., Honerkamp, C., Held, K., Kauch, A.: Phys. Rev. B. **101**, 155104 (2020)
62. Mermin, N.D., Wagner, H.: Phys. Rev. Lett. **17**, 1307 (1966)
63. Krien, F., Valli, A.: Phys. Rev. B. **100**, 245147 (2019)
64. Krien, F., Valli, A., Capone, M.: Phys. Rev. B. **100**, 155149 (2019)
65. van Loon, E.G.C.P., Krien, F., Hafermann, H., Lichtenstein, A.I., Katsnelson, M.I.: Phys. Rev. B. **98**, 205148 (2018)
66. Otsuki, J., Yoshimi, K., Shinaoka, H., Nomura, Y.: Phys. Rev. B. **99**, 165134 (2019)
67. Vilk, Y.M., Tremblay, A.-M.S.: J. Phys. I (France). **7**, 1309 (1997)
68. Schafer, T., Katanin, A.A., Held, K., Toschi, A.: Phys. Rev. Lett. **119**, 046402 (2017)
69. Castellani, C., Castro, C.D., Feinberg, D., Ranninger, J.: Phys. Rev. Lett. **43**, 1957 (1979)
70. LeBlanc, J.P.F., Antipov, A.E., Becca, F., Bulik, I.W., Chan, G.K.-L., Chung, C.-M., Deng, Y., Ferrero, M., Henderson, T.M., Jiménez-Hoyos, C.A., Kozik, E., Liu, X.-W., Millis, A.J., Prokof'ev, N.V., Qin, M., Scuseria, G.E., Shi, H., Svistunov, B.V., Tocchio, L.F., Tupitsyn, I.S., White, S.R., Zhang, S., Zheng, B.-X., Zhu, Z., Gull, E.: Phys. Rev. X. **5**, 041041 (2015)
71. Blumer, N.: "Mott-Hubbard Metal-Insulator Transition and Optical Conductivity in High Dimensions," Ph.D. thesis (Universitat Augsburg, 2002)
72. Park, H., Haule, K., Kotliar, G.: Phys. Rev. Lett. **101**, 186403 (2008)
73. Schafer, T., Toschi, A., Tomczak, J.M.: Phys. Rev. B. **91**, 121107 (2015)
74. Hafermann, H.: "Numerical Approaches to Spatial Correlations in Strongly Interacting Fermion Systems," Ph.D. thesis (University of Hamburg, 2010)
75. van Loon, E.G.C.P., Hafermann, H., Katsnelson, M.: Phys. Rev. B. **97**, 085125 (2018)
76. Kauch, A., Pudleiner, P., Astleithner, K., Thunström, P., Ribic, T., Held, K.: Phys. Rev. Lett. **124**, 047401 (2020)
77. Kitatani, M., Schafer, T., Aoki, H., Held, K.: preprint arXiv:1801.05991 (2018)
78. Vučičević, J., Ayral, T., Parcollet, O.: Phys. Rev. B. **96**, 104504 (2017)
79. Hafermann, H., Kecker, M., Brener, S., Rubtsov, A.N., Katsnelson, M.I., Lichtenstein, A.I.: J. Supercond. Nov. Magn. **22**, 45 (2009)
80. Otsuki, J., Hafermann, H., Lichtenstein, A.I.: Phys. Rev. B. **90**, 235132 (2014)

81. Gunnarsson, O., Schafer, T., LeBlanc, J.P.F., Gull, E., Merino, J., Sangiovanni, G., Rohringer, G., Toschi, A.: Phys. Rev. Lett. **114**, 236402 (2015)
82. Gunnarsson, O., Schafer, T., LeBlanc, J.P.F., Merino, J., Sangiovanni, G., Rohringer, G., Toschi, A.: Phys. Rev. B. **93**, 245102 (2016)
83. Astretsov, G.V., Rohringer, G., Rubtsov, A.N.: Phys. Rev. B. **101**, 075109 (2020)
84. Nourafkan, R., Côté, M., Tremblay, A.-M.S.: Phys. Rev. B. **99**, 035161 (2019)
85. Vučičević, J., Kokalj, J., Žitko, R., Wentzell, N., Tanasković, D., Mravlje, J.: Phys. Rev. Lett. **123**, 036601 (2019)
86. Vilardi, D., Taranto, C., Metzner, W.: Phys. Rev. B. **97**, 235110 (2018)

Part II
A Computational Tool

Chapter 10
DFT + DMFT: Static Properties of Materials

10.1 Density Functional Theory

To solve a problem of the system of N electrons and M nuclei, one needs to solve the multiparticle Schrödinger equation

$$\widehat{H}\Psi\left(\vec{r}_1, \ldots, \vec{r}_N, \vec{R}_1, \ldots, \vec{R}_M\right) = E\Psi\left(\vec{r}_1, \ldots, \vec{r}_N, \vec{R}_1, \ldots, \vec{R}_M\right), \tag{10.1}$$

where

$$\widehat{H} = -\hbar^2 \sum_{i=1}^{N} \frac{\vec{\nabla}^2_{\vec{r}_i}}{2m_e} - \hbar^2 \sum_{I=1}^{M} \frac{\vec{\nabla}^2_{\vec{R}_I}}{2M_I} + \sum_{i=1}^{N} \sum_{j>i}^{N} \frac{e^2}{\left|\vec{r}_i - \vec{r}_j\right|}$$
$$+ \sum_{I=1}^{M} \sum_{I'>I}^{M} \frac{e^2 Z_I Z_{I'}}{\left|\vec{R}_I - \vec{R}_{I'}\right|} - \sum_{i=1}^{N} \sum_{I=1}^{M} \frac{e^2 Z_I}{\left|\vec{r}_i - \vec{R}_I\right|} \tag{10.2}$$

is the Hamiltonian of the system that includes the electron and nuclei kinetic, electron–electron, nuclei–nuclei, and electron–nuclei terms (from left to right). The problem defined by Eqs. (10.1) and (10.2) is a tremendously complicated problem in the case of solids, since it requires to solve equations for $\sim 10^{23} - 10^{25}$ coordinate variables (for a "chunk" of 1 cm^3 solid). Even assuming that the positions of the ions are fixed (they are much heavier than electrons and often can be considered as slow classical particles, thus using the Bohn–Oppenheimer approximation one can assume them to be fixed in space), which reduces the problem to a problem of electrons in the field of ions, the equation does not simplify dramatically:

V. Turkowski, *Dynamical Mean-Field Theory for Strongly Correlated Materials*,
https://doi.org/10.1007/978-3-030-64904-3_10

$$\left(-\hbar^2\sum_{i=1}^{N}\frac{\vec{\nabla}_{\vec{r}_i}^2}{2m_e}+\sum_{i=1}^{N}\sum_{j>i}^{N}\frac{e^2}{\left|\vec{r}_i-\vec{r}_j\right|}-\sum_{i=1}^{N}\sum_{I=1}^{M}\frac{e^2Z_I}{\left|\vec{r}_i-\vec{R}_I\right|}\right)\Psi\left(\vec{r}_1,\ldots,\vec{r}_N\right)$$
$$=E\Psi\left(\vec{r}_1,\ldots,\vec{r}_N\right). \tag{10.3}$$

One still needs to solve the equation with the number of coordinates of the same order as in Eq. (10.1).

To simplify the electron problem (10.3), one can rewrite the Hamiltonian in this equation as

$$\hat{H}=\sum_{i=1}^{N}\left[-\hbar^2\frac{\vec{\nabla}_{\vec{r}_i}^2}{2m_e}-v\left(\vec{r}_{iI}\right)+\sum_{j>i}^{N}U\left(\vec{r}_i-\vec{r}_j\right)\right], \tag{10.4}$$

where $-v\left(\vec{r}_{iI}\right)$ is the potential energy of ith electron in presence of Ith ion and $U\left(\vec{r}_i-\vec{r}_j\right)$ is the Coulomb interaction energy of two electrons. This is not a great simplification so far. The last term in the brackets in Eq. (10.4) (or terms, since one has N such terms in the Hamiltonian) still makes the problem very difficult to solve. Without this term, the problem reduces to a problem of N independent electrons:

$$\left(-\hbar^2\frac{\vec{\nabla}_{\vec{r}_i}^2}{2m_e}-v\left(\vec{r}_{iI}\right)\right)\psi\left(\vec{r}_i\right)=E\psi\left(\vec{r}_i\right) \tag{10.5}$$

that can be easily solved. Using the solution of Eq. (10.5), one can construct the wave function of the system (Slater-determinant) in the ground state:

$$\Psi\left(\vec{r}_1\ldots,\vec{r}_N\right)=\frac{1}{\sqrt{N!}}\begin{vmatrix}\psi_1\left(\vec{r}_1\right) & \psi_2\left(\vec{r}_1\right) & \vdots & \psi_N\left(\vec{r}_1\right)\\ \psi_1\left(\vec{r}_2\right) & \psi_2\left(\vec{r}_2\right) & \vdots & \psi_N\left(\vec{r}_2\right)\\ \vdots & \vdots & \vdots & \vdots\\ \psi_1\left(\vec{r}_N\right) & \psi_2\left(\vec{r}_N\right) & \vdots & \psi_N\left(\vec{r}_N\right)\end{vmatrix}, \tag{10.6}$$

where $\psi_j\left(\vec{r}_i\right)$ is the wave function of the eigenstate with energy ε_j. The wave function of the system in an excited state can be obtained from Eq. (10.6) by substituting there higher-energy eigenfunction(s) (with energy level number(s) larger than N) instead of the ground-state one(s) (with energy levels from 1 to N). Thus, in the noninteracting case, the solution of the many-electron problem in essence consists of the steps to find potential $v\left(\vec{r}_{il}\right)$, solve Eq.(10.5), construct the ground-state wave-function Ψ Eq. (10.6) and to calculate physical observables described by an operator $\widehat{O}$ as $\int \Psi^* \widehat{O} \Psi d\vec{r}_1 \ldots d\vec{r}_N$.

It must be noted that the problem of noninteracting (Coulomb potential neglected) with each other electrons in the field of ions still can be very nontrivial. Two well-known examples are superconductors (where electrons attract to each other due to ions/nuclei vibrations (phonons)) and topological insulators. However, since the main subject of the book is strong electron–electron correlations we omit these cases, and focus on how to deal with systems with the electron–electron interaction.

At the first sight, the electron–electron interaction term in Eq. (10.3) does not allow to formulate the many-particle theory in terms of an equivalent exact one-particle theory, similar to the spirit of Eq. (10.5). However, looking at Eq. (10.4) one can see that it might be possible to incorporate the effects of electron-electron interaction into the (single-electron) ion potential: $v\left(\vec{r}_{il}\right) \to v_s\left(\vec{r}_{il}\right)$. In this case, the equation for each electron in the multi-electron system will become

$$\left(-\hbar^2 \frac{\vec{\nabla}^2_{\vec{r}_i}}{2m_e} - v_s\left(\vec{r}_{il}\right)\right)\psi\left(\vec{r}_i\right) = E\psi\left(\vec{r}_i\right). \tag{10.7}$$

Now, the problem is formally reduced to one-electron problem, but to make Eq. (10.7) useful, one needs to find v_s. It is natural to assume that potential v_s depends on the total electron ground-state charge density—the "main" characteristic of the "non-interacting" electron system, i.e. $v_s = v_s[n_0]\left(\vec{r}\right)$, where

$$n_0\left(\vec{r}\right) = N \int d\vec{r}_1 \int d\vec{r}_2 \ldots \int d\vec{r}_N \, \Psi^*\left(\vec{r}_1, \vec{r}_2, \ldots, \vec{r}_N\right)\Psi\left(\vec{r}_1, \vec{r}_2, \ldots, \vec{r}_N\right). \tag{10.8}$$

Vice versa, since the ground-state charge density is obtained from the many-body wave-function that obeys Schrödinger equation, the charge density is a functional of the potential v_s: $n = n[v_s]\left(\vec{r}\right)$. This relation between the charge density and single-particle potential is unique. Namely, according to the Hohenberg–Kohn theorem [1]

- in an electron system with a given particle–particle interaction there is one-to-one correspondence between $v_s\left(\vec{r}\right)$ and $n_0\left(\vec{r}\right)$, i.e., the external potential is a unique functional of the ground state density (up to an additive constant).

The theorem was proved for the nondegenerate ground state, generalization on the degenerate case can be also done [2]. As the next step, Hohenberg and Kohn showed [1] that the ground-state charge density of the many-electron system can be obtained by minimizing the energy functional of density

$$E[n] = T[n](r) + \int n(\vec{r})V_{\text{ext}}(\vec{r})d\vec{r} + \int n(\vec{r})V_H[n](\vec{r})dr\vec{r} \\ + E_{\text{XC}}[n](\vec{r}), \quad (10.9)$$

where the first term on the right-hand side is the kinetic energy, the second term is the external potential energy (electron–ion interaction), the third—Hartree energy with potential

$$V_H[n](r) = \int \frac{n(\vec{r}')}{|\vec{r} - \vec{r}'|} d\vec{r}', \quad (10.10)$$

and the last term, $E_{\text{XC}}[n](\vec{r})$, is the exchange–correlation (XC) energy defined by the exchange and correlation effects in the electronic system (all terms in Eq. (10.9) are functionals of the charge density).

On the other hand, in the case of an effective one-electron problem (10.7), an equivalent energy functional is

$$E_s[n] = T_s[n](\vec{r}) + \int n(\vec{r})v_s[n](\vec{r})d\vec{r}. \quad (10.11)$$

Differentiating Eqs. (10.9) and (10.11) with respect to the charge density, equalizing the resulting right parts of the equations and canceling the kinetic energy contributions (the kinetic energy functional is the same for both systems, since it does not depend on the interaction) one gets

$$v_s[n](\vec{r}) = V_{\text{ext}}(\vec{r}) + \int \frac{n(\vec{r}')}{|\vec{r} - \vec{r}'|} d\vec{r}' + v_{\text{XC}}[n](\vec{r}), \quad (10.12)$$

where

$$v_{\text{XC}}[n](\vec{r}) = \frac{\delta E_{\text{XC}}[n]}{\delta n(\vec{r})} \quad (10.13)$$

is the XC potential. Substituting (10.12) into Eq. (10.7) one gets the following eigen-equation for the effective one-electron theory:

$$\left(-\hbar^2\frac{\vec{\nabla}^2}{2m_e}+V_{\text{ext}}\left(\vec{r}\right)+\int\frac{n\left(\vec{r}'\right)}{\left|\vec{r}-\vec{r}'\right|}d\vec{r}'+v_{\text{XC}}[n]\left(\vec{r}\right)\right)\psi_j\left(\vec{r}\right)$$
$$=\varepsilon_j\psi_j\left(\vec{r}\right), \tag{10.14}$$

where the XC potential is a functional of the ground-state density given by

$$n_0\left(\vec{r}\right)=\sum_{j=1}^{N}\left|\psi_j\left(\vec{r}\right)\right|^2. \tag{10.15}$$

Equations (10.14) and (10.15) are the Kohn–Sham equations [3] that form the basis of DFT.

There is a useful relation, which is helpful to better understand the nature of the XC potential defined in Eq. (10.13) as a (functional) derivative of the XC energy with respect to the charge density. A similar, site-dependent, relation connects the energy of a level, total energy of the system, and occupancy of the level:

$$\varepsilon_i=\frac{\partial E}{\partial n_i}, \tag{10.16}$$

where n_i is occupancy of the level with energy ε_i. Thus, one may think about the XC potential at given point as the part of the energy of the electron system at this point (energy density) that comes from the exchange and correlation effects.

In many systems (especially in magnets), spins of electrons play an important role. For such situation, the KS equations were generalized on the spin-dependent case (spin DFT), with the spin-resolved XC potential

$$v_{\text{XC}\sigma}[n]\left(\vec{r}\right)=\frac{\delta E_{\text{XC}}}{\delta n_\sigma\left(\vec{r}\right)}. \tag{10.17}$$

Thus, in general case, E_{XC} is a functional of the spin-up and spin-down densities. As a matter of fact, even in the charge DFT one can put in the XC energy functional $n_0(r) = n_{0\uparrow}(r) + n_{0\downarrow}(r)$ (for details, see Ref. [4]).

To conclude the DFT part, we briefly discuss the most popular XC potentials and how they perform for different systems (for the ladder of DFT approximations and references for different potentials, see, e.g., book [5]). The simplest DFT approximation is $v_{\text{XC}}[n](r) = 0$. This approximation is called Hartree approximation, because the corresponding KS equation includes only Hartree potential (10.10). Needless to say, since in the Hartree approximation exchange and correlation effects are neglected, it cannot be used in the systems where the electron–electron interaction plays a nontrivial role. The next simplest XC potential is the local-density approximation (LDA) potential, or in the spin-polarized case, the local spin-density approximation (LSDA) potential. Often, only the exchange (X) part of this functional is taken into account. The exchange part of the L(S)DA was derived by using the dependence of the exchange energy of the isotropic homogeneous free-electron gas on its charge (spin) density. Namely, the X energy was differentiated with respect to the charge density and in the resulting expression the space-independent charge (spin) density was substituted by $n\left(\vec{r}\right)\left(n_\sigma\left(\vec{r}\right)\right)$. The result is

$$v_X^{\mathrm{LDA}}[n]\left(\vec{r}\right) = -\left(\frac{3}{\pi}\right)^{1/3} n^{1/3}\left(\vec{r}\right) \quad \left(v_{X\sigma}^{\mathrm{LSDA}}[n]\left(\vec{r}\right) \approx -4\left(\frac{3}{\pi}\right)^{1/3} n^{1/3}\left(\vec{r}\right)\right) \tag{10.18}$$

(The LSDA X potential in Eq. (10.18) is one of the used approximations). In the LDA approximation, the electron moves in potential defined by the charge density that consists of other electron charges and its own charge. Such a self-interaction leads to some errors, including an underestimation of the bandgap [6, 7] (for how to solve this problem, see below).

The next order of approximation is generalized gradient approximation (GGA) (with PBE, PW91 and other potentials; for details and references, see book [5]), where one includes the charge gradient corrections (in terms of dimensionless variables $\left|\vec{\nabla} n_\sigma\left(\vec{r}\right)\right|/n_\sigma^{4/3}\left(\vec{r}\right)$ and $\vec{\nabla}^2 n_\sigma\left(\vec{r}\right)/n_\sigma^{5/3}\left(\vec{r}\right)$) to the LDA potential. Though these potentials take into account rapid change of the charge density along the unit cell, and lead to some improvement of the ground-state properties, like the total energy, bulk modulus, and cohesion energy, other very important properties obtained with LDA, like the bandgap value in the case of semiconductors and insulators and the quasiparticle energies in different systems, are not described better with GGA [8]. Also, neither LDA nor GGA works well for strongly the correlated systems.

Similarly, the next three levels of approximation—meta- and hyper-GGA and unoccupied orbital approaches—do not work well for correlated materials, except improvement of the value of the bandgap. In meta-GGA, in addition to the GGA gradient-corrected XC potential, the KS equation includes the kinetic energy functionals (defined by the KS orbitals) $\tau_\sigma\left(\vec{r}\right) = \frac{1}{2}\sum_{i=1}^{N}\left|\nabla\varphi_{j\sigma}\left(\vec{r}\right)\right|^2$. In hyper-GGA (most notable class of potentials—hybrid functionals), one uses the exact exchange energy

$$E_x^{\mathrm{exact}}\left(\vec{r}\right) = -\frac{1}{2}\sum_{\sigma,i,j=1}^{N}\int d^3r' \frac{\varphi_{i\sigma}^*\left(\vec{r}'\right)\varphi_{j\sigma}\left(\vec{r}'\right)\varphi_{i\sigma}\left(\vec{r}\right)\varphi_{j\sigma}^*\left(\vec{r}\right)}{\left|\vec{r}-\vec{r}'\right|} \tag{10.19}$$

to obtain the X part of the XC potential. Very popular are hybrid energy functionals that consist of the exact exchange part and the exchange and correlation parts coming from GGA functionals: $E_{\mathrm{xc}}^{\mathrm{hybrid}}\left(\vec{r}\right) = a\mathrm{E}_x^{\mathrm{exact}}\left(\vec{r}\right) + (1-a)E_x^{\mathrm{GGA}}\left(\vec{r}\right) + E_c^{\mathrm{GGA}}\left(\vec{r}\right)$. Probably, the most popular functional of this type is B3LYP. Another class of the hybrid potentials is the range-separated hybrids, where the Coulomb interaction is separated on a short- and a long-range parts with the separation parameter (inverse length) μ. Then, according to this separation the X energy of the system consists of two exchange parts and the correlation part comes from one of the DFT functionals. Such potentials have correct long-range asymptotic behavior that improves results of other approximations for polarizability, exciton-bound state energies (in time-dependent DFT, discussed in Chap. 12), etc.

The last types of popular potentials are orbital potentials that take into account unoccupied orbitals. Inclusion of empty orbitals improves the hybrid functional results for the correlation part of the potential, in particular its asymptotic behavior at short distances. Since unoccupied KS orbitals are also functionals of density, this approach can be regarded as DFT [5]. Performance of these functionals in the case of strongly correlated systems is basically unknown.

To summarize, there are many evidences that DFT approaches, based on homogeneous charge approximation and charge gradient corrections, and their available extensions are not suitable for strongly correlated materials with narrow bands, i.e., with localized electrons that occupy mostly atomic orbitals and are not spread over the lattice. For several transition metal-oxide insulators with partially filled d-orbitals DFT gives a metallic state [9]. Under doping, these systems become metallic, which can be reproduced by DFT. However, many properties of materials are described wrongly by DFT even in metallic phase (see, e.g., book [4]) and references therein. The self-interaction in L(S)DA that underestimates the gap is another problem of DFT, which is just partially removed by the XC potential. The self-interaction also does not allow to describe charge ordering effects with LSDA (e.g., in Fe_3O_4 [10, 11]), since inhomogeneous charge distribution is unstable in this approach (increased atomic level occupancies make the inter-electron potential repulsive).

To cancel the residual self-interaction for occupied states a self-interaction correction (SIC) method [12–14] was proposed. In this approach, one extracts from the XC potential parts that come from the occupied electron states (since orbital charge is defined by orbital wave functions, it makes the potentials orbital-dependent). However, in SIC the potential acting on the occupied-state electrons is reduced, but it does not change for empty states, resulting in energy splitting between occupied and empty states. While SIC gives better values for the magnetic moments and bandgap as compared to LDA [4, 15], it overestimates the correlation effects due to lack of screening.

Several other alternative to DFT methods to study properties of the materials are in use. One of them is popular *ab initio* + many-body GW approach [16, 17], where the DFT electron energies are corrected by many-body self-energy obtained from the electron GF and screened Coulomb potential W. However, GW fails in the case of strongly correlated materials, since the linear in W approximation for the self-energy is usually not sufficient for these systems. To take into account correlation effects, LDA++ method was proposed [18]. In this approach, regimes with different ratios (small, same-order, and large) of the Coulomb repulsion and bandwidth are treated in different way. We begin our discussion on how the correlation effects are incorporated into DFT in the next section.

10.2 DFT + U

Probably, the first efficient and transparent approach to include correlation effects in the framework of DFT was proposed in Ref. [19]. In this method, an additional Coulomb repulsion term $\frac{1}{2}\sum_{a,b} U^{ab} n_a n_b$ is included into the Hamiltonian used in DFT calculations. In this term, a and b are localized orbitals (the indices include also spin variables). Since in this approach one has U-parameters, it is called DFT + U.

To discuss the main details of the DFT + U approach that may be regarded as an intermediate approach between DFT and DFT + DMFT, let us begin with the general Hubbard Hamiltonian and show how the interaction term in this Hamiltonian is transformed to the one used in DFT + U. In Chap. 2, we have derived general Hubbard Hamiltonian with the local electron–electron repulsion

$$\widehat{H} = -\sum_{i,j,\alpha,\sigma} t_{ij}^{\alpha} \widehat{c}_{\alpha i,\sigma}^{+} \widehat{c}_{\beta j,\sigma} - \mu \sum_{i,\alpha,\sigma} \widehat{n}_{\alpha i,\sigma} + \frac{1}{2} \sum_{i} \sum_{\alpha,\beta,\gamma,\delta} \sum_{\sigma,\sigma'} U_i^{\alpha\beta\gamma\delta} \widehat{c}_{\alpha i,\sigma}^{+} \widehat{c}_{\beta i,\sigma'}^{+} \widehat{c}_{\gamma i,\sigma'} \widehat{c}_{\delta i,\sigma}. \tag{10.20}$$

where the hopping and Coulomb interaction matrices are obtained in terms of integrals over the localized (orbital) Wannier functions ϕ:

$$t_{ij}^{\alpha} = -\int d\vec{r}\, \phi^{\alpha *}\left(\vec{r} - \vec{R}_i\right) h_1\left(\vec{r}\right) \phi^{\alpha}\left(\vec{r} - \vec{R}_j\right) \tag{10.21}$$

($h_1\left(\vec{r}\right)$ is an effective single-electron Hamiltonian that includes V_{XC} and other single-electron Hamiltonian terms), and

$$U_i^{\alpha\beta\gamma\delta} = \int d\vec{r} \int d\vec{r}'\, \phi^{\alpha *}\left(\vec{r} - \vec{R}_i\right) \phi^{\beta *}\left(\vec{r} - \vec{R}_i\right) U\left(\vec{r} - \vec{r}'\right) \phi^{\gamma}\left(\vec{r}' - \vec{R}_i\right) \phi^{\delta}\left(\vec{r}' - \vec{R}_i\right) \tag{10.22}$$

($U\left(\vec{r} - \vec{r}'\right)$ is the Coulomb potential).

To derive the U-part of the DFT + U Hamiltonian from general Hubbard Hamiltonian one makes the static mean-field (unrestricted HF) approximation in the four-operator term in Eq. (10.20). For transparence, we write explicitly the orbital, the orbital projection and the spin indices (but omit the same for all operators site index) in this term. Then, the needed approximation is

$$\widehat{c}^{+}_{lm,\sigma}\widehat{c}^{+}_{lm',\sigma'}\widehat{c}_{lm'',\sigma}\widehat{c}_{lm''',\sigma'} = -\widehat{c}^{+}_{lm,\sigma}\widehat{c}_{lm'',\sigma}\left\langle\widehat{c}^{+}_{lm',\sigma'}\widehat{c}_{lm''',\sigma'}\right\rangle$$
$$- \left\langle\widehat{c}^{+}_{lm,\sigma}\widehat{c}_{lm'',\sigma}\right\rangle\widehat{c}^{+}_{lm',\sigma'}\widehat{c}_{lm''',\sigma'}$$
$$+ \widehat{c}^{+}_{lm,\sigma}\widehat{c}_{lm''',\sigma'}\left\langle\widehat{c}^{+}_{lm',\sigma'}\widehat{c}_{lm'',\sigma}\right\rangle$$
$$+ \left\langle\widehat{c}^{+}_{lm,\sigma}\widehat{c}_{lm''',\sigma'}\right\rangle\widehat{c}^{+}_{lm',\sigma'}\widehat{c}_{lm'',\sigma} \quad (10.23)$$

(we flipped the order of the last two operators on the left part, as compared to Eq. (10.20)). The averages in the first two lines of Eq. (10.23) depend on one spin variable. Moreover, due to spin conservation, the averages of the operators with the opposite spins in the last two lines are zero, and only the averages of the same-spin operators survive. For example, in the second line in Eq. (10.23) the average is

$$\left\langle\widehat{c}^{+}_{lm',\sigma'}\widehat{c}_{lm'',\sigma}\right\rangle = \delta_{\sigma\sigma'}\left\langle\widehat{c}^{+}_{lm',\sigma}\widehat{c}_{lm'',\sigma}\right\rangle \equiv \delta_{\sigma\sigma'}n^{\sigma}_{m'm''}, \quad (10.24)$$

and similar for the last term in this equation. This an important result, since it gives diagonal-in-spin "U-part" of the DFT + U Hamiltonian. Using the approximations above in the interacting part of the Hamiltonian and adding the resulting term to the DFT Hamiltonian one gets the DFT + U, or LDA + U, Hamiltonian:

$$\widehat{H}_{\mathrm{DFT+U}} = \widehat{H}_{\mathrm{DFT}} + \sum_{i,l,m,m'} |ilm\sigma\rangle V^{\sigma}_{mm'}\langle ilm'\sigma|, \quad (10.25)$$

where

$$V^{\sigma}_{mm'} = \sum_{m'',\ m'''}\Big[\langle m,m''|U|m',m'''\rangle n^{\bar{\sigma}}_{m'm''}$$
$$+ \left(\langle m,m''|U|m',m'''\rangle - \langle m,m''|U|m''',m'\rangle\right)n^{\sigma}_{m'm''}\Big]$$
$$-\frac{1}{2}\overline{U}\left(n-\frac{1}{2}\right). \quad (10.26)$$

The last term in Eq. (10.26) is the double-counting correction term that cancels the (most part, but not all) double-counted local interaction effects, since besides the Hubbard term contributions, the correlation energy is already included in the hopping (DFT) part of the Hamiltonian through the DFT XC potential. In Eq. (10.26), $\overline{U}$ is the averaged over orbitals Coulomb repulsion parameter (see, e.g., book [4]; in this and the next section, we partially follow the presentation in the last reference). This last term in Eq. (10.26) can be obtained as follows. One begins with writing the expression for the DFT interaction energy in terms of average $\overline{U}$ and total number of d-electrons n:

$$E_{\rm DC} = \frac{1}{2}\overline{U}n(n-1). \tag{10.27}$$

Then, the atomic orbital energy can be calculated: $\varepsilon = \frac{\partial E_{\rm DFT}}{\partial n} = \overline{U}\left(n - \frac{1}{2}\right)$, which is nothing else but the DFT correlation energy that has to be subtracted from the Hamiltonian. Thus, one has to subtract the following double-counting part

$$\widehat{H}_{\rm DC} = \sum_{i,n,l,m,\sigma} \overline{U}\left(n - \frac{1}{2}\right)\widehat{n}_{i,n,l,m,\sigma} \tag{10.28}$$

from the Hamiltonian. The occupancy matrix $n^{\sigma}_{mm'} = \left\langle \widehat{c}^{+}_{lm,\sigma}\widehat{c}_{lm',\sigma}\right\rangle$, defined in Eq. (10.24), is found from

$$n^{\sigma}_{mm'} = -\frac{1}{\pi}\int ImG^{l,mm'}_{i\sigma}(\omega)d\omega, \tag{10.29}$$

where

$$G^{l,mm'}_{i\sigma}(\omega) = \left\langle ilm\sigma\middle|\left(\omega + i\delta - \widehat{H}_{\rm DFT+U}\right)^{-1}\middle|ilm'\sigma\right\rangle \tag{10.30}$$

is the DFT + U GF.

Equations (10.25), (10.26), (10.29), and (10.30) form the set of the DFT + U equations (Eq. (10.25) defines the DFT + U Hamiltonian used in the wave-function equation). They are solved self-consistently. For example, one can solve the DFT problem and find the occupancies from (10.29), (10.30) by using the DFT Hamiltonian instead of the DFT + U. Then, using the obtained occupancies, one can construct the interaction part of the DFT + U Hamiltonian (10.26), solve it, calculate new GF and new occupancies (Eqs. (10.29), (10.30)) and iterate the steps until the occupancies converge (for more details of the solution, see the next subsection, where a generalization of DFT + U, the DFT + DMFT approach, is discussed).

The DFT + U equations can be also obtained by varying the DFT + U energy functional of charge density $\rho(r)$ and occupancies of the correlated orbitals $\{n\}$ (we use ρ to distinguish these two quantities)

$$E^{\rm DFT+U}[\rho(r), \{n\}] = E^{\rm DFT}[\rho(r)] + E^{U}[\{n\}] - E_{\rm DC}[\{n\}], \tag{10.31}$$

where $E^{\rm DFT}[\rho(r)]$ is the DFT functional, the double-counting energy correction term $E_{\rm DC}[\{n\}]$ is defined in Eq. (10.28) and $E^{U}[\{n\}]$ defined by the Coulomb matrix elements (10.26) is

$$E^U = \frac{1}{2}\sum_{m'',\ m'''}\Bigg[\langle m, m''|U|m', m'''\rangle n^{\sigma}_{mm'} n^{\overline{\sigma}}_{m'm''} + (\langle m, m''|U|m', m'''\rangle - \langle m, m''|U|m''', m'\rangle) n^{\sigma}_{mm'} n^{\sigma}_{m'm''}\Bigg] - \frac{1}{2}\overline{U}\left(n - \frac{1}{2}\right). \tag{10.32}$$

Finally, we would like to mention that the potential operator in Eq. (10.25) and the DFT + U equations significantly simplify in the case of "natural" orbital basis where $n^{\sigma}_{m'm''}$ are diagonal in orbital indices. In this case, the potential energy functional (the interaction coefficients in Eq. (10.26)) has a familiar form that follows from the mean-field approximation for the Hubbard Hamiltonian:

$$V^{\sigma}_{m} = \sum_{m'} U^{mm'} n^{\overline{\sigma}}_{m'} + \sum_{m' \neq m}\left(U^{mm'} - J^{mm'}\right) n^{\sigma}_{m'} - \overline{U}\left(n - \frac{1}{2}\right). \tag{10.33}$$

In the one-orbital approximation, the expression for the last potential becomes $V^{\sigma} = Un^{\overline{\sigma}} - U\left(n - \frac{1}{2}\right)$. Thus, in the case of half-filling ($n = 1$), one has $V^{\sigma} = U/2$ when $n^{\overline{\sigma}} = 1$ and $V^{\sigma} = -U/2$ when $n^{\overline{\sigma}} = 0$, i.e., the DFT + U correction to DFT splits the energy of the empty and of the occupied states by $\sim U$. Similar result holds in the multiorbital case (for more details on the DFT + U approach, see book [4]).

10.3 DFT + DMFT

The DFT + U approach can be straightforwardly generalized on the DFT + DMFT case [20, 21], where instead of a mean-field correction to the electron energy (self-energy) of the type (10.26) (non-diagonal matrix) or (10.33) (diagonal matrix), one assumes that the self-energy is frequency-dependent, incorporating the effects of time-resolved on-site interactions into the theory (for details, see review [21] and a book [4]). Then, the DFT + U GF (10.30) is substituted by

$$G^{l,mm'}_{i\sigma}(\omega) = \left\langle ilm\sigma\right|\left(\omega + i\delta - \widehat{H}_0\left(\vec{k}\right) - \widehat{\Sigma}(\omega)\right)^{-1}\left|ilm'\sigma\right\rangle, \tag{10.34}$$

where $\widehat{H}_0\left(\vec{k}\right)$ is the DFT operator of the Hamiltonian with subtracted double counting term in the momentum representation and $\widehat{\Sigma}(\omega)$ is the DMFT self-energy operator. The other two DMFT equations—the Dyson equation and the impurity problem equation—remain the same as in the multi-orbital DMFT, completing the set of the DFT + DMFT equations.

After this sketchy introduction to DFT + DMFT, let us go over the many-body grounds of this theory. DFT + DMFT can be defined in terms of functional Γ that depends on the following DFT and DMFT quantities: Hartree + XC potential V_{HXC} conjugated to the charge density n and correlated-subspace local GF G_{cor} conjugated to the local self-energy Σ_{cor}, correspondingly [22]:

$$\Gamma(n, G_{\rm cor}; V_{\rm HXC}, \Sigma) = {\rm Tr}[{\rm Ln}G] + \Phi[n, G_{\rm cor}] - {\rm Tr}[V_{\rm HXC}n] - {\rm Tr}[\Sigma_{\rm cor}G_{\rm cor}]. \quad (10.35)$$

In Eq. (10.35), $\Phi[n, G_{\rm cor}]$ is the free-energy functional and

$$G = \left(i\omega_n + \mu + \frac{\vec{\nabla}^2}{2} - V_{\rm ext} - V_{\rm HXC} - P^+_{\rm cor}\Sigma_{\rm cor}P_{\rm cor} \right)^{-1}, \quad (10.36)$$

is the inverse continuum GF. In the last equation, $\Sigma_{\rm cor}$ has nonzero elements only in correlated subspace, which is guaranteed by the presence of the projection $P_{\rm cor}$ and $P^+_{\rm cor}$ operators, where $P_{\rm cor}$ is the projection operator on correlated subspace from the space of G, and $P^+_{\rm cor}$ is the corresponding upfold operator. These operators are built as follows: if the correlated subspace is built on functions $|\phi_i\rangle$, one has $P_{\rm cor} = \int d^3x \sum_i |\phi_i\rangle\langle\phi_i|x\rangle\langle x|$, and similar for $P^+_{\rm cor}$. Functional $\Phi[\rho, G_{\rm cor}]$ in Eq. (10.35) contains information about dependence of the free-energy on the correlations. When $\Sigma_{\rm cor} = G_{\rm cor} = 0$ it corresponds to the Kohn–Sham free-energy functional, and when $V_{HXC} = n = 0$—to the Ward–Luttinger functional.

From the extremum conditions for Γ with respect to its function arguments, one can find

$$V_{\rm HXC} = \frac{\delta\Phi[n, G_{\rm cor}]}{\delta n}, \qquad n = {\rm Tr}[G], \quad (10.37)$$

$$\Sigma_{\rm cor} = \frac{\delta\Phi[n, G_{\rm cor}]}{\delta G_{\rm cor}}, \qquad G_{\rm cor} = P_{\rm cor}GP^+_{\rm cor}. \quad (10.38)$$

Equations (10.35)–(10.38) define the DFT + DMFT ingredients, provided the functional $\Phi[n, G_{\rm cor}]$ is known. In DFT + DMFT Φ is approximated by

$$\Phi[n, G_{\rm cor}] = \Phi^{\rm DFT}[n] + \Phi^{\rm DMFT}[G_{\rm cor}] - E_{\rm DC}(N_d), \quad (10.39)$$

where $\Phi^{\rm DFT}[n]$ is Hartree energy plus DFT XC energy (e.g., PBE), $\Phi^{\rm DMFT}[G_{\rm cor}]$ is the DMFT free-energy functional (details of its structure can be found in subsection 3.3.2), and the last term in Eq. (10.39) is the double-counting correction term. After approximations for the functionals are made, one can calculate $V_{\rm HXC} = \frac{\delta\Phi^{\rm DFT}[n]}{\delta n}$ and $\Sigma_{\rm cor} \to \Sigma_{\rm cor} - V_{\rm DC} = \frac{\delta\Phi^{\rm DMFT}[G_{\rm cor}]}{\delta G_{\rm cor}} - \frac{\delta E_{\rm DC}}{\delta N_d}$ and the corresponding conjugated quantities in Eqs. (10.37) and (10.38) [22].

To discuss how to use DFT + DMFT in practice let us begin with a brief summary on how one can calculate the Hubbard model local repulsion and exchange parameters (used in both DFT+U and DFT+DMFT) with DFT, i.e., in an *ab initio* way. For this, let us expand the Coulomb interaction matrix in Eq. (10.26) in terms of spherical harmonics [4, 23]:

$$\langle mm'|U|m''m'''\rangle = \sum_k a_k(mm''m'm''')F^k, \tag{10.40}$$

where $k = 0, 2, \ldots, 2l$,

$$a_k(mm''m'm''') = \frac{4\pi}{2k+1}\sum_{q=-k}^{k}\langle lm|Y_{kq}|lm''\rangle\left\langle lm'|Y^*_{kq}|lm'''\right\rangle, \tag{10.41}$$

and Y_{kq} are spherical harmonics.

Let us present the interaction part of the Hamiltonian operator in the natural-orbital representation, where it has the simplest form, and connect the interaction parameters with its matrix elements (Eq. (10.40)). In the natural-orbital representation, the non-diagonal density matrix elements $n^{\sigma}_{m'm''}$ are much smaller than the diagonal ones and the interaction part of the Hamiltonian operator is (see Eq. (10.33))

$$\begin{aligned} \widehat{H}_{\text{Coulomb}} &= \frac{1}{2}\sum_{i,l,m,m',\sigma}\left(U^{mm'}\widehat{n}^{\sigma}_{ilm}\widehat{n}^{\bar{\sigma}}_{ilm'} + \left(U^{mm'} - J^{mm'}\right)\widehat{n}^{\sigma}_{ilm}\widehat{n}^{\sigma}_{ilm'}\right) \\ &\quad - \frac{1}{2}\sum_{i,l,m'\neq m,\sigma} J^{mm'}\left(\widehat{c}^{\sigma+}_{ilm}\widehat{c}^{\bar{\sigma}}_{ilm}\widehat{c}^{\bar{\sigma}+}_{ilm'}\widehat{c}^{\sigma}_{ilm'} - \widehat{c}^{\sigma+}_{ilm}\,\widehat{c}^{\bar{\sigma}+}_{ilm}\widehat{c}^{\sigma}_{ilm'}\widehat{c}^{\bar{\sigma}}_{ilm'}\right), \end{aligned} \tag{10.42}$$

where

$$U^{mm'} = \langle mm'|U|mm'\rangle \tag{10.43}$$

and

$$J^{mm'} = \langle mm'|U|m'm\rangle \tag{10.44}$$

are the direct Coulomb and the exchange interactions, correspondingly. In Eq. (10.42), the last two ("non-density") terms describe the processes of the simultaneous flip of spins of electrons on different orbitals and of the transition of two electrons with opposite spin values from one orbital to another, respectively. Usually, these terms are neglected to keep the Hamiltonian in the form of density operators (this approximation can be justified by using the fact that in most cases $U \gg J$). Then, the Hamiltonian becomes:

$$\widehat{H}_{\text{Coulomb}} = \frac{1}{2} \sum_{i,l,m,m',\sigma} \left(U^{mm'} n^{\sigma}_{ilm} n^{\bar{\sigma}}_{ilm'} + \left(U^{mm'} - J^{mm'} \right) n^{\sigma}_{ilm} n^{\sigma}_{ilm'} \right). \tag{10.45}$$

One of the most popular choices of the interaction energy parameters for the Hamiltonian (10.45) are Kanamori parameters [24] (mentioned already in Sect. 5.1). They are chosen as follows: U for same-orbital opposite-spin interaction, U-$2J$ for different-orbital opposite-spin interaction, and U-$3J$ for different orbitals same-spin interactions (see Eq. (10.51) where U and J are defined). This gives:

$$\widehat{H}_{\text{Coulomb}} = \frac{1}{2} \sum_{i,l,m,m',\sigma} \left(U\delta_{mm'} n^{\sigma}_{ilm} n^{\bar{\sigma}}_{ilm'} + (U-2J)(1-\delta_{mm'}) n^{\sigma}_{ilm} n^{\bar{\sigma}}_{ilm'} + (U-3J) n^{\sigma}_{ilm} n^{\sigma}_{ilm'} \right). \tag{10.46}$$

Let us express the interaction parameters in Eq. (10.46) in terms of the harmonics components F (Eq. (10.40)) and show how these components can be calculated by using DFT. The direct and exchange parameters are connected with harmonic components [25, 26]. The formula for the Coulomb repulsion parameters (10.40) has the following explicit form:

$$\begin{aligned} U^{mm'} &= \langle mm' | U_{ee} | mm' \rangle \\ &= \int d\vec{r} \int d\vec{r}' \left| \varphi\left(\left| \vec{r} \right| \right) Y_{lm}(\hat{r}) \right|^2 \frac{1}{\left| \vec{r} - \vec{r}' \right|} \left| \varphi\left(\left| \vec{r}' \right| \right) Y_{lm'}(\hat{r}') \right|^2, \end{aligned} \tag{10.47}$$

and similar for $J^{mm'}$ by using Eq. (10.44). Since the integrals over the modulus and over the angle of the radius vector can be separated, it is possible to expand the matrix elements in harmonics:

$$U^{mm'} = \sum_k a_k F^k, \qquad J^{mm'} = \sum_k b_k F^k, \tag{10.48}$$

where

$$F^k = \int r^2 dr \int r'^2 dr' \left| \varphi\left(\left| \vec{r} \right| \right) \right|^2 \frac{r_<^k}{r_>^{k+1}} \left| \varphi\left(\left| \vec{r}' \right| \right) \right|^2, \tag{10.49}$$

$$r_< = r\theta(r' - r) + r'\theta(r - r'), \ \ r_> = r'\theta(r' - r) + r\theta(r - r')$$

and

$$a_k = \frac{4\pi}{2k+1}\sum_{q=-k}^{k}\langle lm|Y_{kq}|lm\rangle\left\langle lm'|Y_{kq}^*|lm'\right\rangle, \tag{10.50}$$

$$b_k = \frac{4\pi}{2k+1}\sum_{q=-k}^{k}|\langle lm|Y_{kq}|lm'\rangle|^2.$$

In the last equation, the integration over the product of averages (Clebsch–Gordan coefficients) can be performed analytically [4, 23]. Substituting the results into Eq. (10.48) one can use the obtained expressions for the interaction matrix elements to calculate the Kanamori parameters by means of

$$U = \frac{1}{(2l+1)^2}\sum_{m,m'} U_{mm'}, \quad U - J = \frac{1}{2l(2l+1)}\sum_{m,m'}(U_{mm'} - J_{mm'}). \tag{10.51}$$

From the Eqs. (10.48) one can obtain

$$U = F^0, \tag{10.52}$$

$$J = \begin{cases} \dfrac{F^2 + F^4}{14}, & 1 = 2\ (d - \text{orbitals}) \\ \dfrac{286F^2 + 195F^4 + 250F^6}{6435}, & 1 = 3\ (\ f - \text{orbitals}) \end{cases}. \tag{10.53}$$

It appears that the ratio F^4/F^2 for many d-electron systems is almost the same: $F^4/F^2 \approx 0.625$. The hydrogen wave functions give similar ratios for different-order Fs for the f-orbital systems: $\frac{F^4}{F^2} \approx 0.668$ and $\frac{F^6}{F^2} \approx 0.495$. These results show that one can estimate J as ~$0.1F^2$.

In the case of Wannier orbitals, U can be calculated as

$$U = \int d\vec{r}\int d\vec{r}'\left|W_n\left(\vec{r}\right)\right|^2 U\left(\vec{r},\vec{r}'\right)\left|W_n\left(\vec{r}'\right)\right|^2, \tag{10.54}$$

where

$$U\left(\vec{r},\vec{r}'\right) = \frac{v\left(\vec{r} - \vec{r}'\right)}{1 - v\left(\vec{r} - \vec{r}'\right)P\left(\vec{r},\vec{r}'\right)}. \tag{10.55}$$

is the screened Coulomb potential. In the last equation, $v\left(\vec{r} - \vec{r}'\right)$ is Coulomb potential and

$$P\left(\vec{r},\vec{r}'\right)=\sum_{i,j}f(\varepsilon_i)\left[1-f(\varepsilon_j)\right]\psi_i\left(\vec{r}\right)\psi_i^*\left(\vec{r}'\right)\psi_j^*\left(\vec{r}\right)\psi_j\left(\vec{r}'\right)$$
$$\times\left(\frac{1}{\varepsilon_i-\varepsilon_j+i\delta}-\frac{1}{\varepsilon_j-\varepsilon_i-i\delta}\right) \tag{10.56}$$

is polarization (i -occupied and j - unoccupied states).

It must be noted that Eq. (10.54) gives different results for U in the case of different states included into polarization (10.56). These states define the screening effects in the system, and hence affect the electron–electron Coulomb interaction (10.55), and as result - the value of U in (10.54). As it follows from the results of calculations, the transitions between d- or f-orbitals should not be included into polarization since the charges in these states are localized and they do not screen each other (otherwise one might get an unphysically small U). Another important finding is that the result for U strongly depends on how many unoccupied states are included (for more details, see book [4]).

Another way to calculate U and J is to use DFT to find the energy of the system E_{DFT} at different fixed d- (or f-orbital) occupancies n_d (or n_f) (constrained calculations), and then to perform numerical differentiation:

$$U=\frac{\partial^2 E_{\mathrm{DFT}}}{\partial^2 n_{d,f}}=\frac{\partial\varepsilon_{d,f}}{\partial n_{d,f}} \tag{10.57}$$

(on how to obtain the last equality in Eq. (10.57), see Eq. (10.25) and the text below). In a similar way, one can calculate J by using formula (10.57), with substitutions $\varepsilon_{d,f}\rightarrow\varepsilon_{d,f\uparrow}-\varepsilon_{d,f\downarrow}$, $n_{d,f}\rightarrow n_{d,f\uparrow}-n_{d,f\downarrow}$, i.e., by differentiating the spin polarization energy with respect to the spin polarization density. For practical implementation of this procedure, we refer the reader to book [4], Section 2.2 (we refer the reader also to work [27] for more details on the construction of a real-material multiorbital Hubbard Hamiltonian).

Let us complete this subsection with some important practical details of the DFT + DMFT calculations. In DFT + DMFT approach, the DFT and DMFT equations have to be solved self-consistently due to the following reason. Assume that one has solved the DFT problem. Then, solving consequent DMFT problem will give changed orbital occupancies and charge distribution. Using these results in new DFT calculations will give new charge spectrum, etc. Moreover, change of the DFT energy and occupancies changes U, which changes the Hubbard Hamiltonian. Thus, the problem should be solved self-consistently:

$$n(r)\rightarrow\widehat{H}_{\mathrm{DFT}},U\rightarrow(\mathrm{DMFT})\rightarrow n_{ilm}\rightarrow n\left(\vec{r}\right), \tag{10.58}$$

and back to $\widehat{H}_{\mathrm{DFT}},U$ until charge density converges [28].

For practical calculations, one also needs to define the correlated subspace and the "hybridization window," the states that hybridize with correlated subspace (in this part, we mainly follow presentation in Ref. [22]). The correlated subspace is usually constructed from local orbitals. Since in solids atomic local orbitals are significantly modified as compared to the isolated-atom orbitals, one needs to define these orbitals. Often, atomic-like orbitals in solids are chosen in the form of Wannier functions [4, 29] that are centered and localized at atoms and form a complete basis of states. The Wannier functions can be constructed as Fourier transforms of the Bloch functions:

$$\left|W_n^{\vec{R}}\right\rangle = \frac{1}{\sqrt{N}}\sum_{\vec{k}} e^{-i\vec{R}\vec{k}}\left|\Psi_{n\vec{k}}\right\rangle, \tag{10.59}$$

where $\vec{R}$ is the unit cell vector, n is combined index of the number of atom in the unit cell and orbital, and $\vec{k}$ is momentum. Wannier functions are not uniquely defined, since there is a freedom in choosing the Bloch functions that can be transformed keeping their orthogonality as $\left|\Psi_{n\vec{k}}\right\rangle = \sum_m U_{mn}^{(\vec{k})}\left|\Psi_{m\vec{k}}\right\rangle$, where $U_{mn}^{(\vec{k})}$ is a unitary transformation matrix. It was proposed by Marzari and Vanderbilt in Ref. [30] to vary $U_{mn}^{(\vec{k})}$ in order to get maximum localized Wannier functions (MLWF procedure). In this approach, the localized states are constructed as linear combinations of band states (with appropriate phase factors) within the hybridization energy window:

$$\left|W_n^{\vec{R}}\right\rangle = \frac{1}{\sqrt{N}}\sum_{\vec{k}} e^{-i\vec{R}\vec{k}}U_{mn}^{(\vec{k})}\left|\Psi_{m\vec{k}}\right\rangle. \tag{10.60}$$

The energy window is chosen to be sufficiently wide so the *d*- or *f*-state correlated subspace (i.e., a subset of the Wannier functions) is sufficiently localized. In this case, the total space is divided on the Wannier function and the remaining states. Then, after Wannier functions (10.60) are found, another unitary transformation, that corresponds to a rotation of orbitals, is made: $\overline{U}_{nm}^{(\vec{k})} = \sum_l \Lambda_{nl}U_{lm}^{(\vec{k})}$ (for details, see Ref. [22]). This gives new wave functions:

$$\left|\overline{W}_n^{\vec{R}}\right\rangle = \frac{1}{\sqrt{N}}\sum_{\vec{k}} e^{-i\vec{R}\vec{k}}\overline{U}_{nm}^{(\vec{k})}\left|\Psi_{m\vec{k}}\right\rangle. \tag{10.61}$$

Transformation (10.61) makes the wavefunctions as local as possible, to maximally decrease the non-diagonal elements of the self-energy and the hybridization matrices. This significantly simplifies numerical simulations.

An important step in the consistent implementation of the DFT + DMFT method is construction of the full charge density. Modern plane-wave codes use either an ultrasoft pseudopotential (PP) or a projector augmented wave (PAW) [31] approaches, where the local charge density consists of two parts: soft PP wave functions term and an "augmentation charge term" defined by difference between the PP wavefunctions and the KS wave functions defined by the full potential. Both charges are expressed in the Wannier representation. The charge density can be expressed in terms of density matrix in the band basis:

$$n\left(\vec{r}\right) = \frac{1}{N_{\vec{k}}}\sum_{\vec{k},i,j} n_{\vec{k},i,j}\left\langle \psi_{\vec{k}i}^{\text{KS}}|\vec{r}\right\rangle\left\langle \vec{r}|\psi_{\vec{k}i}^{\text{KS}}\right\rangle, \tag{10.62}$$

where

$$n_{\vec{k},i,j} = T\sum_{n} G_{\vec{k},i,j}(i\omega_n)e^{-i\omega_n\delta} \tag{10.63}$$

is the density matrix ($\vec{k}$ and i, j are the momentum and the band indices, correspondingly). Outside the hybridization window W, the density-matrix elements are equal to the Fermi function: $n_{\vec{k},i,j} = f_{\vec{k},i}$. For the hybridization window, the easiest way to obtain the density-matrix is to use Wannier representation:

$$n_{\vec{k},i,j} = T\sum_{n} e^{-i\omega_n\delta}\sum_{m,l} \overline{U}_{mi}^{(\vec{k})} G_{\vec{k},m,l}(i\omega_n)\overline{U}_{lj}^{(\vec{k})*} \tag{10.64}$$

($G_{\vec{k},m,l}(i\omega_n)$ are the GFs). This gives

$$\begin{aligned} n\left(\vec{r}\right) &= \sum_{i\notin W} n_i^{\text{DFT}}\left(\vec{r}\right) + \sum_{i,j\epsilon W} n_{ij}^{\text{DMFT}}\left(\vec{r}\right) \\ &\equiv \frac{1}{N_{\vec{k}}}\sum_{\vec{k},i} f_{\vec{k},i}\left\langle \psi_{\vec{k}i}^{\text{KS}}|\vec{r}\right\rangle\left\langle \vec{r}|\psi_{\vec{k}i}^{\text{KS}}\right\rangle + \frac{1}{N_{\vec{k}}}\sum_{\vec{k},i,j} n_{\vec{k},i,j}\left\langle \psi_{\vec{k}i}^{\text{KS}}|\vec{r}\right\rangle\left\langle \vec{r}|\psi_{\vec{k}j}^{\text{KS}}\right\rangle. \end{aligned} \tag{10.65}$$

In the PAW formalism, the KS (ψ^{KS}) and PP ($\widetilde{\psi}$) wavefunctions are connected by the linear transformation $\psi^{\text{KS}} = \widehat{T}\widetilde{\psi}$. Then, the charge density becomes

$$\widehat{T}n\widehat{T} = \widetilde{n} + n^1 - \widetilde{n}^1, \tag{10.66}$$

where $\widetilde{n}$ is the soft charge-density, n^1 is the on-site all-electron charge density and $\widetilde{n}^1$ is the on-site PP charge density. Since in the DMFT part the charge-density is calculated by using the charge-density matrix, one has:

$$\widetilde{n}^{\mathrm{DMFT}}\left(\vec{r}\right)=\frac{1}{N_{\vec{k}}}\sum_{\vec{k},i,j}n_{\vec{k},i,j}\left\langle\widetilde{\psi}_{\vec{k}i}|\vec{r}\right\rangle\left\langle\vec{r}|\widetilde{\psi}_{\vec{k}j}\right\rangle, \tag{10.67}$$

$$n^{1\,\mathrm{DMFT}}\left(\vec{r}\right)=\frac{1}{N_{\vec{k}}}\sum_{\vec{k},i,j}\overline{n}_{nm}\left\langle\phi_m|\vec{r}\right\rangle\left\langle\vec{r}|\phi_n\right\rangle, \tag{10.68}$$

$$\widetilde{n}^{1\,\mathrm{DMFT}}\left(\vec{r}\right)=\frac{1}{N_{\vec{k}}}\sum_{\vec{k},i,j}\overline{n}_{nm}\left\langle\widetilde{\phi}_m|\vec{r}\right\rangle\left\langle\vec{r}|\widetilde{\phi}_n\right\rangle, \tag{10.69}$$

where $|\phi_n\rangle$ and $\left|\widetilde{\phi}_n\right\rangle$ are the all-electrons and the PP partial waves, and

$$\overline{n}_{nm}=\frac{1}{N_{\vec{k}}}\sum_{\vec{k},i,j}n_{k,i,j}\left\langle\widetilde{p}_n|\widetilde{\psi}_{\vec{k}j}\right\rangle\left\langle\widetilde{\psi}_{\vec{k}i}|\widetilde{p}_m\right\rangle \tag{10.70}$$

are the occupancies in the augmentation channel (n,m); $\left|\widetilde{\phi}_n\right\rangle$ are the projector functions which are dual to the PS partial waves projector functions. The self-consistent solution of the DMFT equations is obtained when both the charge density and G_{loc} are converged (for more details, including possible algorithm of the solution, see Ref. [22]).

Once the DMFT solution is obtained, one can calculate different physical quantities, including the energy of the system:

$$E[n,G_{\mathrm{cor}}]=E^{\mathrm{DFT}}[n]+E^{\mathrm{KS}}[n,G_{\mathrm{cor}}]+E^{\mathrm{pot}}[G_{\mathrm{cor}}]-E^{\mathrm{DC}}[N_d], \tag{10.71}$$

where

$$\begin{aligned}E^{\mathrm{DFT}}[n]=&-\sum_i\int d\vec{r}\left\langle\psi_i|\hbar^2\frac{\vec{\nabla}^2}{2m_e}|\psi_i\right\rangle+\int d\vec{r}V_{\mathrm{ext}}\left(\vec{r}\right)n\left(\vec{r}\right)\\&+\int d\vec{r}\int d\vec{r}'\frac{n\left(\vec{r}\right)n\left(\vec{r}'\right)}{\left|\vec{r}-\vec{r}'\right|}+\int d\vec{r}V_{\mathrm{XC}}\left(\vec{r}\right)n\left(\vec{r}\right)\end{aligned} \tag{10.72}$$

is the DFT energy,

$$E^{\mathrm{KS}}[n,G_{\mathrm{cor}}]=\frac{1}{N_{\vec{k}}}\sum_{\vec{k},i,j}\varepsilon^{\mathrm{KS}}_{\vec{k}i}\left(n_{\vec{k},ii}-f_{\vec{k},i}\right) \tag{10.73}$$

is a correction to the band energy, added to correct the error that comes from the difference between the density matrix and the Fermi function in the hybridization window,

$$E^{\text{pot}}[G_{\text{cor}}] = \frac{1}{2} T \sum_n e^{-i\omega_n \delta} Tr \left[\hat{\Sigma}_{\vec{k},i,j}(i\omega_n) \hat{G}_{\vec{k},i,j}(i\omega_n) \right] \tag{10.74}$$

is the potential energy of the "non-DFT" local–subspace interaction and $E^{\text{DC}}[N_d]$ is the double-counting energy.

For the last quantity, different approximations can be made. Let us mention some of them in the case of the interaction part of the Hubbard Hamiltonian with the Kanamori parameters (10.46). Often, to calculate the double-counting energy this part of the Hamiltonian is approximated by the diagonal in the density-matrix part (obtained above by decoupling the products of operators), that corresponds to a mean-field approximation:

$$\begin{aligned} \hat{H}^{\text{cor}}_{\text{MF}} = \frac{1}{2} U \left(N_d^2 - \sum_{\alpha,\sigma} \langle n^\sigma_\alpha \rangle^2 \right) - \frac{3}{2} J \left(\sum_\sigma N^2_{d\sigma} - \sum_{\alpha,\sigma} \langle n^\sigma_\alpha \rangle^2 \right) \\ - J \left(\sum_\sigma N_{d\sigma} N_{d\bar{\sigma}} - \sum_{\alpha,\sigma} \langle n^\sigma_\alpha \rangle \langle n^{\bar{\sigma}}_\alpha \rangle \right), \end{aligned} \tag{10.75}$$

where $N_{d\sigma}$ is total number of d-electrons with spin σ. In the fully localized-limit (FLL) approximation [32–34], it is assumed that $\langle n^\sigma_\alpha \rangle$ is either 0 or 1, so one can put in Eq. (10.75) $\langle n^\sigma_\alpha \rangle^2 = \langle n^\sigma_\alpha \rangle$, which gives $\sum_{\alpha,\sigma} \langle n^\sigma_\alpha \rangle^2 = N_{d\sigma}$. Next, in the PM phase $\langle n^\sigma_\alpha \rangle = \langle n^{\bar{\sigma}}_\alpha \rangle = N_{d\sigma}/2$ and one gets the FLL expression for the double-counting energy:

$$E^{\text{FLL}}_{\text{DC}} = \frac{1}{2} U N_d (N_d - 1) - \frac{5}{4} J N_d (N_d - 2) \tag{10.76}$$

(compare with the DFT + U result (10.27)).

In another popular, "around mean-field" (AMF) [35] approximation, it is assumed that each orbital has an average occupation $\langle n \rangle = \frac{N_d}{10}$, which gives

$$E^{\text{AMF}}_{\text{DC}} = \frac{1}{2} U N_d (N_d - \langle n \rangle) - \frac{5}{4} J N_d (N_d - 2\langle n \rangle). \tag{10.77}$$

To conclude this section, we mention that beginning from the pioneering work [20], many self-consistent DFT + DMFT approaches with various double-counting correction schemes were proposed. However, the main idea used in these approaches is similar to the one described above. We refer the reader to some of the works [20, 22, 36–52] and to review articles [21, 28, 53] and move to the next section, where we discuss an extension of the DFT + DMFT approach—GW + DMFT formalism.

10.4 GW + DMFT

GW + DMFT [54, 55] is a combined GW and DMFT approach where one uses GW approximation [16] to take into account effects of the long-range (screened) electron–electron interaction (and to correct the spectrum and other properties of DFT) and DMFT to take into account the local interactions, or strong electron–electron correlation, effects. Namely, in GW + DMFT Hartree, exchange, Random-Phase Approximation (RPA)-screened Coulomb interaction, and local DMFT diagrams are taken into account. Thus, the approximation is well-beyond DMFT; in particular, the electron self-energy in this scheme is momentum-dependent. Not surprising, it is technically much difficult to implement GW + DMFT comparing to DFT + DMFT. Nevertheless, a significant progress in development and application of this tool has been made. The presentation below follows mostly review [56] (we refer the reader for several other excellent reviews [10, 11, 28, 57]).

Let us begin with details of the GW approach which essentially consists of correcting the DFT results by taking into account a many-body self-energy of electrons. Namely, the electron self-energy that corrects the DFT spectrum and other results consist of the Hartree term discussed before and the Coulomb screened interaction term

$$\Sigma^{GW} = i\int d\omega' G\left(\omega+\omega', \vec{r}, \vec{r}'\right) W\left(\omega', \vec{r}, \vec{r}'\right), \tag{10.78}$$

where the screened Coulomb interaction $W\left(\omega', \vec{r}, \vec{r}'\right)$ is calculated with the RPA. Namely, in the momentum space

$$W(\omega, k) = \frac{V\left(\vec{k}\right)}{1 - V\left(\vec{k}\right)P\left(\omega, \vec{k}\right)}, \tag{10.79}$$

where $V\left(\vec{k}\right)$ is the bare Coulomb potential and

$$P\left(\omega, \vec{k}\right) = -2i\int d\omega' G(\omega+\omega')G(\omega') \tag{10.80}$$

is the one-loop polarization that corresponds to a virtual electron–hole pair (for the derivation of the equations, see below). Substituting the last expression in Eq. (10.79) and expanding the denominator in the last equation in powers of $V\left(\vec{k}\right)P\left(\omega, \vec{k}\right)$ one can show that electrons in this case do not interact directly, but the interaction is mediated by virtual electron–hole pairs, i.e., by multiple screening processes. It is important that the virtual processes are dynamical and hence the interaction is dynamical (frequency-dependent). Though GW includes correlation effects through the addition to the HF part of the self-energy, it cannot describe Hubbard sub-bands, MIT, or other important phenomena in the correlated systems. Thus, GW is usually regarded as an approach to weakly-correlated materials.

The GW method was formulated by Hedin as an approximation for his set of five equations for the self-energy, GF, vertices, polarization, and screened interaction [16]. In order to derive the GW equations, let us begin with writing down Hedin's equations. The first equation is the Dyson equation that connects the GF and the self-energy:

$$G^{-1} = G^{(0)-1} - \Sigma. \tag{10.81}$$

The second equation connects the screened Coulomb interaction and the polarization operator:

$$\begin{aligned} W(1,1';2,2') &= V(1,1';2,2') \\ &\quad + W(1,1';3,3')P(3,3';4,4')V(4,4';2,2'). \end{aligned} \tag{10.82}$$

The interaction terms describe scattering of the incoming electron $2'$ and hole 2 and outgoing electron $1'$ and hole 1 (in Eq. (10.82) the indices are combined space-time indices; two pairs of the space indices are equal: $1 = 1'$,$2 = 2'$).

The third equation connects the polarization operator $P(3,3';4,4')$ with the vertex (for the formulas for these quantities, see below):

$$\begin{aligned} P(1,1';2,2') &= T^{-1}G(1,2')G(2,1') \\ &\quad + T^{-2}G(1,3)G(3',1')\Gamma^*(3,3';4,4')G(4',2')G(2,4) \end{aligned} \tag{10.83}$$

(T is temperature and, similar to Eq. (10.82), the space indices are equal: $1 = 1'$, $2 = 2'$). In the last equation, the polarization and the vertex diagrams do not include the diagrams that can be separated by cutting single-interaction line.

The fourth Hedin's equation is the BSE that connects the 2P irreducible vertex $\Gamma^*(3,3';4,4')$ with the particle-hole irreducible vertex Γ^*_{ph} (see Chap. 9):

$$\begin{aligned} \Gamma^*(1,1';2,2') &= \Gamma^*_{ph}(1,1';2,2') \\ &\quad + T^{-1}\Gamma^*(1,1';3,3')G(3',4)G(4',3)\Gamma^*_{ph}(4,4';2,2'). \end{aligned} \tag{10.84}$$

The fifth equation, for the electron self-energy, can be derived from the equation of motion for the GF where the average of the product of four operators (that come from the Coulomb interaction part of the Hamiltonian) is expressed in terms of the irreducible vertex. It reads as:

$$\Sigma(1,1') = -T^{-1}W(1,3';2,2')G(4',2)G(2',4)G(3',3)\Gamma^*(3,1';4,4') + W(1,2';2,1')G(2',2). \quad (10.85)$$

Equation (10.85) completes the set of Hedin's equations (10.81)–(10.85). In the GW approximation, one puts in these equations $\Gamma^*_{ph} = 0$. Then, the Hedin's equations transform into the following five equations:

$$G^{-1} = G^{(0)-1} - \Sigma^{GW}, \quad (10.86)$$

$$W(1,1';2,2') = V(1,1';2,2') + W(1,1';3,3')P^{GW}(3,3';4,4')V(4,4';2,2'), \quad (10.87)$$

$$P^{GW}(1,1';2,2') = T^{-1}G(1,2')G(2,1'), \quad (10.88)$$

$$\Gamma^*(1,1';2,2') = 0, \quad (10.89)$$

$$\Sigma^{GW}(1,1') = W(1,2';2,1')G(2',2). \quad (10.90)$$

The system of GW Eqs. (10.86)–(10.90) is much simpler comparing to Hedin's equations. Nevertheless, it is still not easy to solve GW equations exactly and some approximations are made. In particular, often the LDA GF $G^{(0)}$ is used to calculate the screened interaction W_0 from Eqs. (10.87) and (10.88). This gives for the self-energy Eq. (10.90): $\Sigma = W_0G^{(0)}$. Such an approximation is sufficient in many cases, in particular to correct the DFT energy bandgap in semiconductors and to renormalize the bandwidth (by changing the quasiparticle parameters, see, e.g., Refs. [58, 59] for the case of Na). Another example of successful application of GW approximation is correct electron scattering rate in Ag and some other systems obtained from the imaginary part of the self-energy (for more details and references, see review [56]).

On the other hand, as mentioned above, GW often fails in the case of correlated systems. In the case of the discussed spectral properties, a typical example is missing satellite peak at −6 eV (a lower Hubbard sub-band) in the spectrum of Ni [60]. At the same time, both LDA + DMFT [61] and GW + DMFT [55] give this peak. Thus, using GW instead of DFT in the DFT + DMFT scheme would have some advantages: taken into account screening effects, a more controllable correction for the double counting (since both GW and DMFT solutions are based on Feynman diagrams and it is easy to track the diagrams which are counted twice). Therefore, similar to combining DFT and DMFT, it was proposed to combine GW and DMFT [54, 55] by using the total functional that consists of the GW and the DMFT parts and to obtain the corresponding equations by differentiating the total functional with respect to the GF and the other unknown quantities. This gave, in particular, the self-energy that is sum of the local DMFT and the nonlocal GW parts. A detailed algorithm of the iterative GW + DMFT approach is shown in Fig. 10.1. The flow is self-explanatory, the only new function is the local dynamic screened interaction $U(\omega)$ used in the impurity action (see, e.g., Ref. [62]) instead of the frequency-independent bare local Coulomb interaction, as in DMFT. In calculation of $U(\omega)$, the

Do LDA calculation, figure 4, yielding $\mathbf{G}_{\mathbf{k}}(\omega) = [\omega\mathbf{1} + \mu\mathbf{1} - \epsilon^{\mathrm{LDA}}(\mathbf{k})]^{-1}$ (eq. (85)).

Calculate GW polarization $\mathbf{P}^{\mathrm{GW}}(\omega) = -2i\int\frac{d\omega'}{2\pi}\mathbf{G}(\omega+\omega')\mathbf{G}(\omega')$ (eq. (78)).

If DMFT polarization $\mathbf{P}^{\mathrm{DMFT}}$ is known (after the 1. iteration), include it
$\mathbf{P}^{\mathrm{GW+DMFT}}(\mathbf{k},\omega) = \mathbf{P}^{\mathrm{GW}}(\mathbf{k},\omega) - \frac{1}{V_{\mathrm{BZ}}}\int d^3k\,\mathbf{P}^{\mathrm{GW}}(\mathbf{k},\omega) + \mathbf{P}^{\mathrm{DMFT}}(\omega)$ (eq. (79)).

With this polarization, calculate the screened interaction (eqs (76), (77)):

$$\mathbf{W}(\mathbf{k};\omega) = \mathbf{V}_{\mathrm{cc}}(\mathbf{k})[\mathbf{1} - \mathbf{V}_{\mathrm{cc}}(\mathbf{k})\mathbf{P}(\mathbf{k};\omega)]^{-1}.$$

Calculate $\mathbf{\Sigma}^{\mathrm{Hartree}}_{\mathbf{k}}$ (eq. (71)) and $\mathbf{\Sigma}^{\mathrm{Hartree}}_{\mathrm{dc}}$ (equation (72)).

Calculate $\Sigma^{\mathrm{GW}}(\mathbf{r},\mathbf{r}';\omega) = i\int\frac{d\omega'}{2\pi}G(\mathbf{r},\mathbf{r}';\omega+\omega')W(\mathbf{r},\mathbf{r}';\omega')$ (equation (75)].

Calculate the DMFT self-energy $\mathbf{\Sigma}^{\mathrm{DMFT}}$ and polarization $\mathbf{P}^{\mathrm{DMFT}}$ as follows:

From the local Green function $\mathbf{G}$ and old self-energy $\mathbf{\Sigma}^{\mathrm{DMFT}}$ calculate $(\mathcal{G}^0)^{-1}(\omega) = \mathbf{G}^{-1}(\omega) + \mathbf{\Sigma}^{\mathrm{DMFT}}(\omega)$ (equation (31)); $\mathbf{\Sigma}^{\mathrm{DMFT}} = 0$ in 1. iteration.

Extract the local screening contributions from $\mathbf{W}$ (equation (83)):

$$\mathbf{U}(\omega) = [\mathbf{W}^{-1}(\omega) - \mathbf{P}^{\mathrm{DMFT}}(\omega)]^{-1}.$$

With $\mathbf{U}$ and $\mathcal{G}^0$, solve impurity problem with effective action (equation (84))

$$\mathcal{A} = \sum_{\nu\sigma\, lm}\psi^{\sigma *}_{\nu m}(\mathcal{G}^{\sigma 0}_{\nu mn})^{-1}\psi^{\sigma}_{\nu n} + \sum_{lm\sigma\sigma'}\int_0^\beta d\tau\,\psi^{\sigma *}_{l}(\tau)\psi^{\sigma}_{l}(\tau)\,U_{lm}(\tau-\tau')\psi^{\sigma' *}_{m}(\tau')\psi^{\sigma'}_{m}(\tau'),$$

resulting in $\mathbf{G}$ and susceptibility χ (equation (82)).

From $\mathbf{G}$ and χ, calculate $\mathbf{\Sigma}^{\mathrm{DMFT}}(\omega) = (\mathcal{G}^0)^{-1}(\omega) - \mathbf{G}^{-1}(\omega)$ (eq. (31)),
$\mathbf{P}^{\mathrm{DMFT}}(\omega) = \mathbf{U}^{-1}(\omega) - [\mathbf{U} - \mathbf{U}\chi\mathbf{U}]^{-1}(\omega)$ (eqs (81), (83)).

Combine this to the total GW self-energy (equation (80)):
$\mathbf{\Sigma}^{\mathrm{GW+DMFT}}(\mathbf{k},\omega) = \mathbf{\Sigma}^{\mathrm{GW}}(\mathbf{k},\omega) - \int d^3k\,\mathbf{\Sigma}^{\mathrm{GW}}(\mathbf{k},\omega) + \mathbf{\Sigma}^{\mathrm{Hartree}}(\mathbf{k}) - \mathbf{\Sigma}^{\mathrm{Hartree\ dc}} + \mathbf{\Sigma}^{\mathrm{DMFT}}(\omega)$.

From this and $\mathbf{G}^0$, calculate $\mathbf{G}^{\mathrm{new}}_{\mathbf{k}}(\omega)^{-1} = \mathbf{G}^0_{\mathbf{k}}(\omega)^{-1} - \mathbf{\Sigma}_{\mathbf{k}}(\omega)$.

Iterate with $\mathbf{G}_{\mathbf{k}} = \mathbf{G}^{\mathrm{new}}_{\mathbf{k}}$ until convergence, i.e. $\|\mathbf{G}_{\mathbf{k}} - \mathbf{G}^{\mathrm{new}}_{\mathbf{k}}\| < \epsilon$.

Fig. 10.1 Flow of the GW + DMFT algorithm. From Ref. [28] (see this work for more details).

local screening is not included, since it is included in the DMFT part. Frequency-dependent U makes the DMFT problem much more difficult to solve, so often the approximation $W(\omega) = W(\omega = 0)$ is used [55].

Thus, GW + DMFT is an approach that takes into account both local (DMFT) and long-range (GW) interaction effects, that in particular results in a momentum-dependent self-energy. Several other developments in the direction of building such a "universal" tool are made [65–71]. In particular, another approach to include the momentum-dependence into the DMFT self-energy was proposed in Ref. [63], where an additional momentum-dependent part of Σ was introduced. This part describes nonlocal dynamical correlations induced by the short-ranged collective spin or charge density wave fluctuations. In Ref. [64], the GW + DMFT was adopted to study superconductivity in 2D Hubbard model, and the results were compared to results obtained with another, proposed in this paper, approach—fluctuation exchange (FLEX) + DMFT method. In FLEX + DMFT local and nonlocal (long-range) correlations are self-consistently incorporated. Nonequilibrium properties of 2D U+V Hubbard were studied by using an implemented nonequilibrium GW + extended DMFT approach [65]. In particular, it was shown that in the case of strong nonlocal interactions the excited system shows a self-sustained population inversion of the doublons and holons that results in a (dynamical) anti-screening. A self-energy embedding theory with CI impurity solver (SEET/CI) + GW was proposed and implemented in Ref. [66], where it was applied for several molecular (H- and N-) systems (see also Ref. [67], where GW + DMFT was applied to the H_2 molecule). The multitier self-consistent GW + EDMFT method, where different degrees of freedom (high- and low-energy bands, local and nonlocal interactions, etc.) are treated within appropriate levels of approximation, was proposed and tested for $SrVO_3$ in Ref. [68]. A full-cell GW + DMFT approach, where instead of small impurities defined in a low-energy subspace, impurities that consist of all atoms in the unit cell or supercell, was proposed in Ref. [69]. The resulting theory does not have empirical truncations and parameters, like standard GW + DMFT with "small" impurities defined in a low-energy subspace.

Now, let us mention some of the various applications of the GW-DMFT approach. In Ref. [70], GW + DMFT was used to study properties of the 2D and 3D Hubbard models with up to third-nearest-neighbor long-range interaction. It was shown that momentum-dependent self-energy enhances the correlation effects and the long-range interaction shifts the boundaries between different phases. Thus, the results of the paper [70] suggest that long-range interaction might play a role in strongly correlated materials (for the studies of the 2D Hubbard model, see also Refs. [71, 72]). Analysis of the spectral properties of $SrVO_3$ with GW + DMFT was performed in Refs. [73, 74] and it was found that obtained results are in a good agreement with available photoemission and inverse photoemission spectra. It was also demonstrated that the "three-peak" structure of the spectral function of $SrVO_3$ comes from orbital effects. In Ref. [75], it was shown that the screening effects in $SrVO_3$ lead to a decrease of the strength of the effective local Coulomb repulsion and to strong plasmonic effects. Thus, in agreement with experiment, the Hubbard subbands in the spectrum of this system are weakly pronounced (since the local repulsion is not strong enough to generate them), contrary to the plasmon satellite peak. In Ref. [76], and the GW + DMFT and the LDA + DMFT spectra of $SrVO_3$ were compared with each other and with the experimental data, and it was found that all three results agree rather well. In fact, the position of the GW + DMFT lower Hubbard subband is closer to the experimental result. As the authors showed, the reason for this is balance between the effects of strong electron correlations, reduced bandwidth, and the large RPA screening, all taken into account in GW + DMFT. GW + DMFT was also successfully applied to several other materials, like other systems of the $SrMO_3$ family ($M = V, Cr, Mn, Nb, \ldots$) [77, 78] and Si(111) surface with different adatoms (Sn, Si, C, Pb) [79].

10.5 *Ab initio* DΓA

Another approach that adds nonlocal effects into DFT + DMFT scheme is *Ab initio* DΓA, or *Abinitio*DΓA (see Chap. 9). Using the DΓA ideas proposed in Ref. [80], the authors of works [81–83] (see also Ref. [84]) extended the approach to nonlocal interactions and multiple orbitals and applied it to $SrVO_3$ [81, 83]. The methodology was used in the developed code AbinitioDΓA [82].

Due to the complexity of the multiorbital problem, currently it is difficult to use computationally-accurate full-parquet DΓA. Therefore, in the *Abinitio*DΓA, the irreducible in the *ph* channels-vertex Γ^{ph} is used (similar to the ladder DΓA). This function is sum of the irreducible local vertex and the nonlocal Coulomb interaction terms:

$$\Gamma^{ph}_{\vec{k}\vec{k}' q,\sigma\sigma',lm,m'l'}(\nu,\nu',\omega) = \Gamma^{ph}_{\sigma\sigma',lmm'l'}(\nu,\nu',\omega) + V_{\vec{q},\sigma\sigma',lmm'l'} - \delta_{\sigma\sigma'}\, V_{\vec{k}'-\vec{k},\sigma\sigma',mm'll'}, \quad (10.91)$$

where l, m, m', l' are the orbital indices (see Fig. 10.2 for the corresponding diagram representation). For the orbitals with not large U (like *s*- and *p*-orbitals), the approximation $\Gamma^{ph}_{\sigma\sigma',lmm'l'}(\nu,\nu',\omega) \approx U$ can be used, which significantly speeds up the calculations [84]. Using the vertex (10.91), one obtains the momentum-dependent self-energy, similar to the previous chapter.

Even though the momentum-dependent part of the irreducible vertex is approximated by simple Coulomb potentials, such an approximation is able to describe many effects, like the RPA screening, when the irreducible vertex is substituted by one in the BSE (particle–hole ladder) approximation. Since local Γ gives the DMFT self-energy, *Abinitio*DΓA, besides some cases, like nonlocal spin fluctuations, is capable to describe both DMFT ("local") and GW ("long-range" interaction) limits (for details, see Ref. [84]).

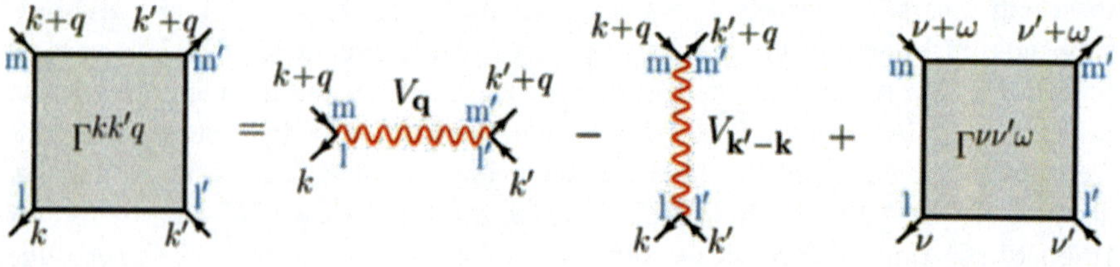

Fig. 10.2 The approximation to the irreducible vertex used in the *Abinitio*DΓA approach. Blue letters are the orbital indices. (Reprinted Fig. 12 with permission from (G. Rohringer, H. Hafermann, A. Toschi, A. A. Katanin, A. E. Antipov, M. I. Katsnelson, A. I. Lichtenstein, A. N. Rubtsov, and K. Held Rev. Mod. Phys. 90, 025003 (2018)). Copyright 2018 by the American Physical Society. https://doi.org/10.1103/RevModPhys.90.025003)

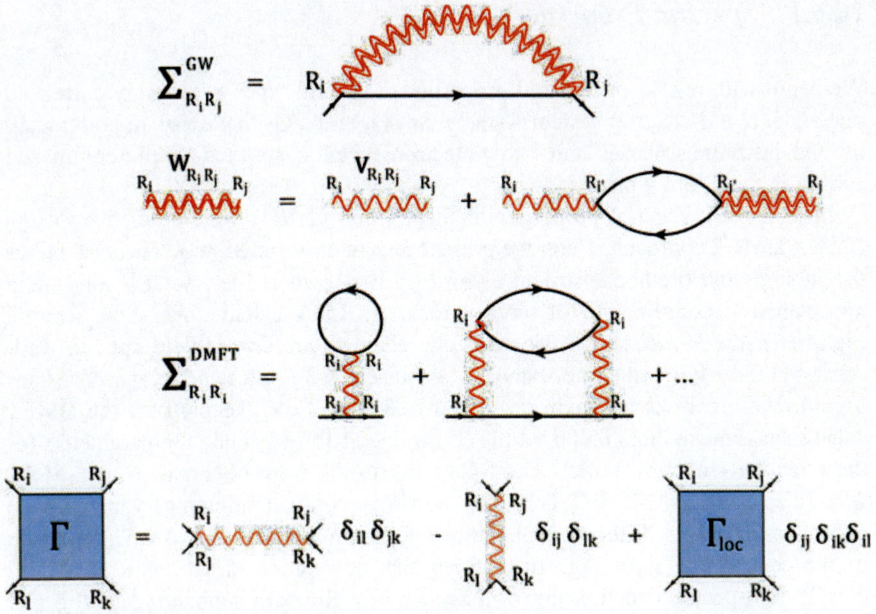

Fig. 10.3 Different *ab initio* schemes, discussed in this chapter, schematically represented in terms of the self-energy, interaction, and vertex diagrams. First and second lines: GW self-energy is defined by the interacting GF and screened interaction, while the screened interaction W is given by the bare interaction and the RPA screening. Third line: the DMFT self-energy is defined by local (on-site) diagrams. Fourth line: In *Abinitio*DΓA, the irreducible vertex Γ is defined by the bare nonlocal Coulomb interaction and the local orbital-dependent vertex. (Reprinted Figs. 1 and 2b with permission from (A. Galler, P. Thunström, P. Gunacker, J.M. Tomczak, and K. Held, Phys. Rev. B 95, 115,107 (2017)). Copyright 2017 by the American Physical Society. https://doi.org/10.1103/PhysRevB.95.115107)

At the end of the summary of the DMFT-based *ab initio* tools, we present Fig. 10.3 that diagrammatically summarizes all three approaches described above: DFT + DMFT, GW + DMFT, and *Abinitio*DΓA. All these approaches can be derived from one general formalism, like the Hedin's equations.

10.6 Application of DFT + DMFT: Properties of Materials

The number of DFT + DMFT papers dramatically increased in the last decade. Thus, it is impossible to over-review all results obtained in these works. Therefore, the goal of this section is rather modest, yet important—to demonstrate how DFT + DMFT works in describing different important properties of materials. For each property, we selected one or a few papers. The choice was made deliberately, with preference to the first study of the material and to a new interesting result.

10.6.1 Spectral Properties

We begin with results on spectral properties of several types of strongly correlated materials: two *d*-electron systems—metal $SrVO_3$ and VO_2 that exists in both metallic and insulating phases and two *f*-electron metallic systems—δ-plutonium and cerium in the α and γ phases.

In Ref. [85], spectral properties of $SrVO_3$ and $CaVO_3$ were studied by using LDA + DMFT approach. Here, we present results only for $SrVO_3$. The system has the cubic perovskite lattice structure (left Fig. 10.4), where the crystal V ions are in an octahedral coordination of oxygen ions. As LDA calculations show (central Fig. 10.4), the V-*3d* states filled with one electron are crystal-field split in well-separated three t_{2g}- and two e_g-bands. The vanadium 3*d* states and oxygen 2*p* states hybridize at energies approximately from −8 to −2 eV. To perform the DMFT calculations, the authors first used the constrained LDA to obtain the parameters for the effective Hubbard model. The following results were obtained: $U = 3.55$eV, $J = 1$eV. The LDA–DMFT DOS for the (low-energy dominating) vanadium t_{2g} orbitals in $SrVO_3$ at different temperatures is shown in right Fig. 10.4. Comparison of the central and right Fig. 10.4 shows that correlation effects split the "non-correlated" spectral function into two asymmetric Hubbard subbands (at −1.5 and 2.5 eV) and produce a high quasiparticle peak at the Fermi energy. The three-peak shape of DOS with a well-pronounced quasiparticle peak suggests that the system is a strongly correlated metal far from MIT. The LDA + DMFT spectrum agrees well with experimental results obtained with high-resolution photoemission [86] and X-ray absorption [87] measurements (not shown here, see Ref. [85]).

Spectral properties of another system—transition-metal oxideVO_2, one of the most popular correlated materials, were studied in Ref. [88]. This system demonstrates change of the crystal structure from the rutile (metallic) to the M_1 monoclinic

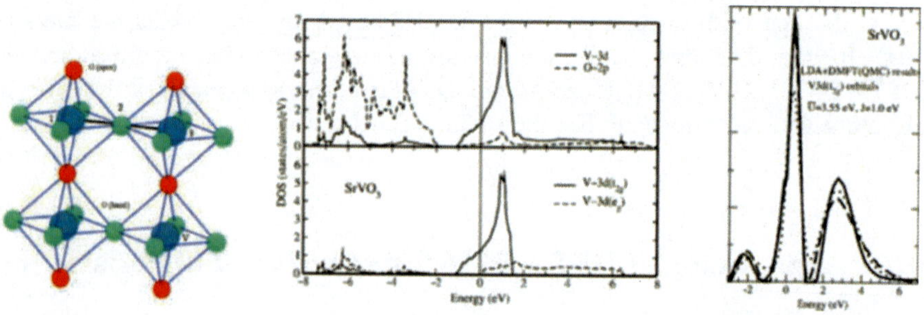

Fig. 10.4 Left: the cubic perovskite lattice structure of $SrVO_3$. Center: the V-*3d*, O-*2p*, and projected V-*3d* DOS of $SrVO_3$ obtained with LDA. Right: The LDA + DMFT results for V-3d DOS of $SrVO_3$ at different temperatures (U = 3.55 eV, J = 1eV, QMC solver). (From Ref. [85] Reprinted Fig. 1, Fig. 3, and right Fig. 6 with permission from (I. A. Nekrasov, G. Keller, D. E. Kondakov, A. V. Kozhevnikov, Th. Pruschke, K. Held, D. Vollhardt, and V. I. Anisimov Phys. Rev. B 72, 155,106 (2005)). Copyright 2005 by the American Physical Society. https://doi.org/10.1103/PhysRevB.72.155106)

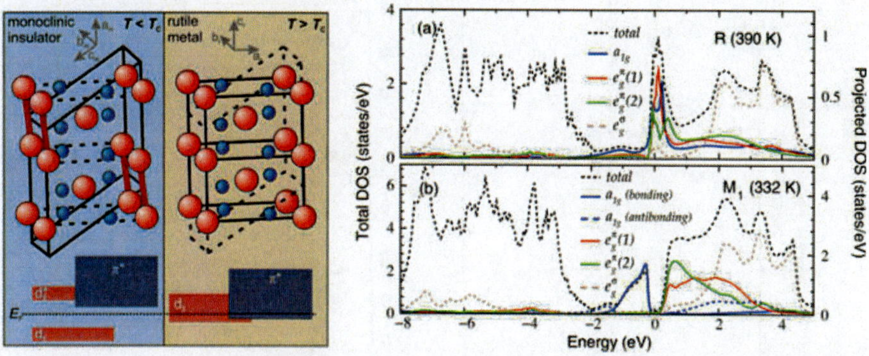

Fig. 10.5 Left: Crystal structures of VO_2 in insulating monoclinic M_1 and metallic rutile phases. From Ref. [89]. Right: Total and projected V d-band DOS of VO_2 in the rutile (top) and monoclinic M_1 (bottom) phases (from Ref. [88], see this paper for more details). (Reprinted Fig. 1 with permission from (W.H. Brito, M.C.O. Aguiar, K. Haule, and G. Kotliar, Phys. Rev. Lett. 117, 056402 (2016)). Copyright 2016 by the American Physical Society. https://doi.org/10.1103/PhysRevLett.117.056402)

(insulating) as temperature goes below the critical value 341 K. The transition to the M_1 monoclinic phase is accompanied by a distortion of the rutile structure with formed V-atom dimers along a_m direction (the crystal structures in both phases are shown in left Fig. 10.5. One of the most important open questions in this system is the mechanism for the gap opening. There are two most popular scenarios that were proposed to explain this phenomenon: the level-splitting mechanism due to V-atom dimerization and the electron-correlation Mott mechanism (see work [89] for references). In paper [88], properties of both metallic and insulating VO_2 systems were studied with DFT + embedded DMFT approach. In the rutile phase (the right top Fig. 10.5), the authors found that the DOS demonstrates a d-orbital lower- (around -1.09 eV) and upper- (around 2.54 eV) Hubbard sub-bands contrary to the DFT results (not shown here) and in agreement with PES measurements [90]. In the M_1 phase, where the V dimers were treated with cluster DMFT, there was found a gap of 0.55 eV (right bottom Fig. 10.5), in agreement with experiment. The gap opening is caused, in particular, by effects of dimerization and antiferroelectric distortion, which lead to a split of the a_{1g} sub-bands in the bonding and the antibonding states. As the authors showed, the bonding–antibonding energy splitting increases upon inclusion of the inter-site dynamic correlations within the dimer. The authors concluded that the gap-opening mechanism can be characterized as "the Mott transition in the presence of strong inter-site exchange." Importantly, DFT + U approach cannot describe the MIT in VO_2 giving insulating phase for both structures.

In work [91], the authors analyzed spectral properties of plutonium with DFT + DMFT and found the $\alpha-$ (simple monoclinic structure) to δ- (fcc structure) phase transformation accompanied by 25% increase in volume of the system (the phases

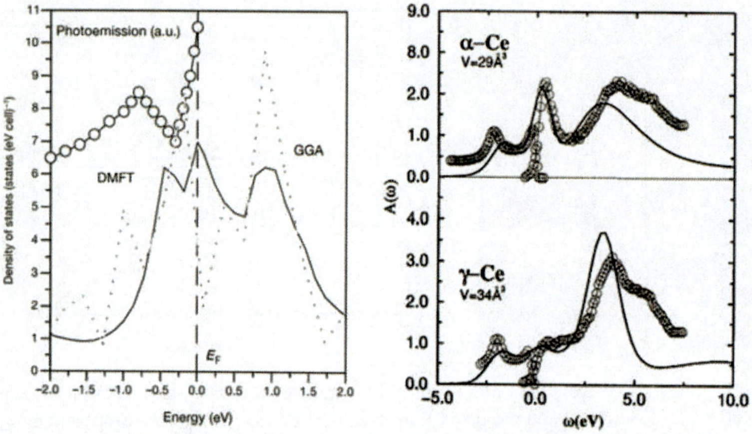

Fig. 10.6 Left: DFT + DMFT (QMC solver), GGA, and experimental (circles) [94] results for the DOS in δ-Pu. From Ref. [91]. Right: DFT + DMFT (QMC solver) and experimental (circles) [98] results for the DOS in α – Ce and γ – Ce. (Reprinted Fig. 11 with permission from (A.K. McMahan, K. Held, and R. T. Scalettar, Phys. Rev. B 67, 075108 (2003)). Copyright 2003 by the American Physical Society. https://doi.org/10.1103/PhysRevB.67.075108)

are stable below 400 K and around 600 K, correspondingly). Understanding of this effect is important for the long-term storage and disposal of the material.

The authors explained the volume expansion between the two phases above, and what is more relevant to this subsection, they predicted the existence of a strong quasiparticle peak near the Fermi level in the δ-phase (left Fig. 10.6), in agreement with experimental data [94]. The DFT + DMFT results are different from the DFT (GGA) (also shown in left Fig. 10.6) and DFT + U results [95, 96] that do not give such a peak. The DFT + DMFT spectrum in left Fig. 10.6 also includes Hubbard sub-bands. From the height of the quasiparticle peak and the distance between the Hubbard sub-bands, one can conclude that in the δ-phase, the *f*-electron subsystem is in a modestly correlated regime, slightly on the side of the "localized-itinerant electron wave function diagram." Since the localization is caused by increased separation between atoms, this explains the larger volume in the δ-phase. This result was also found by DFT + U calculations [95, 96] (that do not give the quasiparticle peak). On the other hand, the α-phase is more metallic (but not weakly correlated, just "slightly on the delocalized side," as the authors mention) with low quasiparticle peak and less pronounced Hubbard sub-bands. Thus, the electrons are less localized in this case and the volume of the system is smaller (the value of the volume and other quantities in this phase are rather well described by DFT).

Spectral properties of the actinide metal Ce with fcc crystal structure at different values of volume and temperature were calculated with DFT + DMFT in Ref. [97] by using QMC and Hubbard-I solvers. It was found that at large volumes, the spectrum has familiar Hubbard–sub-band structure and there is a nonzero local magnetic moment, and in the compressed system for a range of volumes as the system undergoes γ-to-α transition (at 15% volume collapse), there is an enhanced zero-energy quasiparticle (Abrikosov–Suhl resonance) peak and the system is in the PM state. The spectra at the α- and γ-phase volumes agree well with the experimental electron spectroscopy results [98] (right Fig. 10.6). Results for the double occupancy demonstrated that correlations become weaker as the system undergoes the compression-induced transition, yet the system remains in strongly correlated regime that is reflected in the finite Hubbard sub-bands. Agreement between the DFT + DMFT results obtained with the QMC solver (repulsion parameters obtained with constrained LDA $U = 5.72$ eV(α Ce) and $U = 5.98$ eV(γ Ce)) and experiment is rather good for both phases, including positions and weights of the zero-energy quasiparticle peak and of the Hubbard sub-bands, though the theoretical upper sub-band is too narrow (as the authors claim, this is due to not included Hund's exchange coupling). Collapse of the local moments with volume decreasing, as system reached the α-phase, can be explained by increased quasiparticle peak and resulting increase of the hybridization, and hence of the delocalization of *f*-electrons.

10.6.2 Optical Conductivity and Resistivity: Hund's Metal Ruthenates

Now, we move to the transport properties of a few very popular nowadays Hund's metal systems—ruthenates Sr_2RuO_4, $Sr_3Ru_2O_7$, $SrRuO_3$, and $CaRuO_3$. The DFT + DMFT analysis of the transport properties of these systems was performed in Ref. [99]. The DFT + DMFT and DFT results for the optical conductivity and the integrated spectral weight are shown in the left Fig. 10.7. These results demonstrate that correlation effects lead to a strong reduction of the conductivity and of the spectral weight at low frequencies. Since the DFT + DMFT results are in a good agreement with experimental data (see the caption to Fig. 10.7), it suggests that strong electron–electron correlation effects play a dominant role in the transport of ruthenates. Also, in this paper resistivity of the materials was calculated and compared to experimental data. As it follows from the results of the calculations, the agreement between the theoretical and experimental results is rather good for $SrRuO_3$, $CaRuO_3$, while theoretical results for Sr_2RuO_4 and $Sr_3Ru_2O_7$ quite disagree with experiment at high temperatures. In particular, the results for $CaRuO_3$ are in very good agreement with experiment above the Curie temperature $T_c \sim 160$ K, while below T_c there is a discrepancy, probably because the coherence in the FM state that reduces resistivity was not included in the calculations. As the authors suggest, the DFT + DMFT results can be corrected by taking into account vertex corrections or nonlocal interactions.

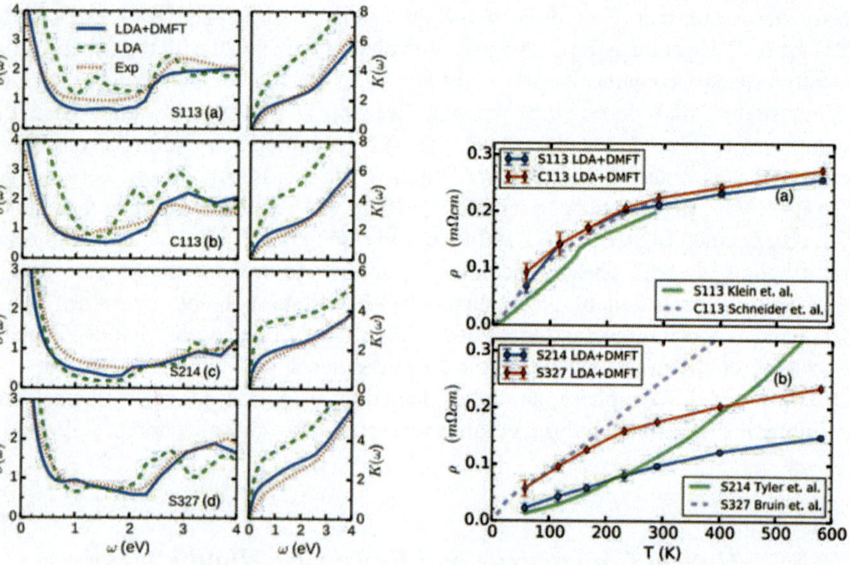

Fig. 10.7 Left: The optical conductivity (left part) and the corresponding integrated spectral weight (right part) of ruthenates obtained from DFT + DMFT and DFT calculations. The red dashed curves correspond to room-temperature experimental data [100]. In the figures, the following notations are used for the materials: S113–$SrRuO_3$, C113–$CaRuO_3$, S214–Sr_2RuO_4, and S327–$Sr_3Ru_2O_7$. Right: The DFT + DMFT and experimental [101–104] results for the resistivity of ruthenates. The results were obtained by using the hybridization expansion CT-HYB solver (for other details of calculations, see Ref. [99]). (Reprinted Figs. 1 and 2 with permission from (X. Deng, K. Haule, and G. Kotliar, Phys. Rev. Lett. **116**, 256401 (2016)). Copyright 2016 by the American Physical Society. https://doi.org/10.1103/PhysRevLett.116.256401)

10.6.3 Optical Properties: Photoemission in Transition-Metal Monoxides

In Ref. [105], *2p* core-level X-ray PES (XPS) for several transition-metal oxides was calculated by using an approach based on DFT + DMFT. To give some more details on the used approach, it is necessary to mention that one can analyze the dynamics of holes in PE spectroscopy by using the emission spectrum. Indeed, since after the absorbed photon creates an electron–hole pair, the electron leaves the system and only the hole affects the emission spectrum. The problem significantly simplifies in the case of core-level XPS since the core-level holes do not move. The hole in this case can be described by single-particle GF in terms of an impurity model that can be constructed with DMFT. As it was formulated in Ref. [105], the core-level spectra can be calculated by performing the following steps: construct strongly correlated Hamiltonian from the DFT results, solve the DMFT problem, calculate the core-level spectra by using the Anderson impurity model with obtained hybridization function (for more details, see work [105]).

As the authors found, the obtained theoretical results are in a good agreement with experiment. We present results for one system—MnO, Fig. 10.8. As it follows

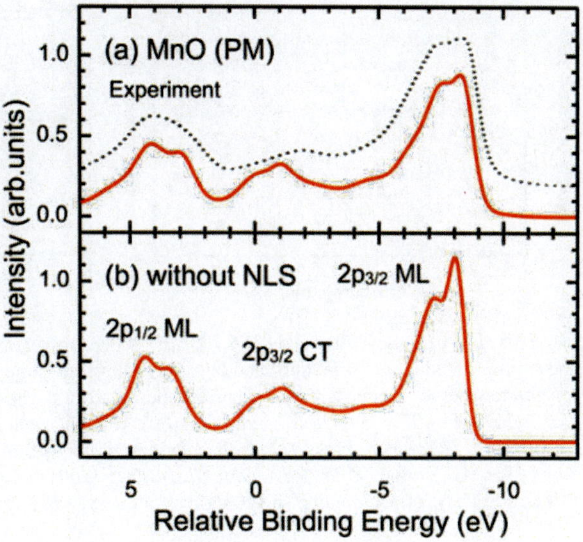

Fig. 10.8 (**a**) The Mn *2p* XPS for MnO obtained with a DMFT-based approach with CT-QMC solver (red curve) and with experimental measurements [121] (dashed curve). (**b**) Theoretical results obtained without nonlocal screening effects. (Reprinted Figs. 5a and 5b with permission from (A. Hariki, T. Uozumi, and J. Kuneš, Phys. Rev. B **96**, 045111 (2017)). Copyright 2017 by the American Physical Society. https://doi.org/10.1103/PhysRevB.96.045111)

from the figure, DMFT-based calculations reproduce well the experimental spectrum (Fig. 10.8a), including the fine features defined in Fig. 10.8b that cannot be obtained with cellular DMFT. Comparison of the calculated spectra obtained with included and not included nonlocal screening (NLS) shows that NLS changes the relative weight of some peaks suggesting that for this system NLS needs to be included in the DMFT calculations.

10.6.4 Magnetic Properties: Fe, Ni, and LaMnPO

We conclude this section with DFT + DMFT results for the magnetic properties of three materials: FM Fe and Ni [61] and AFM LaMnPO [107]. In paper [61], it was shown that LDA + DMFT can describe both the electronic spectrum and the magnetic properties of Fe and Ni below and above the Curie temperature. The results for the LDA and LDA + DMFT spin-resolved spectra for both systems are shown in Fig. 10.9. As it follows from this figure, for Ni LDA + DMFT describes the main correlation effects: satellite peak at 6 eV, 30% narrowing of the occupied part of the *d*-band, and the 50% decrease of exchange splitting, as compared to the LDA results. On the other hand, as the authors found, correlation effects in Fe are less pronounced due to a large spin splitting and a dip in the spin-down DOS near the Fermi energy, which reduces the electron–hole excitations.

LDA + DMFT results also describe well the temperature dependence of the magnetic moment and the slope of Curie–Weiss susceptibility in both systems at high temperatures (Fig. 10.10). As the authors conclude, inclusion of correlation effects is important to describe both the magnetic properties and the spectra of

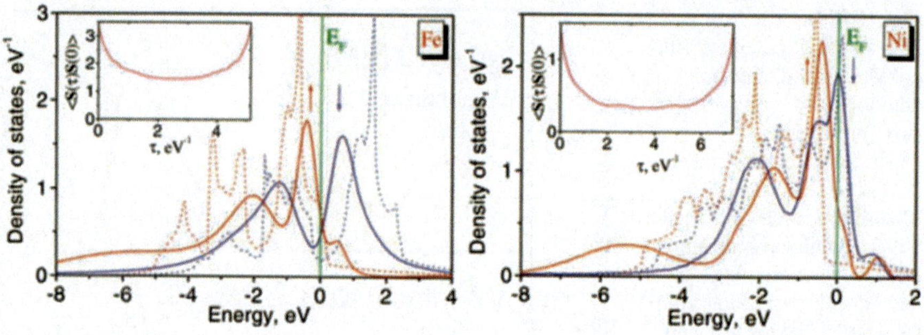

Fig. 10.9 LDA (dashed lines) and LDA + DMFT (QMC solver) (solid lines) results for the orbital- and spin-resolved DOS for Fe (left) and Ni(right) below the Curie temperature. The spin directions for different curves are marked by the corresponding arrows. The used Hubbard model parameters $U = 2.3$ eV (Fe), $U = 3$ eV (Ni), and $J = 0.9$ eV (Fe, Ni) were obtained from constrained LDA calculations. (Reprinted Figs. 1 and 2 with permission from (A. I. Lichtenstein and M. I. Katsnelson, G. Kotliar, Phys. Rev. Lett. **87**, 067205 (2001)). Copyright 2001 by the American Physical Society. https://doi.org/10.1103/PhysRevLett.87.067205)

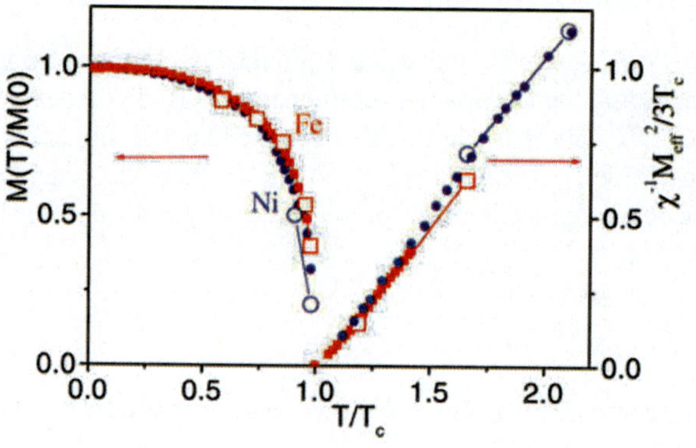

Fig. 10.10 Temperature dependencies of the magnetization and of the inverse FM susceptibility in Fe and Ni obtained with LDA + DMFT and experimental measurements [106, 108] (Fe—squares, Ni—circles). The details of the solution are given in caption in Fig. 10.9. (Reprinted Fig. 3 with permission from (A. I. Lichtenstein and M. I. Katsnelson, G. Kotliar, Phys. Rev. Lett. **87**, 067205 (2001)). Copyright 2001 by the American Physical Society. https://doi.org/10.1103/PhysRevLett.87.067205)

itinerant electrons. Also, the results of the paper show that DMFT describes well physical properties except when systems are close to the transition point, indicating that in this part of the phase diagram fluctuations need to be taken into account.

In work [107], magnetic properties of insulator LaMnPO were studied in both PM and AFM phases by using combined high-temperature inelastic neutron scattering and magnetic susceptibility measurements and DFT + DMFT approach. It was shown that DFT + DMFT calculations can reproduce the direct charge gap (Fig. 10.11) when the Hund coupling $J_H = 0.9$eV (with $U = 8$eV) is taken into account. Thus, it was concluded that LaMnPO is a Mott–Hund's insulator.

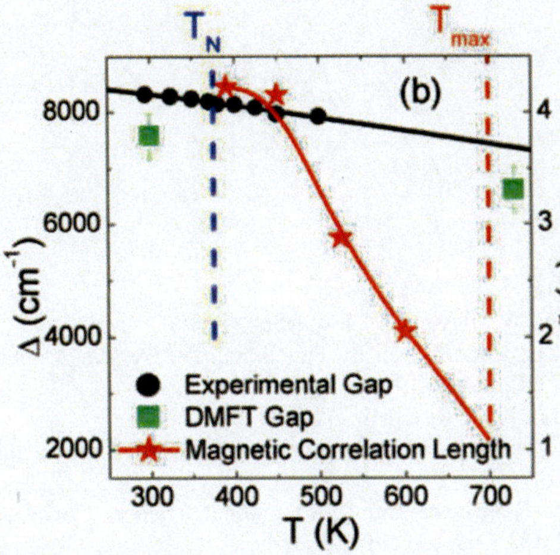

Fig. 10.11 Temperature dependence of the experimental and DFT + DMFT bandgap in the PM and AFM phases of LaMnPO (Néel temperature 375*K*; for details of calculations see the text). From Ref. [107]. (Reprinted Fig. 10 with permission from (D.E. McNally, J.W. Simonson, K. W. Post, Z.P. Yin, M. Pezzoli, G.J. Smith, V. Leyva, C. Marques, L. DeBeer-Schmitt, A.I. Kolesnikov, Y. Zhao, J.W. Lynn, D. N. Basov, G. Kotliar, and M.C. Aronson Phys. Rev. B **90**, 180403 (R) (2014)). Copyright 2014 by the American Physical Society. https://doi.org/10.1103/PhysRevB.90.180403)

As examples of recent applications of DFT + DMFT to study other properties of materials, we mention studies of superconducting properties of iron superconductor LiFeAs [109], of the effects of disorder in VO_2 [110], of the non-Fermi Liquid properties of single SrO layer-doped $SmTiO_3$ [111], and of the nonequilibrium properties of $LaTiO_3$ and $YTiO_3$ [112].

10.7 List of DMFT and DFT + DMFT Codes

At the end of this chapter, we present a list of some of the important static and nonequilibrium DMFT, DFT + DMFT, and other post-DMFT codes, or packages where the DMFT is included (in alphabetic order). Obviously, due to fast developments of the field, this is not complete list, but it may be helpful for newcomers to start to use DMFT.

- Abinit [113]
- AbinitioDGA [114]
- ALPS [115]
- AMULET [116]
- COMSCOPE [92]
- DMFTwDFT [117]
- eDMFT [118]

- NESSi [93]
- TRIQS [119]
- Wien2k [120]

References

1. Hohenberg, P., Kohn, W.: Phys. Rev. **136**, B864 (1964)
2. Dreizler, R.M., Gross, E.K.U.: Density Functional Theory. An Approach to the Quantum Many-Body Problem. Springer (1990)
3. Kohn, W., Sham, L.J.: Phys. Rev. **140**, A1133 (1965)
4. V. Anisimov, Yu. Izyumov, "Electronic Structure of Strongly Correlated Materials" (Springer Series in Solids, 2010)
5. Ullrich, C.A.: Time-Dependent Density-Functional Theory: Concepts and Applications. Oxford University Press, Oxford (2012)
6. Hamada, N., Sawada, H., Terakura, K.: In: Fujimori, A., Tokura, Y. (eds.) Spectroscopy of Mott insulators and correlation metals. Springer-Verlag, Berlin (1995)
7. Sarma, D., Shanthi, N., Barman, S., Hamada, N., Sawada, H., Terakura, K.: Phys. Rev. Lett. **75**, 1126 (1995)
8. Dufek, P., Blaha, P., Sliwko, V., Schwarz, K.: Phys. Rev. B. **49**, 10170 (1994)
9. Mott, N.: Proc. Phys. Soc. London, Ser. A. **49**, 72 (1937)
10. Werner, P., Casula, M.: J. Phys. Condens. Matter. **28**, 383001 (2016)
11. Tomczak, J.M., Liu, P., Toschi, A., Kresse, G., Held, K.: Eur. Phys. J. Special Topics. **226**, 2565 (2017)
12. Svane, A., Gunnarsson, O.: Phys. Rev. Lett. **65**, 1148 (1990)
13. Cowan, R.: Phys. Rev. **163**, 54 (1967)
14. Perdew, J., Zunger, A.: Phys. Rev. B. **23**, 5048 (1981)
15. Fujiwara, T., Arai, M., Ishii, Y.: Chapter 3. In: Anisimov, V. (ed.) Strong Coulomb Correlations in Electronic Structure Calculations: Beyond the Local Density Approximation, vol. 1, pp. 167–202. Gordon and Breach Science Publishers, Singapore (2000)
16. Hedin, L.: Phys. Rev. **139**, A796 (1965)
17. Quinn, J., Ferrell, R.: Phys. Rev. **112**, 812 (1958)
18. Lichtenstein, A., Katsnelson, M.: Phys. Rev. B. **57**, 6884 (1998)
19. Anisimov, V., Zaanen, J., Andersen, O.: Phys. Rev. B. **44**, 943 (1991)
20. Savrasov, S.Y., Kotliar, G.: Phys. Rev. B. **69**, 245101 (2004)
21. Kotliar, G., Savrasov, S.Y., Haule, K., Oudovenko, V.S., Parcollet, O., Marianetti, C.A.: Rev. Mod. Phys. **78**, 865 (2006)
22. Park, H., Millis, A.J., Marianetti, C.A.: Phys. Rev. B. **90**, 235103 (2014)
23. Judd, B.: Operator Techniques in Atomic Spectroscopy. McGraw-Hill, New York (1963)
24. Kanamori, J.: J. Appl. Phys. **31**, S14 (1960)
25. Karolak, M., Ulm, G., Wehling, T., Mazurenko, V., Poteryaev, A., Lichtenstein, A.: J. Electr. Spectr. **181**, 11 (2010)
26. Lichtenstein, A.: In: Pavarini, E., Koch, E., Vollhardt, D., Lichtenstein, A. (eds.) Lecture Notes of the Autumn School on Correlated Electrons. Forschungszentrum Jülich (2014)
27. E. Pavarini, in E. Pavarini, E. Koch, and S. Zhang (eds.) Many-Body Methods for Real Materials Modeling and Simulation, Vol. 9 (Forschungszentrum Juelich, 2019)
28. Held, K.: Adv. Phys. **56**, 829 (2007)
29. Wannier, G.H.: Phys. Rev. **52**, 191 (1937)
30. Marzari, N., Vanderbilt, D.: Phys. Rev. B. **56**, 12847 (1997)
31. Blochl, P.E.: Phys. Rev. B. **50**, 17953 (1994)
32. Wang, X., Dang, H.T., Millis, A.J.: Phys. Rev. B. **84**, 073104 (2011)

33. Park, H., Millis, A.J., Marianetti, C.A.: Phys. Rev. Lett. **109**, 156402 (2012)
34. Dang, H.T., Millis, A.J., Marianetti, C.A.: Phys. Rev. B. **89**, 161113 (2014)
35. Petukhov, A.G., Mazin, I.I., Chioncel, L., Lichtenstein, A.I.: Phys. Rev. B. **67**, 153106 (2003)
36. Minár, J., Chioncel, L., Perlov, A., Ebert, H., Katsnelson, M.I., Lichtenstein, A.I.: Phys. Rev. B. **72**, 045125 (2005)
37. Lechermann, F., Georges, A., Poteryaev, A., Biermann, S., Posternak, M., Yamasaki, A., Andersen, O.K.: Phys. Rev. B. **74**, 125120 (2006)
38. Pourovskii, L.V., Amadon, B., Biermann, S., Georges, A.: Phys. Rev. B. **76**, 235101 (2007)
39. Haule, K., Yee, C.-H., Kim, K.: Phys. Rev. B. **81**, 195107 (2010)
40. Aichhorn, M., Pourovskii, L., Georges, A.: Phys. Rev. B. **84**, 054529 (2011)
41. Grånäs, O., Di Marco, I., Thunström, P., Nordström, L., Eriksson, O., Björkman, T., Wills, J. M.: Comput. Mater. Sci. **55**, 295 (2012)
42. Nekrasov, I.A., Pavlov, V.S., Sadovskii, M.V.: JETP Lett. **95**, 581 (2012)
43. Haule, K., Birol, T.: Phys. Rev. Lett. **115**, 256402 (2015)
44. Haule, K.: Phys. Rev. Lett. **115**, 196403 (2015)
45. Bhandary, S., Assmann, E., Aichhorn, M., Held, K.: Phys. Rev. B. **94**, 155131 (2016)
46. S. Beckes, PhD Thesis (Goethe-Universität, Frankfurt, Germany, 2017)
47. Haule, K.: J. Phys. Soc. Jpn. **87**, 041005 (2018)
48. Schüler, M., Peil, O.E., Kraberger, G.J., Pordzik, R., Marsman, M., Kresse, G., Wehling, T.O., Aichhorn, M.: J. Phys.: Condens. Matter. **30**, 475901 (2018)
49. Adler, R., Kang, C.-J., Yee, C.-H., Kotliar, G.: Rep. Prog. Phys. **82**, 012504 (2019)
50. Bhandary, S., Held, K.: preprint arXiv:1904.02967v2 (2019)
51. Wallerberger, M., Hausoel, A., Gunacker, P., Kowalski, A., Parragh, N., Goth, F., Held, K., Sangiovanni, G.: Comput. Phys. Commun. **235**, 388 (2019)
52. Singh, V., Herath, U., Wah, B., Liao, X., Romero, A.H., Park, H.: preprint arXiv:2002.00068v1 (2019)
53. Minar, J., Ebert, H., Chioncel, L.: Eur. Phys. J. Special Topics. **226**, 2477 (2017)
54. Sun, P., Kotliar, G.: Phys. Rev. B. **66**, 085120 (2002)
55. Biermann, S., Aryasetiawan, F., Georges, A.: Phys. Rev. Lett. **90**, 086402 (2003)
56. K. Held, C. Taranto, G. Rohringer, and A. Toschi, in "Lecture Notes of the Autumn School 2011. Hands-on LDA+DMFT", Eds. E. Pavarini, E. Koch, D. Vollhardt, and A. Lichtenstein (Forschungszentrum, Juelich, 2011)
57. Biermann, S.: J. Phys. Condens. Matter. **26**, 173202 (2014)
58. Northrup, J.E., Hybertsen, M.S., Louie, S.G.: Phys. Rev. Lett. **59**, 819 (1987)
59. Northrup, J.E., Hybertsen, M.S., Louie, S.G.: Phys. Rev. B. **39**, 8198 (1989)
60. Aryasetiawan, F.: Phys. Rev. B. **46**, 13051 (1972)
61. Lichtenstein, A.I., Katsnelson, M.I., Kotliar, G.: Phys. Rev. Lett. **87**, 067205 (2001)
62. Aryasetiawan, F., Imada, M., Georges, A., Kotliar, G., Biermann, S., Lichtenstein, A.I.: Phys. Rev. B. **70**, 195104 (2004)
63. Sadovskii, M.V., Nekrasov, I.A., Kuchinskii, E.Z., Pruschke, T., Anisimov, V.I.: Phys. Rev. B. **72**, 155105 (2005)
64. Kitatani, M., Tsuji, N., Aoki, H.: Phys. Rev. B. **92**, 085104 (2015)
65. Golež, D., Boehnke, L., Strand, H.U.R., Eckstein, M., Werner, P.: Phys. Rev. Lett. **118**, 246402 (2017)
66. Lan, T.N., Shee, A., Li, J., Gull, E., Zgid, D.: Phys. Rev. B. **96**, 155106 (2017)
67. Lee, J., Haule, K.: Phys. Rev. B. **95**, 155104 (2017)
68. Nilsson, F., Boehnke, L., Werner, P., Aryasetiawan, F.: Phys. Rev. Mater. **1**, 043803 (2017)
69. Zhu, T., Chan, G.K.-L: preprint arXiv:2003.01349v1 (2020)
70. Huang, L., Ayral, T., Biermann, S., Werner, P.: Phys. Rev. B. **90**, 195114 (2014)
71. Ayral, T., Werner, P., Biermann, S.: Phys. Rev. Lett. **109**, 226401 (2012)
72. Ayral, T., Biermann, S., Werner, P.: Phys. Rev. B. **87**, 125149 (2013).; Erratum Phys. Rev. B 94, 239906 (2016)

73. Tomczak, J.M., Casula, M., Miyake, T., Aryasetiawan, F., Biermann, S.: Eur. Phys. Lett. **100**, 67001 (2012)
74. Tomczak, J.M., Casula, M., Miyake, T., Biermann, S.: Phys. Rev. B. **90**, 165138 (2014)
75. Boehnke, L., Nilsson, F., Aryasetiawan, F., Werner, P.: Phys. Rev. B. **94**, 201106 (2016)
76. Taranto, C., Kaltak, M., Parragh, N., Sangiovanni, G., Kresse, G., Toschi, A., Held, K.: Phys. Rev. B. **88**, 165119 (2013)
77. Vaugier, L., Jiang, H., Biermann, S.: Phys. Rev. B. **86**, 165105 (2012)
78. Petocchi, F., Nilsson, F., Aryasetiawan, F., Werner, P.: Phys. Rev. Res. **2**, 013191 (2020)
79. Hansmann, P., Ayral, T., Vaugier, L., Werner, P., Biermann, S.: Phys. Rev. Lett. **110**, 166401 (2013)
80. Toschi, A., Rohringer, G., Katanin, A., Held, K.: Ann. Phys. (Berlin). **523**, 698 (2011)
81. Galler, A., Thunström, P., Gunacker, P., Tomczak, J.M., Held, K.: Phys. Rev. B. **95**, 115107 (2017)
82. Galler, A., Thunström, P., Kaufmann, J., Pickem, M., Tomczak, J.M., Held, K.: Comp. Phys. Commun. **245**, 106847 (2017)
83. Galler, A., Kaufmann, J., Gunacker, P., Thunström, P., Tomczak, J.M., Held, K.: J. Phys. Soc. Jpn. **87**, 041004 (2018)
84. Rohringer, G., Hafermann, H., Toschi, A., Katanin, A.A., Antipov, A.E., Katsnelson, M.I., Lichtenstein, A.I., Rubtsov, A.N., Held, K.: Rev. Mod. Phys. **90**, 025003 (2018)
85. Nekrasov, I.A., Keller, G., Kondakov, D.E., Kozhevnikov, A.V., Pruschke, T., Held, K., Vollhardt, D., Anisimov, V.I.: Phys. Rev. B. **72**, 155106 (2005)
86. Sekiyama, A., Fujiwara, H., Imada, S., Eisaki, H., Uchida, S.I., Takegahara, K., Harima, H., Saitoh, Y., Suga, S.: preprint cond-mat/ 0206471 (2002)
87. Inoue, I.H., Hase, I., Aiura, Y., Fujimori, A., Morikawa, K., Mizokawa, T., Haruyama, Y., Maruyama, T., Nishihara, Y.: Physica C. **235–240**, 1007 (1994)
88. Brito, W.H., Aguiar, M.C.O., Haule, K., Kotliar, G.: Phys. Rev. Lett. **117**, 056402 (2016)
89. Wegkamp, D., Stähler, J.: Progr. Surf. Sci. **90**, 464 (2015)
90. Koethe, T.C., Hu, Z., Haverkort, M.W., SchüßlerLangeheine, C., Venturini, F., Brookes, N.B., Tjernberg, O., Reichelt, W., Hsieh, H.H., Lin, H.-J., Chen, C.T., Tjeng, L.H.: Phys. Rev. Lett. **97**, 116402 (2006)
91. Savrasov, S.Y., Kotliar, G., Abrahams, E.: Nature. **410**, 793 (2001)
92. https://www.bnl.gov/comscope/software/comsuite.php
93. M. Schüler, D. Golež, Y. Murakami, N. Bittner, A. Hermann, H.U.R. Strand, P. Werner, M. Eckstein, arXiv:1911.01211 (2019). http://nessi.tuxfamily.org/P2.html
94. Arko, A.J., Joyce, J.J., Morales, L., Wills, J., Jashley, J.: Phys. Rev. B. **62**, 1773 (2000)
95. Savrasov, S.Y., Kotliar, G.: Phys. Rev. Lett. **84**, 3670 (2000)
96. Bouchet, J., Siberchicot, B., Jollet, F., Pasturel, J.: J. Phys. Condens. Matter. **12**, 1723 (2000)
97. McMahan, A.K., Held, K., Scalettar, R.T.: Phys. Rev. B. **67**, 075108 (2003)
98. Liu, L.Z., Allen, J.W., Gunnarsson, O., Christensen, N.E., Andersen, O.K.: Phys. Rev. B. **45**, 8934 (1992)
99. Deng, X., Haule, K., Kotliar, G.: Phys. Rev. Lett. **116**, 256401 (2016)
100. Lee, J.S., Lee, Y.S., Noh, T.W., Nakatsuji, S., Fukazawa, H., Perry, R.S., Maeno, Y., Yoshida, Y., Ikeda, S.I., Yu, J., Eom, C.B.: Phys. Rev. B. **70**, 085103 (2004)
101. Klein, L., Dodge, J.S., Ahn, C.H., Snyder, G.J., Geballe, T.H., Beasley, M.R., Kapitulnik, A.: Phys. Rev. Lett. **77**, 2774 (1996)
102. Tyler, A.W., Mackenzie, A.P., NishiZaki, S., Maeno, Y.: Phys. Rev. B. **58**, R10107 (1998)
103. Schneider, M., Geiger, D., Esser, S., Pracht, U.S., Stingl, C., Tokiwa, Y., Moshnyaga, V., Sheikin, I., Mravlje, J., Scheffler, M., Gegenwart, P.: Phys. Rev. Lett. **112**, 206403 (2014)
104. Bruin, J.A.N., Sakai, H., Perry, R.S., Mackenzie, A.P.: Science. **339**, 804 (2013)
105. Hariki, A., Uozumi, T., Kuneš, J.: Phys. Rev. B. **96**, 045111 (2017)
106. Vonsovsky, S.V.: Magnetism, vol. 2. Wiley, New York (1974)

107. McNally, D.E., Simonson, J.W., Post, K.W., Yin, Z.P., Pezzoli, M., Smith, G.J., Leyva, V., Marques, C., DeBeer-Schmitt, L., Kolesnikov, A.I., Zhao, Y., Lynn, J.W., Basov, D.N., Kotliar, G., Aronson, M.C.: Phys. Rev. B. **90**, 180403 (2014)
108. Altmann, K.N., Petrovykh, D.Y., Mankey, G.J., Shannon, N., Gilman, N., Hochstrasser, M., Willis, R.F., Himpsel, F.J.: Phys. Rev. B. **61**, 15661 (2000)
109. Yin, Z.P., Haule, K., Kotliar, G.: Nat. Phys. **10**, 845 (2014)
110. Weber, C., O'Regan, D.D., Hine, N.D.M., Payne, M.C., Kotliar, G., Littlewood, P.B.: Phys. Rev. Lett. **108**, 256402 (2012)
111. Lechermann, F.: Sci. Rep. **7**, 1565 (2017)
112. Petocchi, F., Beck, S., Ederer, C., Werner, P.: Phys. Rev. B. **100**, 075147 (2019)
113. Gonze, X., et al.: Comput. Phys. Commun. **180**, 2582 (2009). https://www.abinit.org/
114. AbinitioDGA: https://github.com/AbinitioDGA/ADGA
115. B. Bauer et al., J. Stat. Mech. P05001 (2011). https://alpscore.org/
116. http://amulet-code.org/
117. V. Singh et al., arXiv:2002.00068 (2020). https://github.com/DMFTwDFT-project/DMFTwDFT3
118. http://hauleweb.rutgers.edu/tutorials/
119. Parcollet, O., Ferrero, M., Ayral, T., Hafermann, H., Krivenko, I., Messio, L., Seth, P.: Comp. Phys. Comm. **196**, 398 (2015). https://triqs.github.io/triqs/latest/
120. Blaha, P., Schwarz, K., Tran, F., Laskowski, R., Madsen, G.K.H., Marks, L.D.: J. Chem. Phys. **152**, 074101 (2020) http://www.wien2k.at/
121. Bagus, P.S., Broer, R., de Jong, W.A., Nieuwpoort, W.C., Parmigiani, F., Sangaletti, L.: Phys. Rev. Lett. **84**, 2259 (2000)

Chapter 11
DMFT Exchange–Correlation Potentials for Static DFT

11.1 DFT XC Potentials for Strongly Correlated Materials: A Summary

DFT [1, 2] is technically much simpler than many-body theory: DFT equations are equations for one function—space-dependent charge density, while in many-body theory one has to deal with multivariable high-order GFs (see Chap. 9). In DFT, the most complicated part of the problem—effects of electron–electron interaction—is described by the XC potential. Naturally, DFT XC potentials are needed for all types of materials, especially new complex systems, since they allow one to analyze the properties of systems with many atoms and orbitals in the unit cell, including large surfaces and nanoparticles. At present, in all the cases above, DMFT and other many-body approaches have a very limited computational power. Until recent years, theoretical developments of the DFT XC potentials were connected with materials of practical interest—metals and especially semiconductors, i.e., weakly correlated materials (for the basics of the DFT, see the previous chapter). The standard DFT functionals, like LDA, GGA, and hybrid XC potentials, work fairly well for these systems. At the same time, available XC potentials cannot describe properly the ground state properties of strongly correlated materials, e.g., of AFM oxides or orbitally-ordered systems [3]. The reason for this is that not much is known about DFT functionals for these materials, since the general structure of the energy functional for the electron systems with strong interactions, except the homogeneous electron gas [4], is unknown (see, e.g., Refs. [5, 6]).

Reliable energy functionals can be obtained from many-body theory for exactly solvable strongly correlated systems, like 1D and 0D systems [7–13], by using the Bethe ansatz approach [14], or small clusters [5, 15–18], including one- and a few impurity junctions [19, 20] and one-dimensional systems [15, 17, 21–23], that can be solved numerically exactly. Though the obtained results are valid for small systems, they give an idea about the structure of the universal XC potential for strongly correlated systems in all dimensions. To converge to the "universal" all-dimension-

V. Turkowski, *Dynamical Mean-Field Theory for Strongly Correlated Materials*,
https://doi.org/10.1007/978-3-030-64904-3_11

valid XC potential, we need to know the main features of the 2D and 3D potentials. Unfortunately, in high dimensions, analytical or accurate numerical methods for strongly correlated systems are very limited. Analytically, it is possible to calculate the XC energy in the case of low particle density [24–26]. Several numerical DFT approaches for strongly correlated materials were developed and applied, including Gutzwiller DFT [27, 28], where the energy functional is constructed by using projected Gutzwiller wavefunctions, one-body reduced density–matrix functional theory from many-body perturbation theory [29] and lattice DFT [30]. Finally, a machine learning methodology to construct DFT for the Hubbard model was recently developed [31].

Numerically, probably the most powerful approach that covers all regimes of the local Coulomb repulsion is DMFT. It was used to obtain the XC potential for the 3D one-band Hubbard model [32] and the infinite-dimensional Holstein–Hubbard model [33] (see also Ref. [12] for some other potentials obtained with different many-body approximations). In the last two works, the XC energy was calculated for different values of the charge density (doping) n and the XC potential was obtained by numerically differentiating the XC energy with respect to n.

In section 11.2, we summarize DMFT results for the XC potential obtained in Refs. [32, 33] and compare them, where it is possible, with 1D Bethe results. In section 11.3, we discuss details of an alternative way to obtain the DMFT XC potential by using the Sham–Schlueter equation technique [44], that connects the XC potential and a many-body (in our case DMFT) electron self-energy, generalized on the multiorbital case. Using several approximations for the DMFT self-energy, we will discuss possible structures of the DMFT XC potential (for previous (non-DMFT) one-band (one-orbital) applications of this approach, see Refs. [34–42]). It must be noted that obtained in this way potential can be much richer than the pure density (LDA) potentials above obtained by differentiating the XC energy with respect to the charge density (since the potentials obtained from the electron-self energy take into account more subtle effects described by energy, e.g., the interorbital interaction of electrons).

11.2 DMFT XC Potentials: XC Energy Results

In Ref. [32], the authors considered the 3D cubic lattice Hubbard model in presence of an external perturbation. In this chapter, we consider the static part of their solution and the dynamical, TDDFT, solution will be discussed in the next chapter. The used in Ref. [32] static part of the Hamiltonian has the following form:

$$\widehat{H} = -t\sum_{\langle i,j\rangle,\sigma} c_{i\sigma}^{+}c_{j\sigma} + \sum_{i,\sigma} U_i n_{i\uparrow}n_{i\downarrow} + \sum_{i,\sigma} \varepsilon_i n_{i\sigma}, \qquad (11.1)$$

where to include nonhomogeneity effects it was assumed that in general case the site energy ε_i and Coulomb repulsion U_i can be site-dependent; $t = 1$ was used as the energy unit.

The authors of Ref. [32] constructed the energy functional for the Hubbard model that has the structure [43]

$$E_v[n] = T_0[n] + E_H[n] + E_{XC}[n] + \sum_{i,\sigma} \varepsilon_i n_{i\sigma}, \tag{11.2}$$

where $T_0[n]$ is the kinetic energy of the noninteracting system (that can be easily evaluated numerically for each lattice) and $E_H[n] = \frac{1}{4}\sum_i U_i n_i^2$ is the Hartree energy of the noninteracting system (charge density is sum of the spin densities, $n_i = n_{i\uparrow} + n_{i\downarrow}$), and the XC energy E_{XC} was calculated as described below. Using the Lanczos exact diagonalization algorithm for the impurity problem, the authors calculated the energy of the system in DMFT approximation. Namely, the kinetic energy was calculated as

$$K^{DMFT} = 2T\sum_n e^{i\omega_n\delta} \int \frac{d\varepsilon \cdot \varepsilon\rho(\varepsilon)}{i\omega_n - \varepsilon + \mu - \Sigma(i\omega_n)}, \tag{11.3}$$

and the potential energy from

$$V^{DMFT} = UD \equiv U\langle n_\uparrow n_\downarrow \rangle, \tag{11.4}$$

where $\langle n_\uparrow n_\downarrow \rangle$ is the average double occupancy per site (it was calculated for the zeroth site of the impurity model, see below).

Using the DMFT energy $D^{DMFT} = K^{DMFT} + V^{DMFT}$ as the ground state energy of the system, the XC energy was calculated:

$$E_{XC}[n] = E^{DMFT}[n] - T_0[n] - E_H[n]. \tag{11.5}$$

Since the DMFT numerical solution had some noise, especially at low doping $n < 0.2$ and close to half-filling, the authors first smoothed the data and then used a polynomial fitting for the energy function at $0.2 < n < 1$, and used analytical result [25] for the energy for the 3D Hubbard model as function of doping at low doping,

$$E_{XC}[n] = \left(\frac{1}{U^{-1} + 0.1263t^{-1}} - U\right)\frac{n^2}{4} + \lambda n^{\frac{7}{3}}, \tag{11.6}$$

where λ is a fitting parameter (see Ref. [32]).

Differentiating the resulting XC energy with respect to density, the authors obtained DMFT XC potential $V_{XC}[n]$ at $0 < n < 1$. The result for $1 < n < 2$ was obtained by using the particle–hole symmetry relation that gives $V_{XC}[2 - n] = -V_{XC}[n]$ (the XC potential is antisymmetric with respect to the

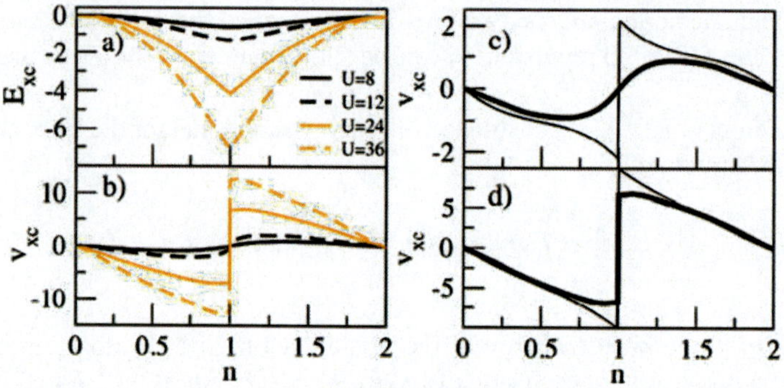

Fig. 11.1 E_{XC} (**a**) and V_{XC} (**b**) as functions of doping for the 3D Hubbard model at different values of U (in units of t). (**c**) DMFT (thick curve) and 1D Bethe ansatz LDA [7, 8] (thin curve) XC potentials at $U = 8t$. (**d**) The same as (**c**) at $U = 24t$. (Reprinted Fig. 1 with permission from (D. Karlsson, A. Privitera, C. Verdozzi, Phys. Rev. Lett. 106, 116401 (2011)). Copyright 2011 by the American Physical Society. https://doi.org/10.1103/PhysRevLett.106.116401)

point $n = 1$). The results for DMFT XC energy and XC potential as functions of doping at different values of U are presented in Fig. 11.1a, b. As it follows from these results, when U crosses the critical (MIT) value of (=14t), the XC energy curve starts to develop a cusp at half-filling. The cusp results in a discontinuity in the XC potential—a DFT sign of MIT. This result is very different from the case of 1D Bethe ansatz LDA potential [7, 8] (Fig. 11.1c, d)), where a much larger discontinuity in the XC potential exists at any U (no MIT).

In Ref. [33], the DMFT XC energy and XC potentials were obtained for a more general case that includes electrons and phonons—the Hubbard–Holstein model. Two cases were considered—one-site and infinite dimensions. We will discuss only the last case, relevant to this book. In the case of infinite dimensions, the authors used the following Hamiltonian:

$$\widehat{H} = -t\sum_{\langle i,j\rangle,\sigma} c_{i\sigma}^{+}c_{j\sigma} + U\sum_{i,\sigma} n_{i\uparrow}n_{i\downarrow} + \sum_{i,\sigma}(\varepsilon_i - \mu)n_{i\sigma} + \omega\sum_i b_i^{+}b_i$$
$$+ \sum_i \sqrt{2}\,\eta_i\widehat{x}_i + \sqrt{2}g\sum_i (n_{i\uparrow} + n_{i\downarrow} - 1)\widehat{x}_i, \tag{11.7}$$

where the new, comparing to the Hubbard model, terms $b_i^{+}(b_i)$ are the phonon creation (annihilation) operators, ω and g are the phonon frequency and the electron–phonon coupling, η_i is an external site-dependent phonon potential, and $\widehat{x}_i = \left(b_i^{+} + b_i\right)/\sqrt{2}$ are the phonon coordinates (η_i was used to control the phonon coordinates).

After the Lang–Firsov transformation $\widehat{H} \to \widehat{H}' = e^{i\widehat{S}}\widehat{H}e^{-i\widehat{S}}$ [45], the Hamiltonian was transformed by the authors of Ref. [33] to the following form:

$$\widehat{\mathrm{H}}' = -\sum_{\langle i,j\rangle,\sigma} \widehat{t}_{ij}' c_{i\sigma}^{+} c_{j\sigma} + U' \sum_{i,\sigma} n_{i\uparrow} n_{i\downarrow} + \sum_{i,\sigma} (\varepsilon_i' - \mu) n_{i\sigma} + \sum_i \left(\omega b_i^{+} b_i - \frac{(\eta_i - g)^2}{\omega} \right), \quad (11.8)$$

where $\widehat{t}_{ij}' = te^{i\sqrt{2}g(\widehat{p}_i - \widehat{p}_j)/\omega}, \widehat{p}_i = i(b_i^{+} - b_i)/\sqrt{2}$ (phonon momentum), $\varepsilon_i' = \varepsilon_i + (g^2 + 2g\eta)/\omega$, $U' = U - 2g^2/\omega$. The Hamiltonian (11.8) was used to calculate the ground state energy of the system, i.e., to construct the XC potential. As the authors of work [33] proved, the total energy of the Holstein–Hubbard model is a functional of n and x, $E = E[n, x]$ and for the homogeneous system the electron and phonon KS Hamiltonians are:

$$H_s^{(e)} = (V_{\mathrm{KS}}[n, x] - \mu) \sum_{i,\sigma} n_{i\sigma} - \sum_{\langle i,j\rangle,\sigma} t(c_{i\sigma}^{+} c_{j\sigma} + h.c.), \quad (11.9)$$

$$H_s^{(ph)} = \omega \sum_i b_i^{+} b_i + \sqrt{2}\eta_{\mathrm{KS}}[n, x] \sum_i \widehat{x}_i, \quad (11.10)$$

The "electron DFT" KS potential in Eq. (11.9) is

$$V_{\mathrm{KS}}[n, x] = V_{\mathrm{ext}} + V_H[n, x] + V_{\mathrm{XC}}[n, x] \equiv V_{\mathrm{ext}} + U\frac{n}{2} + \sqrt{2}gx + \frac{\delta E_{\mathrm{XC}}}{\delta n}, \quad (11.11)$$

where the XC energy E_{XC} is obtained by subtracting the electron–electron, the electron–phonon Hartree interaction terms, and the total energy of the noninteracting system from the total energy $E[n, x]$. Similarly, for the phonon potential in Eq. (11.10) one has

$$\eta_{\mathrm{KS}}[n, x] = \eta_{\mathrm{ext}} + \eta_H[n] + \eta_{\mathrm{XC}}[n, x] \equiv \eta + g(n - 1) + \frac{\delta E_{\mathrm{XC}}}{\delta x}, \quad (11.12)$$

where $\eta_H[n] = g(n - 1)$. Below, we will show the results of paper [33] for the DMFT XC potential obtained from the electron part of the electronic XC energy, since the electron system is the main topic of this and of the other chapters. Regarding the phonon XC potential, as the authors proposed, it can be obtained by beginning with solving the Heisenberg equation of motion for the phonon momentum operator that gives relation between the phonon coordinate and the external $\eta_{\mathrm{ext}} \equiv \eta$ potential: $x = -\sqrt{2}[g(n - 1) + \eta]/\omega$. From the last equation, one can find $\eta_{\mathrm{ext}} \equiv \eta = -g(n - 1) - \omega x/\sqrt{2}$. As the authors noted, since the phonon KS potential reproduces the interacting value of the phonon coordinate x in the noninteracting system, it can be obtained from Eq. (11.12) at $g = 0$. Then, using the last result and Eq. (11.12) one gets $\eta_{\mathrm{XC}} + \eta_H = \eta_{\mathrm{KS}} - \eta_{\mathrm{ext}} = g(n - 1)$, and since $\eta_H = g(n - 1)$ (Eq. (11.12)), one obtains $\eta_{\mathrm{XC}} = 0$. As the authors also mentioned, these arguments, and hence the result, are independent of dimensionality (for more details, see Ref. [33]).

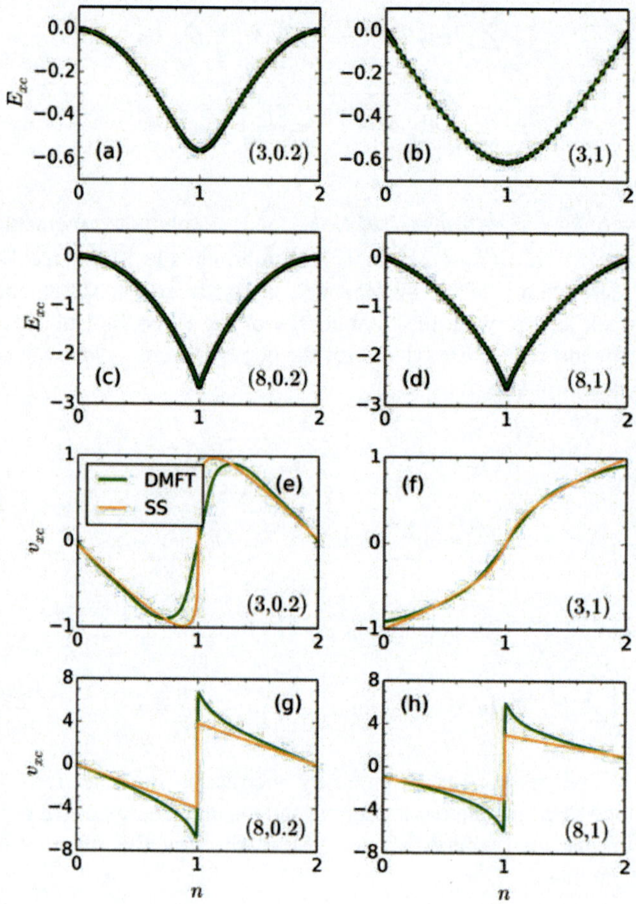

Fig. 11.2 E_{XC} **(a–d)** and V_{XC} **(e–h)** as functions of the electron density at different values of (U, g) (shown in brackets in each figure) in the case of infinite-dimensional Bethe–lattice Hubbard model at $T = 0.005t$. In **(e–h)**, the yellow curves are the single-site (SS) Holstein–Hubbard model results with temperatures that fit the best the DMFT results ($T = 0.2t$ for $U = 3t$ and $T = 0.001t$ for $U = 8t$). Comparison of the infinite-dimensional and the SS potentials is shown to demonstrate to what extent thermal fluctuations can mimic quantum fluctuations. From Ref. [33]

Now, let us discuss in detail the results for the electronic part of the XC potential. In Ref. [33], the authors used DMFT approach to calculate the XC energy and XC potential as functions of doping in the case of infinite-dimensional Bethe lattice Holstein–Hubbard model (bandwidth $4t$, temperature $0.005t$) at different values of U and g. The results are shown in Fig. 11.2. As it follows from Figs. 11.2a–d, similar to the Hubbard model, the XC energy is a smooth function of the charge density at small values of U, and it experiences a kink at half-filling when U is large (the critical (hysteresis) values of U at zero temperature are $U_{c1} = 4.7t$ and $U_{c2} = 5.8t$). Increase in the electron–phonon coupling makes the "half-width" of the energy curve wider.

The results for the XC potential obtained from $V_{XC} = \frac{\delta E_{XC}}{\delta n}$ and fitted by fourth-order polynomials are shown in Figs. 11.2e–h. Again, similar to the Hubbard model case, at large Us, the potential has a cusp at half-filling showing signs of MIT. The electron–phonon interaction delays the inset of discontinuity and decreases the height of the jump.

11.3 DMFT XC Potentials: Sham–Schlueter Equation Approach

In this section, we show an alternative approach to derive the XC potential V_{XC} from the many-body electron self-energy Σ (for details, see, e.g., book [46]). For this, let us write down the many-body Dyson equation

$$\begin{aligned} G\left(t_1\vec{r}_1, t_2\vec{r}_2\right) = &\, G_0\left(t_1\vec{r}_1, t_2\vec{r}_2\right) \\ &+ \iint dt_3 d\vec{r}_3 \iint dt_4 d\vec{r}_4 G_0\left(t_1\vec{r}_1, t_3\vec{r}_3\right) \\ &\times \Sigma[G]\left(t_3\vec{r}_3, t_4\vec{r}_4\right) G\left(t_4\vec{r}_4, t_2\vec{r}_2\right), \end{aligned} \tag{11.13}$$

where the free-electron GF G_0 satisfies the equation

$$\left(\frac{\partial}{\partial t_1} + \frac{\vec{\nabla}^2}{2} - V\left(t_1\vec{r}_1\right)\right) G_0\left(t_1\vec{r}_1, t_2\vec{r}_2\right) = \delta(t_1 - t_2)\delta\left(\vec{r}_1 - \vec{r}_2\right) \tag{11.14}$$

(despite DFT is a static theory, we formally keep all time arguments in the equations below; these dependencies will be used in the next chapter on time-dependent DFT). Similarly, one can write down the nonequilibrium KS GF equation:

$$\left(\frac{\partial}{\partial t_1} + \frac{\vec{\nabla}^2}{2} - V_{HXC}\left(t_1\vec{r}_1\right)\right) G_{KS}\left(t_1\vec{r}_1, t_2\vec{r}_2\right) = \delta(t_1 - t_2)\delta\left(\vec{r}_1 - \vec{r}_2\right). \tag{11.15}$$

From the last three equations, one can derive the Dyson equation

$$\begin{aligned} G\left(t_1\vec{r}_1, t_2\vec{r}_2\right) = &\, G_{KS}\left(t_1\vec{r}_1, t_2\vec{r}_2\right) + \iint dt_3 d\vec{r}_3 \iint dt_4 d\vec{r}_4 G_{KS}\left(t_1\vec{r}_1, t_3\vec{r}_3\right) \\ & \left\{\Sigma_{XC}[G]\left(t_3\vec{r}_3, t_4\vec{r}_4\right) - V_{XC}\left(t_3\vec{r}_3\right)\delta(t_3 - t_4)\delta\left(\vec{r}_3 - \vec{r}_4\right)\right\} G\left(t_4\vec{r}_4, t_2\vec{r}_2\right), \end{aligned} \tag{11.16}$$

where

$$\Sigma_{XC}[G]\left(t_1\vec{r}_1, t_2\vec{r}_2\right) = \Sigma\left(t_1\vec{r}_1, t_2\vec{r}_2\right) - V_H\left(t_1\vec{r}_1\right)\delta(t_1 - t_2)\delta\left(\vec{r}_1 - \vec{r}_2\right). \tag{11.17}$$

Next, using the fact that the many-body and the KS GFs give the same charge density,

$$n\left(t_1\vec{r}_1\right) = -iG\left(r_1t_1, r_1t_1^+\right)\left(t_1\vec{r}_1, t_1^+\vec{r}_1\right) = -iG_{KS}, \tag{11.18}$$

one can obtain from Eq. (11.16) the Sham–Schlueter equation [44] that connects the DFT XC potential $V_{XC}\left(t_2\vec{r}_2\right)$ and the many-body self-energy (see Eq. (11.17)):

$$\iint dr_2 dt_2 G_{KS}\left(t_1\vec{r}_1, t_2\vec{r}_2\right)V_{XC}\left(t_2\vec{r}_2\right)G\left(t_2\vec{r}_2, t_1^+\vec{r}_1\right) = \iint dt_2 d\vec{r}_2 \iint dt_3 d\vec{r}_3 G_{KS}\left(t_1\vec{r}_1, t_2\vec{r}_2\right)\Sigma_{XC}\left(t_2\vec{r}_2, t_3\vec{r}_3\right)G\left(t_3\vec{r}_3, t_1^+\vec{r}_1\right). \tag{11.19}$$

Equation (11.19) can be rewritten as

$$\iint dr_2 dt_2 \tilde{\chi}\left(t_1\vec{r}_1, t_2\vec{r}_2\right)V_{XC}\left(t_2\vec{r}_2\right) = \iint dt_2 d\vec{r}_2 \iint dt_3 d\vec{r}_3 G_{KS}\left(t_1\vec{r}_1, t_2\vec{r}_2\right) \times\Sigma_{XC}\left(t_2\vec{r}_2, t_3\vec{r}_3\right)G\left(t_3\vec{r}_3, t_1^+\vec{r}_1\right) \tag{11.20}$$

where

$$\tilde{\chi}\left(t_1\vec{r}_1, t_2\vec{r}_2\right) = G_{KS}\left(t_1\vec{r}_1, t_2\vec{r}_2\right)G\left(t_2\vec{r}_2, t_1^+\vec{r}_1\right). \tag{11.21}$$

One can rewrite the Sham–Schluter equation (11.20) in another, often more convenient, form:

$$V_{XC}\left(t_1\vec{r}_1\right) = \iint dt_2 d\vec{r}_2 \iint dt_3 d\vec{r}_3 \iint dt_4 d\vec{r}_4 \tilde{\chi}^{-1}\left(t_1\vec{r}_1, t_4\vec{r}_4\right) G_{KS}\left(t_4\vec{r}_4, t_2\vec{r}_2\right) \times\Sigma_{XC}\left(t_2\vec{r}_2, t_3\vec{r}_3\right)G\left(t_3\vec{r}_3, t_4^+\vec{r}_4\right) \tag{11.22}$$

(this is not the solution for V_{XC}, remember that $\tilde{\chi}$ and G_{KS} in the integrals on the right-hand side are functionals of V_{XC}!).

To find the XC potential, one needs to solve self-consistently the system of equations for the many-body and the KS GFs and the many-body self-energy (11.15)–(11.18) and (11.22). This is a very complicated task, thus some approximations have to be made. In the case of not very strong correlations, one can use PT in Eq. (11.16) to find G by iterative expansion in powers of $\left\{\Sigma_{\mathrm{XC}}[G]\left(t_3\vec{r}_3, t_4\vec{r}_4\right) - V_{\mathrm{XC}}\left(t_3\vec{r}_3\right)\delta(t_3 - t_4)\delta\left(\vec{r}_3 - \vec{r}_4\right)\right\}$ with the lowest order approximation $G\left(t_1\vec{r}_1, t_2\vec{r}_2\right) = G_{\mathrm{KS}}\left(t_1\vec{r}_1, t_2\vec{r}_2\right)$, where

$$\begin{aligned} G_{\mathrm{KS}}\left(t_1\vec{r}_1, t_2\vec{r}_2\right) &= \theta(t_1 - t_2)G^{>}_{\mathrm{KS}}\left(t_1\vec{r}_1, t_2\vec{r}_2\right) + \theta(t_2 - t_1)G^{<}_{\mathrm{KS}}\left(t_1\vec{r}_1, t_2\vec{r}_2\right) \\ &= -i\sum_l \left[\phi_l\left(t_1\vec{r}_1\right)\phi_l^*\left(t_2\vec{r}_2\right)\right] \\ &\quad \times \left[\theta(t_1 - t_2)\theta(\varepsilon_l - \varepsilon_F) - \theta(t_2 - t_1)\theta(\varepsilon_F - \varepsilon_l)\right] \end{aligned} \tag{11.23}$$

and $\phi_1\left(t_1\vec{r}_1\right)$ are the "non-correlated" KS wave functions.

The following algorithm to find DMFT XC potential can be used to solve the system of equations (11.15)–(11.18) and (11.22) with PT:

1. Take $\Sigma\left(t_1, t_2, \vec{r}_1\right)$ from the DMFT solution and the Hartree potential $V_H\left(t_1\vec{r}_1\right) = U\frac{n}{2}$ and obtain $\Sigma_{\mathrm{XC}}\left(t_1\vec{r}_1, t_2\vec{r}_2\right)$ from Eq. (11.17).
2. Solve the KS equation with some initial $V_{\mathrm{XC}}\left(t_1\vec{r}_1\right)$ (e.g., $V_{\mathrm{XC}}\left(t_1\vec{r}_1\right) = 0$) and get $\phi_l\left(t_1\vec{r}_1\right)$, and then $G_{\mathrm{KS}}\left(t_1\vec{r}_1, t_2\vec{r}_2\right)$, by using Eq. (11.23).
3. Insert obtained $\Sigma_{\mathrm{XC}}\left(t_1\vec{r}_1, t_2\vec{r}_2\right), V_{\mathrm{XC}}\left(t_1\vec{r}_1\right)$, $G_{\mathrm{KS}}\left(t_1\vec{r}_1, t_2\vec{r}_2\right)$ and $G\left(t_1\vec{r}_1, t_2\vec{r}_2\right) = G_{\mathrm{KS}}\left(t_1\vec{r}_1, t_2\vec{r}_2\right)$ into the right-hand side of Eq. (11.16) and get new $G\left(t_1\vec{r}_1, t_2\vec{r}_2\right)$.
4. Use Eq. (11.22) to calculate new $V_{\mathrm{XC}}\left(t_1\vec{r}_1\right)$.
5. If $V_{\mathrm{XC}}\left(t_1\vec{r}_1\right)$ has not converged, calculate new charge density and new Hartree potential that gives new $\Sigma_{\mathrm{XC}}\left(t_1\vec{r}_1, t_2\vec{r}_2\right)$ from Eq. (11.17) and repeat steps (1)–(4) until $V_{\mathrm{XC}}\left(t_1\vec{r}_1\right)$ converges.

When the correlations are not weak, the system of Eqs. (11.15)–(11.18) and (11.22) has to be solved exactly. Below, we analyze the details of the solution for several cases of an approximated self-energy. We also discuss the structure of an XC potential in these cases.

11.3.1 Local-in-Space and Time Self-Energy

We begin with the simplest case of constant local self-energy

$$\Sigma_{XC}\left(t_1\vec{r}_1, t_2\vec{r}_2\right) = \delta(t_1 - t_2)\delta(t_1 - t_2)\left(\vec{r}_1 - \vec{r}_2\right)\Sigma_{XC}. \tag{11.24}$$

Then, from Eq. (11.22) at $G\left(t_1\vec{r}_1, t_2\vec{r}_2\right) = G_{\mathrm{KS}}\left(t_1\vec{r}_1, t_2\vec{r}_2\right)$ one gets

$$V_{\mathrm{XC}}\left(t_1\vec{r}_1\right) = \iint dt_2 d\vec{r}_2 \iint dt_3 d\vec{r}_3 \iint dt_4 d\vec{r}_4 \chi^{-1}\left(t_1\vec{r}_1, t_4\vec{r}_4\right) G_{\mathrm{KS}}\left(t_4\vec{r}_4, t_2\vec{r}_2\right)$$

$$\times \Sigma_{XC}\left(t_2\vec{r}_2, t_3\vec{r}_3\right) G_{KS}\left(t_3\vec{r}_3, t_4^{+}\vec{r}_4\right)$$

$$= \iint dt_2 d\vec{r}_2 \iint dt_4 d\vec{r}_4 \widetilde{\chi}^{-1}\left(t_1\vec{r}_1, t_4\vec{r}_4\right)\Sigma_{XC}\widetilde{\chi}\left(t_4\vec{r}_4, t_2\vec{r}_2\right) = \Sigma_{XC}. \tag{11.25}$$

Therefore, due to space- and time-invariance of the self-energy Σ_{XC}, XC potential is also space- and time-invariant and is a constant:

$$V_{\mathrm{XC}} = \Sigma_{XC} = \text{constant} \tag{11.26}$$

As it follows from definition Eq. (11.17), $\Sigma_{XC}[G]\left(t_1\vec{r}_1, t_2\vec{r}_2\right) = \Sigma\left(t_1\vec{r}_1, t_2\vec{r}_2\right) - V_H\left(t_1\vec{r}_1\right)\ \delta(t_1 - t_2)\delta\left(\vec{r}_1 - \vec{r}_2\right)$. Thus, in the simplest approximation of $\Sigma\left(t_1\vec{r}_1, t_2\vec{r}_2\right)$ by the static part of the DMFT self-energy $\Sigma^{\mathrm{DMFT}}(\omega = 0)$ (equal $U\frac{n}{2} +$ [higher – order in U terms at $\omega = 0$]) and for the Hartree potential $V_H\left(t_1\vec{r}_1\right) = U\frac{n}{2}$, one gets a simple result

$$V_{\mathrm{XC}}(r_1 t_1) = \Sigma^{\mathrm{DMFT}}(\omega = 0) - U\frac{n}{2}. \tag{11.27}$$

Similarly, for spin DFT one can obtain

$$V_{\mathrm{XC}\sigma}\left(t_1\vec{r}_1\right) = \Sigma_{\sigma}^{\mathrm{DMFT}}(\omega = 0) - U n_{\bar{\sigma}}. \tag{11.28}$$

Since the XC potential is defined up to a constant, such results are meaningless, thus approximation $G\left(t_1\vec{r}_1, t_2\vec{r}_2\right) = G_{\mathrm{KS}}\left(t_1\vec{r}_1, t_2\vec{r}_2\right)$ in Eq. (11.22) is not sufficient to get a nontrivial XC potential from a constant electron self-energy.

11.3.2 Local in Space Self-Energy

Let us consider a more complex case with a frequency-dependent self-energy. In the case of DMFT, one can put:

$$\begin{aligned}\Sigma_{XC}\left(t_1\vec{r}_1,t_2\vec{r}_2\right) &\equiv \delta\left(\vec{r}_1-\vec{r}_2\right)\Sigma_{XC}(t_1-t_2)\\ &= \delta\left(\vec{r}_1-\vec{r}_2\right)\left[\Sigma^{\text{DMFT}}(t_1-t_2)-U\frac{n}{2}\delta(t_1-t_2)\right].\end{aligned} \tag{11.29}$$

Substituting this expression into Eq. (11.22) one gets

$$V_{\text{XC}}\left(t_1\vec{r}_1\right) = \iint dt_2 d\vec{r}_2 \iint dt_3 d\vec{r}_3 \iint dt_4 d\vec{r}_4 \tilde{\chi}^{-1}\left(t_1\vec{r}_1,t_4\vec{r}_4\right) G_{\text{KS}}\left(t_4\vec{r}_4,t_2\vec{r}_2\right)$$
$$\Sigma_{XC}(t_2-t_3)G\left(t_3\vec{r}_2,t_4^{+}\vec{r}_4\right). \tag{11.30}$$

Thus, with fixed time-dependent XC self-energy, the system of equations that define XC potential is dramatically simplified. Making further approximation in Eq. (11.30), $G\left(t_1\vec{r}_1,t_2\vec{r}_2\right)=G_{\text{KS}}\left(t_1\vec{r}_1,t_2\vec{r}_2\right)$, the problem of finding $V_{\text{XC}}\left(t_1\vec{r}_1\right)$ reduces to the solution of the KS equation and Eq. (11.30) (with $G_{\text{KS}}\left(t_1\vec{r}_1,t_2\vec{r}_2\right)$ defined in Eq. (11.23)). This problem is not more complicated than a standard DFT problem with an orbital-dependent XC potential. It is important to note that in the case of a time-dependent self-energy the approximation $G=G_{\text{KS}}$ does not give a trivial XC potential.

11.3.3 Local in Space Self-Energy: An Analytical Hubbard-I Approximation

Finally, let us analyze the structure of the XC potential in the case of the time-(or frequency-) dependent Hubbard-I self-energy

$$\Sigma_{\sigma}^{(\text{at})}(\omega) = Un_{\bar{\sigma}} + \frac{U^2 n_{\bar{\sigma}}(1-n_{\bar{\sigma}})}{\omega-U(1-n_{\bar{\sigma}})} \tag{11.31}$$

that in the spin non-polarized case has the form

$$\Sigma^{(\text{at})}(\omega) = U\frac{n}{2} + \frac{U^2 n(1-n/2)/2}{\omega-U(1-n/2)}. \tag{11.32}$$

First observation that follows from Eq. (11.32) is that in the case of local-in-time approximation (with $\omega = 0$) one gets $\Sigma^{(at)}(\omega = 0) = 0$ and hence, as it follows from Eq. (11.27), one gets a trivial result for the XC potential: $V_{XC}\left(t_1\vec{r}_1\right) = -U\frac{n}{2}$. Thus, frequency (memory) effects are very important in the strongly correlated case.

One can substitute the self-energy (11.32) in time representation in Eq. (11.30) that can be used to solve the system of equations for the XC potential, as in the previous subsection. In the simplest approximation, one can use the local-in-space approximation for the GFs in Eq. (11.30), in the spirit of DMFT. Then, putting there also $G\left(t_1\vec{r}_1, t_2\vec{r}_1\right) = G_{KS}\left(t_1\vec{r}_1, t_2\vec{r}_1\right)$, the equation for the XC potential, becomes

$$V_{XC}\left(t_1\vec{r}_1\right) = \int dt_2 \int dt_3 f\left(t_1, t_2, t_3, \vec{r}_1\right)\Sigma_{XC}(t_2 - t_3), \tag{11.33}$$

where

$$f\left(t_1, t_2, t_3, \vec{r}_1\right) = \int dt_4 \tilde{\chi}^{-1}\left(t_1\vec{r}_1, t_4\vec{r}_1\right) G_{KS}\left(t_4\vec{r}_1, t_2\vec{r}_1\right) G_{KS}\left(t_3\vec{r}_1, t_4^+\vec{r}_1\right). \tag{11.34}$$

We will not make further simplifications to find the XC potential from Eq. (11.33) and mention that due to the structure of the Hubbard-I self-energy, ~$[\omega - U(1 - n/2)]^{-1}$, the major contributions to the XC potential (11.33) come from the GF "pole" energies (frequencies) separated by $\pm U(1 - n/2)$, i.e., the potential is mostly defined by the interaction between electrons in the Hubbard sub-bands.

To summarize, in this section, details of several approximations on how to obtain simple XC potentials from DMFT have been presented. In parallel, it was shown how to self-consistently solve the corresponding DFT + many-body problem. Obviously, these are initial attempts and there is a long road ahead in this direction. In the next chapter, we discuss the generalization of the approaches from this and from the previous section on the time-dependent case.

References

1. Hohenberg, P., Kohn, W.: Phys. Rev. **136**, B864 (1964)
2. Kohn, W., Sham, L.J.: Phys. Rev. **140**, A1133 (1965)
3. Anisimov, V.I., Aryasetiawan, F., Lichtenstein, A.I.: J. Phys. Condens. Matter. **9**, 767 (1997)
4. Ceperley, D.M., Alder, B.J.: Phys. Rev. Lett. **45**, 566 (1980)
5. Capelle, K., Campo Jr., V.L.: Phys. Rep. **528**, 91 (2013)
6. A. Lichtenstein, Lecture Notes of the Autumn School on Correlated Electrons 2014. In E. Pavarini, E. Koch, D. Vollhardt and A. Lichtenstein (Forschungszentrum Jülich, 2014)
7. Lima, N.A., Silva, M.F., Oliveira, L.N., Capelle, K.: Phys. Rev. Lett. **90**, 146402 (2003)
8. Lima, N.A., Oliveira, L.N., Capelle, K.: Europhys. Lett. **60**, 601 (2002)

9. Xianlong, G., Polini, M., Tossi, M.P., Campo Jr., V.L., Capelle, K., Rigol, M.: Phys. Rev. B. **73**, 165120 (2006)
10. Bergfield, J.P., Liu, Z.-F., Burke, K., Staford, C.A.: Phys. Rev. Lett. **108**, 066801 (2012)
11. Campo, V.L.: Phys. Rev. B. **92**, 013614 (2015)
12. Hopjan, M., Karlsson, D., Ydman, S., Verdozzi, C., Almbladh, C.-O.: Phys. Rev. Lett. **116**, 236402 (2016)
13. Senjean, B., Nakatani, N., Tsuchiizu, M., Fromager, E.: Phys. Rev. B. **97**, 235105 (2018)
14. Lieb, E.H., Wu, F.Y.: Phys. Rev. Lett. **20**, 1445 (1968)
15. López-Sandoval, R., Pastor, G.: Phase Transit. **78**, 839 (2005)
16. Carrascal, D., Ferrer, J.: Phys. Rev. B. **85**, 045110 (2012)
17. Xianlong, G., Chen, A.-H., Tokatly, I., Kurth, S., Phys, S.: Rev. B. **86**, 235139 (2012)
18. Carrascal, D., Ferrer, J., Smith, J.C., Burke, K., Phys, J.: Condens. Matter. **27**, 393001 (2015)
19. Liu, Z.-F., Bergfield, J.P., Burke, K., Stafford, C.A.: Phys. Rev. B. **85**, 155117 (2012)
20. Liu, Z.-F., Burke, K.: Phys. Rev. B. **91**, 245158 (2015)
21. Schönhammer, K., Gunnarsson, O., Noack, R.: Phys. Rev. B. **52**, 2504 (1995)
22. Verdozzi, C.: Phys. Rev. Lett. **101**, 166401 (2008)
23. Brosco, V., Ying, Z.-J., Lorenzana, J.: Sci. Rep. **3**, 2172 (2013)
24. Lieb, E.H., Seiringer, R., Solovej, J.P.: Phys. Rev. A. **71**, 053605 (2005)
25. Giuliani, A.: J. Math. Phys. **48**, 023302 (2007)
26. Seiringer, R., Yin, J.: J. Stat. Phys. **131**, 1139 (2008)
27. Bünemann, J., Gebhard, F., Weber, W.: In: Narlikar, A. (ed.) Frontiers in Magnetic Materials. Springer, Berlin (2005)
28. Schickling, T., Bünemann, J., Gebhard, F., Weber, W.: New J. Phys. **16**, 093034 (2014)
29. Giesbertz, K.J.H., Uimonen, A.-M., van Leeuwen, R.: Eur. Phys. J. B. **91**, 282 (2018)
30. Coe, J.P.: Phys. Rev. B. **99**, 165118 (2019)
31. Nelson, J., Tiwari, R., Sanvito, S.: Phys. Rev. B. **99**, 075132 (2019)
32. Karlsson, D., Privitera, A., Verdozzi, C.: Phys. Rev. Lett. **106**, 116401 (2011)
33. Viñas Boström, E., Helmer, P., Werner, P., Verdozzi, C.: Phys. Rev. Res. **1**, 013017 (2019)
34. Sham, L.J.: Phys. Rev. B. **32**, 3876 (1985)
35. Sham, L.J., Schlüter, M.: Phys. Rev. B. **32**, 3883 (1985)
36. Godby, R., Schluter, M., Sham, L.J.: Phys. Rev. Lett. **56**, 2415 (1986)
37. Hanke, W., Sham, L.J.: Phys. Rev. B. **38**, 13361 (1988)
38. Eguiluz, A.G., Hanke, W.: Phys. Rev. B. **39**, 10433 (1989)
39. Eguiluz, A.G., Heinrichsmeier, M., Fleszar, A., Hanke, W.: Phys. Rev. Lett. **68**, 1359 (1992)
40. Niquet, Y.M., Fuchs, M., Gonze, X.: J. Chem. Phys. **118**, 1904 (2003)
41. Engel, E., Dreizler, R.M.: Density Functional Theory: An Advanced Course. Springer Science & Business Media (2011)
42. Hellgren, M., Baguet, L., Calandra, M., Mauri, F., Wirtz, L.: Preprint arXiv. **2005**, 14603 (2010)
43. Schonhammer, K., Gunnarsson, O., Noack, R.M.: Phys. Rev. B. **52**, 2504 (1995)
44. Sham, L.J., Schluter, M.: Phys. Rev. Lett. **14**, 1888 (1983)
45. Lang, I.G., Firsov, Y.A.: Sov. Phys. JETP. **16**, 1301 (1963)
46. Ullrich, C.A.: Time-Dependent Density-Functional Theory: Concepts and Applications. Oxford University Press, Oxford (2011)

Chapter 12
DMFT Exchange-Correlation Potentials for Time-Dependent DFT

12.1 Strongly Correlated Materials Out-of-Equilibrium

As it was mentioned above, materials with strong electron correlations demonstrate many unusual phenomena, e.g., high-temperature superconductivity, enormous heat capacity and giant magnetoresistance, which have important applications in modern technologies. A delicate balance between the role of the itinerant (*s*- and *p*-) and localized (*d*- or *f*-) electronic states allows one to tune the properties of these systems by changing different parameters: chemical composition, pressure, temperature, external fields, system size and shape, etc. (see, e.g., Refs. [1–7]). Even more exotic phenomena can be expected out of equilibrium. Similar to the static properties, dynamical properties of strongly correlated materials may have applications in modern "small and fast" technologies, especially (but not limited to) in the ultrafast optical and electrical switching bulk- and nanotechnologies: switches [8, 9], microelectromechanical system (MEMS) devices [10], smart window applications [11], biosensors [12], and ultrafast lithium storage batteries [13]. In order to fully understand the known and to predict new properties, as well as to describe the experimental results for strongly correlated materials, it is necessary to have a deep understanding of the nonequilibrium properties of these systems, including their ultrafast (femto- and sub-femto-second) spatially-resolved response to external perturbations [14–24].

Such an understanding is challenging, however. Nonequilibrium DMFT, discussed in Chap. 8, and other available many-body approaches cannot be applied directly to many materials because of technical complexity of the problem (many nonequivalent atoms and necessity to consider GFs with several time arguments on a complex time contour), and so far have been used mainly to model the response of simple extended 1–3-orbital systems (see, e.g., Ref. [21]). Similar to the static case (Chap. 11), application of time-dependent *ab initio* approaches looks more promising. Indeed, for weakly correlated systems TDDFT [25–28] can be successfully applied to analyze systems containing hundreds of atoms, similar to DFT [29, 30]. However,

V. Turkowski, *Dynamical Mean-Field Theory for Strongly Correlated Materials*,
https://doi.org/10.1007/978-3-030-64904-3_12

until recent times, direct application of the approach to strongly correlated materials was not possible because of lack of proper TDDFT XC potentials. Many-body or combined *ab initio* and many-body approaches, like DFT + neDMFT (where the static properties are obtained with DFT + DMFT, and the corresponding Hubbard Hamiltonian with an added external perturbation are used to solve the nonequilibrium problem), are computationally very expensive. In particular, in the neDMFT case [21, 33], the analysis is based on calculations of two-time GFs, so that even a four-orbital system is a very challenging task for available computational resources. Needless to say, the case of spatially-nonhomogeneous systems out-of-equilibrium is even more challenging for many-body analysis (where neDMFT, or its cluster extension, has even more limited power due to large supercells). Nevertheless, available neDMFT solutions for rather simple systems already show a rich variety of novel phenomena, including a coupling between the doublon (a doubly occupied-site or -orbital states) and magnetic excitations, Landau–Zener breakdown, Wannier–Stark ladder, and the Bloch oscillations (see Chap. 8 and references therein).

In the last few years, several TDDFT XC potentials for strongly correlated systems have been proposed [31, 32, 34–40], mostly based on exact many-body solutions for low-dimensional strongly correlated models, such as Bethe-ansatz [41] solution for chains [32, 37–40], or exact diagonalization solution for small clusters [31, 42–49]. There, the strategy was to calculate the static dependence of the XC energy on the density of electrons n, to differentiate the result with respect to n and substitute $n \rightarrow n(r, t)$ to obtain the adiabatic XC potential. However, despite a successful description of the Mott insulator transition (MIT) and some nonequilibrium properties, such potentials fail to reproduce the nonequilibrium response in the more complex and important case of systems with strong electron correlations, when the system is close to half-filling (one electron per orbital) and at large values of the local Coulomb repulsion U [39]. One of the main reasons for this in the low-dimensional case might be neglected nonadiabaticity (memory) effects, since the frequency dependence of the XC kernel was not taken into account in these studies.

In the 3D case, similar strategy was used by Karlsson et al. who solved one-band Hubbard model by using DMFT [34]. However, it was also found that obtained adiabatic XC potential gives results that deviate significantly from the exact solution at large values of U or close to half-filling, suggesting importance of the nonadiabatic effects. As a matter of fact, it was demonstrated in the case of Hubbard dimer [50] that nonadiabaticity of the XC kernel is necessary for obtaining an electronic spectrum with the Hubbard satellite peaks and the zero-energy quasiparticle peak. Importance of the nonadiabaticity in the Hubbard dimer was demonstrated also by Fuks and Maitra [44, 45], who showed that the adiabatic approximation gives wrong results for the charge-transfer dynamics. There are also indications that nonadiabatic effects are important in Hubbard chains at some values of doping [51] (On the other hand, as results of recently implemented TDDFT +U algorithms [52, 53] show nonadiabatic approaches can still be useful, especially in calculations of the excitation energies).

To take into account nonadiabatic effects, a more general DMFT - TDDFT approach was proposed [54]. In this approach, the linear-response TDDFT nonadiabatic XC potential (XC kernel) is calculated from the DMFT charge susceptibility. Recently, it was generalized on the spin-dependent, including the noncollinear, case [55] (see also Ref. [31], where the nonadiabatic XC kernel for small Hubbard clusters was calculated).

Thus, at present we have an available adiabatic DMFT XC potential for the general case of nonlinear response for one-band Hubbard model, and a tool to calculate a nonadiabatic XC potential (kernel) for the linear response in general multiorbital case. Below we give details on how to derive potentials in both cases and demonstrate some of their applications. At the end of this chapter, we discuss how to obtain the nonadiabatic DMFT XC potential for the nonlinear case, i.e., the most general form of the XC potential, by using a similar to the static case Sham–Schlueter equation approach [56, 57]. Since all these approaches are based on merging DMFT and TDDFT, we begin with a brief description of TDDFT.

12.2 Time-Dependent Density-Functional Theory

Similar to the static DFT, in TDDFT for the interacting systems a mapping between the time-dependent density and a time-dependent potential is possible. Namely, according to the Runge–Gross theorem [25]

- for a system described by the Hamiltonian

$$\widehat{H} = \widehat{T} + \widehat{H}_{e-e} + \int V\left(t, \vec{r}\right)\widehat{n}\left(t, \vec{r}\right)d\vec{r} \tag{12.1}$$

  where the terms on the right side are the kinetic energy $\widehat{T}$ and the electron-electron interaction $\widehat{H}_{e-e}$, and the last term describes the interaction of the charge density (operator $\widehat{n}\left(t, \vec{r}\right)$) with the external potential *V*(*t*), the charge densities of the system $n\left(t, \vec{r}\right)$ and $n'\left(t, \vec{r}\right)$ that evolve from the same initial state under the action of two external potentials *V*(*r*, *t*) and $\mathrm{V}'\left(t, \vec{r}\right)$ expandable in Taylor series around the initial time, are necessarily different in the case when $V\left(t, \vec{r}\right)$ and $V'\left(t, \vec{r}\right)$ differ by more than a purely time-dependent function, $V\left(t, \vec{r}\right) - V'\left(t, \vec{r}\right) \neq c(t)$.

For details, we refer the reader to books [27, 28]. Here, we just mention that there are important conditions for the theorem to be valid—evolution from the same initial conditions and Taylor expandability of the potentials, i.e., requirement that they are not singular or other "bad" functions. Obviously, if the potentials differ by a time-dependent function, this function can be absorbed into the exponent of the wave function, and the solutions for two wave functions would differ only by a non-important time-dependent phase. In other words, there is one-to-one correspondence between the external potential (up to $c(t)$) and the time-dependent charge density for given initial state. van Leeuwen has proved a theorem that generalizes the Runge–Gross theorem on the case of different initial conditions and different electron–electron interactions [57]:

- Consider two Hamiltonians $\widehat{H}(t)$ and $\widehat{H}'(t)$ of the form (12.1) with different electron–electron interactions $\widehat{H}_{e-e}$ and $\widehat{H}'_{e-e}$ and external $V\left(t,\vec{r}\right)$ and $V'\left(t,\vec{r}\right)$ potentials. Assume that the density $n\left(t,\vec{r}\right)$ evolves from the initial state $|\psi_0\rangle$ with the Hamiltonian $\widehat{H}(t)$. Also, let us assume that the same density $n\left(t,\vec{r}\right)$ with the same current divergence $\vec{\nabla}j\left(t,\vec{r}\right)$ evolves from another initial state $|\psi'_0\rangle$ with the Hamiltonian $\widehat{H}'(t)$. Then, the density $n\left(t,\vec{r}\right)$ uniquely defines the potential $V'\left(t,\vec{r}\right)$ for the system that evolves from $|\psi'_0\rangle$ under $\widehat{H}'(t)$.

Probably, the most important consequence of this theorem that $\widehat{H}_{e-e}$ and $\widehat{H}'_{e-e}$ can be arbitrary, i.e., one can even put $\widehat{H}'_{e-e}=0$, and therefore there is unique $V'\left(t,\vec{r}\right)$ for the noninteracting system that gives the same charge density as in the interacting case with $\widehat{H}_{e-e}\neq 0$ and $V\left(t,\vec{r}\right)$. Thus, there is a unique functional of density $V'\left(t,\vec{r}\right)\equiv V'[n]\left(t,\vec{r}\right)$ called the KS functional that, similar to the static case, has the form (up to a time-dependent function):

$$V_{\mathrm{KS}}[n]\left(t,\vec{r}\right)=V\left(t,\vec{r}\right)+V_H[n]\left(t,\vec{r}\right)+V_{\mathrm{XC}}[n]\left(t,\vec{r}\right). \tag{12.2}$$

Again, similar to the static case, one can write down the TDDFT KS equation [25]:

$$\left[-\frac{\vec{\nabla}^2}{2m_e}+V_{\mathrm{KS}}[n]\left(t,\vec{r}\right)\right]\psi_{\alpha\sigma}\left(t,\vec{r}\right)=i\frac{\partial}{\partial t}\psi_{\alpha\sigma}\left(t,\vec{r}\right), \tag{12.3}$$

where α are the momentum and/or other possible quantum numbers and the Hartree $V_H[n]\left(t,\vec{r}\right)$ and the XC $V_{\mathrm{XC}}[n]$ parts of the KS potential $V_{\mathrm{KS}}[n]\left(t,\vec{r}\right)$ are self-consistently defined by the relation between the charge density and the KS wave functions

$$n\left(t,\vec{r}\right)=\sum_{\alpha\sigma,k<k_F}\left|\psi_{\alpha\sigma}\left(t,\vec{r}\right)\right|^2. \tag{12.4}$$

However, there are essential differences between DFT and TDDFT KS equations. In particular, in DFT the KS potential defines the ground-state density by means of the relation between the charge density and the orbital wave functions, and the solution is obtained by iterations, but in TDDFT the initial density defines the initial KS potential, which is used to propagate the potential and the wave functions as the system evolves in time. Thus, the solution is obtained by propagation in time, not by iterations in the static case. We refer the reader for details on how TDDFT XC potential and KS equations can be derived from the effective action, and for more details on fundamentals of TDDFT to books [27, 28].

After this summary of the main TDDFT theorems and before moving to the discussion on the TDDFT XC potentials for strongly correlated materials, we briefly mention the main TDDFT approximations for "standard," weakly correlated, materials. In fact, much less is known about the TDDFT XC potentials for these systems, as compared to the DFT ones, which is related both to a younger age of the theory and less practical needs of TDDFT so far. Probably, the most often used TDDFT XC potential is the adiabatic LDA potential that can be obtained from the static LDA potential by assuming that the space-dependent charge density is also time-dependent. Other potentials, generalizations of DFT potentials, like GGA and OEP, are also in use. However, most of the potentials are adiabatic, i.e., the potential is defined by the charge density at given moment of time, thus the history of the system does not play any role. As we mentioned above, the memory effects can be very important in many cases.

Before moving to the strongly correlated case, we give details on how the absorption spectrum can be calculated with the linear-response TDDFT. Besides the importance by itself, this material will help to illustrate how nonadiabaticity of the XC potential (kernel) affects the absorption spectrum, one of the most important characteristics of the systems. This analysis is also very helpful for further presentation of the derivation of the DMFT nonadiabatic XC kernel. Let us begin with the many-body formalism by writing down the expression for the fluctuating charge density in the linear-response approximation:

$$n\left(t,\vec{r}\right)=n_0\left(\vec{r}\right)+n_1\left(t,\vec{r}\right), \tag{12.5}$$

where $n_0\left(\vec{r}\right)$ is the initial (static) charge density and $n_1\left(t,\vec{r}\right)$ is the time-dependent part (the fluctuation) due to presence of the time-dependent potential, turned on after time t_0,

$$v\left(t,\vec{r}\right)=v_0\left(\vec{r}\right)+v_1\left(t,\vec{r}\right)\theta(t-t_0). \tag{12.6}$$

The time-dependent part of the potential (12.6) $v_1\left(\vec{r},t\right)$ enters into the perturbation part of the Hamiltonian

$$\widehat{H}_1(t)=\int d^3r' v_1\left(t,\vec{r}'\right)\widehat{n}\left(t,\vec{r}'\right). \tag{12.7}$$

In the linear-response approximation, the fluctuation part of the density (12.5) can be calculated as:

$$n_1\left(t,\vec{r}\right)=\int_{-\infty}^{\infty}dt'\int d^3r'\chi_{nn}\left(t-t',\vec{r},\vec{r}'\right)v_1\left(t',\vec{r}'\right), \tag{12.8}$$

where

$$\chi_{nn}\left(t-t',\vec{r},\vec{r}'\right)=-i\theta(t-t')\left\langle\Psi_0\right|\left[\widehat{n}\left(t-t',\vec{r}\right),\widehat{n}\left(0,\vec{r}'\right)\right]\left|\Psi_0\right\rangle \tag{12.9}$$

is the density–density correlation function (retarded GF). Now, let us transform Eq. (12.9) to a form that can be used in practice. For this, using the interaction representation for the density operators $\widehat{n}(t)=e^{i\widehat{H}_0t}\widehat{n}e^{-i\widehat{H}_0t}$, inserting the identity sum over the state vectors $1=\sum_{n=0}^{\infty}|\Psi_n\rangle\langle\Psi_n|$ between the products of the density operators, and performing the Fourier transformation to the frequency domain, one can obtain instead of (12.9)

$$\chi_{nn}\left(\omega,\vec{r},\vec{r}'\right)=\sum_{n=0}^{\infty}\left(\frac{\left\langle\Psi_0|\widehat{n}\left(\vec{r}\right)|\Psi_n\right\rangle\left\langle\Psi_n|\widehat{n}\left(\vec{r}'\right)|\Psi_0\right\rangle}{\omega-\Omega_n+\Omega_0+i\eta}-\frac{\left\langle\Psi_0|\widehat{n}\left(\vec{r}'\right)|\Psi_n\right\rangle\left\langle\Psi_n|\widehat{n}\left(\vec{r}\right)|\Psi_0\right\rangle}{\omega+\Omega_n-\Omega_0+i\eta}\right), \tag{12.10}$$

where Ω_n are the eigenenergies and $|\Psi_n\rangle$ are the corresponding eigenstates. Eq. (12.10) can be used to calculate the correlation function $\chi_{nn}\left(\omega,\vec{r},\vec{r}'\right)$. Using $\chi_{nn}\left(\omega,\vec{r},\vec{r}'\right)$, one can calculate the charge density fluctuation in the frequency domain:

$$n_1\left(\omega,\vec{r}\right)=\int d^3r'\chi_{nn}\left(\omega,\vec{r},\vec{r}'\right)v_1\left(\omega,\vec{r}'\right). \tag{12.11}$$

A similar to the above many-body theory approach can be used in TDDFT, where the system is "noninteracting" and the potential is

$$V_{KS}[n]\left(t,\vec{r}\right)=v\left(t,\vec{r}\right)+\int d^3r'\frac{n\left(t,\vec{r}'\right)}{\left|\vec{r}-\vec{r}'\right|}+V_{XC}[n]\left(t,\vec{r}\right). \tag{12.12}$$

As usual, in Eq. (12.12) $v\left(\vec{r},t\right)$ is an external potential and the last two terms are the Hartree and XC potentials, correspondingly. In the linear-response TDDFT one can write

$$n_1\left(t,\vec{r}\right) = \int dt' \int d^3r' \chi_{KS}\left(t\vec{r},t'\vec{r}'\right) V_{KS1}\left(t',\vec{r}'\right), \tag{12.13}$$

where index "1" corresponds to the linear perturbative part of the KS potential:

$$V_{KS1}\left(t',\vec{r}'\right) = v_1\left(t,\vec{r}\right) + \int d^3r' \frac{n_1\left(t,\vec{r}'\right)}{\left|\vec{r}-\vec{r}'\right|} + V_{XC1}[n]\left(t,\vec{r}\right) \tag{12.14}$$

and

$$\chi_{KS}\left(t\vec{r},t'\vec{r}'\right) = \frac{\delta n[v_{KS}]\left(t,\vec{r}\right)}{\delta v_{KS}\left(t',\vec{r}'\right)}\Bigg|_{v_{KS}=v_{KS}[n_0]\left(\vec{r}\right)} \tag{12.15}$$

The last term in Eq. (12.14) is the linear-response part of the XC potential

$$V_{XC}[n]\left(t,\vec{r}\right) \approx V_{XC1}[n]\left(t=0,\vec{r}\right) + V_{XC1}[n]\left(t,\vec{r}\right). \tag{12.16}$$

It has the following dependence on the excited charge density:

$$V_{XC1}[n]\left(t,\vec{r}\right) = \int dt' \int d^3r' f_{XC}\left(t\vec{r},t'\vec{r}'\right) n_1\left(t',\vec{r}'\right), \tag{12.17}$$

where

$$f_{XC}\left(t\vec{r},t'\vec{r}'\right) = \left(\frac{\delta V_{XC}[n]\left(t,\vec{r}\right)}{\delta n\left(t',\vec{r}'\right)}\right)_{n(0,\vec{r})} \tag{12.18}$$

is the XC kernel.

Substitution of Eqs. (12.14), (12.17) into Eq. (12.13) gives

$$n_1\left(t,\vec{r}\right) = \int dt' \int d^3r' \chi_{KS}\left(t\vec{r},t'\vec{r}'\right) \times \left[v_1\left(t',\vec{r}'\right) + \int d\tau \int d^3x \left\{\frac{\delta(t'-\tau)}{\left|\vec{r}'-\vec{x}\right|} + f_{XC}\left(t'\vec{r}',\tau\vec{x}\right)\right\} n_1\left(\tau,\vec{x}\right)\right]. \tag{12.19}$$

Then, substituting $n_1\left(t,\vec{r}\right) = \int dt' \int d^3r' \chi\left(t\vec{r},t'\vec{r}'\right) V_1\left(t',\vec{r}'\right)$ into the last equation and cancelling $V_1\left(t',\vec{r}'\right)$ under the integrals in both parts of the equation one gets

$$\chi\left(t\vec{r},t'\vec{r}'\right) = \chi_{KS}\left(t\vec{r},t'\vec{r}'\right) + \int d\tau \int d^3x \int_{-\infty}^{\tau} d\tau' \int d^3x' \chi_{KS}\left(t\vec{r},\tau\vec{x}\right) \times \left\{ \frac{\delta(\tau-\tau')}{\left|\vec{x}-\vec{x}'\right|} + f_{XC}\left(\tau\vec{x},\tau'\vec{x}'\right) \right\} \chi\left(\tau'\vec{x}',t'\vec{r}'\right). \tag{12.20}$$

Thus, the XC kernel that defines the XC potential can be calculated from the last equation, which in frequency representation gives

$$f_{XC}\left(\omega,\vec{r},\vec{r}'\right) = \chi_{KS}^{-1}\left(\omega,\vec{r},\vec{r}'\right) - \chi^{-1}\left(\omega,\vec{r},\vec{r}'\right) - \frac{1}{\left|\vec{r}-\vec{r}'\right|}, \tag{12.21}$$

Since the KS susceptibility can be easily calculated by using the DFT eigenenergies and eigenfunctions,

$$\chi_{KS}\left(\omega,\vec{r},\vec{r}'\right) = \sum_{j,k=1}^{\infty} \left(f_k - f_j\right) \frac{\varphi_j^0\left(\vec{r}\right)\varphi_k^{0*}\left(\vec{r}\right)\varphi_j^{0*}\left(\vec{r}'\right)\varphi_k^0\left(\vec{r}'\right)}{\omega - \omega_{jk} + i\eta} \tag{12.22}$$

(ω_{jk} are the excitation/de-excitation energies, i.e., the differences between the j and k state energies, and $f_k = \theta(\varepsilon_F - \varepsilon_k)$ are the Fermi factors), the problem reduces to find the susceptibility χ.

The formalism above was derived for a discrete spectra, i.e., finite systems. Let us also present the main details of the methodology in the case of solids (continuous spectrum). For this, it is more convenient to rewrite Eq. (12.20) in the frequency representation:

$$\chi\left(\omega,\vec{r},\vec{r}'\right) = \chi_{KS}\left(\omega,\vec{r},\vec{r}'\right) + \int d^3x \int d^3x' \chi_{KS}\left(\omega,\vec{r},\vec{x}\right) \times \left\{ \frac{1}{\left|\vec{x}-\vec{x}'\right|} + f_{XC}\left(\omega,\vec{x},\vec{x}'\right) \right\} \chi\left(\omega,\vec{x}',\vec{r}'\right). \tag{12.23}$$

Due to periodicity of the response function,

$$\chi\left(\omega,\vec{r},\vec{r}'\right)=\chi\left(\omega,\vec{r}+\vec{R},\vec{r}'+\vec{R}\right) \tag{12.24}$$

and of other functions ($\vec{R}$ is the lattice vector), one can introduce Fourier transforms

$$\chi\left(\omega,\vec{r},\vec{r}'\right)=\frac{1}{V}\sum_{\vec{k}}\sum_{\vec{G},\vec{G}'}e^{-i\left(\vec{k}+\vec{G}\right)\vec{r}}\chi\left(\omega,\vec{k}+\vec{G},\vec{k}+\vec{G}'\right)e^{i\left(\vec{k}+\vec{G}'\right)\vec{r}'}, \tag{12.25}$$

$$\frac{1}{\left|\vec{r}-\vec{r}'\right|}=\frac{1}{V}\sum_{\vec{k}}\sum_{\vec{G},\vec{G}'}e^{-i\left(\vec{k}+\vec{G}\right)\vec{r}}\delta_{\vec{G},\vec{G}'}\frac{4\pi}{\left|\vec{k}+\vec{G}\right|^2}e^{i\left(\vec{k}+\vec{G}'\right)\vec{r}'}, \tag{12.26}$$

where $\vec{G}$ and $\vec{G}'$ are the reciprocal vectors (and similar for other functions). Then, in the momentum space Eq. (12.23) can be written as

$$\begin{aligned}\chi_{\vec{G}\vec{G}'}\left(\omega,\vec{k}\right)&=\chi_{\mathrm{KS}\vec{G}\vec{G}'}\left(\omega,\vec{k}\right)+\sum_{\vec{G}_1,\vec{G}_2}\chi_{\mathrm{KS}\vec{G}\vec{G}_1}\left(\omega,\vec{k}\right)\\&\times\left\{v_{\vec{G}_1}\left(\vec{k}\right)\delta_{\vec{G}_1\vec{G}_2}+f_{\mathrm{XC}\vec{G}_1\vec{G}_2}\left(\omega,\vec{k}\right)\right\}\chi_{\vec{G}_2\vec{G}'}\left(\omega,\vec{k}\right),\end{aligned} \tag{12.27}$$

where we introduced notations

$$\chi_{\vec{G}\vec{G}'}\left(\omega,\vec{k}\right)=\chi\left(\omega,\vec{k}+\vec{G},\vec{k}+\vec{G}'\right), \tag{12.28}$$

$$v_{\vec{G}}\left(\vec{k}\right)=\frac{4\pi}{\left|\vec{k}+\vec{G}\right|^2}, \tag{12.29}$$

and

$$\begin{aligned}\chi_{\mathrm{KS}\vec{G}\vec{G}'}\left(\omega,\vec{k}\right)&=\frac{1}{V}\sum_{\vec{k}'}\sum_{j,l=1}^{\infty}\frac{\left(f_{l\vec{k}+\vec{k}'}-f_{j\vec{k}'}\right)}{\omega+\varepsilon_{j\vec{k}'}-\varepsilon_{l\vec{k}+\vec{k}'}+i\delta}\\&\times\int d^3r\varphi^{0*}_{j\vec{k}'}\left(\vec{r}\right)e^{-i\left(\vec{k}+\vec{G}\right)\vec{r}}\varphi^0_{l\vec{k}+\vec{k}'}\left(\vec{r}\right)\\&\times\int d^3r'\varphi^{0*}_{l\vec{k}+\vec{k}'}\left(\vec{r}'\right)e^{i\left(\vec{k}+\vec{G}'\right)\vec{r}'}\varphi^0_{j\vec{k}'}\left(\vec{r}'\right)\end{aligned} \tag{12.30}$$

is the KS susceptibility [27].

Provided the matrix elements of the KS susceptibility and the XC kernel are known, one can find from Eq. (12.27) the susceptibility matrix $\chi_{\vec{G}_1\vec{G}_2}\left(\omega, \vec{k}\right)$. Vice versa, if $\chi_{KS\vec{G}\vec{G}'}$ and $\chi_{\vec{G}_1\vec{G}_2}$ are known one can find the XC kernel:

$$f_{XC\vec{G}_1\vec{G}_2}\left(\omega, \vec{k}\right) = \left(\chi_{KS}^{-1}\right)_{\vec{G}_1\vec{G}_2}\left(\omega, \vec{k}\right) - \left(\chi^{-1}\right)_{\vec{G}_1\vec{G}_2}\left(\omega, \vec{k}\right) - V_{\vec{G}_1}\left(\vec{k}\right)\delta_{\vec{G}_1\vec{G}_2}. \tag{12.31}$$

Finally, to conclude the section on TDDFT, we show how to calculate the absorption spectrum $A(\omega)$ [27] that defines the excitation properties of the system and how nonadiabaticity of the XC kernel affects $A(\omega)$. The absorption spectrum can be obtained from the dielectric function $\epsilon\left(\omega, \vec{r}, \vec{r}'\right)$ that in the case of no magnetic effects connects the displacement and the electric fields:

$$\vec{D}\left(\omega, \vec{r}\right) = \int d^3r' \epsilon\left(\omega, \vec{r}, \vec{r}'\right)\vec{E}\left(\omega, \vec{r}'\right). \tag{12.32}$$

In the case of periodic systems, the last relation has the following form in the momentum representation:

$$\vec{D}_{\vec{G}}\left(\omega, \vec{k}\right) = \sum_{\vec{G}'} \epsilon_{\vec{G}\vec{G}'}\left(\omega, \vec{k}\right)\vec{E}_{\vec{G}'}\left(\omega, \vec{k}\right). \tag{12.33}$$

Averaging the last equation over the unit cell gives equation for the macroscopic functions:

$$\vec{D}_{\text{mac}}(\omega) = \epsilon_{\text{mac}}(\omega)\vec{E}_{\text{mac}}(\omega) \tag{12.34}$$

Thus, provided the averaged (macroscopic) dielectric function $\epsilon_{\text{mac}}(\omega)$ is found, one can obtain the absorption spectrum from its imaginary part $A(\omega) = \frac{\omega}{n_b c} Im\epsilon_{\text{mac}}(\omega)$, where n_b is the refractive index and c is the speed of light in vacuum (for more subtleties in the calculations of $A(\omega)$, see book [58]).

In the homogeneous case, $\epsilon\left(\omega, \vec{r}, \vec{r}'\right) = \epsilon\left(\omega, \vec{r} - \vec{r}'\right)$ and $\epsilon\left(\omega, \vec{r}, \vec{r}'\right)$ depends on the coordinates difference, not on individual $\vec{r}$ and $\vec{r}'$. Then,

$$\epsilon_{\text{mac}}^{\text{hom}}(\omega) = \epsilon^{\text{hom}}\left(\omega, \vec{k} \to 0\right). \tag{12.35}$$

However, in general nonhomogeneous case, due to local-field effects/spatial fluctuations, calculation of the macroscopic dielectric function $\epsilon_{\text{mac}}(\omega)$ is a nontrivial task.

Namely, since $\epsilon_{\vec{G}\vec{G}'}\left(\omega, \vec{k}\right)$ is non-diagonal in reciprocal vector indices, one needs to invert large matrices to calculate $\epsilon_{\text{mac}}(\omega)$:

$$\epsilon_{\text{mac}}(\omega) = \left[\epsilon^{-1}_{\vec{G}\vec{G}'}\left(\omega, \vec{k} \to 0\right)\Big|_{\vec{G}=0,\vec{G}'=0}\right]^{-1}. \tag{12.36}$$

To express $\epsilon^{-1}_{\vec{G}\vec{G}'}\left(\omega, \vec{k} \to 0\right)$ in terms of the TDDFT susceptibility, one can rewrite Eq. (12.32) as

$$V_1\left(\omega, \vec{r}\right) = \int d^3r' \epsilon\left(\omega, \vec{r}, \vec{r}'\right)\left[V_1\left(\omega, \vec{r}'\right) + \int d^3r'' \frac{n_1\left(\omega, \vec{r}''\right)}{\left|\vec{r}' - \vec{r}''\right|}\right]. \tag{12.37}$$

Then, using the last equation and $n_1\left(\omega, \vec{r}''\right) = \int \chi\left(\omega, \vec{r}'', \vec{r}'''\right) V_1\left(\omega, \vec{r}'''\right) d^3r'''$, one can obtain

$$\epsilon^{-1}\left(\omega, \vec{r}, \vec{r}'\right) = \delta\left(\vec{r} - \vec{r}'\right) + \int d^3r'' \frac{\chi\left(\omega, \vec{r}'', \vec{r}'\right)}{\left|\vec{r} - \vec{r}''\right|}, \tag{12.38}$$

and for the periodic systems in the momentum representation:

$$\epsilon^{-1}_{\vec{G}\vec{G}'}\left(\omega, \vec{k}\right) = \delta_{\vec{G}\vec{G}'} + V_{\vec{G}}\left(\vec{k}\right)\chi_{\vec{G}\vec{G}'}\left(\omega, \vec{k}\right). \tag{12.39}$$

Once again, calculation of the macroscopic dielectric function with Eqs. (12.36), (12.39) is a rather complicated task, since it involves inversion of very large matrices. It can be shown that for cubic systems one can use a simple formula [27, 75]

$$\epsilon_{\text{mac}}(\omega) = 1 - V_0\left(\vec{k} \to 0\right)\overline{\chi}_{00}\left(\omega, \vec{k} \to 0\right), \tag{12.40}$$

where $\overline{\chi}_{\vec{G}\vec{G}'}\left(\omega, \vec{\mathrm{k}}\right)$ is calculated by using Eq. (12.27) with

$$V_{\vec{G}}\left(\vec{k}\right) = \left(1 - \delta_{\vec{G}0}\right)\frac{4\pi}{\left|\vec{k} + \vec{G}\right|^2}. \tag{12.41}$$

From Eq. (12.27) for the susceptibility that defines the absorption spectrum it is easy to see how nonadiabaticity, or frequency-dependence, of the XC kernel affects the excitation energies and the absorption spectrum. For a qualitative arguments, let us consider the case with one reciprocal vector and rewrite Eq. (12.27) as $\chi\left(\omega,\vec{k}\right)=\frac{1}{\chi_{\mathrm{KS}}^{-1}\left(\omega,\vec{k}\right)-f_{\mathrm{XC}}\left(\omega,\vec{k}\right)}$. Since the excitation spectrum is defined by imaginary part of χ, i.e., by poles of this function, in absence of the XC potential, the excitation spectrum consists of the transition between the DFT states that enter in the KS susceptibility $\chi_{\mathrm{KS}}\left(\omega,\vec{k}\right)$. The frequency-independent XC kernel just shifts the DFT excitation spectrum levels, while the frequency-dependent one may result in new excitation energies that can be found from $\chi_{\mathrm{KS}}^{-1}\left(\omega,\vec{k}\right)-f_{\mathrm{XC}}\left(\omega,\vec{k}\right)=0$ (instead of the DFT equation $\chi_{\mathrm{KS}}^{-1}\left(\omega,\vec{k}\right)=0$). Thus, new excited states definitely affect the system response, and the nonadiabaticity of the XC kernel should be taken into account in reliable TDDFT approaches.

Now, after we have summarized the main features of TDDFT, we move to the DMFT results for the XC potentials/kernels obtained from the XC energy and by other means.

12.3 Adiabatic DMFT XC Potential

The derived from the XC energy for the 3D Hubbard model adiabatic DMFT XC potential [34], discussed in detail in the static case in Chap. 11, was used to study the charge response in the case of a cubic cluster with 125 atoms, where the Coulomb interaction was assumed to be different from zero only on the central ("interacting impurity") atom, $i=0$. It was also assumed that the energy level on this atom is shifted by energy ε_0 comparing to the energy on the other atoms (to allow non-homogeneity effects) and that the perturbation (Gaussian and rectangular pulses) is also applied only to the charges on this atom. Such a choice of the system was made, besides the analysis of the nonequilibrium properties, to test the DMFT - TDDFT results versus the exact solution that can be found for this rather small cluster. Two different values of U ($8t$ and $24t$ corresponding to gapless and gapped systems) and two different charge concentrations (0.32 and 0.56) were used. It was found that the agreement between the exact and the DMFT - TDDFT results at times of the pulse duration is rather good (Fig. 12.1c–f). After the pulse, the results differ, especially in the case of large Us and/or closer to half-filling (Fig. 12.1c–f), i.e., when the correlations play more important role. As the authors suggested, the reason for this could be absent in the XC potential memory and nonlocal interaction effects. In particular, constructing the exact static XC potential by reverse engineering, the authors showed that the potential is essentially nonlocal in the case of large Us and ns (the results correspond to the case of Fig. 12.1f). Another reason for the deviation of the Fig. 12.1f DMFT result from the exact one is that in the finite cluster approximation the discontinuity in the XC potential at half-filling for 3D Hubbard

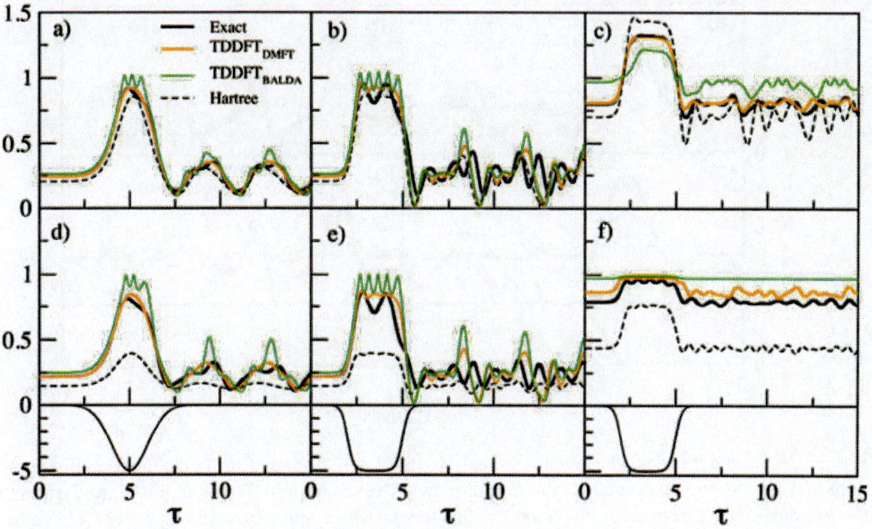

Fig. 12.1 Time dependence of the excited charge density on the central atom of the 125-atom cluster obtained with the exact approach, TDDFT, and Hartree approximation at different values of U ($U = 8t$ (**a–c**), and $24t$ (**d–f**)) and of the density of electrons (0.32 (**a**, **b**, **d**, **e**) and 0.56 (**e**, **f**)). It was assumed that the atomic-level energy on the central atom is shifted in the (**c**) and (**f**) cases (by $\varepsilon_0 = -2.66t$ and $-4t$, correspondingly). (Reprinted Fig. 2 with permission from (D. Karlsson, A. Privitera, C. Verdozzi, Phys. Rev. Lett. 106, 116401 (2011)). Copyright 2011 by the American Physical Society. https://doi.org/10.1103/PhysRevLett.106.116401)

model is significantly modified, but the modification is not included in the dynamic calculations. Thus, the processes when the excited charge density crosses the discontinuity in the XC potential are not described properly.

The authors also compared the DMFT XC potential results with those obtained with adiabatic Bethe ansatz LDA (BALDA) XC potential and with no XC potential (Hartree approximation) (Fig. 12.1). The BALDA results are much worse comparing to the DMFT due to much "stronger" BALDA XC potential (larger discontinuity and presence of gap at any value of U). Similarly, the Hartree approximation gives charge density more different from the exact solution than DMFT-TDDFT one due to missed correlation effects. Thus, DMFT - TDDFT XC potential in the bulk case is much more accurate than other available potentials.

To mimic experiments in ultracold-atom systems, the authors also analyzed dynamics of the system with the DMFT XC potential in the case of an asymmetric confinement and obtained Bloch oscillations of the current, in agreement with experiment [59].

More recently, it was shown [60] that combining a nonlocal XC self-energy (obtained with a time-perturbative conserving many-body approach) with a nonperturbative DMFT XC potential (hybrid TDDFT–nonequilibrium GF approach) gives much better agreement with exact results comparing to the DMFT

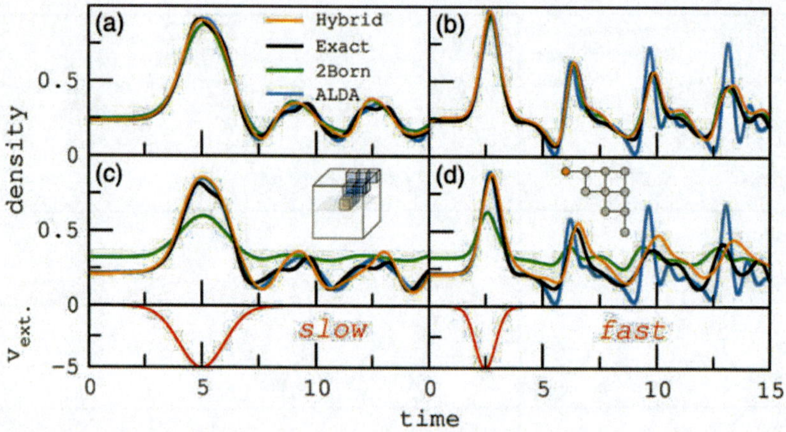

Fig. 12.2 Time-dependent density at the central site of a 125-site cluster at different approximations and $U = 8t$ (top row) and $U = 24t$ (middle row). The shapes of the applied pulses are shown in the bottom row. "2Born approximation" is the second Born approximation (see Ref. [60]) and "ALDA" is the DMFT XC approximation used to obtain the results in Fig. 12.1. (Reprinted Fig. 2 with permission from (M. Hopjan, D. Karlsson, S. Ydman, C. Verdozzi, and C.-O. Almbladh, Phys. Rev. Lett. 116, 236402 (2016)). Copyright 2016 by the American Physical Society. https://doi.org/10.1103/PhysRevLett.116.236402)

XC (ALDA) approach (Fig. 12.2). This suggests that, indeed, the nonlocal-in-space part of the XC potential can be important.

12.4 Nonadiabatic DMFT XC Kernel

Thus, as it was shown in the previous section, nonadiabatic effects may be very important in the case of strong correlations. Here, we demonstrate how one can obtain the nonadiabatic DMFT XC potential in the linear-response regime, i.e., how one can find the DMFT XC kernel. We begin with the PM case in Sect. 12.4.1, and will proceed with general non-colinear spin-polarized case in Sect. 12.4.2.

12.4.1 PM case

To find the nonadiabatic DMFT XC kernel $f_{XC}(r,t;r't')$, we begin with the multiorbital Hubbard Hamiltonian in the Kanamori approximation, with the parameters derived from DFT as described in Chap. 10 (though the formalism can be used for other forms of the Hubbard Hamiltonian without complications):

$$H = -\sum_{i,j,l,m,\sigma} t_{ij,\sigma}^{lm} c_{i\sigma}^{l+} c_{j\sigma}^{m} - \mu \sum_{i,l,\sigma} c_{i\sigma}^{l+} c_{i\sigma}^{l} + U \sum_{i,l} n_{i\uparrow}^{l} n_{i\downarrow}^{l} + (U - 2J) \sum_{i,l\neq m} n_{i\uparrow}^{l} n_{i\downarrow}^{m}$$
$$+ (U - 3J) \sum_{i,l\neq m,\sigma} n_{i\sigma}^{l} n_{i\sigma}^{m}. \tag{12.42}$$

Since the XC kernel can be found from the charge susceptibility as defined in Eq. (12.21), we need to find the single-particle orbital GF

$$G_{\sigma\sigma'}^{ll'}\left(t\vec{r}, t'\vec{r}'\right) = -\left\langle \widehat{T} c_{\sigma}^{l}\left(t, \vec{r}\right) c_{\sigma'}^{l'+}\left(t', \vec{r}'\right)\right\rangle \tag{12.43}$$

and the 2P orbital susceptibility

$$\chi_{\sigma\sigma'}^{ll'}\left(t\vec{r}, t'\vec{r}'\right) = -\left\langle \widehat{T} n_{\sigma}^{l}\left(t, \vec{r}\right) n_{\sigma'}^{l'}\left(t', \vec{r}'\right)\right\rangle, \tag{12.44}$$

where $\widehat{T}$ is the time-ordering operator. Then, the total charge susceptibility can be calculated as

$$\chi\left(t\vec{r}, t'\vec{r}'\right) = -\left\langle \widehat{T} n\left(t, \vec{r}\right) n\left(t', \vec{r}'\right)\right\rangle = \sum_{l,l',\sigma,\sigma'} \chi_{\sigma\sigma'}^{ll'}\left(t\vec{r}, t'\vec{r}'\right) \tag{12.45}$$

In DMFT - TDDFT [54], the functions (12.44) and (12.45) are local in space. We will not stop on details of the calculations of the single-particle GF with DMFT, described in detail in the first part of the book, and will discuss only the 2P susceptibility case. One can obtain the 2P susceptibility by using the BSE approach discussed in detail in Chap. 9. Here, we repeat the main steps. In the BSE approach, one begins with calculating a generalized four-time orbital susceptibility (most conveniently, in the imaginary time representation)

$$\begin{aligned}
\chi^{abcd}(\tau_i, \tau_j, \tau_k, 0) = & -\left\langle \widehat{T} c_i^{a+}(\tau_i) c_i^{b}(\tau_j) c_i^{c+}(\tau_k) c_i^{d}(0)\right\rangle \\
& + \left\langle \widehat{T} c_i^{a+}(\tau_i) c_i^{b}(\tau_j)\right\rangle \left\langle \widehat{T} c_i^{c+}(\tau_k) c_i^{d}(0)\right\rangle \\
& = \frac{1}{Z} \int D[\psi] D[\psi^*] \psi_a^*(\tau_i) \psi_b(\tau_j)\, \psi_c^*(\tau_k) \psi_d(0) \\
& \times exp\left[-\int_0^\beta d\tau_3 \int_0^\beta d\tau_4 \psi_e^*(\tau_3) \mathcal{G}_{ef}^{-1}(\tau_3 - \tau_4) \psi_f(\tau_4) \right. \\
& \left. + \int_0^\beta d\tau_5 U_{ef} \psi_e^*(\tau_5) \psi_e(\tau_5) \psi_f^*(\tau_5) \psi_f(\tau_5) \right],
\end{aligned} \tag{12.46}$$

where the ("non-time") index i corresponds to the (impurity) site and the indices a–f are the spin–orbital indices. We put the last time argument in the correlators equal zero using the time-translation invariance.

Once the susceptibility (12.46) is calculated, one needs to substitute its frequency transform

$$\chi^{abcd}(\omega_m, \omega_l, \omega_n) = T^2 \int_0^\beta d\tau_i \int_0^\beta d\tau_j \int_0^\beta d\tau_k e^{-i\tau_i\omega_m} e^{i\tau_j(\omega_m+\omega_n)} e^{-i\tau_k(\omega_l+\omega_n)} \chi^{abcd}(\tau_i, \tau_j, \tau_k, 0), \tag{12.47}$$

into the equation for the local vertex function $\Gamma(\omega_m, \omega_l, \omega_n)$

$$\begin{aligned} \chi(\omega_m, \omega_l, \omega_n) = & \frac{\delta_{ml}}{T} \chi^{(0)}(\omega_m, \omega_n) \\ & - \chi^{(0)}(\omega_m, \omega_n) \Gamma(\omega_m, \omega_l, \omega_n) \chi^{(0)}(\omega_l, \omega_n), \end{aligned} \tag{12.48}$$

where

$$\chi^{abcd(0)}(\omega_m, \omega_n) = \mathcal{G}_{ac}(\omega_m + \omega_n) \mathcal{G}_{bd}(\omega_m) \tag{12.49}$$

($\mathcal{G}_{ac}$ are the orbital-dependent dynamical mean-field functions, and there is no summation over ω_m!). From Eq. (12.48) one can find the vertex function $\Gamma(\omega_m, \omega_l, \omega_n)$ and use it to calculate the DMFT one-frequency susceptibility:

$$\chi(\omega_n) = 2T \sum_m \chi^0(\omega_m, \omega_n) - 2T^2 \sum_{m,l} \chi^0(\omega_m, \omega_n) \Gamma(\omega_m, \omega_l, \omega_n) \chi^0(\omega_l, \omega_n). \tag{12.50}$$

The last function is used to calculate the XC kernel (see Eq. (12.31) for the general nonlocal case and Eq. (12.56) for the case of DMFT).

Let us recall that in the DΓA formalism the local-vertex correction to susceptibility is used to calculate the momentum-dependent correction to the DMFT self-energy. Here, similar to DΓA, we use it to calculate the correction to the (local) one-loop susceptibility. In Ref. [54], it was proposed to use the ladder approximation for the vertex function [61–63]:

$$\Gamma(\omega_m, \omega_l, \omega_n) \cong U_{ch}, \tag{12.51}$$

where U_{ch} is an effective interaction parameter ("ch" stands for the "charge susceptibility," see also Chap. 9). This gives the ladder susceptibility

$$\chi(\omega_n) = \frac{\chi^{(0)}(\omega_n)}{1 + U_{ch}\chi^{(0)}(\omega_n)}, \tag{12.52}$$

where

$$\chi^{(0)}(\omega_n) = -T\sum_{a,b,m} G_{ab}(\omega_n + \omega_m)G_{ba}(\omega_m). \tag{12.53}$$

is the one-loop susceptibility ($G_{ab}(\omega_m)$ are local single-particle GFs). The parameter U_{ch} for Eq. (12.52) is found from the sum rule for the susceptibility

$$T\sum_m \chi(\omega_n) = n(1 - n) + 2D, \tag{12.54}$$

where

$$D = \langle n_{i\uparrow}n_{i\downarrow}\rangle = \left(\frac{n}{2}\right)^2 + T\sum_{a,n}\Sigma_{aa}(\omega_n)G_{aa}(\omega_n) \tag{12.55}$$

is the one-site electron double occupancy. Despite its simplicity, the ladder approximation for the vertex function is capable to describe many important effects, including superconductivity in the 2D Hubbard model (for details, see Ref. [62] for the one-orbital and Ref. [63] for the multiorbital case).

Thus, in the proposed DMFT - TDDFT approach, the XC kernel has the form

$$f_{XC}\left(\omega, \vec{r}, \vec{r}'\right) = \delta\left(\vec{r} - \vec{r}'\right)\left[\chi^{(0)-1}(\omega) - \chi^{-1}(\omega)\right] - \frac{1}{\left|\vec{r} - \vec{r}'\right|}, \tag{12.56}$$

where in brackets the first and the second terms are the local noninteracting (KS) and interacting (12.52) susceptibilities, correspondingly. The formalism was applied to the one-band Hubbard model at half-filling with Gaussian free-electron DOS $\rho(\varepsilon) = 1/(\sqrt{2\pi}t) \cdot \exp\left(-\frac{\varepsilon^2}{t^2}\right)$ [54]. As it was discussed in the first part of the book, the $U = 0$ (DFT) DOS starts to deviate from the Gaussian one with U increase, and the Gaussian free-electron peak splits into three peaks: a central (zero-frequency) quasiparticle peak and two Hubbard sub-band peaks at $\omega \sim \pm U/2$. The central peak becomes suppressed at large values of U. The central, zero-energy quasiparticle, peak is missed in the static mean-field approximations, like DFT + U. This feature of the DMFT DOS makes the DMFT excitation spectrum, and hence the susceptibility and the XC kernel, different from the static solutions. In fact, at intermediate values of U the DMFT excitation spectrum has a much longer low-frequency shoulder (that corresponds to transitions that involve states in the quasiparticle part of the DOS) as compared to the static mean-field HF one, while at larger Us both the static and the dynamic (frequency-dependent self-energy)

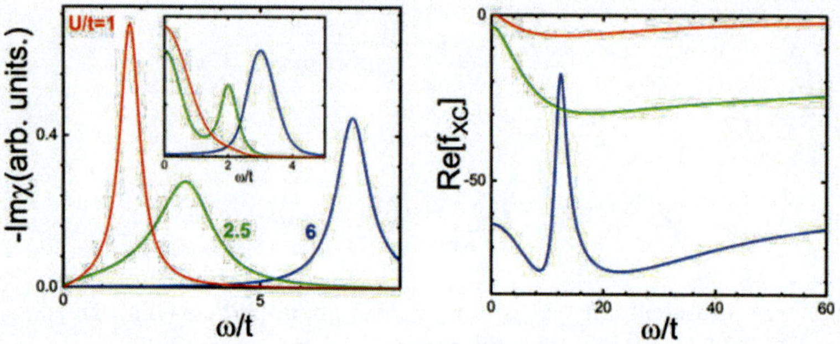

Fig. 12.3 Left: The DMFT results for the imaginary parts of the vertex-corrected susceptibility (excitation spectrum) for the infinite-dimensional one-band Hubbard model at half-filling and different values of U (Gaussian DOS, $T = 0.01t$, IPT solver). Insert: the corresponding DOS. Right: the frequency dependence of the real part of the DMFT nonadiabatic XC kernel at the same values of U as in the left figure (for more details and results, see Ref. [54])

solutions are similar, with a relatively narrow excitation peak at frequency ~ U (the DMFT solution for the DOS and the imaginary part of the susceptibility is shown in the left Fig. 12.3).

The results for the DMFT XC kernels for the infinite-dimensional Hubbard model at different values of U are shown in the right Fig. 12.3. As it follows from this figure, the shape of the XC kernel is defined by the excitation spectrum from the left figure, in particular showing peaks at frequencies of order of the excitation frequency. The values of the frequencies at the peaks of the excitation spectrum are the frequencies of the oscillations of the XC kernel and the width of the peaks defines the memory time of the kernel in time domain.

Similar structures of the excitation spectrum and the XC kernel, with peaks in both functions that come from Coulomb-interaction effects, were found in the cases of the 2D 3 × 3 Hubbard cluster (Fig. 12.4 and left Fig. 12.5) and the Hubbard dimer in Ref. [31] (for the case of an asymmetric Hubbard dimer kernel, see Ref. [64]). It must be noted, however, that as the results for the kernel in systems with small number of neighbor atoms show, the nondiagonal part of the kernel is not negligible (right Fig. 12.5). This is an important question that needs studies in the case of DMFT XC kernel.

The kernel discussed in this subsection corresponds to the PM case. In the next subsection, we consider the spin-polarized case in the most general case of noncollinear systems.

12.4.2 Spin-Polarized Case

Nonequilibrium, especially ultrafast, properties of spin-polarized strongly correlated materials is also of a special interest from both fundamental physics and technological (magnetic recording, spintronics) points of view. An example of important

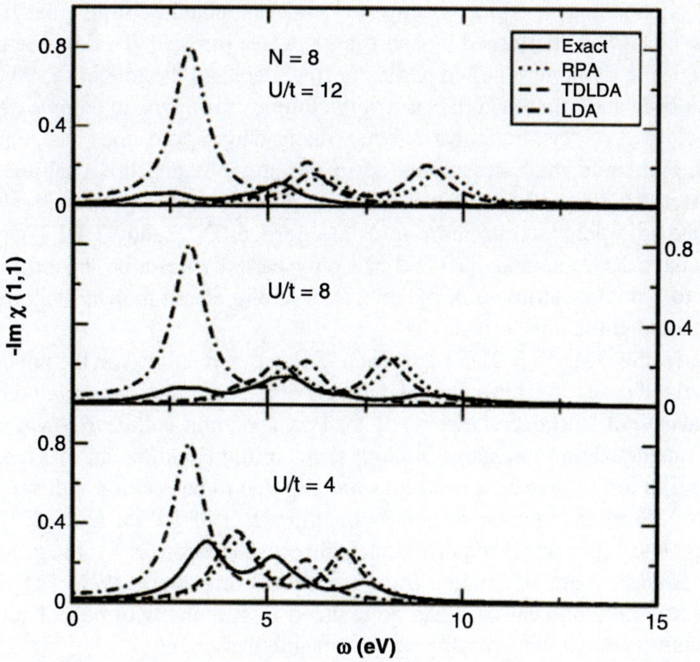

Fig. 12.4 The imaginary part of the diagonal part of the susceptibility for 3×3 Hubbard cluster with eight electrons at different values of U obtained by Lanczos diagonalization and other approaches. (Reprinted Fig. 2 with permission from (F. Aryasetiawan and O. Gunnarsson, Phys. Rev. B **66**, 165119 (2002)). Copyright 2002 by the American Physical Society. https://doi.org/10.1103/PhysRevB.66.165119)

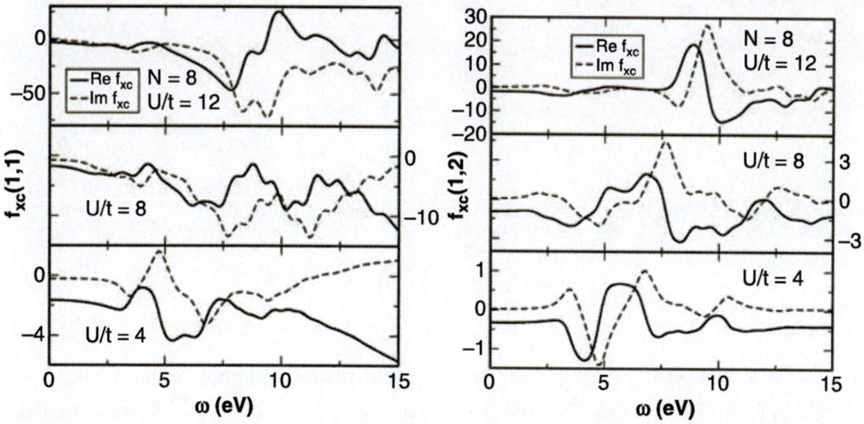

Fig. 12.5 The same as in Fig. 12.4 for the diagonal (left) and off-diagonal diagonal (right) components of the XC kernel. (Reprinted Figs. 4 and 5 with permission from (F. Aryasetiawan and O. Gunnarsson, Phys. Rev. B **66**, 165119 (2002)). Copyright 2002 by the American Physical Society. https://doi.org/10.1103/PhysRevB.66.165119)

ultrafast phenomena in spin systems is 10–50-fs change of magnetization in transition-metal FMs perturbed by short laser pulses [65, 66]. To describe ultrafast charge and spin dynamics, and in particular to understand the role of the correlation effects, in FMs and other materials is a very complex task. Due to complexity of the problem (large systems, nonlinear effects), many-body approaches (see, e.g., Refs. [67–69]), similar to the cases discussed in the previous and this chapter, are not always useful here and have to be used with some severe approximations. Thus, appropriate spin-polarized *ab initio* tools are needed. DFT and TDDFT with standard adiabatic XC potentials [70–72] can only partially describe the experimental data due to a poor description of the effects of strong electron-electron correlations and not included memory effects.

Recently, the DMFT - TDDFT approach [54] was generalized on the noncollinear spin-polarized case [55] (the standard collinear spin case (with conserved total spin) is a partial case of this general theory). In the last work, non-collinearity was required to allow magnetization to change without terms in the Hamiltonian that break total spin conservation (it was assumed that spins flip due to correlation-induced scattering effects). Similar ("non-correlated") noncollinear TDDFT for molecules, which allows spins to flip without requiring total spin conservation, is a topic of an active research in last years beginning from the pioneering works [73, 74]. In bulk materials, the noncollinear case can be realized in systems with helical spin-wave ground state, systems with varying surface magnetization, etc.

To give details of the noncollinear spin (spin-flip) DMFT - TDDFT approach, we begin with spin-flip TDDFT Kohn–Sham equations (non-diagonal in spin XC and external potentials):

$$\left[\left(-\frac{\vec{\nabla}^2}{2m}+V_H[n]\left(t,\vec{r}\right)\right)\delta_{\sigma\sigma'}+V_{XC\sigma\sigma'}[n]\left(t,\vec{r}\right)+V_{ext\sigma\sigma'}\left(t,\vec{r}\right)\right]\Psi_{\vec{k}\sigma'}\left(t,\vec{r}\right)$$
$$=i\frac{\partial\Psi_{\vec{k}\sigma}\left(t,\vec{r}\right)}{\partial t}, \tag{12.57}$$

where $V_{XC\sigma\sigma'}[n](r,t)$ is a functional of the spin-density matrix

$$n_{\sigma\sigma'}(r,t)=\sum_{k\leq k_F}\Psi_{\vec{k}\sigma}\left(t,\vec{r}\right)\Psi^*_{\vec{k}\sigma'}\left(t,\vec{r}\right). \tag{12.58}$$

Similar to spin non-polarized case, in the linear-response approximation, the XC spin-matrix potential can be expressed in terms of the XC kernel matrix $f_{XC\sigma\sigma'\overline{\sigma}\overline{\sigma}'}\left(t\vec{r},t'\vec{r}'\right)$:

$$V_{XC\sigma\sigma'}[n]\left(t,\vec{r}\right) = V_{XC\sigma\sigma'}[n]\left(t=0,\vec{r}\right) + \sum_{\overline{\sigma},\overline{\sigma}'} \int dt' \int d\vec{r}' f_{XC\sigma\sigma'\overline{\sigma}\overline{\sigma}'}\left(t\vec{r},t'\vec{r}'\right) \delta n_{\overline{\sigma}\overline{\sigma}'}\left(t',\vec{r}'\right), \quad (12.59)$$

where

$$f_{XC\sigma\sigma'\overline{\sigma}\overline{\sigma}'}\left(t\vec{r},t'\vec{r}'\right) = \frac{\delta v_{XC\sigma\sigma'}[n]\left(t,\vec{r}\right)}{\delta n_{\overline{\sigma}\overline{\sigma}'}\left(t',\vec{r}'\right)}. \quad (12.60)$$

In the DMFT (local-in-space) approximation, $f_{XC\sigma\sigma'\overline{\sigma}\overline{\sigma}'}\left(t\vec{r},t'\vec{r}'\right)$ has the space- and time-dependencies similar to the PM case:

$$f_{XC\sigma\sigma'\overline{\sigma}\overline{\sigma}'}\left(t\vec{r},t'\vec{r}'\right) = \delta\left(\vec{r}-\vec{r}'\right) f^{DMFT}_{XC\sigma\sigma'\overline{\sigma}\overline{\sigma}'}(t-t') - \delta_{\sigma\sigma'}\delta_{\overline{\sigma}\overline{\sigma}'}\frac{1}{\left|\vec{r}-\vec{r}'\right|} \quad (12.61)$$

where $f^{DMFT}_{XC\sigma\sigma'\overline{\sigma}\overline{\sigma}'}(t-t')$ in the frequency representation satisfies the equation

$$\chi_{\alpha\beta}(\omega) = \chi^{(0)}_{\alpha\beta}(\omega) + \sum_{\gamma,\delta}\chi^{(0)}_{\alpha\gamma}(\omega) f^{DMFT}_{XC\gamma\delta}(\omega)\chi_{\delta\beta}(\omega). \quad (12.62)$$

To shorten notations, in Eq. (12.62) we expressed the XC kernel, the interacting

$$\chi_{\sigma\sigma'\overline{\sigma}\overline{\sigma}'}(t) = -\sum_{a,b}\left\langle \widehat{T} c^{a+}_{\sigma}(t) c^{a}_{\sigma'}(t) c^{b+}_{\overline{\sigma}}(0) c^{b}_{\overline{\sigma}'}(0)\right\rangle \quad (12.63)$$

and the corresponding noninteracting $\chi^{(0)}_{\sigma\sigma'\overline{\sigma}\overline{\sigma}'}(t)$ susceptibilities (defined here in time domain) in the 4×4 matrix representation with indices $1 = \uparrow\uparrow$, $2 = \uparrow\downarrow$, $3 = \downarrow\uparrow$, and $4 = \downarrow\downarrow$ (the independent combinations of the spin-index pairs). Using the one-loop approximation for the DMFT susceptibility (12.63) (with diagonal in orbital indices single-particle GFs) and substituting it into Eq. (12.62), one can obtain the following XC matrix:

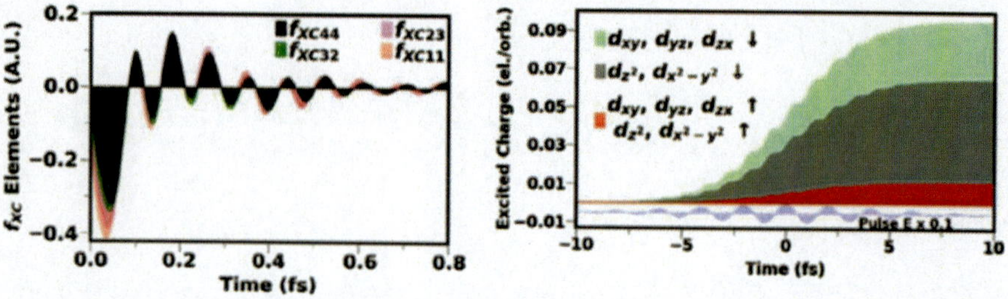

Fig. 12.6 The components of the XC kernel matrix (left) and excited orbital-projected spin density (right, the electric pulse shape is shown in the bottom panel) for Ni (see Ref. [55] for more details). (Reprinted Figs. 3 and 4 with permission from (S.R. Acharya, V. Turkowski, G.P. Zhang, and T.S. Rahman, Phys. Rev. Lett. **125**, 017202 (2020)). Copyright 2020 by the American Physical Society. https://doi.org/10.1103/PhysRevLett.125.017202)

$$\hat{f}_{XC}^{DMFT}(\omega) = \begin{pmatrix} f_{XC11}(\omega) & 0 & 0 & 0 \\ 0 & 0 & f_{XC23}(\omega) & 0 \\ 0 & f_{XC32}(\omega) & 0 & 0 \\ 0 & 0 & 0 & f_{XC44}(\omega) \end{pmatrix}. \quad (12.64)$$

The non-diagonal XC elements in Eq. (12.64) are responsible for the spin-flipping processes (for details, see Ref. [55]). In left Fig. 12.6, we present results for the components of the XC kernel matrix for Ni obtained in the approximation above. Different components have a rather similar time dependence, including the memory time ~ 1 fs. The non-diagonal elements lead to a dramatic decrease of the magnetization in Ni. Indeed, since most of the *d*-electron spin-up states are below and most of the spin-down states are above the Fermi energy, to transfer charges to the spin-down states requires flipping of spins. As it is shown in Fig. 12.6, after the pulse is applied the excited charge density of the spin-down electrons grows from zero to a large value, resulting in a large demagnetization.

12.5 Nonlinear Nonadiabatic DMFT XC Potentials: Sham–Schlueter Equation Approach

In this section, we describe an algorithm of derivation of the most general—nonlinear nonadiabatic multi-orbital system—DMFT XC potential by using the Sham–Schlueter equation approach [56, 57].

Similar to the static case, we begin with the equations for the KS $G_{KS}\left(t_1\vec{r}_1, t_2\vec{r}_2\right)$ and many-body orbital $G_{lm}\left(t_1\vec{r}_1, t_2\vec{r}_2\right)$ GFs:

$$\left(i\frac{\partial}{\partial t_1}+\frac{\vec{\nabla}^2}{2m}-V_H\left(t_1,\vec{r}_1\right)-V_{XC}\left(t_1,\vec{r}_1\right)\right)G_{KS}\left(t_1\vec{r}_1,t_2\vec{r}_2\right)$$
$$=\delta(t_1-t_2)\delta\left(\vec{r}_1-\vec{r}_2\right), \tag{12.65}$$

$$\left(\left(i\frac{\partial}{\partial t_1}+\frac{\vec{\nabla}^2}{2m}\right)\delta(t_1-t_3)\delta\left(\vec{r}_1-\vec{r}_3\right)\delta_{ll'}-\delta_{ll'}\Sigma_{ll'}\left(t_1\vec{r}_1,t_3\vec{r}_3\right)\right)G_{l'm}\left(t_3\vec{r}_3,t_2\vec{r}_2\right)$$
$$=\delta_{lm}\delta(t_1-t_2)\delta\left(\vec{r}_1-\vec{r}_2\right). \tag{12.66}$$

In these equations, the time arguments are defined on the complex Kadanoff–Baym time contour (the "physical" quantities are obtained by using the time arguments on the first branch). Then, similar to the static case in Chap. 11, combining the last two equations (multiplying Eq. (12.65) from the left by $G_{lm}\left(t_4\vec{r}_4,t_1\vec{r}_1\right)$ and Eq. (12.66) from the left by $G_{KS}\left(t_4\vec{r}_4,t_1\vec{r}_1\right)$, integrating both equations over $\left(t_1,\vec{r}_1\right)$ and subtracting one equation from the other) gives the equation that connects two GFs:

$$G_{lm}\left(t_1\vec{r}_1,t_2\vec{r}_2\right)=\delta_{lm}G_{KS}\left(t_1\vec{r}_1,t_2\vec{r}_2\right)$$
$$+\iint dt_3d\vec{r}_3\iint dt_4d\vec{r}_4\Big\{G_{KS}\left(t_1\vec{r}_1,t_3\vec{r}_3\right)$$
$$\times\Sigma_{ll'}\left(t_3\vec{r}_3,t_4\vec{r}_4\right)G_{l'm}\left(t_4\vec{r}_4,t_2\vec{r}_2\right)-G_{lm}\left(t_1\vec{r}_1,t_3\vec{r}_3\right)$$
$$\times\left[V_H\left(t_3\vec{r}_3\right)+V_{XC}\left(t_3\vec{r}_3\right)\right]\delta(t_3-t_4)\delta\left(\vec{r}_3-\vec{r}_4\right)G_{KS}\left(t_4\vec{r}_4,t_2\vec{r}_2\right)\Big\}. \tag{12.67}$$

Next, putting in Eq. (12.67) $t_2=t_1^+,\vec{r}_2=\vec{r}_1,l=m$, performing summation over the orbital index l and using $n\left(t_1,\vec{r}_1\right)=G_{KS}\left(t_1\vec{r}_1,t_1^+\vec{r}_1\right)=\sum_l G_{ll}\left(t_1\vec{r}_1,t_1^+\vec{r}_1\right)$, one gets the time-dependent Sham–Schlueter equation for the multiorbital systems (a generalization of the one-orbital equation [56]):

$$\int dt_2 \int d\vec{r}_2 \left(\sum_l G_{ll}\left(t_1\vec{r}_1, t_2\vec{r}_2\right) \right)$$
$$\times \left[V_H\left(t_2, \vec{r}_2\right) + V_{XC}\left(t_2, \vec{r}_2\right) \right] G\left(t_2\vec{r}_2, t_1^+\vec{r}_1\right)$$
$$= (N_{\text{orb}} - 1) n\left(t_1, \vec{r}_1\right)$$
$$+ \iint dt_3 d\vec{r}_3 \iint dt_4 d\vec{r}_4 G_{KS}\left(t_1\vec{r}_1, t_3\vec{r}_3\right) \Sigma_{ll'}\left(t_3\vec{r}_3, t_4\vec{r}_4\right) G_{l'm}\left(t_4\vec{r}_4, t_1^+\vec{r}_1\right) \tag{12.68}$$

(N_{orb} is the number of orbitals).
Equation (12.68) can be rewritten in the following form:

$$V_{XC}\left(t_1\vec{r}_1\right) = (N_{\text{orb}} - 1) \int dt_2 \int d\vec{r}_2 \widetilde{\chi}^{-1}\left(t_1\vec{r}_1, t_2\vec{r}_2\right) n\left(t_2, \vec{r}_2\right)$$
$$+ \iint dt_2 d\vec{r}_2 \iint dt_3 d\vec{r}_3 \iint dt_4 d\vec{r}_4 \widetilde{\chi}^{-1}\left(t_1\vec{r}_1, t_4\vec{r}_4\right) G_{KS}\left(t_4\vec{r}_4, t_2\vec{r}_2\right)$$
$$\times \Sigma_{ll'}\left(t_2\vec{r}_2, t_3\vec{r}_3\right) G_{l'l}\left(t_3\vec{r}_3, t_1^+\vec{r}_1\right) - V_H(r_1 t_1), \tag{12.69}$$

where

$$\widetilde{\chi}\left(t_1\vec{r}_1, t_2\vec{r}_2\right) = \left(\sum_l G_{ll}\left(t_1\vec{r}_1, t_2\vec{r}_2\right) \right) G_{KS}\left(t_2\vec{r}_2, t_1^+\vec{r}_1\right). \tag{12.70}$$

Thus, to find V_{XC} one needs the KS GF and the many-body multiorbital GF and self-energy (the last two functions are connected by Eq. (12.66)). In the DMFT approximation, we assume that $\Sigma_{lm}\left(t_1\vec{r}_1, t_2\vec{r}_2\right)$ is a local and a space-independent function:

$$\Sigma_{lm}\left(t_1\vec{r}_1, t_2\vec{r}_2\right) \approx \delta\left(\vec{r}_1 - \vec{r}_2\right) \Sigma_{lm}(t_1, t_2), \tag{12.71}$$

where $\Sigma_{lm}(t_1, t_2)$ is obtained by solving the nonequilibrium DMFT problem when the system is perturbed by an external potential (e.g., by a pulse). Substituting the solution for the self-energy (12.71) into Eq. (12.66) one obtains the many-body multiorbital GF $G_{lm}\left(t_1\vec{r}_1, t_2\vec{r}_2\right)$.

Now, the problem is reduced to a pure TDDFT problem: one needs to self-consistently solve the TDDFT KS Eq. (12.3) for the functions $\phi_1\left(t_1, \vec{r}_1\right)$

(index l describes the state (orbital, momentum, spin)) with XC potential defined by Eq. (12.69), where on the right-hand side unknown KS GFs are defined by the KS wave functions:

$$G_{\mathrm{KS}}\left(t_1\vec{r}_1, t_2\vec{r}_2\right) = -i\sum_l \left[\phi_l\left(t_1, \vec{r}_1\right)\phi_l^*\left(t_1, \vec{r}_1\right)\right]$$
$$\times\left[\theta(t_1 - t_2)\theta(\varepsilon_l - \varepsilon_F) - \theta(t_2 - t_1)\theta(\varepsilon_F - \varepsilon_l)\right] \quad (12.72)$$

(ε_F is the Fermi energy). The external potential in the TDDFT case has to be the same as in the preceding neDMFT calculations.

Obviously, such a strategy to solve the problem is a reasonable one: first, one solves less expensive local-in-space neDMFT problem (the inhomogeneity effects are neglected, but it might be regarded as a meaningful first approximation, since the correlations are predominantly local in space and can be well-estimated by the local self-energy). Then, once the self-energy is found one solves self-consistently TDDFT KS Eqs. (12.3), (12.69), and (12.72) to describe often very important spatially non-homogeneous effects.

Once a sufficient amount of data on the dependence of the XC potential on $n\left(t, \vec{r}\right)$ in the case of different pulses is accumulated, one can try to obtain a universal (pulse-independent) functional $V_{\mathrm{XC}}[n]\left(t, \vec{r}\right)$ (as a matter of fact, simple analytical approximations for the XC potential used at the end of the previous chapter in the static case, will not give sufficiently meaningful results in a much more complicated time-dependent case). It must be mentioned that current computational resources do not allow one to solve the neDMFT problems with more than two- to three-orbitals, thus an effective two- or three-orbital model needs to be formulated and solved with neDMFT.

To conclude this chapter, we mention two other possible ways to solve the DMFT - TDDFT problem with the Sham–Schlueter approach. One is to use Eq. (12.67) and apply the PT expansion in the electron self-energy and the XC potential, similar to the static case discussed in the previous chapter. Another possibility is to take advantage of our knowledge of the structure of the DMFT self-energy in the equilibrium case $\Sigma_{lm}^{eq}(\omega)$, i.e., its dependence on frequency and, hence, on the relative time $t = t_1 - t_2$. Then, in the case of not very fast response, separating the relative t and the average $T = \frac{t_1+t_2}{2}$ time coordinates, one can approximate the self-energies in Eq. (12.69) by product of the two functions, the known equilibrium relative-time part $\Sigma_{lm}^{\mathrm{eq}}(t_1 - t_2)$ and unknown slowly-changing average time part $\Sigma_{lm}^{av}(T)$, i.e., $\Sigma_{lm}(t_1, t_2) = \Sigma_{lm}^{\mathrm{eq}}(t)\Sigma_{lm}^{av}(T)$. Since $\Sigma_{lm}^{av}(T)$ changes slowly with time, it can be approximated by a simple smooth function with the decay time obtained from the time scales of the problem. In this case, one avoids solving the neDMFT problem. However, the disadvantages and the advantages of these two approximations are not known so far.

Finally, to take into account nonlocal effects more accurately, one can go beyond neDMFT by including the momentum dependence of the self-energy with DΓA or another approach described in Chap. 9. Thus, we conclude the discussion of implementation of DMFT in *ab initio* approaches, a very promising area of condensed matter physics and material science.

References

1. Fulde, P.: Electron Correlations in Molecules and Solids. Springer-Verlag, Berlin (2002)
2. Tokura, Y.: Phys. Today. **56**, 60 (2003)
3. Yamada, K.: Electron Correlations in Metals. Cambridge University Press (2004)
4. Quintanilla, J., Hooley, C.: Phys. World. **22** (2009)
5. Anisimov, V., Izyumov, Y.: Electronic Structure of Strongly Correlated Materials. Springer (2010)
6. Imada, M., Fujimori, A., Tokura, Y.: Rev. Mod. Phys. **70**, 1039 (1998)
7. Wei, J., Natelson, D.: Nanoscale. **3**, 3509 (2011)
8. Servin, R., Jin-Hyung, P., In-yeal, L., Jeong Min, B., Kyung Soo, Y., Gil-Ho, K.: J. Phys. D: Appl. Phys. **47**, 295101 (2014)
9. Markov, P., Marvel, R.E., Conley, H.J., Miller, K.J., Haglund Jr., R.F., Weiss, S.M.: ACS Photonics. **2**, 1175 (2015)
10. Merced, E., Torres, D., Tan, X., Sepúlveda, N.: J. Microelecromech. Syst. **24**, 100 (2015)
11. Liu, C., Balin, I., Magdassi, S., Abdulhalim, I., Long, Y.: Opt. Express. **3**, A124 (2015)
12. Nie, G., Zhang, L., Lei, J., Yang, L., Zhang, Z., Lu, X., Wang, C.: J. Mat. Chem. A. **2**, 2910 (2014)
13. Yang, S., Gong, Y., Liu, Z., Zhan, L., Hashim, D.P., Ma, L., Vajtai, R., Ajayan, P.M.: Nano Lett. **13**, 1596 (2013)
14. Chemla, D.S., Shah, J.: Nature. **411**, 549 (2001)
15. Onida, G., Reining, L., Rubio, A.: Rev. Mod. Phys. **74**, 601 (2002)
16. Rossi, F., Kuhn, T.: Rev. Mod. Phys. **74**, 895 (2002)
17. Polkovnikov, A., Sengupta, K., Silva, A., Vengalattore, M.: Rev. Mod. Phys. **83**, 863 (2011)
18. Bigot, J.Y., Vomir, M.: Ann. Phys. **525**, 2 (2013)
19. Bigot, J.Y.: Nat. Mat. **12**, 283 (2013)
20. Rana, K.G., Yajima, T., Parui, S., Kemper, A.F., Devereaux, T.P., Hikita, Y., Hwang, H.Y., Banerjee, T.: Sci. Rep. **3**, 1274 (2013)
21. Aoki, H., Tsuji, N., Eckstein, M., Kollar, M., Oka, T., Werner, P.: Rev. Mod. Phys. **86**, 779 (2014)
22. Hopjan, M., Verdozzi, C.: Probing Strongly Correlated Materials in Non-equilibrium: Basic Concepts and Possible Future Trends in First Principle Approaches. In: Di Valentin, C., Bottiand, S., Cococcioni, M. (eds.) First Principles Approaches to Spectroscopic Properties of Complex Materials, vol. 347, pp. 347–384. Springer, Berlin (2014)
23. Kim, S.-H., Kim, B.-J., Jeong, T.-Y., Lee, Y.-S., Yee, K.-J.: J. Appl. Phys. **117**, 163107 (2015)
24. Thompson, Z.J., Stickel, A., Jeong, Y.-G., Han, S., Son, B.H., Paul, M.J., Lee, B., Mousavian, A., Seo, G., Kim, H.-T., Lee, Y.-S., Kim, D.-S.: Nano Lett. **15**, 5893 (2015)
25. Runge, E., Gross, E.K.U.: Phys. Rev. Lett. **52**, 997 (1984)
26. Marques, M.A.L., Ullrich, C.A., Nogueira, F., Rubio, A., Burke, K., Gross, E.K.U.: Time-Dependent Density Functional Theory, vol. 706. Springer-Verlag, Berlin (2006)
27. Ullrich, C.A.: Time-Dependent Density-Functional Theory: Concepts and Applications. Oxford University Press, Oxford (2011)
28. Giuliani, G.F., Vignale, G.: Quantum Theory of the Electron Liquid. Cambridge University Press, Cambridge (2010)
29. Wang, L.W.: Ann. Rev. Phys. Chem. **61**, 19 (2010)
30. Gruner, M.E., Entel, P.: Int. J. Quant. Chem. **112**, 277 (2012)

31. Aryasetiawan, F., Gunnarsson, O.: Phys. Rev. B. **66**, 165119 (2002)
32. Lima, N.A., Silva, M.F., Oliveira, L.N., Capelle, K.: Phys. Rev. Lett. **90**, 146402 (2003)
33. Freericks, J.K., Turkowski, V.M., Zlatic, V.: Phys. Rev. Lett. **97**, 266408 (2006)
34. Karlsson, D., Privitera, A., Verdozzi, C.: Phys. Rev. Lett. **106**, 116401 (2011)
35. Schonhammer, K., Gunnarsson, O., Noack, R.M.: Phys. Rev. B. **52**, 2504 (1995)
36. López-Sandoval, R., Pastor, G.M.: Phys. Rev. B. **61**, 1764 (2000)
37. Lima, N.A., Oliveira, L.N., Capelle, K.: Europhys. Lett. **60**, 601 (2002)
38. Xianlong, G., Polini, M., Tosi, M.P., Campo, V.L., Capelle, K., Rigol, M.: Phys. Rev. B. **73**, 165120 (2006)
39. Verdozzi, C.: Phys. Rev. Lett. **101**, 166401 (2008)
40. Wei, L., Gao, X.L., Kollath, C., Polini, M.: Phys. Rev. B. **78**, 195109 (2008)
41. Lieb, E.H., Wu, F.Y.: Phys. Rev. Lett. **20**, 1445 (1968)
42. López-Sandoval, R., Pastor, G.: Phase Transit. **78**, 839 (2005)
43. Carrascal, D., Ferrer, J.: Phys. Rev. B. **85**(85), 045110 (2012)
44. Fuks, J.I., Maitra, N.T.: Phys. Chem. Chem. Phys. **16**, 14504 (2014)
45. Fuks, J., Maitra, N.: Phys. Rev. A. **89**, 062502 (2014)
46. Lani, G., di Marino, S., Gerolin, A., van Leeuwen, R., Gori-Giorgi, P.: preprint arXiv. **1601**, 03977 (2016)
47. Xianlong, G., Chen, A.-H., Tokatly, I., Kurth, S.: Phys. Rev. B. **86**, 235139 (2012)
48. Capelle, K., Campo Jr., V.L.: Phys. Rep. **528**, 91 (2013)
49. Carrascal, D., Ferrer, J., Smith, J.C., Burke, K.: J. Phys. Condens. Matter. **27**, 393001 (2015)
50. Turkowski, V., Rahman, T.S.: J. Phys. Cond. Matter. **26**, 022201 (2014)
51. Mancini, L., Ramsden, J., Hodgson, M., Godby, R.: Phys. Rev. B. **89**, 195114 (2014)
52. Tancogne-Dejean, N., Oliveira, M.J.T., Rubio, A.: Phys. Rev. B. **96**, 245133 (2017)
53. Orhan, O.K., O'Regan, D.D.: Phys. Rev. B. **99**, 165120 (2019)
54. Turkowski, V., Rahman, T.S.: J. Phys. Condens. Matter. **29**, 455601 (2017)
55. Acharya, S.R., Turkowski, V., Zhang, G.P., Rahman, T.S.: Phys. Rev. Lett. **125**, 017202 (2020)
56. Sham, L., Schlüter, M.: Phys. Rev. Lett. **51**, 1888 (1983)
57. van Leeuwen, R.: Phys. Rev. Lett. **76**, 3610 (1996)
58. Haug, H., Koch, S.W.: Quantum Theory of the Optical and Electronic Properties of Semiconductors. World Scientific, Singapore (1994)
59. Ben Dahan, M., et al.: Phys. Rev. Lett. **76**, 4508 (1996)
60. Hopjan, M., Karlsson, D., Ydman, S., Verdozzi, C., Almbladh, C.-O.: Phys. Rev. Lett. **116**, 236402 (2016)
61. Vilk, Y.M., Tremblay, A.-M.S.: J. Phys. France. **7**, 1309 (1997)
62. Kusunose, H.: J. Phys. Soc. Jpn. **75**, 054713 (2006)
63. Miyahara, H., Arita, R., Ikeda, H.: Phys. Rev. B. **87**, 045113 (2013)
64. Carrascal, D.J., Ferrer, J., Maitra, N., Burke, K.: Eur. Phys. J. B. **91**, 142 (2018)
65. Kirilyuk, A., Kimel, A.V., Rasing, T.: Rev. Mod. Phys. **82**, 2731 (2010)
66. Beaurepaire, E., Merle, J.-C., Daunois, A., Bigot, J.-Y.: Phy. Rev. Lett. **76**, 4250 (1996)
67. Tengdin, P., et al.: Sci. Adv. **4**, eaap9744 (2018)
68. Mueller, B.Y., Baral, A., Vollmar, S., Cinchetti, M., Aeschlimann, M., Schneider, H.C., Rethfeld, B.: Phys. Rev. Lett. **111**, 167204 (2013)
69. Zhang, G.P., Hübner, W.: Phys. Rev. Lett. **85**, 3025 (2000)
70. Zhang, G.P., Bai, Y., Hübner, W., Lefkidis, G., George, T.F.: J. Appl. Phys. **103**, 07B113 (2008)
71. Zhang, G.P., Bai, Y., George, T.F.: J. Phys. Condens. Mat. **28**, 236004 (2016)
72. Krieger, K., Dewhurst, J., Elliott, P., Sharma, S., Gross, E.: J. Chem. Theory Comput. **11**, 4870 (2015)
73. Shao, Y., Head-Gordon, M., Krylov, A.I.: J. Chem. Phys. **118**, 4807 (2003)
74. Gao, J., Liu, W., Song, B., Liu, C.: J. Chem. Phys. **121**, 6658 (2004)
75. Del Sole, F.: Phys. Rev. B. **29**, 4631 (1984)

Index

V. Turkowski, *Dynamical Mean-Field Theory for Strongly Correlated Materials*,
https://doi.org/10.1007/978-3-030-64904-3

D

U

V

W

X

Z